Fundamentals of Atmospheric Physics

This is Volume 61 in the
INTERNATIONAL GEOPHYSICS SERIES
A series of monographs and textbooks
Edited by RENATA DMOWSKA and JAMES R. HOLTON

A complete list of books in this series appears at the end of this volume.

Fundamentals of Atmospheric Physics

Murry L. Salby
CENTER FOR ATMOSPHERIC THEORY AND ANALYSIS
UNIVERSITY OF COLORADO
BOULDER, COLORADO

ACADEMIC PRESS
An Imprint of Elsevier

San Diego   New York   Boston
London   Sydney   Tokyo   Toronto

This book is printed on acid-free paper. ∞

Academic Press
An Imprint of Elsevier
525 B Street, Suite 1900, San Diego, California 92101-4495, USA
http://www.academicpress.com

Academic Press
Harcourt Place, 32 Jamestown Road, London NW1 7BY, UK
http://www.academicpress.com

Library of Congress Cataloging-in-Publication Data

Salby, Murry L.
     Fundamentals of Atmospheric Physics / Murry L. Salby.
          p.   cm. -- (International geophysics series : v. 61)
     Includes bibliographical references and index.
     ISBN-13: 978-0-12-615160-2   ISBN-10: 0-12-615160-1
     1. Atmospheric Physics.   I. Title. II. Series.
QC861861.2S25   1995   95-15419
551.5—dc20

ISBN-13: 978-0-12-615160-2
ISBN-10: 0-12-615160-1

PRINTED IN THE UNITED STATES OF AMERICA
07   08   09   10   11   MM   9   8

DEDICATED TO HARRY

To the resolve you maintained in the face of adversity,
To the quiet wisdom you gave us.

The most fruitful areas for growth of the sciences are those between established fields. Science has been increasingly the task of specialists, in fields which show a tendency to grow progressively narrower. Important work is delayed by the unavailability in one field of results that may have already become classical in the next field. It is these boundary regions of science that offer the richest opportunities to the qualified investigator.

NORBERT WIENER

Contents

Preface   xvii

Chapter 1 A Global View

1.1   Introduction to the Atmosphere   1
    1.1.1   Descriptions of Atmospheric Behavior   1
    1.1.2   Mechanisms Influencing Atmospheric Behavior   2
1.2   Composition and Structure   3
    1.2.1   Description of Air   5
    1.2.2   Stratification of Mass   8
    1.2.3   Thermal and Dynamical Structure   16
    1.2.4   Trace Constituents   22
        Carbon Dioxide   23
        Water Vapor   25
        Ozone   29
        Methane   32
        Chlorofluorocarbons   33
        Nitrogen Compounds   34
        Atmospheric Aerosol   35
    1.2.5   Clouds   36
1.3   Radiative Equilibrium of the Planet   41
1.4   The Global Energy Budget   44
    1.4.1   Global–Mean Energy Balance   44
    1.4.2   Horizontal Distribution of Radiative Transfer   46
1.5   The General Circulation   50
Suggested Reading   53
Problems   53

Chapter 2 Thermodynamics of Gases

2.1   Thermodynamic Concepts   55
    2.1.1   Thermodynamic Properties   55
    2.1.2   Expansion Work   56
    2.1.3   Heat Transfer   58
    2.1.4   State Variables and Thermodynamic Processes   58

2.2 The First Law   63
      2.2.1 Internal Energy   63
      2.2.2 Diabatic Changes of State   63
2.3 Heat Capacity   65
2.4 Adiabatic Processes   68
      2.4.1 Potential Temperature   71
      2.4.2 Thermodynamic Behavior Accompanying Vertical Motion   73
2.5 Diabatic Processes   75
      2.5.1 Polytropic Processes   75
Suggested Reading   76
Problems   77

Chapter 3   The Second Law and Its Implications

3.1 Natural and Reversible Processes   79
      3.1.1 The Carnot Cycle   82
3.2 Entropy and the Second Law   84
3.3 Restricted Forms of the Second Law   87
3.4 The Fundamental Relations   88
      3.4.1 The Maxwell Relations   89
      3.4.2 Noncompensated Heat Transfer   89
3.5 Conditions for Thermodynamic Equilibrium   90
3.6 Relationship of Entropy to Potential Temperature   91
      3.6.1 Implications for Vertical Motion   93
Suggested Reading   96
Problems   96

Chapter 4   Heterogeneous Systems

4.1 Description of a Heterogeneous System   99
4.2 Chemical Equilibrium   102
4.3 Fundamental Relations for a Multicomponent System   104
4.4 Thermodynamic Degrees of Freedom   106
4.5 Thermodynamic Characteristics of Water   107
4.6 Equilibrium Phase Transformations   110
      4.6.1 Latent Heat   110
      4.6.2 Clausius–Clapeyron Equation   112
Suggested Reading   114
Problems   114

Chapter 5   Transformations of Moist Air

5.1 Description of Moist Air   117
      5.1.1 Properties of the Gas Phase   117
      5.1.2 Saturation Properties   120
5.2 Implications for the Distribution of Water Vapor   121

5.3   State Variables of the Two-Component System   123
    5.3.1   Unsaturated Behavior   123
    5.3.2   Saturated Behavior   124
5.4   Thermodynamic Behavior Accompanying Vertical Motion   126
    5.4.1   Condensation and the Release of Latent Heat   126
    5.4.2   The Pseudo-Adiabatic Process   130
    5.4.3   The Saturated Adiabatic Lapse Rate   132
5.5   The Pseudo-Adiabatic Chart   134
    Surface Relative Humidity   135
    Surface Potential Temperature   135
    Surface Dew Point   135
    Cumulus Cloud Base   136
    Equivalent Potential Temperature at the Surface   136
    Freezing Level of Surface Air   136
    Liquid Water Content at the Freezing Level   137
    Temperature inside Cloud at 650 mb   137
    Mixing Ratio inside Cloud at 650 mb   137
Suggested Reading   138
Problems   138

Chapter 6   Hydrostatic Equilibrium

6.1   Effective Gravity   143
6.2   Geopotential Coordinates   145
6.3   Hydrostatic Balance   147
6.4   Stratification   151
    6.4.1   Idealized Stratification   153
    Layer of Constant Lapse Rate   153
    Isothermal Layer   155
    Adiabatic Layer   155
6.5   Lagrangian Interpretation of Stratification   155
    6.5.1   Adiabatic Stratification   156
    6.5.2   Diabatic Stratification   160
Suggested Reading   163
Problems   163

Chapter 7   Hydrostatic Stability

7.1   Reaction to Vertical Displacement   166
7.2   Stability Categories   168
    7.2.1   Stability in Terms of Temperature   169
    7.2.2   Stability in Terms of Potential Temperature   171
    7.2.3   Moisture Dependence   173
7.3   Implications for Vertical Motion   174
7.4   Finite Displacements   176
    7.4.1   Conditional Instability   176
    7.4.2   Entrainment   181
    7.4.3   Potential Instability   184
    7.4.4   Modification of Stability under Unsaturated Conditions   186
7.5   Stabilizing and Destabilizing Influences   188

7.6   Turbulent Dispersion   189
    7.6.1   Convective Mixing   189
    7.6.2   Inversions   191
    7.6.3   Life Cycle of the Nocturnal Inversion   192
Suggested Reading   194
Problems   194

Chapter 8   Atmospheric Radiation

8.1   Shortwave and Longwave Radiation   198
    8.1.1   Spectra of Observed SW and LW Radiation   200
8.2   Description of Radiative Transfer   203
    8.2.1   Radiometric Quantities   203
    8.2.2   Absorption   207
          Lambert's Law   207
    8.2.3   Emission   208
          Planck's Law   209
          Wien's Displacement Law   209
          The Stefan–Boltzmann Law   210
          Kirchhoff's Law   211
    8.2.4   Scattering   213
    8.2.5   The Equation of Radiative Transfer   215
8.3   Absorption Characteristics of Gases   216
    8.3.1   Interactions between Radiation and Molecules 216
    8.3.2   Line Broadening   220
8.4   Radiative Transfer in a Plane Parallel Atmosphere   224
    8.4.1   Transmission Function   228
    8.4.2   Two-Stream Approximation   230
8.5   Thermal Equilibrium   233
    8.5.1   Radiative Equilibrium in a Gray Atmosphere   233
    8.5.2   Radiative–Convective Equilibrium   236
    8.5.3   Radiative Heating   241
8.6   Thermal Relaxation   245
8.7   The Greenhouse Effect   248
Suggested Reading   252
Problems   253

Chapter 9   Aerosol and Clouds

9.1   Morphology of Atmospheric Aerosol   258
    9.1.1   Continental Aerosol   258
    9.1.2   Marine Aerosol   263
    9.1.3   Stratospheric Aerosol   263
9.2   Microphysics of Clouds   264
    9.2.1   Droplet Growth by Condensation   264
    9.2.2   Droplet Growth by Collision   273
    9.2.3   Growth of Ice Particles   275
9.3   Macroscopic Characteristics of Clouds   277
    9.3.1   Formation and Classification of Clouds   277
    9.3.2   Microphysical Properties of Clouds   285
    9.3.3   Cloud Dissipation   286

9.4	Radiative Transfer in Aerosol and Cloud	287
	9.4.1 Scattering by Molecules and Particles	287
	Rayleigh Scattering	287
	Mie Scattering	291
	9.4.2 Radiative Transfer in a Cloudy Atmosphere	295
9.5	Roles of Clouds and Aerosol in Climate	305
	9.5.1 Involvement in the Global Energy Budget	306
	Influence of Cloud Cover	306
	Influence of Aerosol	312
	9.5.2 Involvement in Chemical Processes	314
	Suggested Reading	315
	Problems	315

Chapter 10 Atmospheric Motion

10.1	Descriptions of Atmospheric Motion	321
10.2	Kinematics of Fluid Motion	323
10.3	The Material Derivative	328
10.4	Reynolds' Transport Theorem	329
10.5	Conservation of Mass	330
10.6	The Momentum Budget	331
	10.6.1 Cauchy's Equations of Motion	331
	10.6.2 Momentum Equations in a Rotating Reference Frame	334
10.7	The First Law of Thermodynamics	336
	Suggested Reading	338
	Problems	338

Chapter 11 Atmospheric Equations of Motion

11.1	Curvilinear Coordinates	341
11.2	Spherical Coordinates	344
	11.2.1 The Traditional Approximation	350
11.3	Special Forms of Motion	352
11.4	Prevailing Balances	353
	11.4.1 Motion-Related Stratification	353
	11.4.2 Scale Analysis	354
11.5	Thermodynamic Coordinates	357
	11.5.1 Isobaric Coordinates	357
	11.5.2 Log–Pressure Coordinates	362
	11.5.3 Isentropic Coordinates	364
	Suggested Reading	368
	Problems	369

Chapter 12 Large-Scale Motion

12.1	Geostrophic Equilibrium	371
	12.1.1 Motion on an f Plane	374
	The Helmholtz Theorem	375

12.2	Vertical Shear of the Geostrophic Wind	377
	12.2.1 Classes of Stratification	378
	12.2.2 Thermal Wind Balance	378
12.3	Frictional Geostrophic Motion	381
12.4	Curvilinear Motion	382
	12.4.1 Inertial Motion	385
	12.4.2 Cyclostrophic Motion	385
	12.4.3 Gradient Motion	386
12.5	Weakly Divergent Motion	387
	12.5.1 Barotropic Nondivergent Motion	387
	12.5.2 Vorticity Budget under Baroclinic Stratification	389
	12.5.3 Quasi-Geostrophic Motion	394
	Suggested Reading	397
	Problems	398

Chapter 13   The Planetary Boundary Layer

13.1	Description of Turbulence	406
	13.1.1 Reynolds Decomposition	408
	13.1.2 Turbulent Diffusion	410
13.2	Structure of the Boundary Layer	412
	13.2.1 The Ekman Layer	412
	13.2.2 The Surface Layer	413
13.3	Influence of Stratification	415
13.4	Ekman Pumping	420
	Suggested Reading	423
	Problems	423

Chapter 14   Atmospheric Waves

14.1	Description of Wave Propagation	426
	14.1.1 Surface Water Waves	426
	14.1.2 Fourier Synthesis	429
	14.1.3 Limiting Behavior	432
	14.1.4 Wave Dispersion	434
14.2	Acoustic Waves	439
14.3	Buoyancy Waves	440
	14.3.1 Shortwave Limit	447
	14.3.2 Propagation of Gravity Waves in a Nonhomogeneous Medium	448
	14.3.3 The WKB Approximation	450
	14.3.4 Method of Geometric Optics	451
	Turning Level	451
	Critical Level	455
14.4	The Lamb Wave	456
14.5	Rossby Waves	459
	14.5.1 Barotropic Nondivergent Rossby Waves	459
	14.5.2 Rossby Wave Propagation in Three Dimensions	461
	14.5.3 Planetary Wave Propagation in Sheared Mean Flow	465

14.6   Wave Absorption   468
14.7   Nonlinear Considerations   472
Suggested Reading   479
Problems   479

Chapter 15   The General Circulation

15.1   Forms of Atmospheric Energy   487
       15.1.1   Moist Static Energy   487
       15.1.2   Total Potential Energy   489
       15.1.3   Available Potential Energy   491
                Adiabatic Adjustment   493
15.2   Heat Transfer in an Axisymmetric Circulation   495
15.3   Heat Transfer in a Laboratory Analogue   503
15.4   Tropical Circulations   506
Suggested Reading   513
Problems   513

Chapter 16   Hydrodynamic Instability

16.1   Inertial Instability   517
16.2   Shear Instability   519
       16.2.1   Necessary Conditions for Instability   519
       16.2.2   Barotropic and Baroclinic Instability   521
16.3   The Eady Problem   523
16.4   Nonlinear Considerations   529
Suggested Reading   532
Problems   533

Chapter 17   The Middle Atmosphere

17.1   Ozone Photochemistry   536
       17.1.1   The Chemical Family   537
       17.1.2   Photochemical Equilibrium   538
17.2   Involvement of Other Species   540
       17.2.1   Nitrous Oxide   541
       17.2.2   Chlorofluorocarbons   542
       17.2.3   Methane   544
17.3   Air Motion   546
       17.3.1   The Brewer–Dobson Circulation   547
       17.3.2   Wave Driving of the Mean Meridional Circulation   548
17.4   Sudden Stratospheric Warmings   553
17.5   The Quasi-Biennial Oscillation   557
17.6   Direct Interactions with the Troposphere   560
17.7   Heterogeneous Chemical Reactions   564
Suggested Reading   573
Problems   573

Appendix A: Conversion to SI Units　　577
Appendix B: Thermodynamic Properties of Air and Water　　578
Appendix C: Physical Constants　　579
Appendix D: Vector Identities　　580
Appendix E: Curvilinear Coordinates　　581
Appendix F: Pseudo-Adiabatic Chart　　583
References　　587
Answers to Selected Problems　　595
Index　　603

Preface

Global satellite observations coupled with increasingly sophisticated computer simulation have led to rapid advances in our understanding of the atmosphere. While opening many doors, these tools have also raised new and increasingly complex questions about atmospheric behavior. At the same time, environmental issues have brought atmospheric science to the center of science and technology, where it now plays a key role in shaping national and international policy.

An amalgam of several disciplines, atmospheric science is, by its very nature, interdisciplinary. It therefore requires one to master a wide range of skills and attracts students from varied backgrounds. Current environmental issues, like global warming and ozone depletion, have drawn increased attention to interdisciplinary problems, which involve interactions between traditional subjects like dynamics, radiation, and chemistry. Yet, the demands of specialization often make the introductory course in atmospheric science the only formal opportunity to develop a broad foundation in basic physical principles and how they interact to shape atmospheric behavior.

This book is intended to serve as a text for a graduate core course in atmospheric science. Modern research problems require a unified treatment of material that, historically, has been separated into physical and dynamical meteorology, one which establishes the interrelationship of these subjects. It is in this spirit that this book develops the fundamental principles of atmospheric physics.

The text concentrates on four major themes:

1. Atmospheric thermodynamics
2. Hydrostatic equilibrium and stability
3. Atmospheric radiation and clouds
4. Atmospheric dynamics,

which represent the cornerstones of modern atmospheric research. The global energy budget, hydrological cycle, and photochemistry are also treated, but as applications of the major themes. This material has been integrated by cross-referencing subject matter, with applications drawn from a wide range of topics in global research.

The text is targeted at first-year graduate students with diverse backgrounds in physics, chemistry, mathematics, or engineering, but not necessarily formal training in thermodynamics, radiation, and fluid mechanics. It emphasizes physical concepts, which are developed from first principles. Therefore, the student without formal exposure to a subject is not at a disadvantage, the only prerequisites being a solid grounding in basic undergraduate physics and advanced calculus and some exposure to partial differential equations. The material can also be digested by undergraduates with similar preparation.

The presentation is from a Lagrangian perspective, which considers transformations of an individual air parcel moving through the circulation. In addition to its conceptual advantages, this framework affords a powerful diagnostic tool for interpreting atmospheric behavior. Each chapter opens with a development of basic principles and closes with applications from topical problems in global research. Many are illustrated with behavior on a particular day, which enables the atmosphere to be dissected in contemporaneous properties, like thermal structure, motion, trace species, and clouds, and therefore brings to light interactions among them.

Chapters 2 through 5 are devoted to atmospheric thermodynamics. Emphasis is placed on heterogeneous systems, which figure centrally in cloud formation, its interaction with radiation, and the roles of water vapor in global energetics and chemistry. Atmospheric hydrostatics are treated in Chapters 6 and 7, which interpret stratification from a Lagrangian perspective and atmospheric heating in terms of stabilizing and destabilizing influences. Chapters 8 and 9 concentrate on atmospheric radiation and cloud processes. After developing the fundamental laws governing radiative transfer, the text considers energetics under radiative and radiative–convective equilibrium. Cloud behavior is discussed in relation to the greenhouse effect, climate feedback mechanisms, and chemical processes. Chapters 10 through 16 are devoted to atmospheric dynamics. The perspective is transformed from the Lagrangian to the Eulerian description of atmospheric behavior via Reynolds' transport theorem, from which the equations of motion follow directly. Large-scale motion is treated first in terms of geostrophic and hydrostatic equilibrium and then extended to higher order to introduce vorticity dynamics and quasi-geostrophic motion. Wave propagation is developed from the paradigm of surface water waves. The general circulation is motivated by a zonally symmetric model of heat transfer in the presence of rotation, which, through its laboratory analogue, the rotating annulus, sets the stage for baroclinic instability. The book closes in Chapter 17 with an overview of the middle atmosphere that synthesizes topics developed in earlier chapters. Problems at the end of each chapter are of varied sophistication, ranging in difficulty from direct applications of the development to small computer projects.

One of the challenges in integrating material of this scope is to unify nomenclature, in which an individual quantity can be referred to by as many as half a dozen different expressions (e.g., in radiation). In defining quantities, I have adopted nomenclature that conforms to their physical meaning, for example, to distinguish scalar quantities, like trace speed, from vector quantities, like group velocity, to distinguish between properties referenced to unit mass versus to a particle, and to clearly distinguish between frames of reference. SI units and terminology have been adopted almost throughout. Exceptional are units of mb, which remain the most familiar specification of pressure, chemical reaction rates, for which cgs units are the accepted standard, and the radiative terms *intensity* and *flux*, which are preferred over *radiance* and *irradiance* to conform to more general concepts discussed elsewhere in the text.

This book has benefited from my interaction with a number of individuals. Constructive criticism on earlier versions of the text was selflessly provided by David G. Andrews, Rolando R. Garcia, Dennis L. Hartmann, Raymond Hide, David J. Hoffman, Brian J. Hoskins, James R. Holton, Julius London, Roland A. Madden, John. A. McGinley, and Gerald R. North. Thanks also go to those who supplied data and illustrations. Figures and the supporting calculations were skillfully prepared by Patrick Callaghan, Jacqueline Gratrix, and Kenneth Tanaka, several of which were organized with the guidance of Harry Hendon while I was away on sabbatical. Many of the problems evolved through interaction with John Bergman, Patrick Callaghan, Gil Compo, Gene Francis, and Andrew Fusco, from whom I learned a lot. I am grateful to the National Aeronautics and Space Administration for its continued support during the book's construction and to the Lady Davis Foundation at Hebrew University, which provided support during a sabbatical when I began this exercise. I am also grateful to Atmospheric Systems and Analysis for support to produce several of the illustrations. Last, but surely not least, I am indebted to my wife Eva for enduring the last years, so that this project could be completed.

MURRY L. SALBY

Chapter 1 | A Global View

1.1 Introduction to the Atmosphere

The earth's atmosphere is the gaseous envelope surrounding the planet. Like other planetary atmospheres, the earth's atmosphere figures centrally in transfers of energy between the sun and the planet's surface and from one region of the globe to another; these transfers maintain thermal equilibrium and determine the planet's climate. However, the earth's atmosphere is unique in that it is related closely to the oceans and to surface processes, which, together with the atmosphere, form the basis for life.

Because it is a fluid system, the atmosphere is capable of supporting a wide spectrum of motions, ranging from turbulent eddies of a few meters to circulations having dimensions of the earth itself. By rearranging air, motions influence other atmospheric components such as water vapor, ozone, and clouds, which figure importantly in radiative and chemical processes and make the atmospheric circulation an important ingredient of the global energy budget.

1.1.1 Descriptions of Atmospheric Behavior

The mobility of fluid systems makes their description complex. Atmospheric motions can redistribute mass and constituents into an infinite variety of complex configurations. Like any fluid system, the atmosphere is governed by the laws of continuum mechanics. These can be derived from the laws of mechanics and thermodynamics governing a discrete fluid body by generalizing those laws to a continuum of such systems. In the atmosphere, the discrete system to which these laws apply is an infinitesimal fluid element or *air parcel*, defined by a fixed collection of matter.

Two frameworks are used to describe atmospheric behavior. The *Eulerian description* represents atmospheric behavior in terms of field properties, like the instantaneous distributions of temperature, motion, and constituents. Governed by partial differential equations, the field description of atmospheric behavior is convenient for numerical purposes. The *Lagrangian description* represents atmospheric behavior in terms of the properties of individual air parcels, for example, in terms of their instantaneous positions, temperatures, and constituent concentrations. Because it focuses on transformations of properties within an air parcel and on interactions between that system and its environment, the Lagrangian description offers conceptual as well as certain diagnostic advantages. For this reason, the basic principles governing atmospheric behavior are developed in this text from a Lagrangian perspective.

In the Lagrangian framework, the system considered is an individual air parcel moving through the circulation. Although it may change in form through deformation by the flow and in composition through thermodynamic and chemical processes operating internally, this system is uniquely identified by the matter comprising it initially. Mass can be transferred across the boundary of an air parcel through molecular diffusion and turbulent mixing, but such transfers are slow enough to be ignored for many applications. Then an individual parcel can change through interaction with its environment and internal transformations that alter its composition and state.

1.1.2   Mechanisms Influencing Atmospheric Behavior

Of the factors influencing atmospheric behavior, gravity is the single most important one. Even though it has no upper boundary, the atmosphere is contained by the gravitational field of the planet, which prevents atmospheric mass from escaping to space. Because it is such a strong body force, gravity determines many atmospheric properties. Most immediate is the geometry of the atmosphere. Atmospheric mass is concentrated in the lowest 10 km—less than 1% of the planet's radius. Gravitational attraction has compressed the atmosphere into a shallow layer above the earth's surface, in which mass and constituents are stratified vertically.

Through stratification of mass, gravity imposes a strong kinematic constraint on atmospheric motion. Circulations with dimensions greater than a few tens of kilometers are quasi-horizontal, so vertical displacements of air are much smaller than horizontal displacements. Under these circumstances, constituents like water vapor and ozone fan out in layers or "strata." Vertical displacements are comparable to horizontal displacements only in small-scale circulations like convective cells and fronts, which have horizontal dimensions comparable to the vertical scale of the mass distribution.

The compressibility of air complicates the description of atmospheric behavior because it allows the volume of a fluid element to change as it experiences changes in surrounding pressure. Therefore, concentrations of mass and constituents for an individual air parcel can change, even though the number of molecules remains fixed. The concentration of a chemical constituent can also change through internal transformations, which alter the number of a particular type of molecule. For example, condensation will decrease the abundance of water vapor in an air parcel that passes through a cloud system. Photodissociation of O_2 will increase the abundance of ozone in a parcel that passes through a region of sunlight.

Exchanges of energy with its environment and transformations between one form of energy and another likewise alter the properties of an air parcel. By expanding, an air parcel exchanges energy mechanically with its environment through work that it performs on the surroundings. Heat transfer, as occurs through absorption of radiant energy and conduction with the earth's surface, represents a thermal exchange of energy with a parcel's environment. Absorption of water vapor by an air parcel (e.g., through contact with a warm ocean surface) has a similar effect. When the vapor condenses, latent heat of vaporization carried by it is released to the surrounding molecules of dry air. If the condensed water then precipitates back to the surface, this process leads to a net exchange of heat between the parcel and its environment, similar to the exchange introduced through thermal conduction with the earth's surface.

Like gravity, the earth's rotation exerts an important influence on atmospheric motion and hence on distributions of atmospheric properties. Because the earth is a noninertial reference frame, the conventional laws of mechanics must be modified to account for its acceleration. Forces introduced by the earth's rotation are responsible for properties of the large-scale circulation like the flow of air around centers of low and high pressure. Those forces also inhibit meridional motion and therefore transfers of heat and constituents between the equator and poles. Consequently, rotation tends to stratify properties meridionally, just as gravity tends to stratify them vertically.

The physical processes just described do not operate independently. Instead, they are woven together into a complex fabric of radiation, chemistry, and dynamics. Interactions among these can be just as important as the individual processes themselves. For instance, radiative transfer controls the thermal structure of the atmosphere, which determines the circulation, which in turn influences the distributions of radiatively active components like water vapor, ozone, and clouds. In view of their interdependence, understanding how one of these processes influences atmospheric behavior requires an understanding of how that process is linked to others. This feature makes the study of the atmosphere an eclectic one, involving the integration of many different physical principles. This book develops the most fundamental of these.

1.2 Composition and Structure

The earth's atmosphere consists of a mixture of gases, mostly molecular nitrogen (78% by volume) and molecular oxygen (21% by volume); (see Table 1.1). Water vapor, carbon dioxide, and ozone, along with other minor constituents, comprise the remaining 1% of the atmosphere. Although they appear in very small abundances, trace species like water vapor and ozone play a key role in

Table 1.1
Atmospheric Composition[a]

Constituent	Tropospheric mixing ratio	Vertical distribution (mixing ratio)	Controlling processes
N_2	0.7808	Homogeneous	Vertical mixing
O_2	0.2095	Homogeneous	Vertical mixing
H_2O[b]	≤0.030	Decreases sharply in troposphere; increases in stratosphere; highly variable	Evaporation, condensation, transport; production by CH_4 oxidation
A	0.0093	Homogeneous	Vertical mixing
CO_2[b]	345 ppmv	Homogeneous	Vertical mixing; production by surface and anthropogenic processes
O_3[b]	10 ppmv[c]	Increases sharply in stratosphere; highly variable	Photochemical production in stratosphere; destruction at surface transport
CH_4[b]	1.6 ppmv	Homogeneous in troposphere; decreases in middle atmosphere	Production by surface processes; oxidation produces H_2O
N_2O[b]	350 ppbv	Homogeneous in troposphere; decreases in middle atmosphere	Production by surface and anthropogenic processes; dissociation in middle atmosphere; produces NO transport
CO[b]	70 ppbv	Decreases in troposphere; increases in stratosphere	Production anthropogenically and by oxidation of CH_4 transport
NO	0.1 ppbv[c]	Increases vertically	Production by dissociation of N_2O catalytic destruction of O_3
CFC–11[b] CFC–12[b]	0.2 ppbv 0.3 ppbv	Homogeneous in troposphere; decreases in stratosphere	Industrial production; mixing in troposphere; photodissociation in stratosphere

[a]Constituents are listed with volume mixing ratios representative of the Troposphere or Stratosphere, how the latter are distributed vertically, and controlling processes.
[b]Radiatively active.
[c]Stratospheric value.

the energy balance of the earth through involvement in radiative processes. Because they are created and destroyed in particular regions and are linked closely to the circulation through transport, these and other minor species are highly variable. For this reason, such species are treated separately from the primary atmospheric constituents, which are referred to simply as "dry air."

1.2.1 Description of Air

The starting point for describing atmospheric behavior is the ideal gas law

$$pV = nR^*T$$
$$= \frac{m}{M}R^*T$$
$$= mRT, \tag{1.1}$$

which constitutes the equation of state for a pure (single-component) gas. In (1.1), p, T, and M denote the pressure, temperature, and molar weight of the gas, respectively, and V, m, and $n = m/M$ refer to the volume, mass, and molar abundance of a fixed collection of matter (e.g., an air parcel). The specific gas constant R is related to the universal gas constant R^* through

$$R = \frac{R^*}{M}. \tag{1.2}$$

Equivalent forms of the ideal gas law that do not depend on the dimension of the system are

$$p = \rho RT$$
$$pv = RT, \tag{1.3}$$

where ρ and $v = 1/\rho$ (also denoted α) are the density and specific volume of the gas, respectively.

A mixture of gases, air obeys similar relationships, as do its individual components. The *partial pressure* p_i of the ith component—that pressure the ith component would exert in isolation at the same volume and temperature as the mixture—satisfies the equation of state

$$p_iV = m_iR_iT, \tag{1.4.1}$$

where R_i is the specific gas constant of the ith component. Similarly, the *partial volume* V_i—that volume the ith component would occupy in isolation at the same pressure and temperature as the mixture—satisfies the equation of state

$$pV_i = m_iR_iT . \tag{1.4.2}$$

Dalton's law asserts that the pressure of a mixture of gases equals the sum of their partial pressures:

$$p = \sum_i p_i . \tag{1.5}$$

Likewise, the volume of the mixture equals the sum of the partial volumes[1]:

$$V = \sum_i V_i . \tag{1.6}$$

---

[1] These are among several consequences of the Gibbs–Dalton law, which relates the properties of a mixture to properties of the individual components (e.g., Keenan, 1970).

The equation of state for the mixture can be obtained by summing (1.4) over all of the components:

$$pV = T \sum_i m_i R_i .$$

Then, defining the mean specific gas constant

$$\overline{R} = \frac{\sum_i m_i R_i}{m}, \qquad (1.7)$$

yields the equation of state for the mixture:

$$pV = m\overline{R}T . \qquad (1.8)$$

The mean molar weight of the mixture is defined by

$$\overline{M} = \frac{m}{n} . \qquad (1.9)$$

Because the molar abundance of the mixture is equal to the sum of the molar abundances of the individual components,

$$n = \sum_i \frac{m_i}{M_i},$$

(1.9) may be expressed

$$\overline{M} = \frac{R^* m}{\sum_i m_i (R^*/M_i)} .$$

Then applying (1.2) for the ith component together with (1.7) leads to

$$\overline{R} = \frac{R^*}{\overline{M}}, \qquad (1.10)$$

which is analogous to (1.2) for a single-component gas.

Because of their involvement in radiative and chemical processes, variable components of air must be quantified. The "absolute concentration" of the ith species is measured by its density ρ_i or, alternatively, by its *number density*

$$[i] = \left(\frac{N_A}{M_i}\right)\rho_i \qquad (1.11)$$

(also denoted n_i), where N_A is Avogadro's number and M_i is the molar weight of the species. Partial pressure p_i and partial volume V_i are other measures of absolute concentration.

The compressibility of air makes absolute concentration an ambiguous measure of a constituent's abundance. Even if a constituent is passive, namely, if the number of molecules inside an individual parcel is fixed, its absolute concentration can change through changes of volume. For this reason, a constituent's abundance is more faithfully described by the "relative concentration," which is referenced to the overall abundance air or simply dry air. The relative concentration of the ith species is measured by the *molar fraction*

$$N_i = \frac{n_i}{n}. \tag{1.12}$$

Dividing (1.4) by (1.8) and applying (1.2) for the ith component leads to

$$N_i = \frac{p_i}{p} = \frac{V_i}{V}. \tag{1.13}$$

Molar fraction uses as a reference the molar abundance of the mixture, which can vary through changes of individual species. A more convenient measure of relative concentration is mixing ratio. The *mass mixing ratio* of the ith species,

$$r_i = \frac{m_i}{m_d}, \tag{1.14}$$

where the subscript d refers to dry air, is dimensionless and expressed in g kg^{-1} for tropospheric water vapor and in parts per million by mass (ppmm or simply ppm) for stratospheric ozone. Unlike molar fraction, the reference mass m_d is constant for an individual air parcel. If the ith species is passive, namely, if it does not undergo a transformation of phase or a chemical reaction, its mass m_i is also constant, so the mixing ratio r_i is fixed for an individual air parcel.

For a trace species, such as water vapor or ozone, the mixing ratio is closely related to the molar fraction

$$N_i \cong \frac{r_i}{\epsilon_i}, \tag{1.15.1}$$

where

$$\epsilon_i = \frac{M_i}{M_d}, \tag{1.15.2}$$

because the mass of air in the presence of such species is virtually identical to that of dry air. The *volume mixing ratio* provides similar information and is distinguished from mass mixing ratio by dimensions such as parts per million by volume (ppmv) for stratospheric ozone (Problems 1.2 and 1.3). From (1.13) and (1.12), it follows that the volume mixing ratio is approximately equal to the molar fraction, which reflects the relative abundance of molecules of the ith species.

As noted earlier, the mixing ratio of a passive species is fixed for an individual air parcel. A property that is invariant for individual fluid elements is said to be *conserved*. Although constant for individual air parcels, a conserved property is almost surely not constant in space and time. Unless that property happens to be homogeneous, its distribution must vary because parcels having different values exchange positions. A conserved property is a *material tracer* because particular values track the motion of individual fluid elements. Thus, tracking particular values of r_i provides a description of how air is rearranged by the circulation and therefore of how all conserved species are redistributed.

1.2.2 Stratification of Mass

By confining mass to a shallow layer above the earth's surface, gravity exerts a profound influence on atmospheric behavior. If vertical accelerations are ignored, Newton's second law of motion applied to the column of air between some level at pressure p and a level incrementally higher at pressure $p + dp$ (Fig. 1.1) reduces to a balance between the weight of that column and the net pressure force acting on it

$$pdA - (p + dp)dA = \rho g dV,$$

where g denotes the acceleration of gravity, or

$$\frac{dp}{dz} = -\rho g \ . \tag{1.16}$$

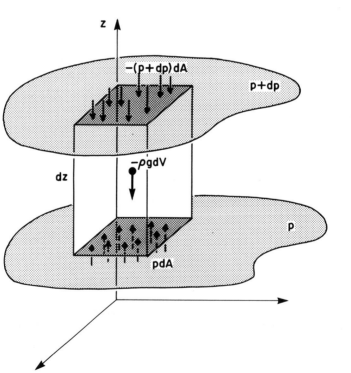

Figure 1.1 Hydrostatic balance for an incremental atmospheric column of cross-sectional area dA and height dz, bounded vertically by isobaric surfaces at pressures p and $p + dp$.

Known as *hydrostatic balance*, this simple form of mechanical equilibrium is a good approximation even if the atmosphere is in motion because vertical displacements of air and their time derivatives are small compared to the forces in (1.16). Applying the same analysis between the pressure p and the top of the atmosphere (where p vanishes) illustrates that the pressure at any level must equal the weight of the atmospheric column of unit cross-sectional area above that level.

The compressibility of air makes the density in (1.16) dependent on the pressure through the gas law. Eliminating ρ with (1.3) and integrating from the surface to an altitude z yields

$$\frac{p}{p_s} = \exp\left[-\int_{z_s}^{z} \frac{dz'}{H(z')}\right], \qquad (1.17.1)$$

where

$$H(z) = \frac{RT(z)}{g} \qquad (1.17.2)$$

is the pressure *scale height* and p_s is the surface pressure. The scale height represents the characteristic vertical dimension of the mass distribution and varies from about 8 km near the surface to 6 km in very cold regions of the atmosphere.

As illustrated by Fig. 1.2, global–mean pressure and density decrease with altitude approximately exponentially. Pressure decreases from about 1000 mb or 10^5 Pascals (Pa) at the surface to only 10% of that value at an altitude of 15 km.[2] According to hydrostatic balance, 90% of the atmosphere's mass then lies beneath this level. Pressure decreases by another factor of 10 for each additional 15 km of altitude. From a surface value of about 1.2 kg m^{-3}, the mean density also decreases with altitude at about the same rate. The sharp decrease with altitude of pressure implies that isobaric surfaces, along which $p = $ const, are quasi-horizontal. Deflections of those surfaces introduce comparatively small horizontal variations of pressure that drive atmospheric motions.

Above 100 km, pressure and density also decrease exponentially (Fig. 1.3), but at a rate which differs from that below and which varies gradually with altitude. The distinct change of behavior near 100 km marks a transition in the processes controlling the stratification of mass and the composition of air. The mean free path of molecules, which is determined by the frequency of collisions, varies inversely with air density. Consequently, the mean free path increases exponentially with altitude from about 10^{-7} m at the surface to of order 1 m at 100 km. Because it controls molecular diffusion, the mean free path determines properties of air such as viscosity and thermal conductivity.

---

[2] See Appendix A for conversions between the Standard International (SI) system of units and others.

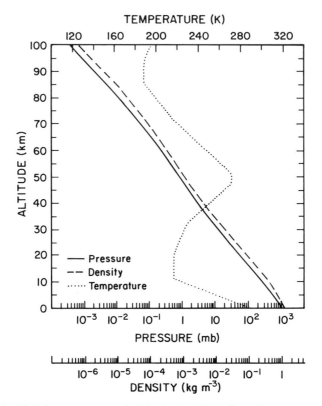

Figure 1.2 Global–mean pressure (solid), density (dashed), and temperature (dotted), as functions of altitude. *Source:* U.S. Standard Atmosphere (1976).

Diffusion of momentum and heat associated with those properties dissipate atmospheric motions by destroying gradients of velocity and temperature.

Below 100 km, the mean free path is small enough for turbulent eddies in the circulation to be only weakly damped by molecular diffusion. At those altitudes, bulk transport by turbulent air motions dominates diffusive transport of atmospheric constituents. Because turbulent air motions stir different gases with equal efficiency, the mixing ratios of passive constituents are homogeneous in this region and the components of air are said to be "well mixed." Turbulent mixing below 100 km makes the densities of passive constituents decrease with altitude at the same exponential rate, which gives air a homogeneous composition with constant mixing ratios $r_{N_2} \cong 0.78$, $r_{O_2} \cong 0.21$ and the constant gas properties[3]

$$M_d = 28.96 \text{ g mol}^{-1} \tag{1.18.1}$$

$$R_d = 287.05 \text{ J kg}^{-1}\text{K}^{-1} . \tag{1.18.2}$$

[3] Properties of dry air are tabulated in Appendix B along with other thermodynamic constants.

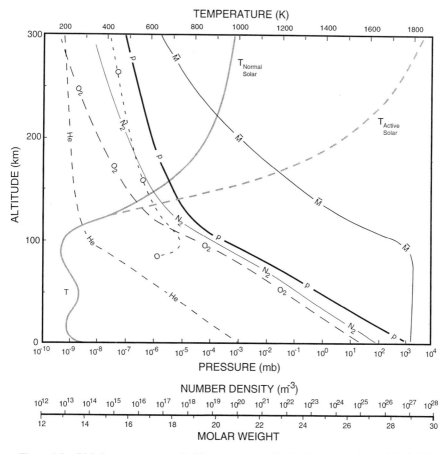

Figure 1.3   Global–mean pressure (bold), temperature (shaded), mean molar weight (solid), and number densities of atmospheric constituents, as functions of altitude. *Source:* U.S. Standard Atmosphere (1976).

The well-mixed region below 100 km is known as the *homosphere* and processes within it form the primary focus of this book.

Above 100 km, the mean free path quickly becomes larger than turbulent displacements of air. As a result, turbulent air motions are strongly damped by diffusion of momentum and heat, and diffusive transport becomes the dominant mechanism for transferring properties vertically. The transition from turbulent transport to diffusive transport occurs at the *homopause* (also known as the *turbopause*), which has an average altitude of about 100 km. Figure 1.4 shows a rocket vapor trail traversing this region. Below 100 km, the trail is marked by turbulent eddies which are produced in the wake of the rocket and which homogenize different constituents with equal efficiency. These eddies are conspicuously absent above 107 km, where molecular diffusion suppresses turbulent air motions and becomes the prevailing form of vertical transport.

The region above the homopause and below 500 km is known as the *heterosphere*. In the heterosphere, air flow is nearly laminar. Because it operates on gases according to their molar weights, molecular diffusion stratifies constituents so that the heaviest species O_2 decreases with altitude more rapidly than the second heaviest species N_2 and so forth (Fig. 1.3). For this reason, composition changes with altitude in the heterosphere, as evidenced by the mean molar weight in Fig. 1.3. Although constant below the homopause, $\overline{M}$ changes abruptly near 100 km, above which it decreases monotonically with altitude.

Diffusive separation is primarily responsible for the stratification of constituents in the heterosphere. However, photodissociation also plays a role. Energetic ultraviolet (UV) radiation incident at the top of the atmosphere dissociates molecular oxygen, providing an important source of atomic oxygen at these altitudes. In fact, O becomes the dominant form of oxygen not far above the homopause. Photodissociation of H_2O at lower altitudes by less energetic UV radiation liberates atomic hydrogen, which is gradually mixed and diffused to higher altitudes. By dissociating such molecules, energetic radiation is filtered from the solar spectrum penetrating to lower levels.

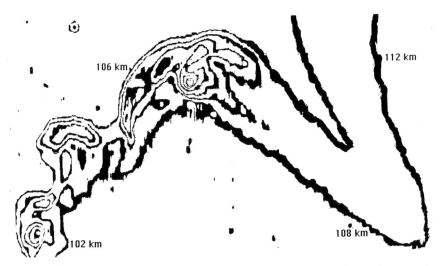

Figure 1.4   Constant-density contours of a chemical vapor trail released by a rocket traversing the turbopause. Beneath 107 km, the vapor trail is distorted by an array of turbulent eddies that form in the wake of the rocket. Above 107 km, the vapor trail remains laminar, reflecting the absence of turbulence, and expands under the action of molecular diffusion. Adapted from Roper (1977).

In the homosphere and heterosphere, atmospheric molecules interact strongly through frequent collisions. Above an altitude of about 500 km, which is referred to as the *critical level*, molecular collisions are so rare that a significant fraction of the molecules passes out of the atmosphere without sustaining a single collision. Known as the *exosphere*, the region above the critical level contains molecules that leave the denser atmosphere and move out into space. As is illustrated in Fig. 1.5, these molecules follow ballistic trajectories that are determined by the molecular velocity at the critical level and the gravitational attraction of the planet. Most of the molecules in the exosphere are captured by the earth's gravitational potential. They return to the denser atmosphere along approximately parabolic trajectories. However, some molecules have velocities great enough to escape the earth's gravitational potential entirely and those are lost to deep space.

The escape velocity v_e is determined by the kinetic energy adequate to liberate a molecule from the potential well of the planet's gravitational field. Since that energy equals the work performed to displace a molecule from the critical level to infinity,

$$\frac{1}{2}mv_e^2 = \int_a^\infty mg_0 \left(\frac{a}{r}\right)^2 dr, \tag{1.19}$$

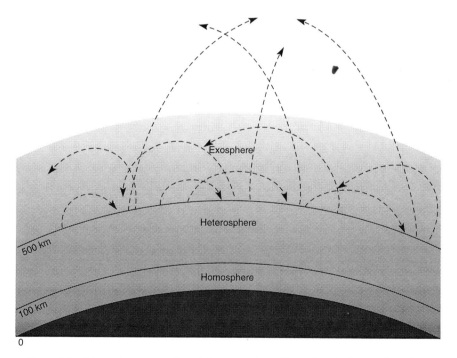

Figure 1.5 Schematic cross section of the atmosphere illustrating the homosphere, heterosphere, and the exosphere, in which molecular trajectories are shown.

where a is the earth's radius, g_0 is the gravitational attraction averaged over the surface of the earth, and the difference between a and the radial distance to the critical level is negligible. The escape velocity follows from (1.19) as

$$v_e = \sqrt{2g_0a}. \tag{1.20}$$

For earth, v_e has a value of about 11 km s^{-1}. Independent of molecular weight, v_e is the same for all molecules. However, different molecules do not all have the same distribution of velocities. Because energy is equipartitioned in a molecular ensemble (e.g., Lee, Sears, and Turcotte, 1973), lighter molecules have greater velocities than heavier ones. Consequently, lighter atmospheric constituents escape to space more readily than do heavier constituents.

The critical level of the earth's atmosphere lies near 500 km. At this altitude, the temperature is about 1000 K under conditions of normal solar activity, but can reach 2000 K during disturbed conditions (Fig. 1.3). Figure 1.6 shows for

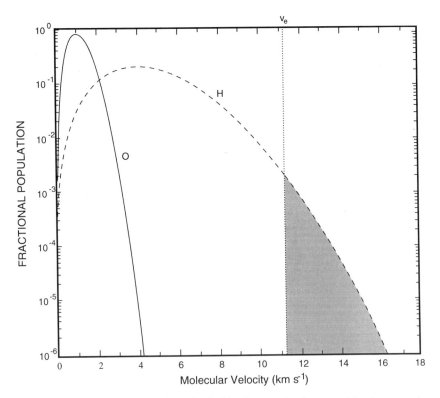

Figure 1.6 Boltzmann distribution of velocities for a molecular ensemble of oxygen atoms and hydrogen atoms. Escape velocity v_e for earth also indicated.

O and H the Boltzmann distributions of a molecular ensemble as functions of molecular velocity

$$\frac{dn}{n} = \frac{4}{\sqrt{\pi}} \frac{v^2}{v_0^3} \exp\left[-\left(\frac{v}{v_0}\right)^2\right] dv, \tag{1.21}$$

where dn/n represents the fractional number of molecules having velocities in the range $(v, v + dv)$,

$$v_0 = \sqrt{\frac{2kT}{m}} \tag{1.22}$$

is the most probably velocity, m is the mass of the molecules, and k is the Boltzmann constant. The fraction of molecules with velocities exceeding the escape velocity is

$$\frac{\Delta n_e}{n} = \int_{v_e}^{\infty} \frac{4}{\sqrt{\pi}} \frac{v^2}{v_0^3} \exp\left[-\left(\frac{v}{v_0}\right)^2\right] dv$$

$$\cong \frac{2}{\sqrt{\pi}} \left(\frac{v_e}{v_0}\right) \exp\left[-\left(\frac{v_e}{v_0}\right)^2\right] \tag{1.23}$$

for $v_0 \ll v_e$.

For atomic oxygen, the most probable velocity is $v_0 = 1.02$ km s^{-1}. The fraction of O molecules having velocities greater than v_e is only about 10^{-45}. A lower bound on the time to deplete all O molecules initially at the critical level (e.g., neglecting production of O molecules locally by photodissociation and diffusive transport from below) is given by the mean time between collisions divided by the fraction of molecules moving upward with $v > v_e$. Near 500 km, the mean time between collisions is of order 10 s. Hence, the time for all O molecules initially at the critical level to escape the earth's gravitational field is greater than 10^{46} s—far greater than the four billion years the planet has existed. Heavier species are captured by the earth's gravitational field even more effectively.

The situation for hydrogen differs sharply. The population of H is distributed over much higher velocities, so many more molecules exceed the escape velocity v_e. The most probable velocity is 4.08 km s^{-1}, whereas the fraction of molecules having velocities greater than v_e is about 10^{-4}. By previous reasoning, a significant fraction of H molecules initially at the critical level is lost to deep space after only 10^5 s or 1 day. During conditions of disturbed solar activity, temperatures at the critical level are substantially higher, so hydrogen molecules are boiled off of the atmosphere even faster. The rapid escape of atomic hydrogen from the planet's gravitational field explains why H is found only in very small abundances in the earth's atmosphere, despite its continual production by photodissociation of H_2O.

1.2.3 Thermal and Dynamical Structure

The atmosphere is categorized according to its thermal structure, which determines the dynamical properties of individual regions. The simplest picture of the atmosphere's thermal structure is provided by the vertical profile of global–mean temperature in Fig. 1.2. From the surface up to about 10 km, temperature decreases with altitude at a nearly constant *lapse rate*, which is defined as the rate of "decrease" of temperature with altitude. This layer immediately above the earth's surface is known as the *troposphere*, which means "turning sphere" and symbolizes the convective overturning that characterizes this region. Having a global–mean lapse rate of about 6.5 K km^{-1}, the troposphere contains most of what is known as weather and is driven ultimately by surface heating. The upper boundary of the troposphere or "tropopause" lies at an altitude of about 10 km (100 mb) and is marked by a sharp change of lapse rate.

The region from the tropopause to an altitude of about 85 km is known as the *middle atmosphere*. Above the tropopause, temperature first remains nearly constant and then increases in the *stratosphere*, which means "layered sphere" and is symbolic of properties at these altitudes. Increasing temperature with altitude (negative lapse rate) in the stratosphere reflects ozone heating, which results from the absorption of solar UV. Contrary to the troposphere, the stratosphere involves only weak vertical motions and is dominated by radiative processes. The upper boundary of the stratosphere or "stratopause" lies at an altitude of about 50 km (1 mb), where temperature reaches a maximum.

Above the stratopause, temperature again decreases with altitude in the *mesosphere*, where ozone heating diminishes. Convective motions and radiative processes are both important in the mesosphere. Meteor trails form in this region of the atmosphere, as do lower layers of the ionosphere during daylight hours. The "mesopause" lies at an altitude of about 85 km (0.01 mb), where a second minimum of temperature is reached.

Above the mesopause, temperature increases steadily in the *thermosphere* (compare Fig. 1.3). Unlike lower regions, the thermosphere cannot be treated as an electrically neutral continuum. Ionization of molecules by energetic solar radiation produces a plasma of free electrons and ions, each of which interacts differently with the earth's electric and magnetic fields. As is apparent in Fig. 1.3, this region of the atmosphere is influenced strongly by variations of solar activity. However, its influence on processes below the mesopause is very limited.

A more complete picture of the thermal structure of the atmosphere is provided by the zonal–mean temperature $\overline{T}$, where the overbar denotes the longitudinal average, which is shown in Fig. 1.7 as a function of latitude and altitude during northern winter. In the troposphere, temperature decreases with altitude and latitude. The tropopause, which is characterized by an abrupt change of lapse rate, is highest in the tropics ($\sim$16 km), where temperatures

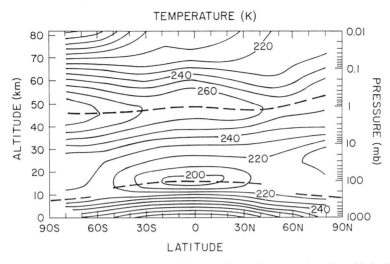

Figure 1.7 Zonal–mean temperature during northern winter as a function of latitude and altitude. Adapted from Fleming *et al.* (1988).

are quite cold, and lowest in polar regions (~8 km). A sharp change of zonal–mean lapse rate is not observed at midlatitudes, which is symbolized by a break in the tropopause. In the stratosphere, temperature increases with altitude. Temperatures are warmest over the summer pole and decrease steadily to coldest values over the winter pole. In the mesosphere, where temperature again decreases with altitude, the horizontal temperature gradient is reversed. Temperatures are actually coldest over the summer pole, which lies in perpetual daylight, and increase steadily to warmest values over the winter pole, which lies in perpetual darkness. This peculiarity of the temperature distribution, which is contrary to radiative considerations, illustrates the importance of dynamics to establishing observed thermal structure.

The thermal structure in Fig. 1.7 is related closely to the zonal-mean circulation $\bar{u}$, which is shown in Fig. 1.8 at the same time of year. In the troposphere, the circulation is characterized by subtropical *jet streams*, which strengthen with altitude up to the tropopause. These jets describe circumpolar motion that is westerly in each hemisphere.[4] Above the subtropical jets, the zonal-mean flow first weakens with altitude and then intensifies with opposite sign in the two hemispheres. In the winter hemisphere, westerlies intensify above the tropopause in the *polar-night jet*, which reaches speeds of 60 m s^{-1} in the lower mesosphere. In the summer hemisphere, westerly flow weakens above the tropopause and is then replaced by easterly flow that intensifies up to the mesosphere. Reaching speeds somewhat stronger than the zonal-mean flow

[4] In meteorological parlance, *westerly* refers to motion from the west and *easterly* to motion from the east.

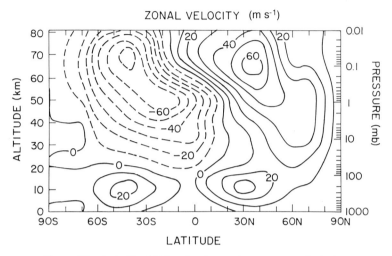

Figure 1.8 As in Fig. 1.7, but for the zonal component of velocity.

in the winter hemisphere, this easterly circulation merges below with weak easterlies in the tropical troposphere.

On individual days, the circulation is more complex and involves a good deal more variability than is represented in the zonally and time-averaged distributions in Figs. 1.7 and 1.8. Figure 1.9a shows for an individual day the instantaneous circulation on the 500-mb isobaric surface: that locus of points where the pressure equals 500 mb. Contoured over the circulation is the altitude of the 500-mb isobaric surface. Because hydrostatic equilibrium (1.17) implies a single-valued relationship between pressure and altitude (Fig. 1.2), the elevation of this isobaric surface may be interpreted analogously to the pressure on a surface of constant altitude. Because pressure decreases monotonically with altitude, low elevation of the 500-mb surface corresponds to low pressure on the surface with a constant altitude of 5 km. Similarly, high elevation of the 500-mb surface corresponds to high pressure on that constant altitude surface. Therefore, centers of low elevation that punctuate the height of the 500-mb surface in Fig. 1.9a imply centers of low pressure on a constant altitude surface, whereas centers of high elevation imply the reverse.

The 500-mb surface slopes downward toward the pole, in the direction of decreasing tropospheric temperature (Fig. 1.7). The circulation at 500 mb is characterized by westerly circumpolar flow, which corresponds to the zonal-mean subtropical jet in Fig. 1.8. But the jet stream on the individual day shown is far from zonally symmetric. Half a dozen depressions of the 500-mb surface, which are associated with *synoptic weather systems*, distort the circulation into a wavy pattern that meanders around the globe. These unsteady disturbances deflect the air stream meridionally and migrate from west to east. The circumpolar flow, although highly disturbed by these features, remains nearly parallel to contours of isobaric elevation.

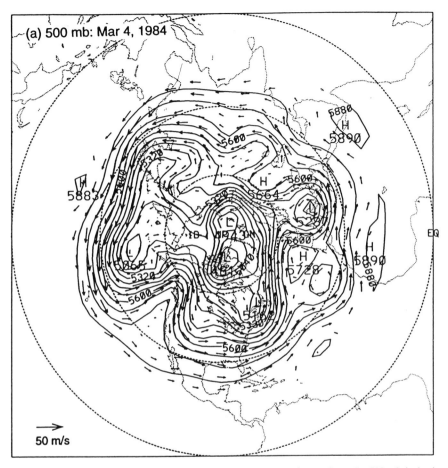

Figure 1.9   (a) Height (contours) of and horizontal velocity (vectors) on the 500-mb isobaric surface for March 4, 1984. (*continues*)

In addition to synoptic weather systems, disturbances of global dimension also appear in the 500-mb circulation. Known as *planetary waves*, these disturbances are manifested by a displacement of the circumpolar flow out of zonal symmetry (e.g., at high latitudes) and by a gradual undulation about latitude circles of the jet stream. Planetary waves are more evident in the time-mean circulation, shown in Fig. 1.9b, in which unsteady synoptic weather systems are filtered out. The time-mean circulation exhibits more zonal symmetry and therefore a greater correspondence to the subtropical jet in Fig. 1.8 than does the instantaneous circulation in Fig. 1.9a. However, the time–mean flow is still disturbed, only on larger scales. Steady planetary waves, which are not removed by time-averaging, displace the circumpolar flow out of zonal symmetry to again deflect the air stream across latitude circles. In addition, the

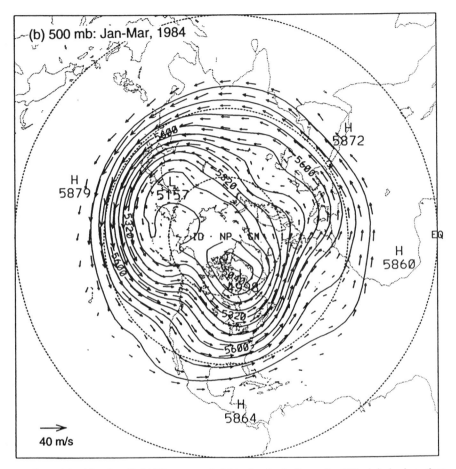

Figure 1.9 (*Continued*) (b) Time–mean height of and velocity on the 500-mb isobaric surface for January–March 1984. Isobaric heights shown in meters. From National Meteorological Center (NMC) analyses.

time–mean circulation contains locally intensified jets that mark the "North Atlantic" and "North Pacific storm tracks." Also marked by a steep meridional gradient of height, these are preferred regions of cyclone development and accompanying convective activity. By determining these and other longitudinally dependent features of the time–mean circulation, planetary waves control regional climates.

The circulation in the stratosphere is also unsteady. However, the synoptic disturbances that prevail in the troposphere are not evident in the instantaneous circulation at 10 mb, which is shown in Fig. 1.10a for the same day as in Fig. 1.9a. Instead, only planetary-scale disturbances appear at this altitude. The time–mean circulation at 10 mb (Fig. 1.10b) is characterized by

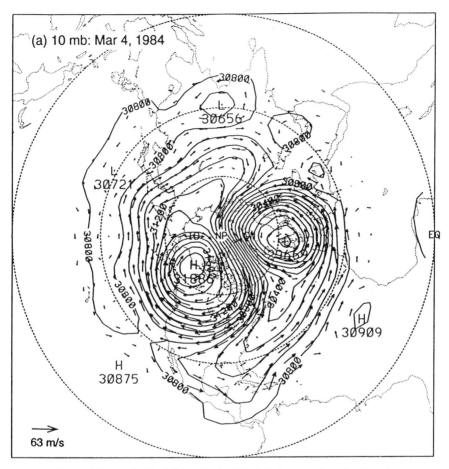

Figure 1.10 As in Fig. 1.9, but for the 10-mb isobaric surface. (*continues*)

strong westerly flow that corresponds to the polar-night jet in Fig. 1.8 and to a circumpolar vortex. Like motion at 500 mb, the time–mean circulation is disturbed from zonal symmetry by steady planetary waves that are not eliminated by averaging over time. The instantaneous circulation in Fig. 1.10a is much more disturbed than the time–mean circulation. Counterclockwise motion associated with the polar-night vortex has been displaced well off the pole and highly distorted by a clockwise circulation that has amplified and temporarily invaded the polar cap. This anomalous feature deflects the air stream through large excursions in latitude, which transport air from one radiative environment to another.

In addition to the large-scale features described, the circulation is also disturbed on smaller dimensions that are not resolved in the global analyses shown in Figs. 1.9 and 1.10. These small-scale disturbances, which are known

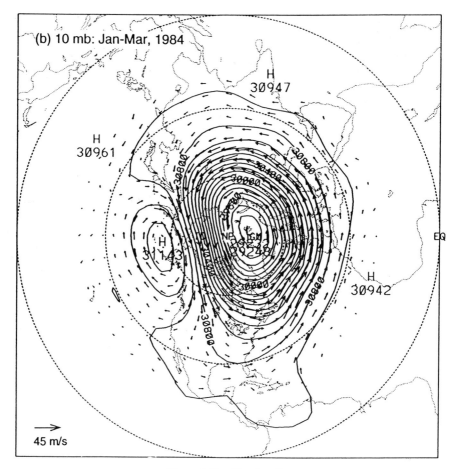

Figure 1.10   (*Continued*)

as *gravity waves*, owe their existence to buoyancy and the stratification of mass. Gravity waves are manifested in wavy patterns that appear in layered clouds, as are often observed from the ground and in satellite imagery like that shown in Fig. 1.11. Like planetary waves, gravity waves contain both transient and steady components, so they are present even in time–mean fields.

1.2.4   Trace Constituents

Beyond its primary constituents, air contains a variety of trace species. Although they exist in relatively minor abundances, several of these play key roles in radiative and chemical processes in the atmosphere. Perhaps the simplest such trace species is CO_2 because it is chemically inert away from the surface and therefore well mixed throughout the homosphere. Like N_2 and O_2,

Figure 1.11 Satellite image of wavy cloud patterns found downwind of mountainous terrain. From Scorer (1986). Reproduced with permission of Ellis Horwood Ltd.

carbon dioxide has a nearly uniform mixing ratio, $r_{CO_2} \cong 350$ ppmv. However, unlike the primary constituents of air, CO_2 is tied to human activities.

CARBON DIOXIDE

Involved in chemical and biological processes, CO_2 is produced naturally near the surface. However, increasing levels of carbon dioxide in historical records point to human activities as an important perturbation to the natural budget of CO_2. Since the dawn of the industrial age, global burning of fossil fuels has steadily increased the amount of carbon dioxide that is introduced into the atmosphere. Even though interactions with the oceans and the biosphere make the budget of CO_2 complex, the influence of human activities is strongly suggested in recent measurements. Figure 1.12 shows the record of r_{CO_2} at Mauna Loa in the latter half of the twentieth century. A rather modest annual variation of about 1%, which is associated with biological production and the growing season of the Northern Hemisphere, is superposed on a steady trend that reflects an increase of nearly 15% in just the last four decades.

The rapid increase of CO_2 in recent years has prompted concerns over global warming because of the role carbon dioxide plays in trapping radiant energy near the earth's surface. Such concerns are supported, in part, by large-

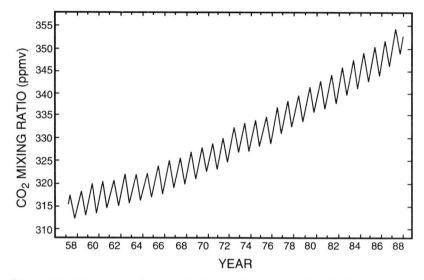

Figure 1.12 Mixing ratio of carbon dioxide over Mauna Loa, Hawaii. After Keeling *et al.* (1989).

scale numerical integrations. General circulation models (GCMs) are used to study climate by including a wide array of physical processes, many of which can be only crudely parameterized. Nevertheless, most GCMs are in agreement in predicting a 2 to 3 K increase of global–mean surface temperature in response to a doubling of CO_2, as appears imminent in the next century. In the absence of mitigating factors, such an increase of global–mean temperature would introduce important changes to the earth's climate, for example, by melting polar ice caps and increasing the level of the oceans, not to mention changes in weather and regional climates. In the stratosphere, where CO_2 plays a dominant role in the radiative energy balance through infrared (IR) cooling to space, temperature decreases as large as 10 K have been suggested, which could alter other radiatively active constituents such as ozone.

Concerns over increasing CO_2 are also supported by historical records of the earth's climate. Glacial ice cores, drilled from great depths, provide a record of atmospheric composition dating far into the past. In combination with geological information on atmospheric temperature, that record suggests a link between atmospheric CO_2 and global temperature. During previous climates of the earth, these quantities varied in a systematically related fashion. Figure 1.13 shows geological records of CO_2 and temperature that extend back 160,000 years. Sharp minima in CO_2 coincide with anomalously cold temperatures that have been identified with ice ages 20,000 and 140,000 years ago. Some caution should be exercised in interpreting these records due to uncertainties in the stability of gases trapped inside ice and in fixing the times of individual events in the records. Equally important is the issue of causality:

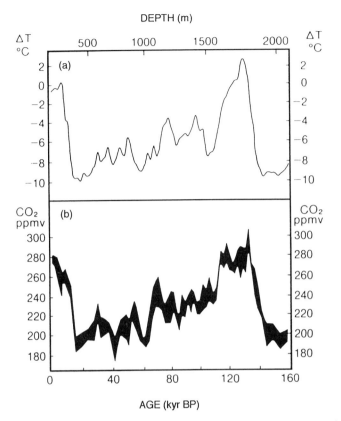

Figure 1.13 (a) Temperature change and (b) atmospheric carbon dioxide as functions of time before present (BP), inferred from an ice core drilled at Vostok, Antarctica. Reprinted with permission from *Nature* (Barnola *et al.*, 1987). Copyright 1987 Macmillan Magazines Limited.

Although they illustrate the interdependence of CO_2 and temperature, these geological records provide no information on which produced changes in the other. Nevertheless, geological records like those in Fig. 1.13 provide important evidence of previous climates of the earth and the roles atmospheric constituents play in the current climate.

WATER VAPOR

The uniform composition of dry air in the homosphere is not to be confused with the distributions of trace species such as water vapor and ozone. These constituents are highly variable because they are not merely redistributed by atmospheric motions, which would eventually homogenize them. Instead, water vapor and ozone are continually produced in some regions and destroyed in others. By transporting them from their source regions to their sink regions, the circulation exerts an important influence on these species and makes their distributions dynamic.

Owing to its involvement in radiative processes, cloud formation, and in exchanges of energy with the oceans, water vapor is the single most important trace species in the atmosphere. The zonal–mean distribution of water vapor is shown in Fig. 1.14 as a function of latitude and altitude. Water vapor is confined almost exclusively to the troposphere. Its zonal–mean mixing ratio $\bar{r}_{H_2O}$ decreases steadily with altitude, from a maximum of about 20 g kg^{-1} at the surface in the tropics to a minimum of a few parts per million at the tropopause. The absolute concentration of water vapor, or *absolute humidity*, $\bar{\rho}_{H_2O}$ (shaded area) decreases with altitude even more rapidly. From (1.14), the density of the ith constituent is just its mixing ratio times the density of dry air

$$\rho_i = r_i \rho_d. \tag{1.24}$$

Because ρ_d decreases exponentially with altitude, water vapor tends to be concentrated in the lowest 2 km of the atmosphere. The zonal–mean mixing ratio also decreases with latitude, falling to under 5 g kg^{-1} poleward of 60°. These characteristics of water vapor reflect its production at the earth's surface, redistribution by the atmospheric circulation, and destruction at altitude and at middle and high latitudes through condensation and precipitation.

Owing to those production and destruction mechanisms and the rapid transport of air between source and sink regions, tropospheric water vapor is short

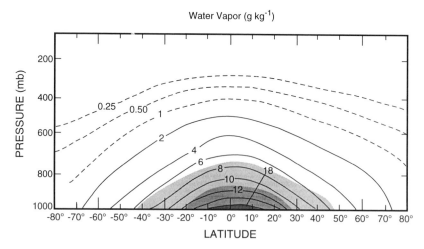

Figure 1.14   Zonal–mean mixing ratio of water vapor (contoured) and density of water vapor or absolute humidity (shaded), as functions of latitude and pressure. The shaded levels correspond to 20, 40, and 60% of the maximum value. *Source:* Oort and Peixoto (1983).

lived. A characteristic lifetime, which may be defined as the time for r_{H_2O} inside an individual parcel to change significantly, is of order days. Every few days, an air parcel encounters a warm ocean surface, where it absorbs moisture through evaporation, or a region of cloudiness, where it loses water vapor through condensation and precipitation.

Most of the water vapor in Fig. 1.14 originates near the equator at warm ocean surfaces. Consequently, transport by the circulation plays a key role in determining the mean distribution $\bar{r}_{H_2O}$. Vertical and horizontal transport, which are referred to as *convection* and *advection*, respectively, each contributes to the redistribution of r_{H_2O}. Introduced at the surface of the tropical atmosphere, water vapor is carried aloft by deep convective cells and horizontally by large-scale eddies that disperse r_{H_2O} across the globe in complex fashion. Some bodies of air escape production and destruction long enough for r_{H_2O} to be rearranged as a tracer.

Figure 1.15 presents for the day shown in Fig. 1.9 an image from the 6.3-μm water vapor channel of the geostationary satellite Meteosat-2, which observes the earth from above the Greenwich meridian (see Fig. 1.24 for geographical landmarks). The gray scale in Fig. 1.15 represents cold emission temperatures (high altitudes) as bright, and warm emission temperatures (low altitudes) as dark. Since the water vapor column is optically thick at this wavelength (i.e., outgoing radiation is emitted by H_2O at the highest levels), behavior in Fig. 1.15 corresponds to the top of the moisture layer. Bright regions indicate moisture at high altitudes and deep convective displacements of surface air, whereas dark regions indicate moisture that remains close to the earth's surface. Thus, Fig. 1.15 reflects the horizontal distribution of r_{H_2O}.

Unlike the mean distribution in Fig. 1.14, which is fairly smooth, the global distribution of water vapor on an individual day is quite variable. The moisture pattern is granular in the tropics, where water vapor has been displaced vertically by deep convective cells that have dimensions of tens to a few hundred kilometers. At middle and high latitudes, the pattern is smoother, but still complex. Swirls of light and dark mark bodies of air that are rich and lean in water vapor, respectively, for example, air that originated in tropical and extratropical regions and has been rearranged by the circulation. The local abundance reflects the history of the air parcel residing at that location, namely, where that parcel has been and what processes influencing water vapor have acted on it. A tongue of water vapor stretches northeastward from deep convection over the Amazon Basin (see Fig. 1.24), across the Atlantic, and into Africa, where it joins a tongue of drier air that is being drawn southward behind a cyclone in the eastern Atlantic (compare Fig. 1.9a). In the Southern Hemisphere, a band of high moisture is sharply delineated from neighboring lower moisture along a front that trails behind a cyclone in the South Atlantic.

More relevant to radiative processes than the local concentration is the total abundance of a species over a position on the earth's surface. The

Figure 1.15 Water vapor image on March 4, 1984, from the 6.3-μm channel of Meteosat-2, which is in geostationary orbit over the Greenwich meridian. Gray scale displays equivalent blackbody temperature from warmest (black) to coldest (white). (See Fig. 1.24 for geographical landmarks.) Supplied by the European Space Agency.

column abundance,

$$\Sigma_i = \int_0^\infty \rho_i \, dz$$
$$= \frac{1}{g} \int_0^{p_s} r_i dp, \tag{1.25}$$

describes the mass of the ith species contained by an atmospheric column of unit cross-sectional area. Figure 1.16 displays the distribution of Σ_{H_2O}, which is referred to as *total precipitable water vapor* and expressed in millimeters of liquid water, for the day shown in Fig. 1.15. Sharply confined to the tropics, Σ_{H_2O} resembles the distribution of temperature. Signatures of deep convection appear in enhanced column abundance (e.g., over tropical Africa, South America, and the eastern Atlantic), but Σ_{H_2O} is distributed more uniformly than r_{H_2O}. In part, this feature of the water vapor distribution follows from

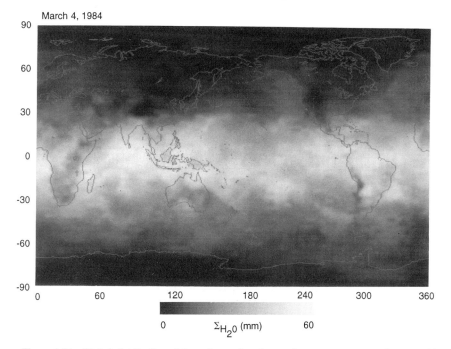

Figure 1.16 Global distribution of the column abundance of water vapor, or *total precipitable water vapor*, on March 4, 1984, derived from the TIROS Operational Vertical Sounder (TOVS). Data courtesy of I. Wittmeyer and T. Vonderharr (CSU).

the concentration of ρ_{H_2O} in the lowest 2 km. But it also indicates that, at higher altitudes, even deep convective towers contain comparatively little water in vapor phase.

Together, the zonal–mean and horizontal distributions illustrate the prevailing mechanisms controlling atmospheric water vapor. Large values of $\bar{r}_{H_2O}$ at the earth's surface in the tropics reflect production of water vapor through evaporation of warm tropical oceans. Only inside convective towers and in analogous features of greater dimension are large mixing ratios found far above the ground. Even there, vertical transport of H_2O is limited by thermodynamic constraints that prevent water vapor from reaching great altitudes, where it would be photodissociated by energetic solar radiation.

OZONE

Another radiatively active trace gas, ozone plays a key role in supporting life at the earth's surface. By intercepting harmful UV radiation, ozone allows life as we know it to exist. In fact, the evolution of the earth's atmosphere and the formation of the ozone layer are thought to be closely related to the development of life on Earth.

Geological evidence suggests that primitive forms of plant life developed *in aqua* deep in the oceans, at a time when the earth's atmosphere contained little or no oxygen and damaging UV radiation passed freely to the planet's surface. Through photosynthesis, these early forms of life are thought to have liberated oxygen, which then passed to the surface where it was photodissociated by UV radiation according to the reaction

$$O_2 + h\nu \rightarrow 2O. \tag{1.26}$$

Atomic oxygen produced by (1.26) could then recombine with O_2 to form ozone in the termolecular reaction

$$O_2 + O + M \rightarrow O_3 + M, \tag{1.27}$$

where M represents a third body needed to carry off excess energy liberated by the combination of O and O_2. Ozone created in (1.27) is dissociated by UV radiation according to the reaction

$$O_3 + h\nu \rightarrow O_2 + O. \tag{1.28}$$

If third bodies are abundant, atomic oxygen produced by (1.28) recombines almost immediately with O_2 in (1.27) to again form ozone. Thus, reactions (1.27) and (1.28) constitute a "closed cycle" that involves no net loss of components. Since the only result is the absorption of solar energy, this cycle can process UV radiation very efficiently. By removing harmful UV from the solar spectrum, ozone is thought to have allowed life to spread upward to the oceans' surfaces, where it had greater access to visible radiation, could produce more oxygen through photosynthesis, and was then able to evolve into more sophisticated forms.

The zonal–mean distribution of ozone mixing ratio is shown in Fig. 1.17 as a function of latitude and altitude. Whereas atmospheric water vapor is confined chiefly to the troposphere, ozone is concentrated in the stratosphere. Ozone mixing ratio increases sharply above the tropopause, reaching a maximum of about 10 ppmv near 30 km (10 mb). The zonal–mean ozone mixing ratio $\bar{r}_{O_3}$ is largest in the tropics, where the flux of solar UV and photodissociation of O_2 are large.

The photochemical lifetime of ozone varies sharply with altitude. In the lower stratosphere, ozone has a photochemical lifetime of several weeks. Since this is long compared to the characteristic timescale of air motion (~ 1 day), r_{O_3} behaves as a tracer at these altitudes and its distribution is controlled by dynamical influences. Should ozone find its way into the troposphere, it is quickly destroyed. Its water solubility makes O_3 readily absorbed by convective systems, which precipitate it to the surface, where it can be destroyed by a variety of oxidation processes. Thus, the troposphere serves as a sink of stratospheric ozone. The photochemical lifetime of ozone also decreases upward, to of order 1 day by 30 km and only 1 h by the stratopause. For this reason, the

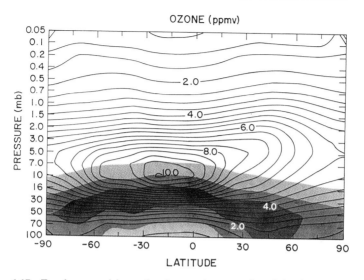

Figure 1.17   Zonal–mean mixing ratio of ozone (contoured) and density of ozone (shaded) averaged over January–February 1979, as functions of latitude and pressure, obtained from the Limb Infrared Monitor of the Stratosphere (LIMS) on board Nimbus-7. The shaded levels correspond to 20, 40, and 60% of the maximum value.

distribution of ozone in the upper stratosphere and mesosphere is controlled chiefly by photochemical influences.

Even though its mixing ratio $\bar{r}_{O_3}$ maximizes near 30 km, atmospheric ozone is concentrated in the lower stratosphere (Fig. 1.17). Because air density decreases exponentially with altitude, the density of ozone $\bar{\rho}_{O_3}$ (shaded area) is concentrated at altitudes of 10 to 20 km (1.24). Largest values are found in a shallow layer near 30 mb in the tropics, which descends and deepens in extratropical regions. The column abundance, or *total ozone*, Σ_{O_3} is expressed in Dobson units (DU), which measure in thousandths of a centimeter the depth the ozone column would assume if brought to standard temperature and pressure. Figure 1.18 shows the zonally averaged column abundance of ozone as a function of latitude and season. Values of $\overline{\Sigma}_{O_3}$ range from about 250 DU near the equator to in excess of 400 DU at high latitudes. The entire ozone column measures less than one-half of 1 cm at standard temperature and pressure! Even though most stratospheric ozone is produced in the tropics, the greatest column abundances are found at middle and high latitudes. Like the temperature distribution in the mesosphere (Fig. 1.7), this peculiarity of the ozone distribution illustrates the importance of dynamics to observed composition and structure.

As is true for water vapor, the circulation plays a key role in determining the mean distribution of O_3 and makes the global distribution on individual days dynamic. Figure 1.19 shows the distribution of Σ_{O_3} over the Northern

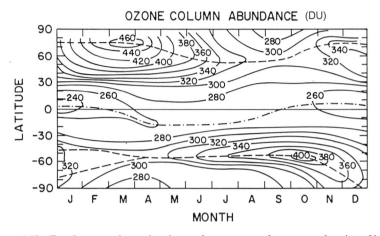

Figure 1.18   Zonal–mean column abundance of ozone, or *total ozone*, as a function of latitude and month. Based on the historical record prior to 1980. From London (1980).

and Southern Hemispheres on individual days, as observed by the Total Ozone Mapping Spectrometer (TOMS). The distribution over the Northern Hemisphere (Fig. 1.19a) has been arranged by the circulation into several anomalies in which Σ_{O_3} varies by as much as 100%. Figure 1.19a, which is contemporaneous with the 500-mb circulation in Fig. 1.9a, reveals a strong correspondence between variations of Σ_{O_3} and synoptic weather systems in the troposphere. Both evolve on a timescale of a day. The distribution over the Southern Hemisphere (Fig. 1.19b) is distinguished by column abundances of less than 200 DU (white) that delineate the "Antarctic ozone hole." Anomalously low Σ_{O_3} appears over the South Pole each year during Austral spring and then disappears some two months later. The formation of the ozone hole, which emerged in the 1980s, is attributed to increasing levels of atmospheric chlorine. Its disappearance each year occurs through dynamics.

METHANE

Several other trace gases also figure importantly in radiative and chemical processes. These include species that are produced naturally, like methane (CH_4), and species that are produced solely by human activities, like chlorofluorocarbons (CFCs). Methane is produced primarily by bacterial and surface processes that occur naturally. However, anthropogenic sources such as mining and industrial activities may constitute as much as 20% of CH_4 production.

Figure 1.19   Distribution of total ozone (color) from the Total Ozone Mapping Spectrometer (TOMS) on board Nimbus-7 and pressure (contours) on the 375 K isentropic surface (see Sec. 2.4.1) over (a) the Northern Hemisphere on March 4, 1984 and (b) the Southern Hemisphere on October 25, 1983.

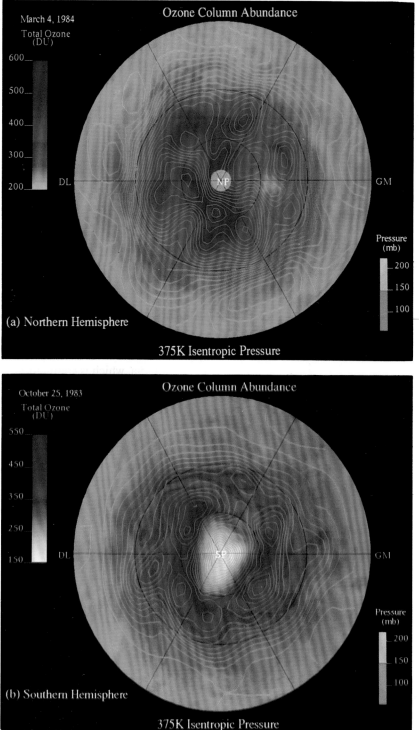

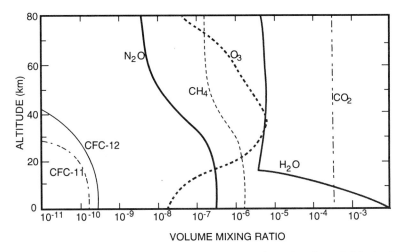

Figure 1.20   Mixing ratios of radiatively active trace species as functions of altitude. *Source:* Goody and Yung (1989).

Methane is long lived and therefore well mixed in the troposphere, where it has a constant mixing ratio of $r_{CH_4} \cong 1.7$ ppmv (Fig. 1.20). In the stratosphere, r_{CH_4} decreases with altitude as a result of oxidation. This process ultimately leads to the formation of stratospheric water vapor and is thought to be responsible for the increase of r_{H_2O} observed in the stratosphere.

Like CO_2, methane concentrations are increasing steadily. Although the exact rate is a matter of some debate, an increase of 1 to 2% per year is generally agreed on. Increasing levels of CH_4 are predicted by GCM studies to lead to increased global temperature, but not as great as those anticipated from increases of CO_2. Methane is also involved in the complex photochemistry of ozone.

CHLOROFLUOROCARBONS

Industrial halocarbons like CCl_4 (CFC-10) and $CFCl_3$ (CFC-11) and CF_2Cl_2 (CFC-12) are used widely as aerosol propellants, in refrigeration, and in a variety of manufacturing processes. As shown in Fig. 1.21, the release of these gases into the atmosphere has increased steadily since the Second World War, although the rate of increase has abated in the wake of heightened environmental concern and reduced demand. These anthropogenic species are stable in the troposphere, where their water insolubility makes them immune to normal scavenging processes associated with precipitation. Because they are long lived, CFCs are well mixed in the troposphere, CFC-11 and CFC-12 having nearly uniform mixing ratios of $r_{CFC-11} \cong 0.2$ ppbv and $r_{CFC-12} \cong 0.3$ ppbv. Their radiative properties involve industrial halocarbons in the energy balance of the earth, making them another consideration for climate change.

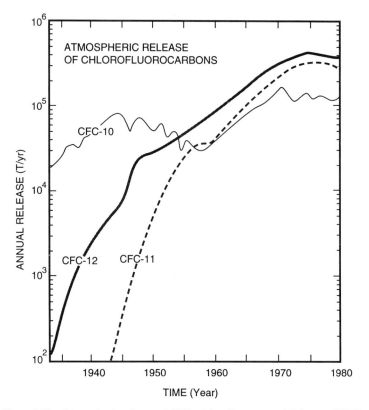

Figure 1.21   Atmospheric release of CFCs. After Brasseur and Solomon (1986).

The greatest interest in CFCs surrounds their impact on the ozone layer. Because they are long lived in the troposphere, CFCs are eventually transported into the stratosphere, where UV radiation photodissociates them. Although constant in the troposphere, mixing ratios of CFC-11 and -12 decrease in the stratosphere (Fig. 1.20). Free chlorine liberated via dissociation of CFCs can destroy ozone. Catalytic destruction of O_3 by Cl is responsible for the formation of the Antarctic ozone hole in Fig. 1.19b. Chlorine is but one piece of the scientific puzzle surrounding the ozone hole. Very high clouds that rarely form except over Antarctica also play a key role, as does the stratospheric circulation during the spring breakdown of the ozone hole.

NITROGEN COMPOUNDS

Oxides of nitrogen, such as nitrous oxide (N_2O) and nitric oxide (NO), are also relevant to the photochemistry of ozone. Nitrous oxide is produced primarily by natural means relating to bacterial processes in soils. Anthropogenic sources of N_2O include nitrogen fertilizers and combustion of fossil fuels, which may account for as much as 25% of the total production. These sources have altered the natural cycle of nitrogen by introducing it in the form of N_2O instead of N_2. Like methane, N_2O is long lived and therefore well mixed in the troposphere, with a nearly uniform mixing ratio of $r_{N_2O} \cong 300$ ppbv.

In the stratosphere, r_{N_2O} decreases with altitude due to dissociation of nitrous oxide, which represents the primary source of stratospheric NO. Like free chlorine, NO can destroy ozone catalytically. Nitric oxide is also produced as a by-product of inefficient combustion, for example, in aircraft exhaust. Nitrous oxide, similar to other anthropogenic gases, has increased steadily in recent years. Beyond its relevance to ozone, the increasing level of N_2O has implications to the energy budget of the earth because nitrous oxide is radiatively active, albeit much less so than CO_2.

Atmospheric Aerosol

Suspensions of liquid and solid particles are relevant to radiative as well as chemical processes. Ranging in size from thousandths of a micron to several hundred microns, aerosol particles are vital to atmospheric behavior because they promote cloud formation. Aerosol particles serve as condensation nuclei for water droplets and ice crystals, which do not form readily in their absence. These small particulates are produced naturally, for example, as dust, sea salt, and volcanic debris. They are also produced anthropogenically through combustion and industrial processes. An important source of aerosols is *gas-to-particle conversion*, which occurs through chemical reactions involving, among other precursors, the gaseous emission sulfur dioxide (SO_2) that is produced by industry and naturally by volcanos. Reflecting anthropogenic sources, aerosol concentrations are high in urban areas and near industrial complexes, for example, as much as an order of magnitude greater than concentrations over maritime regions (Fig. 1.22). Even higher concentrations are found in association with windblown silicates during desert dust storms. In addition to their role in cloud formation, aerosol particles scatter solar radiation at visible wavelengths and absorb IR radiation emitted by the planet's surface and atmosphere. These radiative properties involve aerosols in the energy balance of the earth, wherein their influence is thought to be as great as that of CO_2.

Aerosols play key roles in several chemical processes. One of the more notable is heterogeneous chemistry, which requires the presence of multiple phases. Reactions involving chlorine, which take place on the surfaces of cloud particles, lie at the heart of the chain of events that culminates in the formation of the Antarctic ozone hole in Fig. 1.19b. Beyond this role, aerosols are thought to figure in climatic fluctuations and possibly in the evolution of the atmosphere. Sporadic increases of aerosol produced by major volcanic eruptions have been linked to changes in thermal, optical, and chemical properties of the atmosphere.

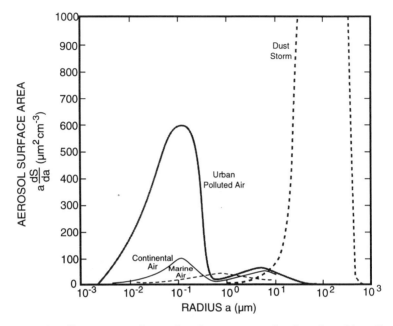

Figure 1.22 Size spectrum of aerosol surface area, as a function of particle radius. The representation is area-preserving: Equal areas under the curves represent equal number densities of aerosol particles. *Source:* Slinn (1975).

1.2.5   Clouds

One of the most striking features of the earth, when viewed from space, is its extensive coverage by clouds. At any instant, about half of the planet is cloud covered. Clouds appear with a wide range of shapes, sizes, and microphysical properties. Like trace species, the cloud field is highly dynamic because of its close relationship to the circulation. Beyond these more obvious characteristics, clouds and related phenomena figure importantly in a variety of atmospheric processes.

The reason clouds are so striking from space is that they reflect back to space a large fraction of incident solar radiation, which is concentrated at visible wavelengths. Owing to their high reflectance at visible wavelengths, clouds shield the planet from solar radiation, which is one important role they play in the earth's energy budget. Clouds also figure importantly in the budget of terrestrial radiation that is emitted in the IR by the planet's surface and atmosphere to offset solar heating. Because they absorb strongly in the IR, cloud particles of water and ice increase the atmosphere's opacity to terrestrial radiation and hence its trapping of radiant energy near the earth's surface.

Their optical properties allow clouds to be observed from space in measurements of visible and IR radiation. Figure 1.23 shows an image from the 11-μm

Figure 1.23   Infrared image at 1200 GMT on March 4, 1984, from the 11-μm channel of Meteosat-2. Gray scale displays equivalent blackbody temperature from warmest (black) to coldest (white). (See Fig. 1.24 for geographical landmarks.) Supplied by the European Space Agency.

channel of Meteosat-2 that is contemporaneous with the water vapor image in Fig. 1.15. The gray scale in Fig. 1.23 emphasizes the highest objects emitting in the IR. Bright areas correspond to cold high clouds, whereas dark areas indicate warm surfaces under cloud-free conditions. Many features in the IR image are well correlated with features in the water vapor image. This is especially true of high convective clouds in the tropics, which extend upward to the tropopause. Having horizontal dimensions of tens to a few hundred kilometers, those clouds mark deep convective towers that displace surface air vertically on timescales of hours to a day. On the other hand, the IR image reveals the subtropical Sahara and Kalahari deserts in dark areas that are far less evident in the water vapor image.

High clouds at midlatitudes also correspond well with features in the water vapor image. Most striking is the band of high cirrus that extends northeastward from South America (Fig. 1.24), across the Atlantic, and into Africa. Overlying the tongue of moisture in Fig. 1.15, this high cloud cover is part of

Figure 1.24 Visible image at 1200 GMT on March 4, 1984, from Meteosat-2. Supplied by the European Space Agency.

a frontal system that accompanies the cyclone in the eastern Atlantic, which appears in Fig. 1.23 as a spiral of lower clouds (compare Fig. 1.9a). As it advances, that cyclone draws moist tropical air poleward ahead of the front and entrains it with drier midlatitude air.

Clouds also appear in visible imagery, like that shown in Fig. 1.24, which is contemporaneous with the IR and water vapor images. Unlike those images, which follow from emission of IR radiation, the visible image represents reflected solar radiation, so it does not depend on the temperature of the objects involved. Consequently, low and comparatively warm stratiform clouds appear just as prominently as high cold convective clouds. The prevalence of stratiform features in Fig. 1.24 indicates that even shallow clouds are efficient reflectors of solar radiation. Their extensive coverage of the globe makes such clouds an important consideration in the earth's radiation budget. Swirls of light and dark in the water vapor image, which are signatures of horizontal motion, are only weakly evident in the IR image and are completely absent from the visible image.

The roles clouds play in the budgets of solar and terrestrial radiation make them a key ingredient of climate. In fact, the influence cloud cover exerts on the earth's energy balance is an order of magnitude greater than that of CO_2. Its dependence on the circulation, thermal structure, and moisture distribution, make the cloud field an especially interactive component of the earth–atmosphere system.

With the exception of shallow stratus, most clouds in Figs. 1.23 and 1.24 develop through vertical motion. Two forms of convection are distinguished in the atmosphere. *Cumulus convection*, which is often implied by the term convection alone, involves thermally driven circulations that operate on horizontal dimensions of order 100 km and smaller. Deep tropical clouds in Fig. 1.23 are a signature of cumulus convection, which displaces surface air vertically on small horizontal dimensions. *Sloping convection* is associated with forced lifting, when one body of air overrides another, and occurs coherently over large horizontal dimensions. The band of high cloud cover preceding the cyclone in the eastern Atlantic is a signature of sloping convection.

Beyond its involvement in radiative processes, convection plays a key role in the dynamics of the atmosphere and in its interaction with the oceans. Deep convection in the tropics liberates large quantities of latent heat that are released when water vapor condenses and precipitates back to the surface. Derived from heat exchange with the oceans, latent heating in the tropics represents a major source of energy for the atmosphere. It is for this reason that deep cumulus clouds are used as a proxy for atmospheric heating.

Figure 1.25a shows a nearly instantaneous image of the global cloud field, as constructed from 11-μm radiances measured aboard six satellites. The highest (brightest) clouds are found in the tropics in a narrow band of cumulus convection. Known as the *Inter Tropical Convergence Zone* (ITCZ), this band of organized convection is oriented parallel to the equator, except over the tropical landmasses: South America, Africa, and the "maritime continent" over Indonesia, where the zone of convection widens. Inside the ITCZ, deep convection is supported by the release of latent heat when moisture condenses. Transfers of moisture and energy make the ITCZ important to the tropical circulation and to interactions between the atmosphere and oceans.

The ITCZ emerges prominently in the time–mean cloud field, shown in Fig. 1.25b. Over maritime regions, the time–mean ITCZ appears as a narrow strip parallel to the equator, which reflects the convergence of surface air from the two hemispheres inside the Hadley circulation. Over tropical landmasses, time–mean cloud cover expands due to the additional influence of surface heating, which triggers convection diurnally. There, as throughout the tropics, the unsteady component of the cloud field is as large as the time–mean component. In addition to operating on timescales of hours to a day, organized convection migrates north and south annually with the sun and involves the monsoons over southeast Asia and northern Australia during the solstices.

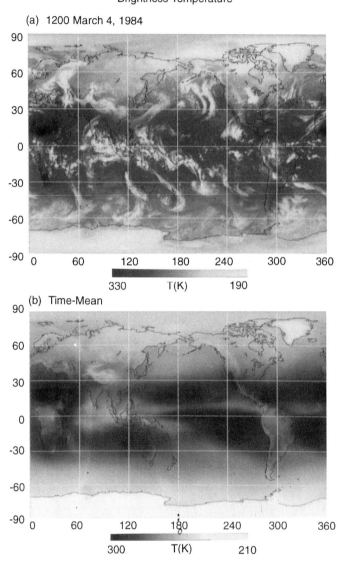

Brightness Temperature

(a) 1200 March 4, 1984

(b) Time-Mean

Figure 1.25   Global cloud image constructed from 11-μm radiances measured aboard six satellites simultaneously viewing the earth. (a) Instantaneous cloud field at 1200 GMT on March 4, 1984. (b) Time-mean cloud field for January–March 1984. Gray scale displays equivalent black-body temperature as indicated.

Time–mean cloudiness also reveals the North Atlantic and North Pacific storm tracks, where convection is organized by synoptic weather systems. Several such systems are evident in the instantaneous cloud field in Fig. 1.25a. In the Southern Hemisphere, they are distributed throughout a nearly continuous storm track. This feature of the tropospheric circulation reflects the relative absence of major orographic features in the Southern Hemisphere, which, by exciting planetary waves, disrupt the zonal circulation of the Northern Hemisphere.

Clouds are also important in chemical processes. Condensation and precipitation constitute the primary removal mechanism for many chemical species. Gaseous pollutants that are water soluble are absorbed in cloud droplets and eliminated when those droplets precipitate to the surface. Referred to as *rain out*, this mechanism also scavenges aerosol pollutants, which serve as condensation nuclei for cloud droplets and ice crystals. Although they improve air quality, these scavenging mechanisms transfer pollutants to the surface, where they can produce "acid rain."

Another chemical process in which clouds figure importantly relates to the ozone hole in Fig. 1.19b. Because moisture is sharply confined to the troposphere, clouds form in the stratosphere only under exceptionally cold conditions. The Antarctic stratosphere is one of the coldest sites in the atmosphere and, as a result, is populated by a rare cloud form. Thin and very high, polar stratospheric clouds (PSCs) are common over the Antarctic. Heterogeneous chlorine chemistry that takes place on the surfaces of cloud particles is responsible for the formation of the ozone hole each year during Austral spring.

1.3 Radiative Equilibrium of the Planet

The driving force for the atmosphere is the absorption of solar energy at the earth's surface. Over timescales long compared to those controlling the redistribution of energy, the earth–atmosphere system is in thermal equilibrium, so the net energy gained must vanish. Consequently, absorption of solar radiation, which is concentrated in the visible and termed *shortwave* (SW) *radiation*, must be balanced by emission to space from the planet's surface and atmosphere of terrestrial radiation, which is concentrated in the IR and termed *longwave* (LW) *radiation*. This basic principle leads to a simple estimate of the mean temperature of the planet.

The earth intercepts a beam of SW radiation of cross-sectional area πa^2 and flux F_s, as illustrated in Fig. 1.26. A fraction of the intercepted radiation, the albedo $\mathcal{A}$, is reflected back to space by the planet's surface and components of the atmosphere. The remainder of the incident SW flux: $(1 - \mathcal{A})F_s$, is then absorbed by the earth–atmosphere system and distributed across the globe as it spins in the line of the beam.

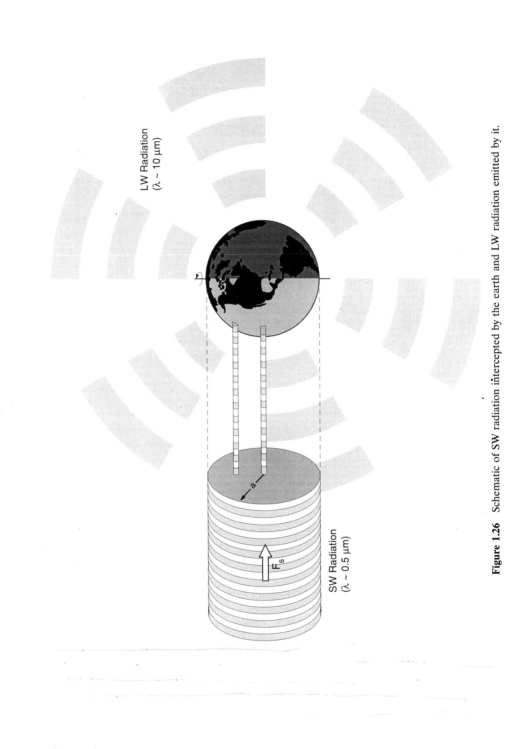

Figure 1.26 Schematic of SW radiation intercepted by the earth and LW radiation emitted by it.

To maintain thermal equilibrium, the earth and atmosphere must re-emit to space LW radiation at exactly the same rate. Also referred to as *outgoing longwave radiation* (OLR), the emission to space of terrestrial radiation is described by the Stefan–Boltzmann law

$$\pi B = \sigma T^4, \tag{1.29}$$

where πB represents the energy flux integrated over wavelength that is emitted by a blackbody at temperature T and where σ is the Stefan–Boltzmann constant. Integrating the emitted LW flux over the surface of the earth and equating the result to the SW energy absorbed by the earth–atmosphere system results in the simple energy balance

$$(1 - \mathscr{A})F_s \pi a^2 = 4\pi a^2 \sigma T_e^4, \tag{1.30.1}$$

where T_e is the equivalent blackbody temperature of the earth. Then

$$T_e = \left[\frac{(1 - \mathscr{A})F_s}{4\sigma} \right]^{\frac{1}{4}} \tag{1.30.2}$$

provides a simple estimate of the planet's temperature. An incident SW flux of $F_s = 1372$ W m^{-2} and an albedo of $\mathscr{A} = 0.30$ lead to an equivalent blackbody temperature for earth of $T_e = 255$ K. This value is some 30 K colder than the global–mean surface temperature, $T_s = 288$ K.

The discrepancy between T_s and T_e follows from the different ways the atmosphere processes SW and LW radiation. Although nearly transparent to SW radiation (wavelengths $\lambda \sim 0.5$ μm), the atmosphere is almost opaque to LW radiation ($\lambda \sim 10$ μm) that is re-emitted by the planet's surface. For this reason, SW radiation passes freely to the earth's surface, where it can be absorbed, but LW radiation emitted by the planet's surface is captured by the overlying air. Energy absorbed in an atmospheric layer is re-emitted, half upward and half back downward. The upwelling radiation is absorbed again in overlying layers, which subsequently re-emit that energy in similar fashion. This process is repeated until LW energy is eventually radiated beyond all absorbing components of the atmosphere and rejected to space. By inhibiting the transfer of energy from the earth's surface, repeated absorption and emission by the atmosphere traps LW energy and elevates the surface temperature over what it would be in the absence of an atmosphere.

The elevation of surface temperature that results from the atmosphere's different transmission characteristics to SW and LW radiation is known as the *greenhouse effect*. The greenhouse effect is controlled by the IR opacity of atmospheric constituents, which radiatively insulate the planet. In the earth's atmosphere, the primary absorbers are water vapor, clouds, and carbon dioxide. Ozone, methane, and nitrous oxide are also radiatively active at wavelengths of terrestrial radiation, as are aerosols and CFCs.

1.4 The Global Energy Budget

Because it follows from a simple energy balance, the equivalent blackbody temperature provides some insight into where LW radiation is ultimately emitted to space. The value $T_e = 255$ K corresponds to the middle troposphere, above most of the water vapor and where clouds are abundant. Most of the energy received by the atmosphere is supplied from the earth's surface, where SW radiation is absorbed. Transfers of energy from the surface constitute a heat source for the atmosphere, whereas IR cooling to space in the middle troposphere constitutes a heat sink for the atmosphere. These energy transfers drive the atmosphere and make its circulation behave as a global heat engine in the energy budget of the earth.

Energy absorbed at the earth's surface must be transmitted to the middle troposphere, where it is rejected to space. From the time it is absorbed at the surface as SW radiation until it is eventually rejected to space, energy assumes a variety of forms. Most of the energy transfer between the earth's surface and the atmosphere occurs through LW radiation. In addition, energy is transferred through thermal conduction, which is referred to as the transfer of *sensible heat*, and through the transfer of *latent heat*, when water vapor absorbed by the atmosphere condenses and precipitates back to the surface.

In addition, energy is stored internally in the atmosphere in both thermal and mechanical forms. The thermal or internal energy of air, which is measured by temperature, represents the random motion of molecules. Mechanical energy is represented in the distribution of atmospheric mass within the earth's gravitational field (potential energy) and in the motion of air (kinetic energy). These internal forms of energy are involved in the redistribution of energy within the atmosphere, but they do not contribute to globally integrated transfers between the Earth's surface, the atmosphere, and deep space.

1.4.1 Global–Mean Energy Balance

The globally averaged energy budget is illustrated in Fig. 1.27. Incoming solar energy is distributed across the earth. Therefore, the global–mean SW flux incident on the top of the atmosphere is given by $\bar{F}_s = F_s/4 = 343$ W m^{-2}, where the factor 4 represents the ratio of the surface area of the earth to the cross-sectional area of the intercepted beam of SW radiation (Fig. 1.26). Of the incident 343 W m^{-2}, a total of 106 W m^{-2} (approximately 30%) is reflected back to space: 21 W m^{-2} by air, 69 W m^{-2} by clouds, and 16 W m^{-2} by the surface. The remaining 237 W m^{-2} is absorbed in the earth–atmosphere system. Of this, 68 W m^{-2} (about 20% of the incident SW flux) is absorbed by the atmosphere: 48 W m^{-2} by atmospheric water vapor, ozone, and aerosol and 20 W m^{-2} by clouds. This leaves 169 W m^{-2} to be absorbed by the surface—nearly 50% of that incident on the top of the atmosphere.

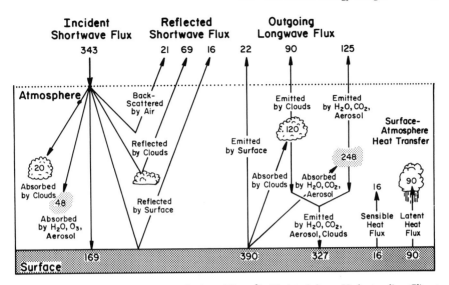

Figure 1.27 Global–mean energy budget (W m^{-2}). Updated from *Understanding Climate Change* (1975) with recent satellite measurements of the earth's radiation budget. *Additional sources:* Ramanathan (1987), Ramanathan *et al.* (1989).

The 169 W m^{-2} absorbed at the ground must be re-emitted to maintain thermal equilibrium of the earth's surface. At a global–mean surface temperature of $T_s = 288$ K, the surface emits 390 W m^{-2} of LW radiation according to (1.29)—far more than it absorbs as SW radiation. Excess LW emission must be balanced by transfers of energy from other sources. Owing to the greenhouse effect, the surface also receives LW radiation that is emitted downward by the atmosphere in the amount 327 W m^{-2}. Collectively, these contributions result in a net transfer of radiative energy to the surface:

SW absorption	+	LW absorption from atmosphere	−	LW emission	=	Net radiative forcing of surface
169 W m^{-2}	+	327 W m^{-2}	−	390 W m^{-2}	=	+106 W m^{-2}.

The surplus of 106 W m^{-2} represents net "radiative heating of the surface." For equilibrium, this must be balanced by transfers of sensible and latent heat to the atmosphere. Sensible heat transfer accounts for 16 W m^{-2} through conduction with the atmosphere. Latent heat transfer accounts for the remaining 90 W m^{-2}, which follows from evaporative cooling of the oceans. Were it not for these supplemental forms of energy transfer, the earth's surface would have to be some 50 K warmer to balance the net absorption of radiant energy.

The energy budget of the atmosphere must also balance to zero to maintain thermal equilibrium. The atmosphere receives 68 W m^{-2} directly through absorption of SW radiation. Of the 390 W m^{-2} of LW radiation emitted by the earth's surface, only 22 W m^{-2} passes freely through the atmosphere and is rejected to space. The remaining 368 W m^{-2} is absorbed by the atmosphere: 120 W m^{-2} by clouds and 248 W m^{-2} by water vapor, CO_2, and aerosol. The atmosphere loses 327 W m^{-2} through LW emission to the earth's surface. Another 215 W m^{-2} is rejected to space: 90 W m^{-2} emitted by clouds and 125 W m^{-2} emitted by water vapor, CO_2, and other minor constituents. Collecting these contributions gives the net flux of radiative energy to the atmosphere:

SW absorption	+	LW absorption from surface	−	LW emission to surface	−	LW emission to space	=	Net radiative forcing of atmosphere
68 W m^{-2}	+	368 W m^{-2}	−	327 W m^{-2}	−	215 W m^{-2}	=	−106 W m^{-2}.

The deficit of 106 W m^{-2} represents net "radiative cooling of the atmosphere," cooling that is balanced by the transfers of sensible and latent heat from the earth's surface.

1.4.2   Horizontal Distribution of Radiative Transfer

The preceding discussion focuses on vertical transfers associated with the global–mean energy balance. Were the absorption of SW energy and the emission of LW energy uniform across the earth, those vertical transfers could accomplish most of the energy exchange needed to preserve thermal equilibrium. However, geometrical considerations, variations in optical properties, and cloud cover make radiant energy transfer nonuniform across the globe.

Even if optical properties of the earth–atmosphere system were homogeneous, the absorption of solar energy would not be distributed uniformly. At low latitudes, SW radiation arrives nearly perpendicular to the earth (see Fig. 1.26). A pencil of radiation incident on the top of the atmosphere is distributed there across a perpendicular cross section. Consequently, the flux of energy crossing the top of the atmosphere at low latitudes equals that passing through the pencil. On the other hand, SW radiation at high latitudes arrives at an oblique angle. A pencil of radiation incident on the top of the atmosphere is distributed there across an oblique cross section. Because the latter has an area greater than that of a perpendicular cross section, the flux of energy crossing the top of the atmosphere is less than that passing through the pencil and therefore less than that arriving at low latitudes.

The daily-averaged SW flux incident on the top of the atmosphere is referred to as the *insolation*. Insolation depends on the *solar zenith angle*, which is measured from local vertical, and on the length of day. Figure 1.28 shows the insolation as a function of latitude and season. During equinox, insolation is a maximum on the equator and decreases poleward like the cosine of latitude. At solstice, the maximum insolation occurs over the summer pole, with values uniformly high across much of the summer hemisphere. In the win-

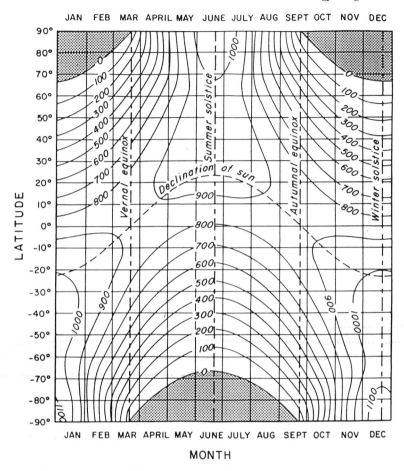

Figure 1.28   Average daily SW flux incident on the top of the atmosphere (cal cm^{-2}day^{-1}), as a function of latitude and time of year. After List (1958).

ter hemisphere, insolation decreases sharply with latitude and vanishes at the *polar-night terminator*, beyond which the earth is not illuminated.[5]

Optical properties of the atmosphere also lead to nonuniform heating of the planet. At low latitudes, SW radiation passes almost vertically through the atmosphere, so the distance traversed through absorbing constituents is minimized. However, at middle and high latitudes, SW radiation traverses the atmosphere along a slant path that involves a much longer distance and therefore results in greater absorption. A similar effect is brought about by scattering of SW radiation by atmospheric aerosol, which increases sharply

---

[5] The slight asymmetry between the hemispheres evident in Fig. 1.28 follows from the eccentricity of the earth's orbit, which brings the planet closest to the sun during January and farthest from the sun during July.

with solar zenith angle, Each of these atmospheric optical effects leads to more SW energy reaching the earth's surface at low latitudes than at high latitudes.

Optical properties of the earth's surface introduce similar effects. These emerge in OLR and reflected SW radiation observed by the Earth Radiation Budget Experiment (ERBE), which are shown in Fig. 1.29 during northern winter. Figure 1.29a shows the distribution of albedo. Low latitudes, which account for much of the earth's surface area, have extensive coverage by ocean.

(a) ALBEDO

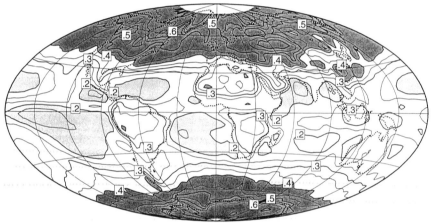

(b) LONGWAVE (Wm⁻²)

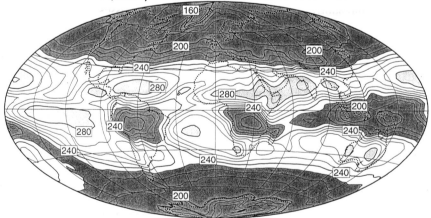

Figure 1.29 Top-of-the-atmosphere radiative properties for December–February derived from the Earth Radiation Budget Experiment (ERBE), which was on board the ERBS and NOAA-9 satellites: (a) albedo, (b) outgoing longwave radiation, (c) net radiation. Adapted from Hartmann (1993). (*continues*)

(c) NET RADIATION (Wm^{-2})

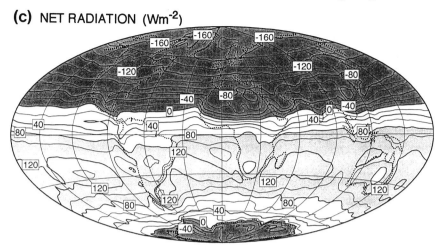

Figure 1.29   *(Continued)*

Those regions are characterized by very low albedo and thus absorb most of the SW energy incident on them. High latitudes, particularly in the winter hemisphere, have extensive coverage by snow and ice. Having a high albedo, those surfaces reflect back to space much of the SW energy incident on them, especially at large solar zenith angles, which increases scattering. Cloud cover adds to the high albedo of extratropical regions, for example, in the storm tracks. A high albedo is also evident in the tropics, where deep convection leads to maxima over the Amazon Basin, tropical Africa, and Indonesia. Flanking these regions are areas of low albedo that correspond to cloud-free oceanic regions in the subtropics and between tropical landmasses.

Outgoing longwave radiation (Fig. 1.29b) is dominated by the poleward decrease of temperature, which results in OLR decreasing steadily with latitude (1.29). At middle and high latitudes, anomalously low OLR is found over the Tibetan Plateau and Rocky Mountains, where cold surface temperatures reduce LW emission. Anomalously low OLR is also found near the equator, over the convective centers of South America, Africa, and Indonesia. High cloud tops in those regions have very low temperatures, so their LW emissions are substantially smaller than those from the surface in neighboring cloud-free regions. In fact, subtropical latitudes are marked by distinct maxima of OLR in regions flanking organized convection. Sinking motion there, which compensates rising motion inside the ITCZ, inhibits cloud formation and precipitation, maintaining arid zones that typify the subtropics.

The *net radiation*, which is defined as the difference between the SW radiation absorbed and the LW radiation emitted by the earth–atmosphere system, is shown in Fig. 1.29c. Although the global–mean net radiation must vanish for thermal equilibrium (1.30), this need not be the case locally. Low latitudes are characterized by a surplus of net radiation (warming), whereas high latitudes have a deficit of net radiation (cooling). During northern winter, the entire Northern Hemisphere poleward of 15° absorbs less energy in the form of SW radiation than it emits as LW radiation. At the same time, most of the Southern Hemisphere absorbs more energy in the form of SW radiation than it emits as LW radiation. Only over Antarctica, where permanent snow cover produces a high albedo, is the net radiation negative. Net radiation is positive and large almost uniformly across the tropics and in the subtropics of the summer hemisphere, where strong insolation and cloud-free conditions prevail. Even though, individually, albedo and OLR exhibit large geographical variations across the tropics, a cancellation between these components in regions of convective activity nearly eliminates those anomalies from the distribution of net radiation. As a result, net radiative forcing of the earth–atmosphere system exhibits a good measure of zonal symmetry.

The optical properties described above lead to nonuniform heating of the earth–atmosphere system. Low latitudes are heated radiatively, whereas middle and high latitudes are cooled radiatively. To preserve thermal equilibrium, the surplus of radiative energy arriving at low latitudes must be transferred to middle and high latitudes, where it offsets a deficit of radiative energy. About 60% of this meridional transfer of energy is accomplished by the general circulation of the atmosphere.

1.5   The General Circulation

The term *general circulation* usually refers to the aggregate of motions controlling transfers of heat, momentum, and constituents. Broadly speaking, the general circulation also includes interactions with the earth's surface (e.g., transfers of latent heat and moisture from the oceans) because atmospheric behavior, especially that operating on timescales longer than a month, is influenced importantly by such interactions. For this reason, properties of the lower boundary, like sea surface temperature (SST), figure importantly in the general circulation of the atmosphere.

The general circulation is maintained against frictional dissipation by a conversion of potential energy, which is associated with the distribution of atmospheric mass, to kinetic energy, which is associated with the motion of air. Radiative heating acts to expand the atmospheric column at low latitudes, according to hydrostatic equilibrium (1.17), and to raise its center of mass. By contrast, radiative cooling acts to compress the atmospheric column at middle and high latitudes and lower its center of mass there. The uneven distribution of mass that results introduces an imbalance of pressure forces that drives a meridional overturning, with air rising at low latitudes and sinking at middle and high latitudes.

The simple meridional circulation implied above is modified importantly by the earth's rotation. As is evident from the instantaneous circulation at

500 mb (Fig. 1.9a), the large-scale circulation remains nearly parallel to contours of isobaric height. Net radiative heating in Fig. 1.29c tends to establish a time–mean thermal structure in which isotherms and contours of isobaric height are oriented parallel to latitude circles. Consequently, the time–mean circulation at 500 mb (Fig. 1.10b) is nearly circumpolar at middle and high latitudes. Characterized by a nearly zonal jet stream, the time–mean circulation possesses only a small meridional component to transfer heat between the equator and poles. A similar conclusion applies to the stratosphere, where time–mean motion is strongly zonal (Fig. 1.10b).

For this reason, asymmetries in the instantaneous circulation that deflect air meridionally play a key role in transferring heat between the equator and poles. In the troposphere, much of the heat transfer is accomplished by unsteady synoptic weather systems, which transport heat in sloping convection that exchanges cold polar air with warm tropical air. Ubiquitous in the troposphere, those disturbances contain much of the kinetic energy at midlatitudes. They develop preferentially in the North Pacific and North Atlantic storm tracks and in the continuous storm track of the Southern Hemisphere. By rearranging air, synoptic disturbances also control the distributions of water vapor and other constituents produced at the earth's surface.

In the stratosphere and mesosphere, synoptic disturbances are absent. Planetary waves, which propagate upward from the troposphere (Fig. 1.10), play a role at these altitudes similar to the one played by synoptic disturbances in the troposphere. Generated near the earth's surface, these global-scale disturbances force the middle atmosphere mechanically. By deflecting air across latitude circles, planetary waves transport heat and constituents between low latitudes and high latitudes. Such transport is behind the largest abundances of ozone being found at middle and high latitudes (Fig. 1.18), despite its production at low latitudes.

The earth's rotation exerts a smaller influence on air motions at low latitudes. Kinetic energy there is associated primarily with *thermally direct circulations*, in which air rises in regions of heating and sinks in a regions of cooling. Thermally direct circulations in the tropics are forced by the geographical distribution of heating (e.g., as is implied by time–mean cloud cover in Fig. 1.25b). Latent heat release inside the ITCZ drives a meridional *Hadley circulation*, in which air rises near the equator and sinks at subtropical latitudes. Subsiding air in the descending branch of the Hadley circulation maintains deserts that prevail at subtropical latitudes; compare Fig. 1.29b.

Nonuniform heating also drives zonal overturning, known as a *Walker circulation*, in which air rises at longitudes of heating and sinks at longitudes of cooling. The nonuniform distribution of land and sea and asymmetries in radiative, conductive, and latent heating that accompany it lead to Walker circulations along the equator. The concentration of latent heating over Indonesia (Fig. 1.29b) forces the Pacific Walker circulation, which is illustrated in Fig. 1.30. This circulation reinforces easterly trade winds across the equa-

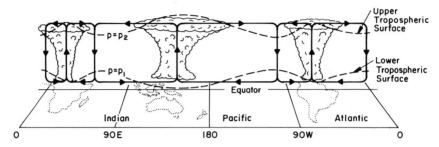

Figure 1.30 Schematic of the equatorial Walker circulation. Levels indicated correspond to isobaric surfaces in the upper and lower troposphere. Adapted from Webster (1983).

torial Pacific, and maintains the arid climate that typifies the eastern Pacific (compare Figs. 1.29a,b).

Latent heat release also excites unsteady wave motions. Another form of asymmetry in the circulation, these disturbances propagate the influence of their source region, which is subsequently deposited in regions where they are absorbed. Wave motions are also excited mechanically at the earth's surface. By displacing atmospheric mass vertically, orographic features like the Alps, the Himalayas, and the Rocky Mountains excite planetary waves and gravity waves that radiate away from those source regions. Like wave activity generated by unsteady heating, those disturbances can propagate the influence of their forcing far from regions where they are excited.

Kinetic energy associated with the general circulation is damped by frictional dissipation, which involves turbulent air motion that operates on a wide range of scales. About half of the large-scale kinetic energy is dissipated in the lowest kilometer of the atmosphere. Inside the *planetary boundary layer*, small-scale turbulence extracts energy from large scales and cascades it to small dimensions, where it is dissipated by molecular diffusion. The remaining large-scale kinetic energy is dissipated in the free atmosphere by convective motions and dynamical instability, which generate turbulence from large-scale organized motion. Large-scale circulations are also damped by thermal dissipation (e.g., by IR cooling to space), which acts on motion indirectly by destroying the accompanying temperature anomalies.

The description of the general circulation completes this overview of the atmosphere, which has been presented in terms of a wide range of behavior. Dynamical behavior comprising the general circulation follows from the stratification of mass and thermal structure, which in turn are shaped by energy transfer in the atmosphere. Because air motion controls transfers of sensible and latent heat as well as distributions of radiatively active species, the processes responsible for atmospheric behavior are interconnected. With these phenomenological ingredients as motivation, we now proceed to develop the fundamental principles of thermodynamics, hydrostatics, radiative transfer, and atmospheric dynamics that form cornerstones of atmospheric behavior.

Suggested Reading

The Boltzmann distribution is developed along with other components of molecular kinetics in *Statistical Thermodynamics* (1973) by Lee, Sears, and Turcotte.

Aeronomy of the Middle Atmosphere (1986) by Brasseur and Solomon includes a comprehensive treatment of anthropogenic trace gases and impacts they may have on the atmosphere.

The Intergovernmental Panel Report on Climate Change (1990), edited by Houghton, Jenkins, and Ephraums, provides an overview of the climate problem, numerical modeling of it, and historical evidence of previous climates.

Atmospheric Science: An Introductory Survey (1977) by Wallace and Hobbs contains an excellent introduction to the general circulation and basic concepts of radiative transfer.

A comprehensive treatment of the global energy budget and its relationship to the general circulations of the atmosphere and oceans is presented in *Global Physical Climatology* (1994) by Hartmann.

Problems

1.1. Derive expression (1.15) for the molar fraction of the ith species in terms of its mass mixing ratio.

1.2. Derive an expression for the volume mixing ratio of the ith species in terms of its mass mixing ratio.

1.3. Relate the volume mixing ratio of the ith species to the fractional abundance of i molecules present.

1.4. Show that the sum of partial volumes equals the total volume occupied by a mixture of gases (1.6).

1.5. Demonstrate that 1 atm of pressure is equivalent to that exerted by (a) a 760-mm column of mercury and (b) a 32-ft column of water. (The density of Hg at 273 K is $1.36 \ 10^4$ kg m^{-3}.)

1.6. Consider moist air at a pressure of 1 atm and a temperature of 20°C. Under these conditions, a relative humidity of 100% corresponds to a vapor pressure of 23.4 mb. At that vapor pressure, air holds the maximum abundance of water vapor possible under the foregoing conditions and is said to be *saturated*. For this mixture, determine (a) the molar fraction N_{H_2O}, (b) the mean molar weight $\overline{M}$, (c) the mean specific gas constant $\overline{R}$, (d) the absolute concentration of vapor ρ_{H_2O}, (e) the mass mixing ratio of vapor r_{H_2O}, and (f) the volume mixing ratio of vapor.

1.7. From the distribution of water vapor mixing ratio shown in Fig. 1.14, plot the vertical profile of water vapor number density $[H_2O](z)$ at (a) the equator and (b) 60°N.

1.8. In terms of the vertical profile of water vapor mixing ratio $r_{H_2O}(z)$, derive an expression for the total precipitable water vapor Σ_{H_2O} in millimeters.

1.9. If
$$r_{H_2O}(\phi, z) = r_0(\phi)e^{-z/h},$$
where $h = 3$ km and $r_0(\phi)$ is the zonal–mean mixing ratio at the surface and latitude ϕ in Fig. 1.14, use the results of Problem 1.8 to compute the total precipitable water vapor at (a) the equator and (b) 60°N.

1.10. From the mixing ratio of ozone r_{O_3} shown in Fig. 1.17, plot the vertical profile of ozone number density $[O_3](z)$ above 100 mb at (a) the equator and (b) 60°N.

1.11. As in Problem 10a, but for the contribution to total ozone $\Sigma_{O_3}(z)$ from 100 mb to an altitude z over the equator.

1.12. Clouds contribute much of the Earth's albedo, which has a value of about 0.30. Calculate the change of equivalent blackbody temperature corresponding to a 5% reduction of albedo.

1.13. The earth's equivalent blackbody temperature, $T_e = 255$ K, corresponds to a level in the middle troposphere, above the layer in which water vapor is concentrated. This also corresponds to the level from which most of the outgoing LW radiation is emitted to space. Suppose atmospheric water vapor increases to raise the top of the (optically thick) water vapor layer by 1 km, while the albedo remains unchanged. (a) What is the corresponding change of the earth's equivalent blackbody temperature? (b) If a lapse rate of 6.5 K km^{-1} is maintained inside the water vapor layer, by how much will the earth's mean surface temperature change?

1.14. By appealing to the hydrostatic relationship (1.17), contrast the vertical spacing of isobaric surfaces ($p = $ const) at a horizontal site where air is cold from one where air is warm.

1.15. Estimate the characteristic timescale for advection (a) in the troposphere, where synoptic disturbances to the subtropical jet have a characteristic zonal wavenumber of 5 and move slowly relative to the zonal-mean flow, and (b) in the stratosphere, where planetary wave disturbances to the polar-night jet have a zonal wavenumber of approximately 1 and are quasi-stationary.

1.16. The mean pressure observed in a planetary atmosphere varies with height as
$$p(z) = \frac{p_0}{1 + (z/h)^2}.$$
Determine the mean variation of temperature with height.

1.17. Venus has a mean distance from the sun of 0.7 that of Earth and an albedo of 0.75. Calculate the radiative-equilibrium temperature of Venus and compare it with the observed surface temperature of 750 K.

1.18. Calculate the fractional exchanges of energy in the global–mean energy budget (Fig. 1.27), referenced to the mean incoming flux.

Chapter 2 | Thermodynamics of Gases

The link between the circulation and transfers of radiative, sensible, and latent heat between the earth's surface and the atmosphere is thermodynamics. Thermodynamics deals with internal transformations of the energy of a system and exchanges of energy between that system and its environment. Here, we develop the principles of thermodynamics for a discrete system, namely, an air parcel moving through the circulation. In Chapter 10, those principles are generalized to a continuum of such systems, which represents the atmosphere as a whole.

2.1 Thermodynamic Concepts

A *thermodynamic system* refers to a specified collection of matter (Fig. 2.1). Such a system is said to be "closed" if no mass is exchanged with its surroundings and "open" otherwise. The air parcel that will serve as our system is, in principle, closed. In practice, mass can be exchanged with the surroundings through entrainment and mixing across the system's boundary, which is referred to as the *control surface*. In addition, trace species like water vapor can be absorbed through diffusion across the control surface. Above the planetary boundary layer, such exchanges are slow compared to other processes influencing an air parcel, so the system may be regarded as closed.

The thermodynamic state of a system is defined by the various properties characterizing it. In a strict sense, all of those properties must be specified to define the system's thermodynamic state. However, that requirement is simplified for many applications, as is discussed below.

2.1.1 Thermodynamic Properties

Two types of properties characterize the state of a system. A property that does not depend on the mass of the system is said to be *intensive*, and *extensive* otherwise. Intensive and extensive properties are usually denoted with lowercase and uppercase symbols, respectively. Pressure and temperature are examples

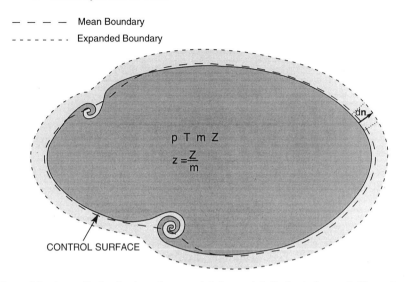

Figure 2.1 A specified collection of matter defining an infinitesimal air parcel. The system's thermodynamic state is characterized by the extensive properties m and Z and by the intensive properties p, T, and z.

of intensive properties, whereas volume is an extensive property. An intensive property z may be defined from an extensive property Z, by referencing the latter to the mass m of the system

$$z = \frac{Z}{m}, \tag{2.1}$$

in which case the intensive property is referred to as a *specific property*. The specific volume $v = V/m$ is an example. A system is said to be *homogeneous* if its properties do not vary in space, and *heterogeneous* otherwise. Because an air parcel is of infinitesimal dimension, it is by definition homogeneous so long as it involves only gas phase. On the other hand, stratification of density and pressure make the atmosphere as a whole a heterogeneous system.

A system can exchange energy with its surroundings through two fundamental mechanisms. It can perform work on the surroundings, which represents a mechanical exchange of energy with its environment. In addition, heat can be transferred across the control surface, which represents a thermal exchange of energy between the system and its environment.

2.1.2 Expansion Work

The system relevant to the atmosphere is a compressible gas, perhaps containing an aerosol of liquid and solid particles. For this reason, the primary mechanical means of exchanging energy with the environment is expansion work (Fig. 2.1). If a pressure imbalance exists across the control surface, it is

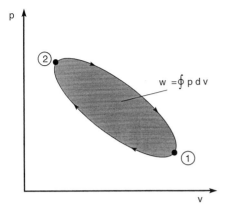

Figure 2.2 Expansion work performed by the system during a cyclic process, wherein it is restored to its initial state.

out of mechanical equilibrium. The system will then expand or contract to relieve the mechanical imbalance. By adjusting to the environmental pressure, the system performs expansion work or has such work performed on it. The incremental expansion work δW performed by displacing a section of control surface $dS = dSn$, with unit normal n, perpendicular to itself by dn is

$$\delta W = (pdS) \cdot dn$$
$$= pdV,$$

(2.2)

where $dV = dSdn$ is the incremental volume displaced by the section dS. Integrating (2.2) between the initial and final volumes yields the net work performed "by" the system

$$W_{12} = \int_{V_1}^{V_2} pdV.$$

(2.3)

Dividing both sides by the mass of the system then gives the specific work:

$$w_{12} = \int_{v_1}^{v_2} pdv.$$

(2.4.1)

For a cyclic variation, namely, one in which the initial and final states of the system are identical,

$$w = \oint pdv.$$

(2.4.2)

The cyclic work (2.4.2) is represented in the area enclosed by the path of the system in the p–v plane (Fig. 2.2). Although the system is restored to its initial state, namely, the net change of properties is zero, the same is not true of the work performed during the cycle. Unless the system returns to its initial state along the same path it followed out of that state, net work is performed during a cyclic variation.

On a fluid system, another form of work can be performed, one analogous to stirring. *Paddle work* corresponds to a rearrangement of fluid, as would occur by turning a paddle immersed in the system. The reaction of the fluid against the paddle must be overcome by a torque, which in turn produces work when exerted across an angular displacement. Although secondary to expansion work, paddle work has applications to disspative processes associated with turbulence (see Fig. 2.1).

2.1.3   Heat Transfer

Energy can also be exchanged thermally via heat transfer Q across the system's control surface. If it moves into an environment of a higher temperature, an air parcel will absorb heat from its surroundings, for example, through diffusion or thermal conduction. If the system is open, a similar process can occur through the absorption of water vapor from the surroundings, condensation and the release of latent heat, and precipitation out of the system of the condensed water. Heat can also be exchanged through radiative transfer. By communicating radiatively with surroundings at a higher temperature, an air parcel will absorb more radiant energy than it emits.

For many applications, heat transfer is secondary to processes introduced through motion, which operates on a timescale of order one day and shorter. This is especially true above the boundary layer, where turbulent mixing between bodies of air is relatively weak. In the free atmosphere and away from clouds, the prevailing form of heat transfer is radiative. Operating on a timescale of order two weeks, radiative transfer is slow by comparison with expansion work, which usually influences an air parcel on timescales of only a day.

If no heat is exchanged between a system and its environment, the control surface is said to be *adiabatic*, and *diabatic* otherwise. Because heat transfer is slow compared to other processes influencing a parcel, adiabatic behavior is a good approximation for many applications. Obviously, heat transfer must be central to processes that operate on long timescales since it is ultimately responsible for driving the atmosphere. For this reason, diabatic effects prove to be important for the long-term maintenance of the general circulation, even though they can be ignored for behavior operating on shorter timescales.

2.1.4   State Variables and Thermodynamic Processes

In general, describing the thermodynamic state of a system requires us to specify all of its properties. However, that requirement can be relaxed for gases and other substances at normal temperatures and pressures. A *pure substance* is one whose thermodynamic state is uniquely determined by any two intensive properties, which are then referred to as *state variables*. From any two state variables z_1 and z_2, a third z_3 can be determined through an *equation of state* of the form

$$f(z_1, z_2, z_3) = 0,$$

or, equivalently,

$$z_3 = g(z_1, z_2).$$

In this sense, a pure substance has two *thermodynamic degrees of freedom.* Any two state variables fix the thermodynamic state. An ideal gas is a pure substance that has as its equation of state the ideal gas law (1.1). Because no more than two intensive properties are independent, the state of such a system can be specified in the plane of any two state variables (z_1, z_2). The latter represents the *state space* of the system, which defines the collection of all possible thermodynamic states. For instance, the state space of an ideal gas may be represented by the $v - T$ plane. Then any third state variable describes a surface as a function of the two independent state variables, for example, $p = p(v, T)$, which is illustrated in Fig. 2.3 for an ideal gas.

A homogeneous system is said to be in *mechanical equilibrium* if there exists at most an infinitesimal pressure difference between the system and its surroundings (unless its control surface happens to be rigid). Likewise, a homogeneous system is said to be in *thermal equilibrium* if there exists at most an infinitesimal temperature difference between the system and its surroundings (unless its control surface happens to be adiabatic). If it is in mechanical equilibrium and in thermal equilibrium, a homogeneous system is in *thermodynamic equilibrium.*

Thermodynamics addresses how a system evolves from one state to another, for example, from one position in state space to another position. The transformation of a system between two states describes a path in state space, which is referred to as a *thermodynamic process.* As depicted in Fig. 2.4, infinitely many paths connect two thermodynamic states. Thus, a process depends on the particular path in state space followed by the system. However, the change of a state variable depends only on the initial and final states of the system, namely, it is "path independent." As illustrated in Fig. 2.2, a cyclic process can be decomposed into two legs: a forward leg, from the initial state to an intermediate state, and a reverse leg, from that intermediate state back to the initial state. Because it is path independent, the change of a state variable z along the reverse leg must equal minus the change along the forward leg. Therefore, the cyclic integral of a state variable vanishes

$$\oint dz = 0. \tag{2.5}$$

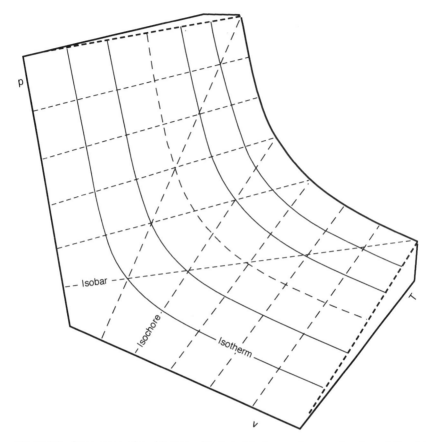

Figure 2.3 State space of an ideal gas, illustrated in terms of p as a function of the state variables v and T. Superposed are contours of constant pressure (*isobars*), constant temperature (*isotherms*), and constant volume (*isochores*).

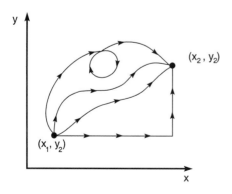

Figure 2.4 Possible thermodynamic processes between two states: (x_1, y_1) and (x_2, y_2).

The incremental change of a quantity $z(x, y)$ satisfying (2.5) may be represented as an exact differential:

$$dz = \frac{\partial z}{\partial x}dx + \frac{\partial z}{\partial y}dy, \tag{2.6}$$

which is true only under certain conditions.

Exact Differential Theorem

Consider two continuously differentiable functions $M(x, y)$ and $N(x, y)$ in a simply connected region of the $x-y$ plane. The contour integral between two points (x_0, y_0) and (x, y)

$$\int_{(x_0,y_0)}^{(x,y)} M(x', y')dx' + N(x', y')dy' \tag{2.7}$$

(e.g., along a specified contour $g(x, y)$ relating x and y) is path independent if and only if

$$\frac{\partial M}{\partial y} = \frac{\partial N}{\partial x}. \tag{2.8}$$

Under these circumstances, the quantity

$$M(x, y)dx + N(x, y)dy = dz$$

represents an exact differential. That is, there exists a function z such that

$$M(x, y) = \frac{\partial z}{\partial x},$$
$$N(x, y) = \frac{\partial z}{\partial y}, \tag{2.9}$$

in which case the contour integral (2.7) reduces to

$$\int_{z(x_0,y_0)}^{z(x,y)} dz = z(x, y) - z(x_0, y_0)$$
$$= \Delta z$$

or

$$z = z(x, y) + c.$$

The variable defined by (2.7) is a *point function*: It depends only on the evaluation point (x, y), and its net change along a contour depends only on the initial and final points of the contour. Because it is unique only up to an additive constant, only changes of a point function are significant. Also referred to as a *potential function*, $z(x, y)$ defines an *irrotational* vector field[1]

$$\boldsymbol{v} = \nabla z$$

$$= M(x, y)\boldsymbol{i} + N(x, y)\boldsymbol{j}, \tag{2.10.1}$$

which satisfies

$$\nabla \times \boldsymbol{v} \equiv 0. \tag{2.10.2}$$

Conversely, an irrotational vector field may be represented as the gradient of a scalar potential z. Gravity $\boldsymbol{g}$ is an irrotational vector field: $\nabla \times \boldsymbol{g} \equiv 0$. The work performed to displace a unit mass between two points in a gravitational field is independent of path and defines the gravitational potential Φ.

Thermodynamic state variables are point functions. As properties of the system, they depend only on the system's state but not on its history. By contrast, the work performed by the system and the heat transferred into it during a thermodynamic process are not properties of the system. Work and heat transfer are, in general, *path functions*. They depend on the path in state space followed by the system. For this reason, the thermodynamic process must be specified to define those quantities unambiguously.

Path-dependent, work and heat transfer can differ along the forward and reverse legs of a cyclic process. Consequently, the net work and heat transfer during a cyclic variation of the system need not vanish, as does the net change of a state variable (2.5). The path dependence of work and heat transfer produces a *hysteresis* in the $w–q$ plane, illustrated in Fig. 2.5. During a cyclic variation of the system, the cumulative work and heat

$$w = \int \delta w \quad \text{and} \quad q = \int \delta q$$

do not return to their original positions after the system has been restored to its initial state. The discrepancy $\Delta w = \oint \delta w$ after one cycle equals the area in the $p–v$ plane enclosed by the contour in Fig. 2.2. Due to hysteresis, successive cycles lead to a drift of the above quantities in the $w–q$ plane, which reflects the net work performed by and the heat transferred into the system.

Under special circumstances, the work performed by a system or the heat transferred into it "is" independent of path, in which case that quantity does vanish for a cyclic process. Because it is then a point function, this special form of work or heat transfer can be used to define a state variable. For instance, the displacement work in a gravitational field is path independent. Hence, that work can then be used to define the gravitational potential Φ, which is a property of the system.

---

[1] If it refers to force, the vector field $\boldsymbol{v}$ is said to be *conservative* because the net work performed along a cyclic path vanishes (2.5).

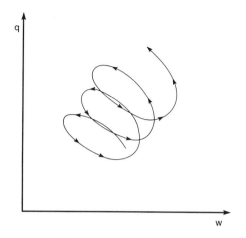

Figure 2.5 Cumulative work and heat transfer during successive thermodynamic cycles of the system. The path dependence of w and q introduces a hysteresis into those quantities, even though the system (i.e., any state variable) is restored to its initial state after each cycle.

2.2   The First Law

2.2.1   Internal Energy

The first law of thermodynamics is inspired by the observation that the work performed on an adiabatic system is independent of the process, that is, it is independent of the path in state space followed by the system. For an adiabatic process, expansion work depends only on the initial and final states of the system, so it behaves as a state variable. The *internal energy u* is defined as that state variable whose difference equals the work performed on the system under adiabatic conditions, or minus the work performed "by" the system under adiabatic conditions

$$\Delta u = -w_{ad}. \qquad (2.11.1)$$

For an incremental process,

$$du = -\delta w_{ad}, \qquad (2.11.2)$$

where δ refers to an incremental change that is, in general, path dependent, and d to one that is path independent (i.e., of a state variable). Under the special circumstances defined earlier, work is path independent, so the net work performed during a cyclic process vanishes. It follows that the system returns to its initial state along the same path it followed out of that state.

2.2.2   Diabatic Changes of State

Consider now a diabatic process, as illustrated in Fig. 2.6. Because heat is exchanged with the environment,

$$w \neq w_{ad} = -\Delta u.$$

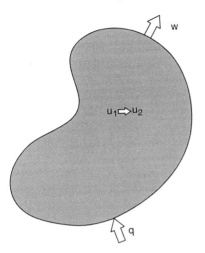

Figure 2.6   An air parcel undergoing a process between states 1 and 2, during which it performs work w and absorbs heat q.

The work performed by the system w will differ from that performed under adiabatic conditions w_{ad} by an amount q, which equals the energy transferred "into" the system through heat exchange. Hence,

$$\Delta u = q - w. \qquad (2.12.1)$$

Equation (2.12.1) constitutes the *first law of thermodynamics*. For an incremental process and for a system capable of expansion work only, the first law can be expressed

$$du = \delta q - p\,dv. \qquad (2.12.2)$$

In (2.12), $p\,dv$ represents the work performed "by" the system on its surroundings and q the heat transfer "into" the system. Although the change of internal energy between two states is path independent, the same is not true of the work performed by the system and the heat transferred into it. Except under adiabatic conditions, the work performed by the system during a change of state depends on the thermodynamic process. Likewise, except under special circumstances, the heat transferred into the system depends on the path followed in state space. Since each is path dependent, neither the work nor the heat transfer vanishes for a cyclic process. However, the change of internal energy does vanish for a cyclic process, in which case the first law reduces to

$$\oint p\,dv = \oint \delta q. \qquad (2.13)$$

The net work performed by the system during a cyclic process is balanced by the net heat absorbed by the system.

A closed system that performs work through a conversion of heat absorbed by it is said to be a *heat engine*. Conversely, a system that rejects heat through a conversion of work performed on it is a *refrigerator*.[2] We will see in Chapter 6 that individual air parcels comprising the circulation of the troposphere behave as a heat engine. By absorbing heat at the earth's surface through transfers of radiative, sensible, and latent heat, and by rejecting heat in the upper troposphere through LW emission to space, air parcels perform net work as they evolve through a thermodynamic cycle (refer to Fig. 1.27). According to (2.13), the net work performed during such a cycle equals the net heat absorbed by the parcel. That work is ultimately realized as kinetic energy, which maintains the circulation of the troposphere against frictional dissipation and makes it thermally driven.

In contrast, the circulation of the stratosphere behaves as a radiative refrigerator. For vertical motion to occur, individual air parcels must have work performed on them, which is eventually rejected to space in the form of LW radiation. In that sense, the circulation of the stratosphere is mechanically driven. Gravity waves and planetary waves that propagate upward from the troposphere and are dissipated in the stratosphere exert an influence analogous to paddle work. By rearranging stratospheric air, those disturbances drive the circulation out of radiative equilibrium, which results in net IR cooling to space.

It is convenient to introduce the state variable *enthalpy*

$$h = u + pv. \tag{2.14}$$

In terms of enthalpy, the first law becomes

$$dh = \delta q + v dp. \tag{2.15}$$

Enthalpy is useful for diagnosing processes that occur at constant pressure. Under those circumstances, the first law reduces to a statement that the change in enthalpy equals the heat transferred into the system.

2.3 Heat Capacity

Observations indicate that the heat absorbed by a homogeneous system which is maintained at constant pressure or at constant volume is proportional to the change of the system's temperature. The constants of proportionality between heat absorption and temperature change define the *specific heat capacity at constant pressure*

$$c_p = \frac{\delta q_p}{dT} \tag{2.16.1}$$

---

[2] In some texts, the term *refrigerator* refers to a particular type of heat engine. However, we reserve this term for a system that converts work into heat and use the term *heat engine* exclusively for a system that converts heat into work.

and the *specific heat capacity at constant volume*

$$c_v = \frac{\delta q_v}{dT},\tag{2.16.2}$$

where the subscripts denote *isobaric* (p = const) and *isochoric* (v = const) processes, respectively.

The specific heat capacities are related closely to the internal energy and enthalpy of the system. Because they are state variables, u and h can be expressed in terms of any two other state variables. For instance, $u = u(v, T)$, in which case

$$du = \left(\frac{\partial u}{\partial v}\right)_T dv + \left(\frac{\partial u}{\partial T}\right)_v dT.$$

Incorporating this transforms the first law (2.12.2) into

$$\left(\frac{\partial u}{\partial T}\right)_v dT + \left[\left(\frac{\partial u}{\partial v}\right)_T + p\right] dv = \delta q.$$

For an isochoric process, this reduces to

$$\left(\frac{\partial u}{\partial T}\right)_v = \frac{\delta q_v}{dT}$$

$$= c_v.\tag{2.17.1}$$

Thus, the specific heat capacity at constant volume measures the rate internal energy increases with temperature during an isochoric process. In a similar fashion, the enthalpy may be expressed $h = h(p, T)$, with which the first law (2.15) becomes

$$\left[\left(\frac{\partial h}{\partial T}\right)_T - v\right] dp + \left(\frac{\partial h}{\partial T}\right)_p dT = \delta q.$$

For an isobaric process, this reduces to

$$\left(\frac{\partial h}{\partial T}\right)_p = \frac{\delta q_p}{dT}$$

$$= c_p.\tag{2.17.2}$$

The specific heat capacity at constant pressure measures the rate enthalpy increases with temperature during an isobaric process.

In a strict sense, c_p and c_v are state variables, so they depend on pressure and temperature. However, over ranges of pressure and temperature relevant to the atmosphere, the specific heats may be regarded as constant. Therefore, the change of internal energy during an isochoric process is proportional to the change of temperature alone and similarly for the change of enthalpy during an isobaric process. It turns out that, for an ideal gas, the relationships (2.17) hold irrespective of process.

Consider the internal energy of an ideal gas in the form $u = u(p, T)$. Joule's experiment, which is pictured in Fig. 2.7, demonstrates that u is then a function of temperature alone. Two ideal gases that are initially isolated and at pressures p_1 and p_2 are brought into contact and allowed to equilibrate, for example, by rupturing a diaphragm that separates them. Observations indicate that no heat transfer takes place with the environment during this process. Then the first law reduces to a statement that the change of internal energy equals minus the work performed. However, the volume of the system (that occupied by both gases) does not change, so the work also vanishes and $\Delta u = 0$. Yet, the final equilibrated pressure clearly differs from the initial pressures of the two gases in isolation. Since the total internal energy of the system equals the sum of the contributions from the two cells, it follows that u is not a function of pressure. Therefore, $u(p, T)$ reduces to

$$u = u(T),$$

which must hold regardless of process for an ideal gas. Incorporating this result and the ideal gas law into (2.14) demonstrates that the enthalpy is likewise a function of temperature alone

$$h = h(T),$$

which again holds irrespective of process.

Based on these conclusions, equations (2.17) are valid in general for an ideal gas. Integrating with respect to temperature yields finite values of internal energy and enthalpy, which are unique up to constants of integration. It is customary to define u and h so that they vanish at a temperature of absolute zero. With that convention,

$$u = c_v T, \tag{2.18.1}$$

$$h = c_p T. \tag{2.18.2}$$

It follows that

$$c_p - c_v = R. \tag{2.19}$$

According to statistical mechanics, the specific heat at constant volume is given by

$$c_v = \frac{3}{2} R \tag{2.20.1}$$

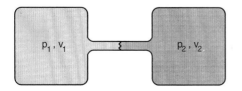

Figure 2.7 Schematic of Joule's experiment: Two ideal gases in states 1 and 2 are brought into contact by rupturing a diaphragm that separates them initially. No heat transfer with the surroundings is observed.

for a monotomic gas, and by

$$c_v = \frac{5}{2}R \tag{2.20.2}$$

for a diatomic gas (e.g., Lee, Sears, and Turcotte, 1973). These values are confirmed experimentally over a wide range of pressure and temperature relevant to the atmosphere. Taking air to be chiefly diatomic together with the value of R_d in (1.18) yields the specific heats for dry air

$$c_{vd} = 717.5 \text{ J kg}^{-1}\text{K}^{-1},$$
$$c_{pd} = 1004.5 \text{ J kg}^{-1}\text{K}^{-1}, \tag{2.21.1}$$

and the dimensionless constants

$$\gamma = c_p/c_v = 1.4, \tag{2.21.2}$$
$$\kappa = R/c_p$$
$$= (\gamma - 1)/\gamma$$
$$\cong 0.286. \tag{2.21.3}$$

With the aforementioned definitions, the first law can be expressed in the two equivalent forms:

$$c_v dT + p dv = \delta q, \tag{2.22.1}$$

$$c_p dT - v dp = \delta q. \tag{2.22.2}$$

For isochoric and isobaric processes, these expressions reduce to

$$\delta q_v = c_v dT, \tag{2.23.1}$$

$$\delta q_p = c_p dT, \tag{2.23.2}$$

respectively. Because the right-hand sides of (2.23) involve only state variables, the same must be true of the left-hand sides. Thus, under these special circumstances, heat transfer behaves as a state variable. Although generally path dependent, heat transfer during an isochoric process or during an isobaric process is uniquely determined by the change of temperature.

2.4 Adiabatic Processes

For an adiabatic process, the first law reduces to

$$c_v dT + p dv = 0, \tag{2.24.1}$$

$$c_p dT - v dp = 0. \tag{2.24.2}$$

Dividing through by T and introducing the the gas law transforms (2.24) into

$$c_v d \ln T + R d \ln v = 0, \tag{2.25.1}$$

$$c_p d \ln T - R d \ln p = 0, \tag{2.25.2}$$

which may be integrated to obtain the identities

$$T^{c_v} v^R = \text{const} \tag{2.26.1}$$

$$T^{c_p} p^{-R} = \text{const.} \tag{2.26.2}$$

A third identity, which relates p and v, can be derived from (2.25) with the aid of a differential form of the ideal gas law

$$d \ln p + d \ln v = d \ln T, \tag{2.27}$$

which follows from (1.1). Using (2.27) to eliminate T from (2.25.2) gives

$$c_v d \ln p + c_p d \ln v = 0, \tag{2.28}$$

which on integration yields

$$p^{c_v} v^{c_p} = \text{const.} \tag{2.29}$$

The three identities, (2.26.1), (2.26.2), and (2.29), can be cast in terms of dimensionless constants as

$$T v^{\gamma - 1} = \text{const}, \tag{2.30.1}$$

$$T p^{-\kappa} = \text{const}, \tag{2.30.2}$$

$$p v^{\gamma} = \text{const.} \tag{2.30.3}$$

Known as *Poisson's equations*, (2.30) define adiabatic paths in the state space of an ideal gas. Each describes the evolution of a state variable during an adiabatic process in terms of only one other state variable. Thus, the change of a single state variable together with the condition that the process be adiabatic is sufficient to determine the change of a second state variable and hence the change of thermodynamic state. For this reason, an adiabatic system possesses only one independent state variable and thus only one thermodynamic degree of freedom.

Because the state space of a pure substance is represented by the plane of any two intensive properties z_1 and z_2, a thermodynamic process describes a contour

$$g(z_1, z_2) = \text{const.}$$

Poisson's equations (2.30) are of this form and describe a family of contours, known as *adiabats*, in the plane of any two of the state variables p, T, and v (Fig. 2.8). In a similar fashion, isobaric processes describe a family of *isobars* ($p = \text{const}$), isothermal processes describe a family of *isotherms* ($T = \text{const}$),

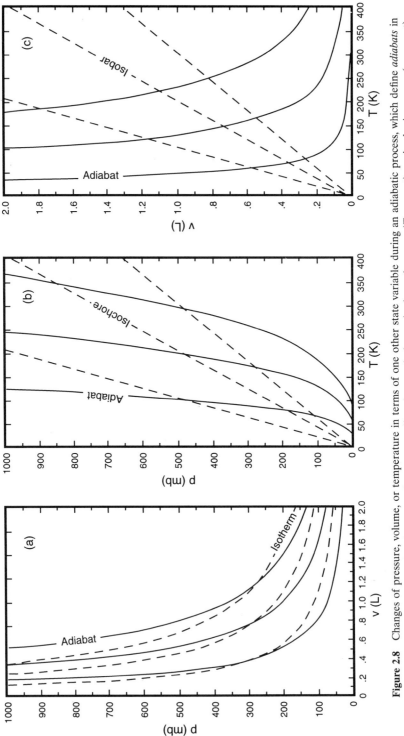

Figure 2.8 Changes of pressure, volume, or temperature in terms of one other state variable during an adiabatic process, which define *adiabats* in the state space of an ideal gas. Also shown are the corresponding changes during an isothermal process (T = const), an isochoric process (v = const), and an isobaric process (p = const), which define *isotherms*, *isochores*, and *isobars*, respectively.

and isochoric processes describe a family of *isochores* ($v =$ const), which are superposed in Fig. 2.8. Just as each of the foregoing paths is characterized by invariance of a certain state variable, so too are adiabats.

2.4.1 Potential Temperature

Poisson's relation between pressure and temperature motivates the introduction of a new state variable that is preserved during an adiabatic process. The *potential temperature* θ is defined as that temperature assumed by the system when compressed or expanded adiabatically to a reference pressure of $p_0 = 1000$ mb. According to (2.30.2), an adiabatic process from the state (p, T) to the reference state (p_0, θ) satisfies

$$\theta p_0^{-\kappa} = T p^{-\kappa}.$$

Hence, the potential temperature is described by

$$\frac{\theta}{T} = \left(\frac{p_0}{p}\right)^{\kappa}. \qquad (2.31)$$

A function of pressure and temperature, θ is a state variable. According to (2.31) and Poisson's relation (2.30.2), θ is invariant along an adiabatic path in state space.

Adiabatic behavior of individual air parcels is a good approximation for many atmospheric applications. Above the boundary layer and outside of clouds, the timescale of heat transfer is of order two weeks and thus long compared to the characteristic timescale of displacements, which influence an air parcel through changes of pressure and expansion work. For instance, vertical displacements of air and accompanying changes of pressure and volume occur in cumulus convection on a timescale of minutes to hours. Even in motions of large horizontal dimension (e.g., in sloping convection associated with synoptic disturbances), air displacements occur on a characteristic timescale of order one day. Thus, over a fairly wide range of motions, the timescale for an air parcel to adjust to changes of pressure and to perform expansion work is short compared to the characteristic timescale of heat transfer.

Under these circumstances, the potential temperature of individual air parcels is approximately conserved. An air parcel descending to greater pressure experiences an increase of temperature according to (2.30.2) due to compression work performed on it, but in such proportion to its increase of pressure as to preserve the parcel's potential temperature through (2.31). Similar considerations apply to an air parcel that is ascending. It follows that, under adiabatic conditions, θ is a conserved quantity and therefore behaves as a tracer of air motion. On timescales for which individual parcels can be regarded as adiabatic, particular values of θ track the movement of those bodies of air. Conversely, a collection of air parcels that has a particular value of θ

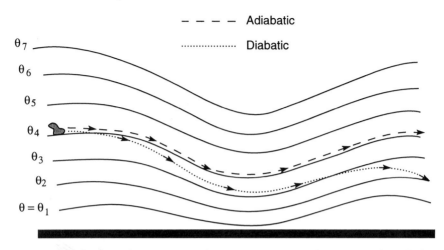

Figure 2.9 Surfaces of constant potential temperature θ. An air parcel remains coincident with a particular θ surface under adiabatic conditions, whereas it drifts across θ surfaces under diabatic conditions.

moves through space coincidentally with the corresponding isopleth of potential temperature.

The distribution of θ in the atmosphere is determined by the distributions of pressure and temperature. Because pressure decreases sharply with altitude, (2.31) implies that surfaces of constant θ tend to be quasi-horizontal like isobaric surfaces (Fig. 2.9). On timescales characteristic of air motion, an air parcel initially coincident with a certain θ surface must remain coincident with that surface, even though the position of that surface may change with time. Hence, deflections of these quasi-horizontal surfaces describe the vertical motion of individual bodies of air, which are constrained to move along them under adiabatic conditions.

Magnified values of ozone column abundance that appear in Fig. 1.10 are attributable, in part, to vertical motion along θ surfaces. Contoured over Σ_{O_3} is the pressure on the 375 K potential temperature surface, which lies just above the tropopause. Above tropospheric cyclones (see Fig. 1.9a), that surface is deflected downward to greater pressure. Air moving along the θ surface then descends and undergoes compression, which in turn increases the absolute concentration ρ_{O_3} and hence Σ_{O_3} (1.25).

Another example is provided by the distribution of nitric acid on the 470 K potential temperature surface (Fig. 2.10), which also lies in the lower stratosphere. Like ozone, HNO_3 is long lived at this level, so its mixing ratio is conserved. Therefore, r_{HNO_3} is passively rearranged by the circulation on potential temperature surfaces, along which air moves. The distribution of r_{HNO_3} on the 470 K surface has been distorted into a pentagonal pattern by tropospheric cyclones that are distributed almost uniformly along the continuous

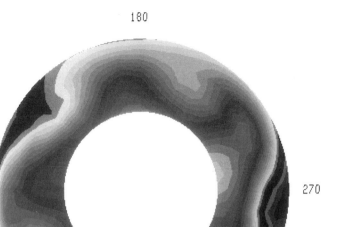

Figure 2.10 Distribution of nitric acid mixing ratio r_{HNO_3} (ppbv) on the $\theta = 470$ K potential temperature surface over the Southern Hemisphere on January 3, 1979, as observed by Nimbus-7 LIMS. Adapted from Miles and Grose (1986).

storm track of the Southern Hemisphere (see Fig. 1.25a). In fact, the pattern of r_{HNO_3} suggests that air is being overturned horizontally by those synoptic disturbances.

2.4.2 Thermodynamic Behavior Accompanying Vertical Motion

According to (2.30.2), the temperature of an air parcel moving vertically changes due to expansion work, but in such proportion to its pressure as to preserve its potential temperature. An expression for the rate that a parcel's temperature T' changes with its altitude z' under adiabatic conditions follows from (2.25.2) in combination with hydrostatic equilibrium (1.16). With the gas law, the change with altitude of environmental pressure p is described by

$$d \ln p = -\frac{dz}{H}, \tag{2.32}$$

where H is given by (1.17.2). Since the parcel's pressure p' adjusts automatically to the environmental pressure to preserve mechanical equilibrium, (2.32) also describes how the parcel's pressure changes with altitude. Then, for the

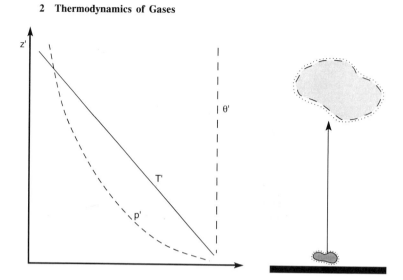

Figure 2.11 Profiles of pressure, temperature, and potential temperature for an individual air parcel ascending adiabatically.

parcel, (2.25.2) becomes

$$c_p dT' + g dz' = 0.$$

Thus, the temperature of a displaced air parcel evolving adiabatically decreases with its altitude at a constant rate

$$-\frac{dT'}{dz'} = \frac{g}{c_p}$$

$$= \Gamma_d, \tag{2.33}$$

which defines the *dry adiabatic lapse rate*. The linear profile of temperature followed by a parcel moving vertically under adiabatic conditions (Fig. 2.11) is associated with a uniform profile of potential temperature, namely, along which θ' for an individual parcel remains constant. Both describe the evolution of a displaced air parcel that does not communicate thermally with its environment, namely, one for which the only interaction with its surroundings is mechanical.

Having a value of approximately 9.8 K km^{-1}, Γ_d applies to a parcel of dry air undergoing vertical motion in a hydrostatically stratified environment. For reasons developed in Chapter 5, (2.33) also applies to an air parcel that is moist, but outside of clouds, because heat transfer remains slow compared to expansion work. Inside clouds, the simple behavior predicted by adiabatic considerations breaks down due to the release of latent heat, which accompanies condensation of water vapor and occurs on the same timescale as vertical motion. Although the behavior of an air parcel is then more complex, a generalization of (2.33) holds under those circumstances and is developed in Chapter 5.

2.5 Diabatic Processes

Whereas for many purposes the behavior of an air parcel can be regarded as adiabatic, that condition is violated in certain locations and over long timescales. Near the surface, thermal conduction and turbulent mixing become important on timescales of order one day. A similar conclusion holds inside clouds, where the release of latent heat operates on the same timescale as vertical motion. Over timescales longer than a week, radiative transfer becomes important.

Under diabatic conditions, the system interacts with its environment thermally as well as mechanically. The potential temperature of an air parcel is then no longer conserved. Instead, θ changes in proportion to heat transferred into the system, where θ and other unprimed variables are hereafter understood to refer to an individual air parcel. Taking the logarithm of (2.31) gives a differential relation among the variables θ, p, and T

$$d \ln \theta - d \ln T = -\kappa d \ln p. \tag{2.34}$$

Likewise, dividing the first law (2.22.2) by T and incorporating the gas law leads to a similar relation

$$d \ln T - \kappa d \ln p = \frac{\delta q}{c_p T}. \tag{2.35}$$

Combining (2.34) and (2.35) then yields

$$d \ln \theta = \frac{\delta q}{c_p T}. \tag{2.36}$$

Thus, the increase of potential temperature is a direct measure of the heat transferred into the system. For this reason, an air parcel will drift across potential temperature surfaces in proportion to the net heat exchanged with its environment (Fig. 2.9). Equation (2.36) can be regarded as an alternate and more compact expression of the first law, one in which expansion work has been absorbed into the state variable θ. In this light, θ for a compressible fluid like air is analogous to T for an incompressible fluid like water. Each increases in direct proportion to the heat absorbed by the system.

2.5.1 Polytropic Processes

Most of the energy exchanged between the earth's surface and the atmosphere and between one atmospheric layer and another is accomplished through radiative transfer (refer to Fig. 1.27). In fact, radiative transfer is the primary diabatic influence outside of the boundary layer and clouds. It is sometimes convenient to model radiative transfer as a *polytropic process*

$$\delta q = cdT, \tag{2.37}$$

wherein the heat transferred into the system is proportional to the system's change of temperature. The constant of proportionality c is the *polytropic specific heat capacity*. *Newtonian cooling*, in which heat transfer is proportional to the departure from an equilibrium temperature, can be regarded as a polytropic process. For deep temperature anomalies, as is characteristic of planetary waves, Newtonian cooling is a useful approximation to IR cooling to space.

For a polytropic process, the two expressions of the first law (2.22) reduce to

$$(c_v - c)dT + p\,dv = 0, \tag{2.38.1}$$

$$(c_p - c)dT - v\,dp = 0. \tag{2.38.2}$$

Equations (2.38) resemble the first law for an adiabatic process, but with modified specific heats. Consequently, previous formulas valid for an adiabatic process hold for a polytropic process with the transformation

$$\begin{aligned} c_p &\to (c_p - c), \\ c_v &\to (c_v - c). \end{aligned} \tag{2.39}$$

Then (2.39) may be used in (2.31) to define a *polytropic potential temperature* $\hat{\theta}$, which is conserved during a polytropic process (Problem 2.9). Analogous to potential temperature, $\hat{\theta}$ is the temperature assumed by an air parcel if compressed polytropically to 1000 mb, and it characterizes a family of polytropes in state space.

In terms of potential temperature, the first law becomes

$$d\ln\theta = \left(\frac{c}{c_p}\right)d\ln T. \tag{2.40}$$

If c is positive, the effective specific heats $(c_p - c)$ and $(c_v - c)$ are reduced from their adiabatic values. Therefore, a given change of pressure or volume in (2.38) leads to a greater change of temperature. For an increase of temperature, the additional warming implies heat absorption from the environment and an increase of θ by (2.40). This would be the situation for an air parcel that is cooler than its surroundings. For a decrease of temperature, the additional cooling implies heat rejection to the environment and a decrease of θ, as would occur for an air parcel that is warmer than its environment. The reverse behavior follows if c is negative.

Suggested Reading

A thorough discussion of thermodynamic properties of gases is given in *Atmospheric Thermodynamics* (1981) by Iribarne and Godson.

Statistical Thermodynamics (1973) by Lee, Sears, and Turcotte provides a clear treatment of specific heats and other gas properties from the perspective of statistical mechanics.

Problems

2.1. A plume of heated air leaves the cooling tower of a power plant at 1000 mb with a temperature of 30°C. If the air may be treated as dry, to what level will the plume ascend if the ambient temperature varies with altitude as: (a) $T(z) = 20 - 8z$(C) and (b) $T(z) = 20 + z$(C), with z in kilometers.

2.2. Suppose that the air in Problem 2.1 contains moisture, but its effect is negligible beneath a cumulus cloud, which forms overhead between 900 and 700 mb. (a) What is the potential temperature inside the plume at 950 mb? (b) How is the potential temperature inside the plume at 800 mb related to that at 950 mb? (c) How is the buoyancy of air at 800 mb related to that if the air were perfectly dry?

2.3. One mole of water is vaporized at 100°C and 1 atm of pressure (1.013×10^5 Pa), with an observed increase in volume of 3.02×10^{-2} m³. (a) How much work is performed by the water? (b) How much work would be performed if the water behaved as an ideal gas and accomplished the same change of volume isobarically? (c) The enthalpy of vaporization for water is 4.06×10^4 J mol^{-1}. What heat input is necessary to accomplish the above change of state? Why does the heat input differ from the work performed?

2.4. Demonstrate (a) that $h = h(T)$ and (b) relation (2.19).

2.5. Demonstrate that θ is conserved during an adiabatic process.

2.6. A large helium-filled balloon carries an instrumented payload from sea level into the stratosphere. If the balloon ascends rapidly enough for heat transfer across its surface to be negligible, what relative increase in volume must be accommodated for the balloon to reach 10 mb? (*Note:* The balloon exerts negligible surface tension—it serves merely to contain the helium.)

2.7. Derive (2.27) for an ideal gas.

2.8. Describe physical circumstances under which polytropic heat transfer would be a useful approximation to actual heat transfer.

2.9. Use transformation (2.29) to define a *polytropic potential temperature* $\hat{\theta}$, which is conserved during a polytropic process. How is $\hat{\theta}$ related to θ?

2.10. Through sloping convection, dry air initially at 20°C ascends from sea level to 700 mb. Calculate (a) its initial and final specific volumes, (b) its final temperature, (c) the specific work performed, and (d) changes in its specific energy and enthalpy.

2.11. A commercial aircraft is en route to New York City from Miami, where the surface pressure was 1013 mb and its altimeter was adjusted at takeoff to indicate 0 ft above sea level. If the surface pressure at New

York is 990 mb and the temperature is $-3°C$, estimate the altimeter reading when the aircraft lands if no subsequent adjustments are made.

2.12. Commercial aircraft normally cruise near 200 mb, where $T = -60°C$. (a) Calculate the temperature of air if compressed adiabatically to the cabin pressure of 800 mb. (b) How much specific heat must then be added/removed to maintain a cabin temperature of 25°C?

2.13. A system comprised of pure water is maintained at 0°C and 6.1 mb, at which the specific volumes of ice, water, and vapor are $v_i = 1.09 \times 10^{-3}$ m^{3}kg^{-1}, $v_w = 1.00 \times 10^{-3}$ m^3 kg^{-1}, $v_v = 206$ m^3 kg^{-1}. (a) How much work is performed by the water substance if 1 kg of ice melts? (b) How much work is performed by the water substance if 1 kg of water evaporates? (c) Contrast the values in (a) and (b) in light of the heat transfers required to accomplish the preceding transformations of phase: $Q_{fusion} = 0.334 \times 10^6$ J and $Q_{vaporization} = 2.50 \times 10^6$ J.

2.14. Surface winds blow down a mountain range during a *Chinook*, which is an Indian term for "snow eater." If the temperature at 14,000 ft along the continental divide is $-10°C$ and if the sky is cloud-free leeward of the divide, what is the surface temperature at Denver, which lies at approximately 5000 ft?

2.15. A scuba diver releases a bubble at 30°C at a depth of 32 ft, where the absolute pressure is 2 atm. Presume that the bubble remains intact, is large enough for heat transfer across its surface to be negligible, and condensational heating in its interior can be ignored. Beginning from hydrostatic equilibrium and the first law, (a) derive an expression for the bubble's temperature as a function of its elevation z above its initial elevation and (b) determine the bubble's temperature upon reaching the surface under the foregoing conditions. (c) How would these results be modified under fully realistic conditions? (*Note*: Normal respiration processes only about 30% of the oxygen available, so the bubble can be treated as air.)

2.16. An air parcel descends from 10 to 100 mb. (a) How much specific work is performed on the parcel if its change of state is accomplished adiabatically? (b) If its change of state is accomplished isothermally at 220 K? (c) Under the conditions of (b), what heat transfer must occur to achieve the change of state? (d) What is the parcel's change of potential temperature under the same conditions?

Chapter 3 | The Second Law and Its Implications

The first law of thermodynamics describes how the state of a system changes in response to work it performs and heat absorbed by it. The second law of thermodynamics deals with the direction of thermodynamic processes and the efficiency with which they occur. Since these characteristics control how a system evolves out of a given state, the second law also underlies the stability of thermodynamic equilibrium.

3.1 Natural and Reversible Processes

A process for which the system can be restored to its initial state, without leaving a net influence on the system or on its environment, is said to be *reversible*. A reversible process is actually an idealization: one that is devoid of friction and for which changes of state occur slowly enough for the system to remain in thermodynamic equilibrium. By contrast, a *natural process* is one that proceeds freely, for example, a spontaneous process which can be stimulated by a small perturbation to the system. Because the system is then out of thermodynamic equilibrium with its surroundings, a natural process cannot be reversed entirely, that is, without leaving a net influence on either the system or its environment. Thus, a natural process is inherently "irreversible."

Irreversibility is introduced by any process that drives the system out of thermodynamic equilibrium, as occurs during rapid changes of state. Performing work across a finite pressure difference, wherein the system is out of mechanical equilibrium, and transferring heat across a finite temperature difference, wherein the system is out of thermal equilibrium, are both irreversible. A process involving friction (e.g., turbulent mixing, which is accompanied by frictional dissipation) is also irreversible, as are transformations of phase when different phases are not at equilibrium with one another (e.g., in states away from saturation).

An example of irreversible work follows from a gas that is acted on by a piston and is maintained at a constant temperature through contact with a heat reservoir (Fig. 3.1). Suppose the gas is expanded isothermally at temperature T_{12} from state 1 to state 2 and subsequently restored to state 1 through

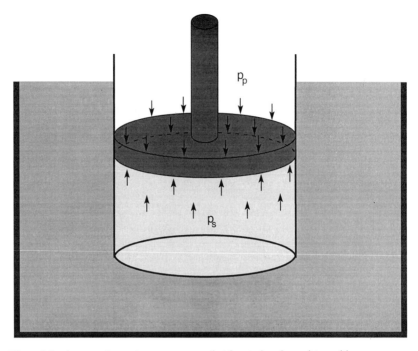

Figure 3.1 A gas at the system pressure p_s that is acted on by a piston with pressure p_p and is maintained at constant temperature through contact with a heat reservoir.

isothermal compression at the same temperature (Fig. 3.2). If the cycle is executed very slowly and without friction, the system pressure p_s is uniform throughout the gas and equals that exerted by the piston p_p at each stage of the process. Therefore, the work performed by the system during expansion is given by

$$w_{12} = \int_1^2 p_p dv$$

$$= \int_1^2 \frac{RT_{12}}{v} dv$$

$$= RT_{12} \ln\left(\frac{v_2}{v_1}\right). \tag{3.1.1}$$

Likewise, the work performed on the system during compression is

$$-w_{21} = -\int_2^1 p_p dv$$

$$= RT_{12} \ln\left(\frac{v_2}{v_1}\right), \tag{3.1.1}$$

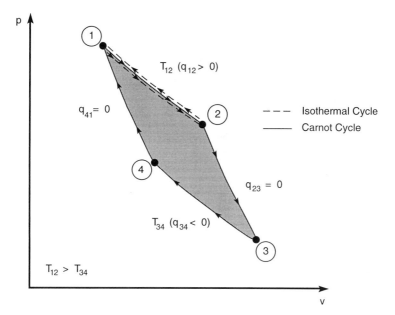

Figure 3.2   Isothermal cycle (dashed) and Carnot cycle (solid) in the p–v plane.

which equals the work performed by the system during expansion. Hence, the net work performed during the cycle vanishes. Since $\oint du = 0$, (2.13) implies

$$\oint \delta q = \oint p\,dv = 0$$

and the net heat absorbed during the cycle also vanishes. Consequently, both the system and the environment are restored to their original states, so the compression from state 1 to state 2 is entirely reversible. The same follows for the expansion and for the cycle as a whole.

If the cycle is executed rapidly, some of the gas is accelerated. Pressure is then no longer uniform across the system and p_s does not equal p_p. During rapid compression, the piston must exert a pressure which exceeds that exerted when the compression is executed slowly (e.g., to offset the reaction of the gas being accelerated). Therefore, the work performed on the system exceeds that performed under reversible conditions.[1] Since Δu vanishes, the first law implies that heat transferred out of the system during rapid compression likewise exceeds that when the compression is executed slowly. During rapid expansion, the piston must exert a pressure which is smaller than that exerted when the expansion is executed slowly. The work performed by the system is then less than that performed under reversible conditions and, since $\Delta u = 0$, so is the

---

[1] Kinetic energy generated by the excess work is eventually dissipated and lost through heat transfer to the environment.

heat transferred into the system. Thus, executing the isothermal cycle rapidly results in

$$\oint p\,dv = \oint \delta q < 0.$$

Net work is performed on the system, which must be compensated by net rejection of heat to the environment.

This example illustrates that completing a transition between two states reversibly minimizes the work that must be performed on a system and maximizes the work that is performed by the system. These results can be extended to a system that performs work through a conversion of heat transferred into it, namely, to a heat engine. With (2.13), it follows that the cyclic work performed by such a system is maximized when the cycle is executed reversibly. Irreversibility reduces the net work performed by the system over a cycle, which must be associated with some heat rejection to the environment.

3.1.1 The Carnot Cycle

Consider a cyclic process comprised of two isothermal legs and two adiabatic legs (Fig. 3.2). During isothermal expansion and compression, the system is maintained at constant temperature through contact with a thermal reservoir, which serves as a heat source or heat sink. During adiabatic compression and expansion, the system is thermally isolated. If executed reversibly, this process describes a *Carnot cycle*. The leg from state 1 to state 2 is isothermal expansion at temperature T_{12}, so

$$\Delta u_{12} = 0.$$

Then the heat absorbed during that leg is given by

$$q_{12} = w_{12}$$
$$= \int_1^2 p\,dv$$
$$= RT_{12}\ln\left(\frac{v_2}{v_1}\right). \tag{3.2.1}$$

The leg from state 2 to state 3 is adiabatic, so

$$q_{23} = 0$$

and

$$-w_{23} = \Delta u_{23}$$
$$= c_v(T_{34} - T_{12}). \tag{3.2.2}$$

Similarly, the leg between states 3 and 4 is isothermal compression, so the heat transfer and work are given by

$$q_{34} = w_{34} = RT_{34} \ln \left(\frac{v_4}{v_3} \right), \tag{3.2.3}$$

whereas the leg between states 4 and 1 is adiabatic, so it involves work

$$w_{41} = c_v(T_{12} - T_{34}). \tag{3.2.4}$$

If the cycle is executed in reverse, the work and heat transfer during each leg are exactly cancelled by those during the corresponding leg of the forward cycle. Therefore, the Carnot cycle is reversible.

During the two adiabatic legs of (3.2.2) and (3.2.4), the work performed cancels, while the heat transfer vanishes identically. Hence, the net work and heat transfer over the cycle follow from the isothermal legs alone. The volumes in (3.2.1) and (3.2.3) are related to the changes of temperature along the adiabatic legs, which follow from Poisson's identity (2.30.1)

$$\frac{T_{12}}{T_{34}} = \left(\frac{v_3}{v_2} \right)^{\gamma-1} = \left(\frac{v_4}{v_1} \right)^{\gamma-1}.$$

Thus,

$$\frac{v_2}{v_1} = \frac{v_3}{v_4}. \tag{3.3}$$

Collecting the heat transferred during the isothermal legs yields

$$\oint \delta q = RT_{12} \ln \left(\frac{v_2}{v_1} \right) + RT_{34} \ln \left(\frac{v_4}{v_3} \right)$$

$$= R(T_{12} - T_{34}) \ln \left(\frac{v_2}{v_1} \right), \tag{3.4}$$

which is positive for $T_{12} > T_{34}$ if $v_2 > v_1$. Then

$$\oint pdv = \oint \delta q > 0. \tag{3.5}$$

The system performs net work, which is converted from heat absorbed during the cycle.

The Carnot cycle is a paradigm of a heat engine. For net work to be performed by the system, heat must be absorbed at high temperature T_{12} (during isothermal expansion) and rejected at low temperature T_{34} (during isothermal compression). According to (3.2), more heat is absorbed at high temperature than is rejected at low temperature. The first law (2.13) then implies that net heat absorbed during the cycle is balanced by net work performed by the system. If the preceding cycle is executed in reverse, the system constitutes a refrigerator. More heat is rejected at high temperature than is absorbed at low temperature. Then (2.13) asserts that net heat rejected during the cycle must be compensated by work performed on the system.

Net work and heat transfer over the cycle are proportional to the temperature difference $T_{12} - T_{34}$ between the heat source and heat sink. Not all of the heat absorbed by the system during expansion is converted into work. Some of that heat is rejected during compression. Therefore, the efficiency of a heat engine is limited—even for a reversible cycle.

3.2 Entropy and the Second Law

In the development of the first law, we observed that work is independent of path under adiabatic conditions, which allowed the internal energy be introduced as a state variable. The second law of thermodynamics is inspired by the observation that the quantity q/T is independent of path under reversible conditions. As for the first law, this allows the introduction of a state variable, *entropy*, which is defined in terms of a quantity that is, in general, path dependent.

Consider the Carnot cycle. According to (3.4),

$$\oint \frac{\delta q}{T} = R \left[\ln \left(\frac{v_2}{v_1} \right) + \ln \left(\frac{v_4}{v_3} \right) \right] = 0,$$

which is equivalent to the identity

$$\frac{q_{12}}{T_{12}} + \frac{q_{41}}{T_{41}} = 0. \tag{3.6}$$

It turns out this relationship holds under fairly general circumstances.

Carnot's Theorem

The identity (3.6) holds for any reversible cycle between two heat reservoirs at temperatures T_{12} and T_{41}, irrespective of details of the cycle.

It can be shown that any reversible cycle can be represented as a succession of infinitesimal Carnot cycles (e.g., Keenan, 1970). Therefore, Carnot's theorem implies that the identity

$$\oint \left(\frac{\delta q}{T} \right)_{rev} = 0 \tag{3.7}$$

holds irrespective of path.

By the exact differential theorem (Sec. 2.1.4), it follows that the quantity $(\delta q/T)_{rev}$ represents a point function. That is, $\int_{(x_0, y_0)}^{(x, y)} (\delta q/T)_{rev}$ depends only on the thermodynamic state (x, y) and not on the path along which the system evolved to that state. In the spirit that internal energy was introduced (Sec. 2.2), we define the state variable *entropy*

$$ds = \left(\frac{\delta q}{T}\right)_{rev}, \tag{3.8}$$

which constitutes a property of the system.

Carnot's theorem and the identity (3.7) apply to a reversible process. Under more general circumstances, the following inequality holds.

The Clausius Inequality

For a cyclic process,

$$\oint \frac{\delta q}{T} \leq 0, \tag{3.9}$$

where equality applies if the cycle is executed reversibly.

One of several statements of the second law, the Clausius inequality has the following consequences that pertain to the direction of thermodynamic processes:

1. Heat must be rejected to the environment somewhere during a cycle.
2. Under reversible conditions, more heat is exchanged at high temperature than at low temperature.
3. Irreversibility reduces the net heat absorbed during a cycle.

The first consequence precludes the possibility of a process that converts heat from a single source entirely into work: a *perpetual motion machine of the second kind*. Some of the heat absorbed by a system that performs work must be rejected. Representing a thermal loss, that heat rejection limits the efficiency of any heat engine, even one operated reversibly. The second consequence of (3.9) implies that net work is performed by the system during a cycle (namely, it behaves as a heat engine) if heat is absorbed at high temperature and rejected at low temperature. Conversely, net work must be performed on the system during the cycle (namely, it behaves as a refrigerator) if heat is rejected at high temperature and absorbed at low temperature. The third consequence of (3.9) implies that irreversibility reduces the net work performed by the system, in the case of a heat engine, and increases the net work that must be performed on the system, in the case of a refrigerator.

A differential form of the second law, which applies to an incremental process, may be derived from the Clausius inequality. Consider a cycle comprised of a reversible and an irreversible leg between two states 1 and 2 (Fig. 3.3). A cycle comprised of two reversible legs may also be constructed between those states. For the first cycle, the Clausius inequality yields

$$\int_1^2 \left(\frac{\delta q}{T}\right)_{rev} + \int_2^1 \left(\frac{\delta q}{T}\right)_{irrev} \leq 0,$$

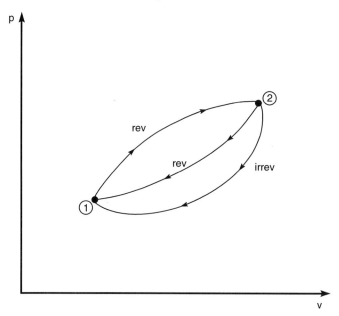

Figure 3.3 Cyclic processes between two states comprised of (*i*) two reversible legs and (*ii*) one reversible and one irreversible leg.

whereas for the second cycle

$$\int_1^2 \left(\frac{\delta q}{T} \right)_{\text{rev}} + \int_2^1 \left(\frac{\delta q}{T} \right)_{\text{rev}} = 0.$$

Subtracting gives

$$\int_2^1 \left(\frac{\delta q}{T} \right)_{\text{irrev}} - \int_2^1 \left(\frac{\delta q}{T} \right)_{\text{rev}} \leq 0$$

or with (3.8)

$$\int_2^1 ds \geq \int_2^1 \left(\frac{\delta q}{T} \right)_{\text{irrev}}.$$

Since it holds between two arbitrary states, the inequality must apply to the integrands too. Thus,

$$ds \geq \frac{\delta q}{T}, \tag{3.10}$$

where equality holds if the process is reversible.

Equation (3.10) is the most common form of the second law. In combination with (3.8), it implies an upper bound to the heat that can be absorbed by the system during a given change of state:

$$\delta q \leq T ds,$$

namely, the heat absorbed when that process is executed reversibly. If the process is executed irreversibly, additional heat is rejected to the environment. This reduces δq and the net heat absorbed, which in turn reduces the net work performed by the system if the process is cyclic (2.13).

Also represented in (3.10) is the direction of thermodynamic processes. Through the inequality, the second law asserts whether or not a system is capable of evolving along a given path. A process for which the change of entropy satisfies (3.10) is possible. If (3.10) is satisfied through equality, that process is reversible, whereas if it is satisfied through inequality that process is irreversible (e.g., a natural process). By contrast, a process that satisfies the reverse inequality is impossible.

3.3   Restricted Forms of the Second Law

The entropy of a system can either increase or decrease, depending on the heat transfer needed to achieve the same change of state under reversible conditions. For certain processes, the change of entropy implied by the second law is simplified.

For an adiabatic process, (3.10) reduces to

$$ds_{ad} \geq 0, \tag{3.11}$$

so the entropy can then only increase. It follows that irreversible work increases a system's entropy. Letting the control surface of a hypothetical system pass to infinity eliminates heat transfer to the environment and leads to the conclusion that the entropy of the universe can only increase. For a reversible adiabatic process, (3.11) implies $ds = 0$, so such a process is *isentropic*.

For an isochoric process, wherein expansion work vanishes, the first law (2.23.1) transforms (3.8) into

$$ds = c_v \left(\frac{dT}{T} \right)_{rev}.$$

Since it involves only state variables, this expression must hold whether or not the process is reversible. Hence,

$$ds_v = c_v \left(\frac{dT}{T} \right)_v. \tag{3.12}$$

In the absence of work, entropy can either increase or decrease, depending on the sign of dT. It follows that heat transfer can either increase or decrease a system's entropy.

According to (3.11) and (3.12), changes of entropy follow from

1. Irreversible work and
2. Heat transfer.

Irreversible work only increases s, whereas heat transfer (irreversible or otherwise) can either increase or decrease s. In the absence of work, the change of entropy equals $\delta q/T$, irrespective of path (e.g., whether or not the process is reversible).

3.4   The Fundamental Relations

Substituting the second law (3.10) into the two forms of the first law (2.22) leads to the inequalities

$$du \leq Tds - pdv, \tag{3.13.1}$$

$$dh \leq Tds + vdp. \tag{3.13.2}$$

For a reversible process, these reduce to

$$du = Tds - pdv, \tag{3.14.1}$$

$$dh = Tds + vdp. \tag{3.14.2}$$

Since they involve only state variables, the equalities (3.14) cannot depend on path. Consequently, they must hold whether or not the process is reversible. Known as the *fundamental relations*, these identities describe the change in one state variable in terms of the changes in two other state variables. Even though generally valid, the fundamental relations cannot be evaluated easily under irreversible conditions. The values of p and T in (3.14) refer to the pressure and temperature "of the system," which can be specified only under reversible conditions. Under irreversible conditions, the relationship among these variables reverts to the inequalities (3.13), wherein p and T denote "applied values," which can be specified.

It is convenient to introduce two new state variables, that are referred to as auxiliary functions. The *Helmholtz function* is defined by

$$f = u - Ts. \tag{3.15.1}$$

The *Gibbs function* is defined by

$$g = h - Ts$$
$$= u + pv - Ts. \tag{3.15.2}$$

In terms of these variables, the fundamental relations become

$$df = -sdT - pdv, \tag{3.16.1}$$

$$dg = -sdT + vdp. \tag{3.16.2}$$

The Helmholtz and Gibbs functions are each referred to as the *free energy* of the system: f for an isothermal process and g for an isothermal–isobaric process, because they reflect the energy available for conversion into work under those conditions (Problem 3.10).

3.4.1 The Maxwell Relations

Variables appearing in (3.14) and (3.16) are not entirely independent. Involving only state variables, each fundamental relation has the form of an exact differential. According to the exact differential theorem (2.8), these relations can hold only if the cross derivatives of the coefficients on the right-hand sides are equal. Applying that condition to (3.14) and (3.16) yields the identities

$$\left(\frac{\partial T}{\partial v}\right)_s = -\left(\frac{\partial p}{\partial s}\right)_v$$

$$\left(\frac{\partial s}{\partial v}\right)_T = \left(\frac{\partial p}{\partial T}\right)_v$$

$$\left(\frac{\partial T}{\partial p}\right)_s = \left(\frac{\partial v}{\partial s}\right)_p \tag{3.17}$$

$$\left(\frac{\partial s}{\partial p}\right)_T = -\left(\frac{\partial v}{\partial T}\right)_p,$$

which are known as the *Maxwell relations*.

3.4.2 Noncompensated Heat Transfer

The inequalities in (3.13) account for additional heat rejection to the environment that occurs through irreversibility. Those inequalities can be eliminated in favor of equalities by introducing the *noncompensated heat transfer* $\delta q'$, defined by

$$\delta q = T ds - \delta q', \tag{3.18}$$

where $\delta q' > 0$. The noncompensated heat transfer represents the additional heat that must be rejected to the environment as a result of irreversibility, for example, that associated with frictional dissipation of kinetic energy.

Substituting (3.18) into the first law (2.22.1) yields

$$du = T ds - p dv - \delta q', \tag{3.19}$$

where p and T are understood to refer to applied values. An expression for the noncompensated heat transfer can be derived from (3.14) and (3.19), which hold for applied values under reversible and irreversible conditions, respectively. Subtracting gives

$$\delta q' = (T - T_{rev})ds - (p - p_{rev})\,dv, \tag{3.20}$$

where T_{rev} and p_{rev} refer to applied values under equilibrium conditions (i.e., those assumed by the system when the process is executed reversibly). According to (3.20), noncompensated heat transfer results from the thermal disequilibrium of the system, which is represented in the difference $T - T_{rev}$, and from the mechanical disequilibrium of the system, which is represented in the difference $p - p_{rev}$.

3.5 Conditions for Thermodynamic Equilibrium

By defining the direction of thermodynamic processes, the second law implies whether or not a path out of a given thermodynamic state is possible. Since this determines the likelihood of the system remaining in that state, the second law characterizes the stability of thermodynamic equilibrium.

Consider a system in a given thermodynamic state. An arbitrary infinitesimal process emanating from that state is referred to as a *virtual process*. The system is said to be in *stable* or *true thermodynamic equilibrium* if no virtual process emanating from that state is a natural process, i.e., if all virtual paths out of that state are either reversible or impossible. If all virtual paths out of the state are natural processes, the system is said to be in *unstable equilibrium*. A small perturbation will then result in a finite change of state. If only some of the virtual processes out of the state are natural, the system is said to be in *metastable equilibrium*. A small perturbation then may or may not result in a finite change of state, depending on the details of the perturbation.

These definitions can be combined with the second law to determine conditions that characterize thermodynamic equilibrium. The relations

$$du \leq Tds - pdv,$$
$$dh \leq Tds + vdp,$$
$$df \leq -sdT - pdv,$$
$$dg \leq -sdT + vdp,$$

hold for a reversible process, in the case of equality, and for an irreversible (e.g., natural) process, in the case of inequality. For a state to correspond to thermodynamic equilibrium, the reverse must be true, that is,

$$du \geq Tds - pdv,$$
$$dh \geq Tds + vdp,$$
$$df \geq -sdT - pdv, \tag{3.21}$$
$$dg \geq -sdT + vdp,$$

where inequality describes an impossible process. Inequalities (3.21) provide criteria for thermodynamic equilibrium. If they are satisfied by all virtual processes out of the current state, the system is in true thermodynamic equilibrium. If only some virtual paths emanating from the state satisfy (3.21), the system is in metastable equilibrium.

Under special circumstances, simpler criteria for thermodynamic equilibrium exist. For an adiabatic enclosure, (3.10) reduces to

$$ds \geq 0,$$

where inequality corresponds to a natural process. Then a criterion for thermodynamic equilibrium is

$$ds_{ad} \leq 0, \tag{3.22}$$

which describes reversible and impossible processes. According to (3.22), a state of thermodynamic equilibrium for an adiabatic system coincides with a local maximum of entropy (Fig. 3.4). Therefore, an adiabatic system's entropy must increase as it approaches thermodynamic equilibrium.

Choosing processes for which the right-hand sides of (3.21) vanish yields other criteria for thermodynamic equilibrium:

$$du_{s,v} \geq 0,$$
$$dh_{s,p} \geq 0,$$
$$df_{T,v} \geq 0, \tag{3.23}$$
$$dg_{T,p} \geq 0,$$

which must be satisfied by all virtual paths out of the current state for the system to be in equilibrium. For the particular processes just discussed, (3.23) implies that a state of thermodynamic equilibrium coincides with local minima in the properties u, h, f, and g, respectively. Thus, for those processes the internal energy, enthalpy, Helmholtz function, and Gibbs function must all decrease as a system approaches thermodynamic equilibrium.

3.6 Relationship of Entropy to Potential Temperature

In Chapter 2, we saw that the change of potential temperature is proportional to the heat absorbed by an air parcel. Under reversible conditions, (3.10) implies that the same is true of entropy. Substituting (2.36) into the second

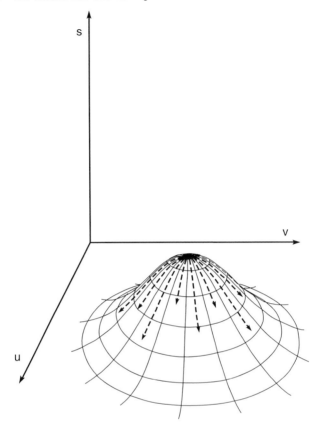

Figure 3.4 Local entropy maximum in the u–v plane, symbolizing a state that corresponds to thermodynamic equilibrium.

law yields

$$d \ln \theta \leq \frac{ds}{c_p}, \tag{3.24}$$

where equality holds for a reversible process. Because it involves only state variables, that equality must hold irrespective of whether or not the process is reversible. Hence,

$$d \ln \theta = \frac{ds}{c_p}, \tag{3.25}$$

and the change of potential temperature is related directly to the change of entropy. Despite its validity, (3.25) applies to properties of the system (e.g., to

p and T through θ). Thus, like the fundamental relations, it can be evaluated only under reversible conditions.

3.6.1 Implications for Vertical Motion

If a process is adiabatic, $d\theta = 0$ and $ds \geq 0$. The entropy remains constant or it can increase through irreversible work (e.g., that associated with frictional dissipation of kinetic energy). In the case of an air parcel, the conditions for adiabatic behavior are closely related to those for reversibility. Adiabatic behavior requires not only that no heat be transferred across the control surface, but also that no heat be exchanged between one part of the system and another (e.g., Landau *et al.*, 1980). The latter requirement excludes turbulent mixing, which is the principal form of mechanical irreversibility in the atmosphere. It also excludes irreversible expansion work because such work introduces internal motions that eventually result in mixing.

Since they exclude the important sources of irreversibility, the conditions for adiabatic behavior are tantamount to conditions for isentropic behavior. Thus, adiabatic conditions for the atmosphere are equivalent to requiring isentropic behavior for individual air parcels. Under these circumstances, potential temperature surfaces, $\theta = $ const, coincide with isentropic surfaces, $s = $ const. An air parcel coincident initially with a certain isentropic surface remains on that surface. Because those surfaces tend to be quasi-horizontal, adiabatic behavior implies no net vertical motion (see Fig. 2.9). Air parcels can ascend and descend along isentropic surfaces, but they undergo no systematic vertical motion.

Under diabatic conditions, an air parcel moves across isentropic surfaces according to the heat exchanged with its environment (2.36). Consider an air parcel advected horizontally through different thermal environments, like those represented in the distribution of net radiation in Fig. 1.27. Because radiative transfer varies sharply with latitude, this occurs whenever the parcel's motion is deflected across latitude circles. Figure 3.5 shows a wavy trajectory followed by an air parcel that is initially at latitude ϕ_0. Symbolic of the disturbed circulations in Figs. 1.9 and 1.10, that motion advects air through different radiative environments. Also indicated in Fig. 3.5 is the distribution of radiative-equilibrium temperature $T_{RE}(\phi)$, at which air emits radiant energy at the same rate as it absorbs it. That thermal structure is achieved if the motion is everywhere parallel to latitude circles, because air parcels then have infinite time to adjust to local thermal equilibrium.

Suppose the displaced motion in Fig. 3.5 is sufficiently slow for the parcel to equilibrate with its surroundings at each point along the trajectory. The parcel's temperature then differs from T_{RE} only infinitesimally (Fig. 3.6), so the parcel remains in thermal equilibrium and heat transfer along the trajectory occurs reversibly. Between two successive crossings of the latitude ϕ_0, the

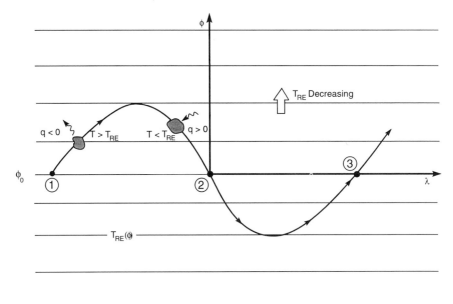

Figure 3.5   A wavy trajectory followed by an air parcel, which symbolizes the disturbed circulations in Figs 1.9a and 1.10a. When displaced poleward, the parcel is warmer than the local radiative–equilibrium temperature (contours), so it rejects heat. The reverse process occurs when the parcel moves equatorward.

parcel absorbs heat such that

$$\int_1^2 c_p d \ln \theta = \int_1^2 \frac{\delta q}{T}. \tag{3.26}$$

If the heat exchange depends only on the parcel's temperature, for example,

$$\delta q = T df(T),$$

which is symbolic of radiative transfer, then (3.26) reduces to

$$c_p \ln \left(\frac{\theta_2}{\theta_1} \right) = \Delta f = 0,$$

since $T_2 = T_1 = T_{RE}(\phi_0)$. Thus, $\theta_2 = \theta_1$ and the parcel is restored to its initial thermodynamic state when it returns to latitude ϕ_0.

  While moving poleward, the parcel is infinitesimally warmer than the local radiative–equilibrium temperature, so it emits more radiant energy than it absorbs. Rejection of heat results in the parcel drifting off its initial isentropic surface toward lower θ, which, for reasons developed in Chapter 7, corresponds to lower altitude. While moving equatorward, the parcel is in-

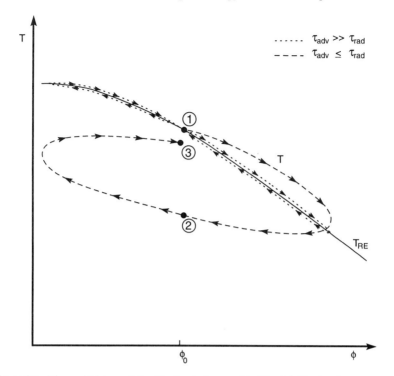

Figure 3.6 Thermal history of the idealized air parcel in Fig. 3.5. If advection occurs on a timescale much longer than radiative transfer (dotted), the parcel remains in thermal equilibrium with its surroundings (i.e., it assumes the local radiative-equilibrium temperature). The heat it absorbs and rejects is then reversible. If advection occurs on a timescale comparable to or shorter than radiative transfer (dashed), the parcel is driven out of thermal equilibrium with its surroundings. The heat it absorbs and rejects is then irreversible, which introduces a hysteresis into the parcel's thermodynamic state as it is advected through successive cycles of the disturbance (refer to Fig. 2.5).

finitesimally colder than the local radiative-equilibrium temperature, so it absorbs more radiant energy than it emits. Absorption of heat then results in the parcel ascending to higher θ, just enough to restore the parcel to its initial isentropic surface when it returns to the latitude ϕ_0. Thus, successive crossings of the latitude ϕ_0 result in no net vertical motion and the parcel's evolution is perfectly cyclic.

Now suppose the motion is sufficiently fast to carry the parcel between radiative environments before it has equilibrated to the local radiative-equilibrium temperature. During the excursion poleward of ϕ_0, the parcel is out of thermal equilibrium, so heat transfer along the trajectory occurs irreversibly. Because its temperature then lags that of its surroundings, the parcel returns to the latitude ϕ_0 with a temperature different from that initially (Fig. 3.6). By the foregoing analysis, the parcel's potential temperature θ_2 also differs from that initially

$$c_p \ln \left(\frac{\theta_2}{\theta_1} \right) = \Delta f \neq 0,$$

since $T_2 \neq T_1$. Thus, the parcel is not restored to its initial isentropic surface, but rather remains displaced vertically after returning to the latitude ϕ_0. Similar reasoning shows that heat transfer during the excursion equatorward of ϕ_0 (e.g., between positions 2 and 3 in Fig. 3.5) does not exactly cancel net heat transfer during the poleward excursion. Hence, a complete cycle results in net heat transfer and therefore a net vertical displacement of the parcel from its initial isentropic surface.

Whether the parcel returns above or below that isentropic surface depends on the radiative-equilibrium temperature and on details of the motion, which control the history of heating and cooling. In either event, irreversible heat transfer introduces a hysteresis (refer to Fig. 2.5), through which successive cycles produce a vertical drift of air across isentropic surfaces. Because advection operates on a timescale much shorter than radiative transfer in the atmosphere, disturbed horizontal motion invariably drives air out of thermal equilibrium, introducing irreversible heat transfer, which in turn drives vertical motion. Vertical motions generated in this fashion play an important role in the mean meridional circulation of the atmosphere, which transfers heat, moisture, and chemical constituents.

Suggested Reading

An illuminating treatment of the second law and its consequences to mechanical and chemical systems is given in *The Principles of Chemical Equilibrium* (1971) by K. Denbigh.

Statistical Physics (1980) by Landau *et al.* includes a clear discussion of reversibility and its implications for fluid systems.

Problems

3.1. Two hundred grams of mercury at $100°C$ is added to 100 g of water at $20°C$. If the specific heat capacities of water and mercury are 4.18 and $0.14 \text{ J kg}^{-1}\text{K}^{-1}$, respectively, determine (a) the limiting temperature of the mixture, (b) the change of entropy for the mercury, (c) the change of entropy for the water, and (d) the change of entropy for the system as a whole.

3.2. For reasons developed in Chapter 9, many clouds are supercooled: They contain droplets at temperatures below $0°C$. Consider 1 mol of supercooled water in the metastable state: $T = -10°C$ and $p = 1$ atm. Following a perturbation, the water freezes spontaneously, with the ice and its surroundings eventually returning to the original temperature. If the heat capacities of water and ice remain approximately constant with the values 75 and 38 J K^{-1} mol^{-1}, respectively, and if the enthalpy of fusion at 0°C is 6026 J mol^{-1}, calculate (a) the change of entropy for the water and (b) the change of entropy for its environment.

3.3. Consider a parcel crossing isentropic surfaces. If radiative cooling is just large enough to maintain the parcel on a fixed isobaric surface, what is the relative change of its temperature when the parcel's potential temperature has decreased to 90% of its original value?

3.4. Air moving inland from a cooler maritime region warms through conduction with the ground. If the temperature and potential temperature increase by 5 and 6%, respectively, determine (a) the fractional change of surface pressure between the parcel's initial and final positions and (b) the heat absorbed by the parcel.

3.5. During a cloud-free evening, LW heat transfer with the surface causes an air parcel to descend from 900 to 910 mb and its entropy to decrease by 15 J kg^{-1} K^{-1}. If its initial temperature is 280 K, determine the parcel's (a) final temperature and (b) final potential temperature.

3.6. Derive expression (3.20) for the noncompensated heat transfer.

3.7. Air initially at 20°C and 1 atm is allowed to expand freely into an evacuated chamber to assume twice its original volume. Calculate the change of specific entropy. (*Hint*: How much work is performed by the air?)

3.8. The thermodynamic state of an air parcel is represented conveniently on the *pseudo-adiabatic chart* (Chapter 5), which displays altitude in terms of the variable p^κ. Show that the work performed during a cyclic process equals the area circumscribed by that process in the θ–p^κ plane.

3.9. One mole of water at 0°C and 1 atm is transformed into vapor at 200°C and 3 atm. If the enthalpy of vaporization for water is 4.06×10^4 J mol^{-1} and if the specific heat of vapor is approximated by

$$c_p = 8.8 - 1.9 \times 10^{-3}T + 2.2 \times 10^{-6}T^2 \qquad \text{cal } K^{-1}mol^{-1},$$

calculate the change of (a) entropy and (b) enthalpy.

3.10. The Helmholtz and Gibbs functions are referred to as *free energies* or *thermodynamic potentials*. Use the definitions (3.15) together with the first and second laws for a closed system to show that (a) under isothermal conditions, the decrease of Helmholtz function describes the maximum total work which can be performed by the system:

$$w_{\max} = -\Delta f,$$

(b) under isothermal–isobaric conditions, the decrease of Gibbs function describes the maximum work—exclusive of expansion work, which can be performed by the system:

$$w'_{\max} = -\Delta g,$$

where $\delta w' = \delta w - p\,dv$.

Chapter 4 | Heterogeneous Systems

The thermodynamic principles developed in Chapters 2 and 3 apply to a homogeneous system, which can involve only a single phase. For it to be in thermodynamic equilibrium, a homogeneous system must be in thermal equilibrium: at most an infinitesimal temperature difference exists between the system and its environment, and also in mechanical equilibrium: at most an infinitesimal pressure difference exists between it and its environment. A heterogeneous system can involve more than one phase and, for it, thermodynamic equilibrium requires an additional criterion. The system must also be in *chemical equilibrium*: No conversion of mass occurs from one phase to another. Analogous to thermal and mechanical equilibrium, chemical equilibrium requires a certain state variable to have virtually no difference between the phases present.

4.1 Description of a Heterogeneous System

For a homogeneous system, two intensive properties describe the thermodynamic state. Conversely, only two state variables may be varied independently, so a homogeneous system has two thermodynamic degrees of freedom. For a heterogeneous system, each phase may be regarded as a homogeneous subsystem, one that is "open" due to exchanges with the other phases present. Consequently, the number of intensive properties that describes the thermodynamic state of a heterogeneous system is proportional to the number of phases present. Were they independent, those properties would constitute additional degrees of freedom for a heterogeneous system. However, thermodynamic equilibrium between phases introduces additional constraints that actually reduce the degrees of freedom of a heterogeneous system below those of a homogeneous system.

The system we consider is a two-component mixture of dry air and water, with the latter existing in vapor[1] and possibly one condensed phase (Fig. 4.1). This heterogeneous system is described conveniently in terms of extensive

---

[1] The term *vapor* denotes water in gas phase and should not be confused with aerosol or other forms of condensate.

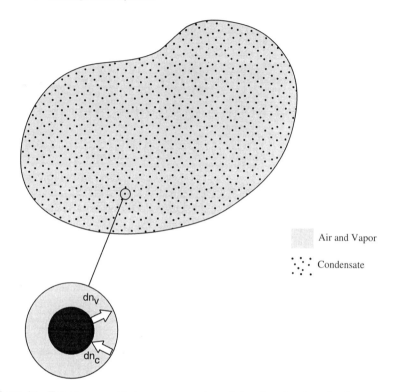

Figure 4.1 Two-component heterogeneous system of dry air and water, with the water component present in vapor and one condensed phase. The water component is transformed incrementally from condensate to vapor in amount dn_v and from vapor to condensate in amount dn_c.

properties, which depend on the abundance of each species present. An extensive property Z for the total system follows from the contributions from the individual phases

$$Z_{tot} = Z_g + Z_c, \tag{4.1}$$

where the subscripts g and c refer to the gas and condensate subsystems and the former includes both dry air and vapor. The extensive property for the gas phase subsystem is specified by its pressure and temperature and the molar abundances of dry air and vapor

$$Z_g = Z_g(p, T, n_d, n_v).$$

If the condensed phase is a pure substance, the extensive property for it is determined by the pressure and temperature and the molar abundance of condensate

$$Z_c = Z_c(p, T, n_c).$$

Then changes of the extensive property Z or the individual subsystems are described by

$$dZ_g = \left(\frac{\partial Z_g}{\partial T}\right)_{pn} dT + \left(\frac{\partial Z_g}{\partial p}\right)_{Tn} dp + \left(\frac{\partial Z_g}{\partial n_d}\right)_{pTn} dn_d + \left(\frac{\partial Z_g}{\partial n_v}\right)_{pTn} dn_v, \quad (4.2.1)$$

$$dZ_c = \left(\frac{\partial Z_c}{\partial T}\right)_{pn} dT + \left(\frac{\partial Z_c}{\partial p}\right)_{Tn} dp + \left(\frac{\partial Z_c}{\partial n_c}\right)_{pT} dn_c, \quad (4.2.2)$$

where the subscript n denotes that all molar values are held fixed except, possibly, the one being varied.

It is convenient to introduce a state variable that measures how the extensive property Z of the total system changes with an increase of one of the components, for example, through a conversion of mass from one phase to another. We will see in Sec. 4.5 that an isobaric change of phase also occurs isothermally. Consequently, the foregoing state variable is expressed most conveniently for processes that occur at constant pressure and temperature. The *partial molar property* is defined as the rate at which an extensive property changes with a change in the molar abundance of the kth species under isobaric and isothermal conditions

$$\overline{Z}_k = \left(\frac{\partial Z}{\partial n_k}\right)_{pTn}, \quad (4.3.1)$$

where the subscript k refers to a particular component and phase. Similarly, the *partial specific property* is defined as

$$\overline{z}_k = \left(\frac{\partial Z}{\partial m_k}\right)_{pTm}. \quad (4.3.2)$$

In general, the partial molar and partial specific properties differ from the molar and specific properties for a pure substance:

$$\tilde{z}_k = \frac{Z}{n_k},$$
$$z_k = \frac{Z}{m_k}, \quad (4.4)$$

due to interactions with the other components present.[2] However, for a system of dry air and water in which the latter appears only in trace abundance, such

---

[2] For example, adding air to a mixture of air and water at constant pressure and temperature results in a change of the extensive property Z which differs from that brought about by adding the same amount of air to a system of pure air if, in the former case, some of that air passes into solution with the water.

differences are small enough to be ignored. Hence, to a good approximation, the partial properties may be replaced by the molar and specific properties:

$$\overline{Z}_d \cong \tilde{z}_d \quad \overline{z}_d \cong z_d,$$

$$\overline{Z}_v \cong \tilde{z}_v \quad \overline{z}_v \cong z_v, \tag{4.5}$$

$$\overline{Z}_c \cong \tilde{z}_c \quad \overline{z}_c \cong z_c.$$

If the system is closed, the abundance of individual components is also preserved, so

$$dn_d = 0, \tag{4.6.1}$$

$$d(n_v + n_c) = 0. \tag{4.6.2}$$

Adding (4.2) and incorporating (4.6) and (4.5) yields the change of the extensive property Z for the total system

$$dZ_{\text{tot}} = \left(\frac{\partial Z_{\text{tot}}}{\partial T}\right)_{pn} dT + \left(\frac{\partial Z_{\text{tot}}}{\partial p}\right)_{Tn} dp + (\tilde{z}_v - \tilde{z}_c)dn_v, \tag{4.7.1}$$

where

$$Z_{\text{tot}} = n_d\tilde{z}_d + n_v\tilde{z}_v + n_c\tilde{z}_c. \tag{4.7.2}$$

In terms of specific properties, the change of Z for the total system is given by

$$dZ_{\text{tot}} = \left(\frac{\partial Z_{\text{tot}}}{\partial T}\right)_{pm} dT + \left(\frac{\partial Z_{\text{tot}}}{\partial p}\right)_{Tm} dp + (z_v - z_c)dm_v, \tag{4.8.1}$$

where

$$Z_{\text{tot}} = m_d z_d + m_v z_v + m_c z_c. \tag{4.8.2}$$

These expressions represent generalizations of the exact differential relation (2.6) for a homogeneous system to account for exchanges of mass between the phases of a heterogeneous closed system.

4.2   Chemical Equilibrium

We now establish a criterion for two phases of the system to be at equilibrium with one another. In addition to thermal and mechanical equilibrium, those phases must also be in chemical equilibrium. The latter is determined by the diffusion of mass from one phase to the other, which, for reasons to become apparent, is closely related to the Gibbs function. The *chemical potential* for the kth species μ_k is defined as the partial molar Gibbs function:

$$\mu_k = \overline{G}_k = \left(\frac{\partial G}{\partial n_k}\right)_{pTn}, \tag{4.9}$$

which is tantamount to the molar Gibbs function $\tilde{g}_k$. For the gas phase, (4.2.1) gives the change of Gibbs function

$$dG_g = \left(\frac{\partial G_g}{\partial T}\right)_{pn_v n_d} dT + \left(\frac{\partial G_g}{\partial p}\right)_{Tn_v n_d} dp + \mu_d dn_d + \mu_v dn_v.$$

In a constant n_v, n_d process (e.g., one not involving a phase transformation), this expression must reduce to the fundamental relation (3.16.2) for a homogeneous closed system. Accordingly, we identify

$$\begin{aligned}
\left(\frac{\partial G_g}{\partial T}\right)_{pn_v n_d} &= -S_g, \\
\left(\frac{\partial G_g}{\partial p}\right)_{Tn_v n_d} &= V_g.
\end{aligned} \qquad (4.10)$$

Since they involve only state variables, expressions (4.10) must hold irrespective of path (e.g., whether or not a phase transformation is involved). Thus,

$$dG_g = -S_g dT + V_g dp + \mu_d dn_d + \mu_v dn_v \qquad (4.11)$$

describes the change of Gibbs function for the gas phase. A similar analysis leads to the relation

$$dG_c = -S_c dT + V_c dp + \mu_c dn_c \qquad (4.12)$$

for the condensed phase. Adding (4.11) and (4.12) and incorporating (4.6.2) yields the change of Gibbs function for the total system

$$dG_{tot} = -S_{tot} dT + V_{tot} dp + (\mu_v - \mu_c) dn_v, \qquad (4.13.1)$$

where

$$G_{tot} = n_d \tilde{g}_d + n_v \tilde{g}_v + n_c \tilde{g}_c. \qquad (4.13.2)$$

For the heterogeneous system to be in thermodynamic equilibrium, the pressures and temperatures of the different phases present must be equal. Further, there can be no conversion of mass from one phase to another. Corresponding to the fundamental relation (4.13) is the inequality

$$dG_{tot} \leq -S_{tot} dT + V_{tot} dp + (\mu_v - \mu_c) dn_v, \qquad (4.14)$$

where p, T, and n_v refer to applied values and where equality and inequality correspond to reversible and irreversible processes, respectively. For the system to be in equilibrium, all virtual processes emanating from the state under consideration must be either reversible or impossible. Hence, thermodynamic equilibrium is characterized by the reverse inequality

$$dG_{tot} \geq -S_{tot} dT + V_{tot} dp + (\mu_v - \mu_c) dn_v, \qquad (4.15)$$

which must hold for all virtual paths out of the state. Consider a virtual process involving a transformation of phase that occurs at constant pressure and temperature. Then

$$dG_{tot} \geq (\mu_v - \mu_c)dn_v$$

must hold for equilibrium. Because this expression must apply irrespective of the sign of dn_v, it follows that

$$\mu_v = \mu_c \tag{4.16}$$

is a criterion for chemical equilibrium. The chemical potential of different phases of the water component must be equal. This is analogous to the temperatures and pressures of the phases being equal under thermal and mechanical equilibrium.

The chemical potential determines the diffusive flux of mass from one species to another. In that respect, μ_k may be regarded as a diffusion potential from the kth species.[3] Under the circumstances described by (4.16), the flux of mass from one phase of water to another is exactly balanced by a flux in the opposite sense, so the net diffusion of mass vanishes. If the chemical potential differs between the phases, there will exist a net diffusion of mass from the phase of high μ to the phase of low μ. This transformation of mass will continue until the difference of chemical potential between the phases has been eliminated and chemical equilibrium has been restored.

4.3   Fundamental Relations for a Multicomponent System

In a manner similar to that used to develop the change of Gibbs function, all of the fundamental relations for a homogeneous system can be generalized to a heterogeneous system of C chemically distinct components and P phases. Expanding U, H, F and G as was done above leads to the relations

$$dU = TdS - pdV + \sum_{j=1}^{P}\sum_{i=1}^{C}\mu_{ij}dn_{ij},$$

$$dH = TdS + Vdp + \sum_{j=1}^{P}\sum_{i=1}^{C}\mu_{ij}dn_{ij},$$

$$dU = -SdT - pdV + \sum_{j=1}^{P}\sum_{i=1}^{C}\mu_{ij}dn_{ij}, \tag{4.17}$$

$$dG = -SdT + Vdp + \sum_{j=1}^{P}\sum_{i=1}^{C}\mu_{ij}dn_{ij},$$

---

[3] Chemical potential is the ultimate determinant of diffusion and supersedes concentration in certain applications; see Denbigh (1971).

where we have introduced alternate expressions for the chemical potential of the ith component and jth phase:

$$\mu_{ij} = \left(\frac{\partial U}{\partial n_{ij}}\right)_{SVn} = \left(\frac{\partial H}{\partial n_{ij}}\right)_{Spn} = \left(\frac{\partial F}{\partial n_{ij}}\right)_{TVn} = \left(\frac{\partial G}{\partial n_{ij}}\right)_{pTn} \tag{4.18}$$

in terms of the variables in (4.17). As will be seen later, the definitions in (4.18) are equivalent. However, only the last corresponds to an isobaric–isothermal process. Consequently, the latter three do not represent partial molar properties according to the definition (4.3). Because transformations of phase at constant pressure also occur at constant temperature, the last expression in (4.18) is the most convenient form of chemical potential and, for the same reason, the last of the fundamental relations in (4.17) is the most convenient description of such transformations.

For a closed system, the mass of each component is preserved, so

$$\sum_{j=1}^{P} dn_{ij} = 0, \qquad i = 1, 2, 3, ...C. \tag{4.19}$$

For the ith component, (4.19) implies

$$\sum_{j=1}^{P} \mu_{ij} dn_{ij} = \sum_{j=2}^{P} \mu_{ij} dn_{ij} + \mu_{i1} dn_{i1}$$

$$= \sum_{j=2}^{P} (\mu_{ij} - \mu_{i1}) dn_{ij}, \tag{4.20}$$

where $j = 1$ defines an arbitrary reference phase for that component. Then (4.17) may be expressed

$$dU = TdS - pdV + \sum_{j=2}^{P} \sum_{i=1}^{C} (\mu_{ij} - \mu_{i1}) dn_{ij},$$

$$dH = TdS + Vdp + \sum_{j=2}^{P} \sum_{i=1}^{C} (\mu_{ij} - \mu_{i1}) dn_{ij},$$

$$\tag{4.21}$$

$$dF = -SdT - pdV + \sum_{j=2}^{P} \sum_{i=1}^{C} (\mu_{ij} - \mu_{i1}) dn_{ij},$$

$$dG = -SdT + Vdp + \sum_{j=2}^{P} \sum_{i=1}^{C} (\mu_{ij} - \mu_{i1}) dn_{ij},$$

which represent generalizations of the fundamental relations (3.14) and (3.16) to a heterogeneous closed system. According to (4.21), each of the definitions of chemical potential in (4.18) implies the same statement of chemical equilibrium (4.16). However, only the expression involving Gibbs function applies to a process conducted at constant pressure and temperature, for example, to an isobaric transformation of phase.

4.4 Thermodynamic Degrees of Freedom

Each phase of a heterogeneous system constitutes a homogeneous subsystem, so its state is specified by two intensive properties. Thus, the number of intensive properties that describes the thermodynamic state of a heterogeneous system is proportional to the number of phases present. Consider a single-component system involving two phases. The state of each homogeneous subsystem is specified by two intensive properties, so four intensive properties describe the state of the total system. However, thermodynamic equilibrium requires each subsystem to be in thermal, mechanical, and chemical equilibrium with the other subsystem. Consequently, the heterogeneous system must also satisfy three constraints:

$$T_1 = T_2,$$
$$p_1 = p_2, \qquad (4.22)$$
$$\mu_1 = \mu_2,$$

which leave only one independent state variable.

Thus, a one-component system involving two phases at equilibrium with one another possesses only one thermodynamic degree of freedom. Such a system must therefore possess an equation of state of the form

$$p = p(T), \qquad (4.23)$$

which is the same form that describes adiabatic changes of a homogeneous system (2.30). The equation of state (4.23) describes a family of curves, along which the heterogeneous system evolves reversibly (i.e., during which it remains in thermodynamic equilibrium). According to (4.23), fixing the temperature of a single-component mixture of two phases also fixes its pressure and vice versa. Consequently, an equilibrium transformation of phase that occurs at constant pressure also occurs at constant temperature. This feature makes the first expression in (4.18) the most useful definition of chemical potential and the change of Gibbs function the most convenient of the fundamental relations (4.21).

If all three phases are present, six intensive properties describe the state of a single-component heterogeneous system. For the system to be in thermodynamic equilibrium, those properties must also satisfy six independent constraints like those in (4.22). Consequently, a single-component system involving all three phases possesses no thermodynamic degrees of freedom. Such a system can exist only in a single state, which is referred to as the *triple point*.

In general, the number of thermodynamic degrees of freedom possessed by a heterogeneous system is described by the following principle.

Gibbs' Phase Rule

The number of independent state variables for a heterogeneous system involving C chemically distinct but nonreactive components[4] and P phases is given by

$$N = C + 2 - P. \tag{4.24}$$

According to (4.24), the number of degrees of freedom possessed by a heterogeneous system increases with the number of chemically distinct components but decreases with the number of phases present.

4.5 Thermodynamic Characteristics of Water

The remainder of this chapter focuses on the thermodynamic behavior of a single-component system of pure water involving one, two, or possibly all three phases at equilibrium with one another (Fig. 4.2). Because water is a pure substance, its equation of state can be expressed

$$p = p(v, T), \tag{4.25}$$

regardless of how many phases are present. However, the particular form of (4.25) depends on which phases are present. If only vapor is present, (4.25) is given by the ideal gas law (1.1). If two phases are present, the equation of state must reduce to the form of (4.23).

[4]A generalization of the phase rule and its application to reactive components can be found in Denbigh (1971).

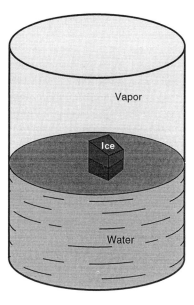

Figure 4.2 Single-component heterogeneous system of pure water.

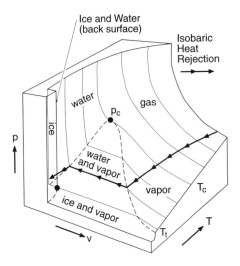

Figure 4.3 State space of a single-component system of pure water, illustrated in terms of its pressure as a function of its specific volume and temperature. Heterogeneous states occur below the *critical point* (p_c, T_c). A process corresponding to isobaric heat rejection (bold arrows) is also indicated. In the heterogeneous region of vapor and water, isobars coincide with isotherms. Adapted from Iribarne and Godson (1981) by permission of Kluwer Academic Publishers.

Like any pure substance, water has an equation of state that describes a surface over the plane of two intensive properties, as shown in Fig. 4.3 for p as a function of v and T. Portions of that surface describe states in which only a single phase is present and the system is homogeneous. In those regions of state space, the system possesses two thermodynamic degrees of freedom and any two state variables may be varied independently. For instance, both v and T are required to specify p and the thermodynamic state in the region of vapor, where the behavior is that of an ideal gas (compare Fig. 2.3).

At pressures and temperatures below the *critical point*: (p_c, T_c), the system can assume heterogeneous states, wherein multiple phases coexist at equilibrium with one another.[5] If two phases are present, the system possesses only one thermodynamic degree of freedom (4.24). Consequently, in such regions, the surface in Fig. 4.3 assumes a simpler form and is discontinuous in slope at the boundary with regions where the system is homogeneous. For instance, in the region of vapor and water, isotherms coincide with isobars. Specifying one of those properties determines the other and thus the thermodynamic state of the pure substance. Similar behavior is found in the regions of vapor and ice and water and ice.

At temperatures below the critical point, the system can evolve from a homogeneous state of one phase to a homogeneous state of another phase only by passing through intermediate states that are heterogeneous. Due to a change in the number of degrees of freedom, the path describing this process

---

[5] At temperatures above T_c, the system remains homogeneous and its properties vary smoothly across the entire range of pressure, for example, condensed phases are not possible for any pressure. For those temperatures, water substance is referred to as *gas*, to distinguish it from vapor, which can coexist with condensed phases.

is discontinuous in slope at the boundary between homogeneous and heterogeneous states. Consider isobaric heat rejection, indicated in Fig. 4.3. From an initial temperature above the critical point, where the system is entirely gaseous, temperature and volume both decrease, reflecting the two degrees of freedom possessed by a homogeneous system. Eventually, the process encounters the boundary separating homogeneous states, wherein only vapor is present, from heterogeneous states, wherein vapor and water coexist at equilibrium. At that state, the slope of the process changes discontinuously because the temperature of the system can no longer decrease. Instead, isobaric heat rejection results in condensation of vapor, which is attended by a sharp reduction of volume, all at constant temperature. This simplified behavior continues until the vapor has been converted entirely into water, at which point the system is again homogeneous and the slope of the process changes discontinuously a second time. Beyond that state, heat rejection results in a decrease of both temperature and volume, reflecting the two degrees of freedom again possessed by the system.

In a heterogeneous state, different phases coexist at equilibrium only if their temperatures, pressures, and chemical potentials are equal. The individual phases are then said to be *saturated* because the net flux of mass from one phase to another vanishes. If one of those phases is vapor, the pressure of the heterogeneous system represents the *equilibrium vapor pressure* with respect to water or ice, denoted p_w and p_i, respectively. Should the heterogeneous system have a pressure below the equilibrium vapor pressure, the chemical potential of the vapor will be less than that of the condensed phase. Mass will then diffuse from the condensed phase to the vapor phase until chemical equilibrium has been restored, at which point the net flux of mass between the two subsystems vanishes. Conversely, a pressure above the equilibrium vapor pressure will result in a conversion of mass from vapor to condensate, again until the difference of chemical potential between the phases has been eliminated and chemical equilibrium has been restored.

According to Gibbs' phase rule, there exists a single state at which all three phases coexist at equilibrium. Defined by the intersection of surfaces where water and vapor coexist, where vapor and ice coexist, and where water and ice coexist, the triple point for water is given by

$$p_T = 6.1 \text{ mb},$$
$$T_T = 273 \text{ K},$$
$$v_{Tv} = 2.06 \times 10^5 \text{ m}^3 \text{ kg}^{-1}, \qquad (4.26)$$
$$v_{Tw} = 1.00 \times 10^{-3} \text{ m}^3 \text{ kg}^{-1},$$
$$v_{Ti} = 1.09 \times 10^{-3} \text{ m}^3 \text{ kg}^{-1},$$

where the subscripts v, w, and i refer to vapor, water, and ice, respectively.

4.6   Equilibrium Phase Transformations

According to Sec. 2.3, heat transferred during an isobaric process between two homogeneous states of the same phase is proportional to the change of temperature. Under those circumstances, the constant of proportionality is the specific heat at constant pressure c_p. By contrast, heat transfer during an isobaric process between two heterogeneous states of the same two phases involves no change of temperature (Fig. 4.3). Instead, heat transfer results in a conversion of mass from one phase to the other (which is associated with a change of internal energy) and in work being performed when the system's volume changes. Conversely, an isobaric transformation of phase can occur only if it is accompanied by a transfer of heat to support the change of internal energy and the work performed by the system during its change of volume.

4.6.1   Latent Heat

The preceding effects are collected in the change of enthalpy, which equals the heat transfer into the system during an isobaric process (2.15). Analogous to the specific heat capacity at constant pressure, the specific *latent heat* of transformation is defined as the heat absorbed by the system during an isobaric phase transformation

$$l = \delta q_p$$

$$= dh, \tag{4.27.1}$$

which, by the first law, equals the change in enthalpy of phase transformation. In (4.27.1), the mass in l refers only to the substance undergoing the transformation of phase (e.g., to the water component in a mixture with dry air). The specific latent heats of vaporization, fusion (solid $\rightarrow$ liquid), and sublimation (solid $\rightarrow$ vapor) are denoted l_v, l_f, and l_s, respectively, and are related as

$$l_s = l_f + l_v. \tag{4.27.2}$$

Like specific heat capacity, latent heat is a property of the system. Thus, l depends on the thermodynamic state, which may be expressed $l = l(T)$ for a heterogeneous system of two phases. The dependence on temperature of l can be established by considering a transformation from one phase to another and how that transformation varies with temperature. Consider a homogeneous state wherein the system is entirely in phase a and another homogeneous state wherein it is entirely in phase b. By (2.23.2), an isobaric process between two homogeneous states that involve only phase a results in a change of enthalpy

$$dh_a = \left(\frac{\partial h_a}{\partial T}\right)_p dT$$

$$= c_{pa} dT.$$

Likewise, an isobaric process between two homogeneous states that involve only phase b results in

$$dh_b = \left(\frac{\partial h_b}{\partial T}\right)_p dT$$
$$= c_{pb} dT.$$

Subtracting gives an expression for how the difference of enthalpy between phases a and b changes with temperature

$$d(\Delta h) = (c_{pb} - c_{pa})dT.$$

Since the enthalpy difference equals the latent heat of transformation between phase a and phase b (4.27.1),

$$\frac{dl}{dT} = \Delta c_p, \tag{4.28}$$

where Δc_p refers to the difference of specific heat between the phases.

Known as *Kirchhoff's equation*, (4.28) provides a formula for calculating the latent heat, as a function of temperature, in terms of the difference of specific heat between the two phases. Of the three latent heats, only l_f varies significantly over a range of temperature relevant to the atmosphere. At 0°C, the latent heats of water have the values

$$l_v = 2.50 \times 10^6 \, \text{J kg}^{-1},$$
$$l_f = 3.34 \times 10^5 \, \text{J kg}^{-1}, \tag{4.29}$$
$$l_s = 2.83 \times 10^6 \, \text{J kg}^{-1}.$$

Since it is defined for an isobaric process, l describes the change of enthalpy during a transformation of phase at constant pressure. The corresponding change of internal energy follows from the first law. Consider the system in a heterogeneous state and undergoing an isobaric transformation of phase. Then (2.12.2) becomes

$$du = l - pdv. \tag{4.30}$$

For fusion, dv is negligible, so $du = l$. The change of internal energy is then just the latent heat, which also represents the change of enthalpy under those circumstances. For vaporization and sublimation,

$$dv \cong v_v, \tag{4.31}$$

where v_v refers to the volume of vapor produced. Incorporating the ideal gas law for the vapor gives the change of internal energy

$$du = l - R_v T \tag{4.32}$$

during isobaric vaporization and sublimation.

4.6.2   Clausius–Clapeyron Equation

In states involving two phases, the system of pure water possesses only one thermodynamic degree of freedom. Thus, specifying its temperature determines the system's pressure and hence its thermodynamic state. Under those circumstances, the equation of state (4.25) reduces to the form of (4.23), which describes the simplified behavior of the water surface in Fig. 4.3. The equation of state describing those heterogeneous regions may be derived from the fundamental relations, subject to conditions of chemical equilibrium.

Consider two phases a and b and a transformation between them that occurs reversibly (e.g., wherein the system remains in thermodynamic equilibrium). The heat transfer during such a process equals the latent heat of transformation. Then with (4.27), the second law becomes

$$ds = \frac{l}{T}. \tag{4.33}$$

For the system to be in chemical equilibrium, the chemical potential of phase a must equal that of phase b, so by (4.5)

$$g_a = g_b.$$

Since this condition is satisfied throughout the process, the change of Gibbs function for one phase must track that of the other

$$dg_a = dg_b.$$

Applying the fundamental relation (3.16.2) to each subsystem then implies

$$-(s_b - s_a)dT + (v_b - v_a)dp = 0$$

or

$$\frac{dp}{dT} = \frac{\Delta s}{\Delta v}, \tag{4.34}$$

where Δ refers to the change between the phases. Incorporating (4.33) yields a relationship between the equilibrium pressure and temperature:

$$\frac{dp}{dT} = \frac{l}{T\Delta v}, \tag{4.35}$$

where l is the latent heat appropriate to the phases present. Known as the *Clausius–Clapeyron equation*, (4.35) relates the equilibrium vapor pressure (e.g., $p = p_w$ or p_i) to the temperature of the heterogeneous system. It thus constitutes an equation of state for the heterogeneous system when two phases are present and describes the simplified surfaces in Fig. 4.3 corresponding to such states.

The Claussius–Clapeyron equation may be specialized to each of the heterogeneous regions in Fig. 4.3. For water and ice, l corresponds to fusion.

Under those circumstances, (4.35) is expressed most conveniently in inverted form

$$\frac{dT}{dp} = \frac{T\Delta v}{l},$$

which describes the influence on melting temperature exerted by a change of pressure. Because the change of volume during fusion is negligible, the equation of state in the region of water and ice reduces to

$$\left(\frac{dT}{dp}\right)_{\text{fusion}} \cong 0. \tag{4.36}$$

Changing the pressure has only a negligible effect on the temperature at which water is at equilibrium with ice. Consequently, the surface of water and ice in Fig. 4.3 is vertical.

For vapor and a condensed phase, l corresponds to the latent heat of vaporization or sublimation. Under those circumstances, the change of volume is approximately equal to that of the vapor produced (4.31). Thus,

$$\Delta v \cong \frac{R_v T}{p},$$

which transforms (4.35) into

$$\left(\frac{d \ln p}{dT}\right)_{\substack{\text{vaporization} \\ \text{sublimation}}} = \frac{l}{R_v T^2}. \tag{4.37}$$

For $l \cong \text{const}$, (4.37) gives

$$\ln\left(\frac{p_2}{p_1}\right) = \frac{l}{R_v}\left(\frac{1}{T_1} - \frac{1}{T_2}\right). \tag{4.38}$$

Using the value of l_v in (4.29) yields the equilibrium vapor pressure with respect to water:

$$\log_{10} p_w \cong 9.4041 - \frac{2.354 \times 10^3}{T}, \tag{4.39}$$

where p_w is in millibars and $T = 0°C$ is used as a reference state. Similarly, the value of l_s in (4.29) yields the equilibrium vapor pressure with respect to ice

$$\log_{10} p_i \cong 10.55 - \frac{2.667 \times 10^3}{T}, \tag{4.40}$$

which differs from (4.39) only modestly.

Equations (4.39) and (4.40) describe the simplified surfaces in Fig. 4.3 that correspond to vapor being in chemical equilibrium with a condensed phase and to the system pressure equaling the equilibrium vapor pressure p_w or p_i. If the system's pressure is below the equilibrium vapor pressure for the temperature of the system, water will evaporate or sublimate until the system's pressure reaches the equilibrium vapor pressure. A pressure above the equilibrium vapor pressure will result in the reverse transformations. Owing to the exponential dependence in (4.39) and (4.40), p_w and p_i vary sharply with temperature. In the presence of a condensed phase, substantially more water can exist in vapor phase at high temperature than at low temperature. We will see in Chapter 5 that the principles governing a single-component heterogeneous system carry over to a two-component system of dry air and water. The equilibrium vapor pressure is then the maximum amount of vapor that can be supported by air at a given temperature.

The exponential dependence of p_w on T has an important implication for exchanges of water between the earth's surface and the atmosphere. According to (4.39), warm tropical oceans with a high sea surface temperature can transfer substantially more water into the atmosphere than can colder extra-tropical oceans. For this reason, tropical oceans serve as the primary source of water vapor for the atmosphere, which is subsequently redistributed over the globe by the circulation (refer to Fig. 1.15). Much of the water vapor absorbed by the tropical atmosphere is precipitated back to the earth's surface in organized convection inside the Inter Tropical Convergence Zone (ITCZ) (Fig. 1.25). However, latent heat that is released during condensation remains in the overlying atmosphere. Thus, cyclic transfer of moisture between ocean surfaces and the tropical troposphere results in a net transfer of heat to the atmosphere. Eventually converted into work, that heat generates kinetic energy, which, along with radiative transfer from the earth's surface (Fig. 1.27), maintains the general circulation against frictional dissipation.

Suggested Reading

The Principles of Chemical Equilibrium (1971) by Denbigh includes a thorough development of phase equilibria in heterogeneous systems.

A detailed treatment of water substance and accompanying thermodynamic properties is presented in *Atmospheric Thermodynamics* (1981) by Iribarne and Godson.

Problems

4.1. The Gibbs–Dalton law implies that the partial pressure of vapor at equilibrium with a condensed phase of water is the same in a mixture with dry air as the equilibrium vapor pressure if the water component were in isolation. Since it corresponds to the abundance of vapor at which no mass is transformed from one phase to another, this vapor pressure describes the state at which air is *saturated*. For a lapse rate of 6.5 K km^{-1}, which is representative of thermal structure in the troposphere (Fig. 1.2), calculate (a) the equilibrium vapor pressure as a function of altitude and

(b) the corresponding mixing ratio of water vapor as a function of altitude.

4.2. Derive the fundamental relations (4.17) for a mixture involving multiple phases.

4.3. Derive alternate expressions for chemical equilibrium under (a) isothermal–isochoric conditions and (b) reversible-adiabatic and isobaric conditions.

4.4. Consider a mixture of dry air and water. Describe the state space of this generally heterogeneous system and the geometry its graphical representation assumes, noting the number of thermodynamic degrees of freedom in regions where one, two, and three phases are present.

4.5. Use (4.39) to estimate the latent heat of vaporization for water.

4.6. A more accurate version of (4.39) is given by

$$\log_{10} p_w = -\frac{2937.4}{T} - 4.9283 \log_{10} T + 23.5471 \qquad \text{(mb)}.$$

Estimate the boiling temperature for water at the altitude of (a) Denver, 5000 ft; (b) the continental divide, 14,000 ft; and (c) the summit of Mt. Everest, 29,000 ft.

4.7. Present-day Venus contains little water, most of which is thought to have been absorbed by its atmosphere and eventually destroyed during the planet's evolution (Sec. 8.7). (a) For present conditions, wherein Venus has a surface pressure of 9×10^6 Pa, at what surface temperature does water vapor become the atmosphere's primary constituent? Compare this to Venus' present surface temperature of 750 K. (b) Were these conditions to prevail to altitudes where energetic UV radiation is present, what would be the implication for the abundance of water on the planet? (c) Contrast these circumstances with present-day conditions on Earth.

4.8. Clouds seldom form in the stratosphere because air there is very dry. However, polar stratospheric clouds (PSCs) are known to form over the Antarctic and, less frequently, over the Arctic. The thicker of these clouds (Type II PSCs), which are still quite tenuous compared to tropospheric clouds, are known to be composed of ice. (a) For a mixing ratio at 80 mb of 3 ppmv, which is representative of water vapor in the lower stratosphere (refer to Fig. 17.8), calculate the temperature at which ice cloud forms. (b) Referring to Fig. 1.7 and to the discussion in Sec. 17.3.2, where are such clouds likely to form?

4.9. The average precipitation rate inside the ITCZ is of order 10 mm day^{-1} (Fig. 9.38). (a) In watts per square meter, calculate the average column heating rate inside the ITCZ. (b) Compare this value with the longwave radiative flux emitted to (and largely absorbed by) the atmosphere if the surface behaves as a blackbody at a temperature of 300 K (1.29).

4.10. The Gibbs–Dalton law implies that equilibrium properties of water vapor in solution with dry air are identical to those of pure water. In light of this result, explain why the vertical profile of atmospheric water vapor is distinguished from other trace gases in Fig. 1.20. In particular, use the global–mean temperature (Fig. 1.2) to demonstrate why the mixing ratio of water vapor decreases sharply with height in the troposphere.

Chapter 5 | Transformations of Moist Air

The atmosphere is a mixture of dry air and water in varying proportions. Although its abundance varies widely, water vapor seldom represents more than a few percent of air by mass. We shall consider a two-component system comprised of these species, with water appearing in possibly one condensed phase. According to the Gibbs–Dalton law (which is accurate at pressures below the critical point), an individual component of a mixture of gases behaves the same as if the other components were absent. Consequently, the abundance of vapor at equilibrium with a condensed phase in a mixture of water and dry air is the same as if the water component were in isolation.[1] For this reason, concepts established in Chapter 4 for a single-component system of pure water carry over to a two-component system of dry air and water.

5.1 Description of Moist Air

5.1.1 Properties of the Gas Phase

For the moment, we focus on the gas phase of this system, irrespective of whether condensate happens to be present. The vapor in solution with dry air is represented by its partial pressure e, which obeys the equation of state

$$ev_v = R_v T. \qquad (5.1.1)$$

In (5.1.1),

$$v_v = \frac{V}{m_v} \qquad (5.1.2)$$

is the specific volume of vapor (not to be confused with the partial volume),

$$R_v = \frac{R^*}{M_v}$$

$$= \frac{1}{\epsilon_v} R_d, \qquad (5.1.3)$$

and $\epsilon_v = M_v/M_d \cong 0.622$ is the ratio of molar weights defined by (1.15.2).

[1]This and other implications of the Gibbs–Dalton law can be found in Keenan (1970).

The *absolute humidity* $\rho_v = 1/v_v$ measures the absolute concentration of vapor. The relative concentration of vapor is measured by the *specific humidity*:

$$q = \frac{\rho_v}{\rho} = \frac{m_v}{m},$$
(5.2)

which equals the ratio of the masses of vapor and mixture. Closely related is the mass *mixing ratio r*, which is referenced to to the mass of dry air (1.14). Since

$$m = m_d + m_v,$$
(5.3)

the mixing ratio is approximately equal to the specific humidity

$$r = \frac{q}{1-q}$$
$$\cong q$$
(5.4)

because vapor exists only in trace abundance (Table 1.1). Hence, to a good approximation, q and r can be used interchangeably. Both are conserved for an individual air parcel outside regions of condensation. By contrast, measures of absolute concentration like e and ρ_v change for an individual air parcel through changes of its pressure, even if the mass of vapor remains fixed.

Despite the advantages of relative concentration, chemical equilibrium of the water component is controlled by the absolute concentration of vapor. For this reason, it is convenient to express the mixing ratio r in terms of the vapor pressure e. The dry air component of the mixture obeys the equation of state

$$p_d v_d = R_d T,$$
(5.5)

where p_d and v_d denote the partial pressure and specific volume of dry air. Dividing (5.1) by (5.5) obtains

$$\left(\frac{e}{p_d}\right)\left(\frac{v_v}{v_d}\right) = \frac{1}{\epsilon},$$
(5.6)

where $\epsilon = \epsilon_v$ is understood. Now

$$\left(\frac{v_v}{v_d}\right) = \left(\frac{m_d}{m_v}\right)$$
$$= \frac{1}{r}$$

and, since vapor exists only in trace abundance,

$$p = p_d + e$$
$$\cong p_d.$$

Then (5.6) reduces to

$$\frac{r}{\epsilon} = \frac{e}{p}$$
$$= N_v, \tag{5.7}$$

where the molar fraction of vapor N_v is defined by (1.12).

Because the composition of air varies with the abundance of water vapor, so too do composition-dependent properties like the specific gas constant R. It is convenient to absorb such variations into state variables and deal with the fixed properties of dry air. From (1.7) and (5.3), the specific gas constant of the mixture may be expressed

$$R = (1 - q)R_d + qR_v$$
$$= (1 - q)R_d + \frac{q}{\epsilon}R_d$$
$$= \left[1 + \left(\frac{1}{\epsilon} - 1\right)q\right]R_d$$

or

$$R = (1 + 0.61q)R_d. \tag{5.8}$$

Then the equation of state for the mixture becomes

$$pv = (1 + 0.61q)R_d T. \tag{5.9}$$

The moisture dependence in (5.9) can be absorbed into the *virtual temperature*

$$T_v = (1 + 0.61q)T. \tag{5.10}$$

Then the equation of state for the gas phase of the system is simply

$$pv = R_d T_v, \tag{5.11}$$

which permits the fixed specific gas constant of dry air to be used for the mixture. In practice, $q = O(0.01)$, so T can be used in place of the T_v to a good approximation.

The specific heats of air also depend on moisture content:

$$c_v = (1 + 0.97q)c_{vd},$$
$$c_p = (1 + 0.87q)c_{pd}, \tag{5.12}$$

from which the dimensionless quantities γ and κ (2.21) follow directly. Like virtual temperature, c_v, c_p, γ, and κ differ only slightly from the constant values for dry air.

5.1.2 Saturation Properties

Consider now the gas phase of the system in the presence of a condensed phase of water. If the vapor is in chemical equilibrium with the condensed phase, it is said to be *saturated*. Corresponding to this condition and for a given pressure and temperature are particular values of the foregoing moisture variables, which are referred to as *saturation values*. According to the Gibbs–Dalton law, the *saturation vapor pressure* with respect to water e_w is identical to the equilibrium vapor pressure p_w of a single-component system of vapor and water (Sec. 4.6). Likewise, the saturation vapor pressure with respect to ice e_i is identical to p_i for a single-component system.[2]

The saturation vapor pressure e_c, where c denotes either of the condensed phases, is a function of temperature alone and described by the Clausius–Clapeyron relations (4.39) and (4.40). The *saturation specific humidity* q_c and the *saturation mixing ratio* r_c, which follows from e_c through (5.7), also describe the abundance of vapor at equilibrium with a condensed phase. Like the saturation vapor pressure, these quantities are state variables. But, because they refer to the mixture and not just the vapor, q_c and r_c also depend on pressure, in accord with Gibbs' phase rule (4.24) for a two-component system involving two phases. However, the strong temperature dependence of $e_c(T)$ in the Clausius–Clapeyron equation is the dominant influence on $q_c(p, T)$ and $r_c(p, T)$. Therefore, a decrease of temperature following from adiabatic expansion sharply reduces the saturation values q_c and r_c. Just the reverse results from an increase of temperature following from adiabatic compression.

Contrary to saturation values, which change with the thermodynamic state of the system, the abundance of vapor actually present changes only through a transformation of phase. If no condensed phase is present (e.g., under unsaturated conditions), the abundance of vapor is preserved. A decrease of temperature then results in a decrease of r_c, but no change of r. On the other hand, if the system is saturated, $r = r_c(p, T)$. A change of state in which the system remains saturated must then result in a change of both $r_c(p, T)$ and r, that is, the vapor and condensate must adjust to preserve chemical equilibrium between those phases.

---

[2] Strictly, the water component of the two-component system does not behave exactly as it would in isolation. Discrepancies in that idealized behavior stem from (1) near saturation, departures of the vapor from the behavior of an ideal gas, (2) the condensed phase being acted on by the total pressure and not just that of the vapor, and (3) some of the air passing into solution with the water. However, these effects introduce discrepancies that are smaller than 1%, so they can be ignored for most applications; see Iribarne and Godson (1981) for a detailed treatment.

Two other quantities are used to describe the abundance of water vapor. The *relative humidity* is defined as

$$RH = \frac{N_v}{N_{vc}}$$
$$= \frac{e}{e_c} \simeq \frac{r}{r_c}, \tag{5.13}$$

where N_{vc} denotes the saturation molar abundance of vapor, and equilibrium with respect to water is usually implied. The *dew point* T_d is defined as that temperature to which the system must be cooled "isobarically" to achieve saturation. If saturation occurs below 0°C, that temperature is the *frost point* T_f. The *dew point spread* is given by the difference $(T - T_d)$. For a given temperature T, a high dew point implies a small dew point spread. Each corresponds to a large abundance of vapor, which requires only a small depression of temperature to achieve saturation.

Neither relative humidity nor dew point spread are direct measures of vapor concentration but, rather, describe how far the system is from saturation. Because saturation values increase sharply with temperature, inferring moisture content from the aforementioned quantities is misleading. For instance, r_w at 1000 mb and 30°C is nearly 30 g kg^{-1}, but only 4 g kg^{-1} at the same pressure and 0°C. Therefore, a relative humidity of 50% implies an abundance of vapor of 2 g kg^{-1} in the latter but of 15 g kg^{-1} in the former—more than seven times as large.

5.2 Implications for the Distribution of Water Vapor

Saturation values describe the maximum abundance of vapor that can be supported by air for a given temperature and pressure. At that abundance, diffusion of mass from vapor to condensate is balanced by diffusion of mass in the opposite sense. If the system is heterogeneous and has a vapor abundance below the saturation value (e.g., an unsaturated air parcel in contact with a warm ocean surface), vapor will be absorbed until the difference of chemical potential between the phases of water has been eliminated. For this transformation to occur, the water component must absorb heat equal to the latent heat of vaporization. Conversely, if the system is heterogeneous and has a vapor abundance slightly above the saturation value (e.g., a supersaturated air parcel containing an aerosol of droplets), vapor will condense to relieve the imbalance of chemical potential. For this transformation to occur, the water component must reject heat equal to the latent heat of vaporization.

At temperatures and pressures representative of the atmosphere, the saturation vapor pressure seldom exceeds 60 mb and the saturation mixing ratio seldom exceeds 30 g kg^{-1} or 0.030. It is for this reason that water vapor exists only in trace abundance in the atmosphere. One should note that the foregoing moisture properties refer only to vapor—not to the total water content of the system. Condensation results in a reduction of q and r, but a commensurate increase of condensate according to (4.6.2). Unless condensate precipitates out of the system, the total water content of an air parcel is preserved.

According to the Clausius–Clapeyron equation, saturation vapor pressure depends exponentially on temperature. Thus, air can support substantially more vapor in solution at high temperature than at low temperature. For this reason, water vapor is produced efficiently in the tropics, where warm sea surface temperature (SST) corresponds to high equilibrium vapor pressure (Fig. 5.1), and its mixing ratio decreases poleward (Fig. 1.14). Conversely, water vapor is destroyed aloft through condensation and precipitation, after displaced air parcels have cooled through adiabatic expansion work (2.33) and suffered a sharp reduction of saturation mixing ratio. Condensation also occurs at high latitudes, after displaced air parcels have cooled through radiative and conductive heat transfer.

Production of vapor at an ocean surface can occur only if the water component absorbs latent heat to support the transformation of phase. Absorbed from the ocean or directly from shortwave radiation, that latent heat is then transferred to the air with which the vapor passes into solution. When the water recondenses, the latent heat is released to the air surrounding the condensate (Fig. 5.2) and remains in the atmosphere after the condensate has precipitated back to the surface. Therefore, the preceding cycle results in no permanent exchange of mass but a net transfer of heat from the ocean to the atmosphere (see Fig. 1.27).

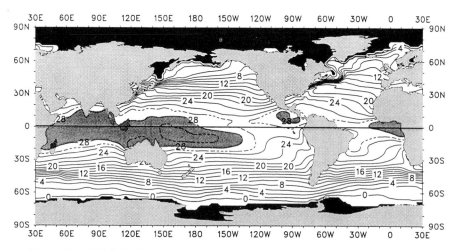

Figure 5.1 Global distribution of sea surface temperature (SST) for March. Temperatures warmer than 28°C are shaded. From Shea *et al.* (1990).

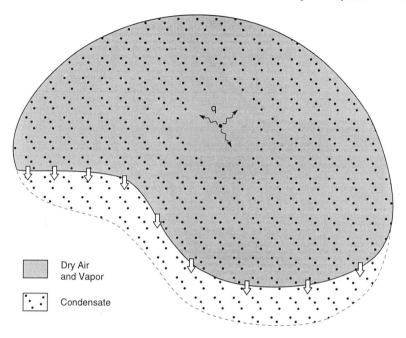

Figure 5.2 Schematic illustration of a saturated air parcel undergoing condensation, in which latent heat q is released to the gas phase and some of the condensate precipitates out of the parcel.

5.3 State Variables of the Two-Component System

Under unsaturated conditions, the two-component system involves only a single phase. By Gibbs' phase rule (4.24), the system then possesses three thermodynamic degrees of freedom. Thus, three intensive properties are required to specify the system's state if no condensate is present. Usually, the state of moist air is specified by pressure, temperature, and a humidity variable like mixing ratio, for example, $\theta = \theta(p, T, r)$. Under saturated conditions, two phases are present and one of the independent state variables is eliminated by the constraint of chemical equilibrium [e.g., $r = r_c(p, T)$]. This leaves two degrees of freedom—the same number as for a homogeneous single-component system. Thus, under saturated conditions, a state variable can be specified as $\theta = \theta(p, T)$.

5.3.1 Unsaturated Behavior

Under unsaturated conditions, thermodynamic processes for moist air occur much as they would for dry air because thermal properties are modified only slightly by the trace abundance of vapor. A change of temperature leads to changes of internal energy and enthalpy given by (2.18), but with slightly modified specific heats (5.12). The saturation mixing ratio r_c, which measures the capacity of air to support vapor in solution, varies with the system's pressure and temperature. By contrast, the mixing ratio r of vapor actually present remains constant so long as the system is unsaturated.

The same is approximately true of the potential temperature. Under adiabatic conditions, pressure and virtual temperature change in such proportion to preserve the *virtual potential temperature*

$$\frac{\theta_v}{T_v} = \left(\frac{p_0}{p}\right)^{\kappa_d}, \tag{5.14}$$

where κ_d denotes the value for dry air. Like virtual temperature, θ_v is nearly identical to its counterpart for dry air, θ. Therefore, virtual properties of moist air like T_v and θ_v will hereafter be referred to by their counterparts for dry air, but the former will be understood to apply in a strict sense.

5.3.2 Saturated Behavior

Under saturated conditions (e.g., in the presence of an aerosol of droplets), the aforementioned relationships no longer hold. Because the vapor must then be at equilibrium with a condensed phase, $e = e_c$. A change of thermodynamic state that alters e_c then also alters e, which must result in a transformation of mass from one phase of the water component to another. Accompanying that transformation of mass is an exchange of latent heat between the condensed and gas phases of the heterogeneous system (Fig. 5.2), one that alters the potential temperature of the gas phase through (2.36).

State variables describing the two-component heterogeneous system must account for these changes. For a closed system, (4.7) gives the change of total enthalpy

$$dH = \left(\frac{\partial H}{\partial T}\right)_{pm} dT + \left(\frac{\partial H}{\partial p}\right)_{Tm} dp + (h_v - h_c)dm_v. \tag{5.15}$$

Because

$$H = m_d h_d + m_v h_v + m_c h_c,$$

$$\left(\frac{\partial H}{\partial T}\right)_{pm} = m_d \left(\frac{\partial h_d}{\partial T}\right)_{pm} + m_v \left(\frac{\partial h_v}{\partial T}\right)_{pm} + m_c \left(\frac{\partial h_c}{\partial T}\right)_{pm}$$

$$= m_d c_{pd} + m_v c_{pv} + m_c c_{pc}. \tag{5.16}$$

Similarly,

$$\left(\frac{\partial H}{\partial p}\right)_{Tm} = m_d \left(\frac{\partial h_d}{\partial p}\right)_{Tm} + m_v \left(\frac{\partial h_v}{\partial p}\right)_{Tm} + m_c \left(\frac{\partial h_c}{\partial p}\right)_{Tm}.$$

For the gas phase,

$$\left(\frac{\partial h_d}{\partial p}\right)_{Tm} = \left(\frac{\partial h_v}{\partial p}\right)_{Tm} = 0$$

by (2.18). For the condensed phase, the corresponding change of enthalpy can be expressed

$$\left(\frac{\partial h_c}{\partial p}\right)_{Tm} = v_c(1 - T\alpha_p),$$

where

$$\alpha_p = \frac{1}{v}\left(\frac{\partial v}{\partial T}\right)_p$$

defines the *isobaric expansion coefficient* (e.g., Denbigh, 1971). Because α_p is small for condensed phases,

$$\left(\frac{\partial h_c}{\partial p}\right)_{Tm} \cong v_c.$$

The contribution to (5.15) from this pressure term can be shown to be negligible compared to the corresponding contribution from temperature (5.16). Incorporating the above into (5.15) and identifying the specific heat of the heterogeneous system as

$$c_p = \frac{m_d c_{pd} + m_v c_{pv} + m_c c_{pc}}{m} \tag{5.17}$$

and the latent heat as the difference of enthalpy between the phases of water

$$l = h_v - h_c, \tag{5.18}$$

obtains the change of enthalpy for the system:

$$dh = c_p dT + l\frac{dm_v}{m}$$
$$\cong c_p dT + ldr. \tag{5.19}$$

The last expression holds exactly if the specific enthalpy refers to a unit mass of dry air, derivation of which is left as an exercise. In (5.19), the term

$$\delta q = -ldr$$

represents the heat transferred to the gas phase (which is chiefly dry air) from the water component when the latter undergoes a transformation of phase. If l is treated as constant (Sec. 4.6.1), (5.19) may be integrated to yield an expression for the absolute enthalpy of the two-component heterogeneous system:

$$h = c_p T + lr + h_0, \tag{5.20}$$

where h_0 denotes the enthalpy at a suitably-defined reference state.

Expressions for the internal energy and entropy of the system follow in similar fashion:

$$u = c_v T + lr + u_0, \tag{5.21}$$

$$\frac{s}{c_p} = \ln T - \kappa_d \ln p + \frac{lr}{c_p T} + \frac{s_0}{c_p}$$

$$= \ln \theta + \frac{lr}{c_p T} + \frac{s_0}{c_p}. \tag{5.22}$$

Like (5.19), the expressions for absolute enthalpy, internal energy, and entropy of the two-component system are exact if referenced to a unit mass of dry air.

5.4 Thermodynamic Behavior Accompanying Vertical Motion

The thermodynamic state of a moist air parcel changes through vertical motion. Vertical displacement alters the environmental pressure, which varies hydrostatically according to (1.17). To preserve mechanical equilibrium, the parcel expands or contracts, which results in work being performed. Compensating that work is a change of internal energy, which alters the temperature and hence the saturation vapor pressure of the two-component system.

5.4.1 Condensation and the Release of Latent Heat

From (4.38), the change of saturation vapor pressure between a reference temperature T_0 and a temperature T can be expressed

$$\ln \left(\frac{e_c}{e_{c0}} \right) = -\frac{l}{R_v} \left(\frac{1}{T} - \frac{1}{T_0} \right). \tag{5.23}$$

Then (5.7) implies that the saturation mixing ratio varies with pressure and temperature as

$$\frac{r_c}{r_{c0}} = \frac{\exp \left[-\frac{l}{R_v} \left(\frac{1}{T} - \frac{1}{T_0} \right) \right]}{\left(\frac{p}{p_0} \right)}. \tag{5.24}$$

According to (5.24), the saturation mixing ratio increases with decreasing pressure. However, r_c decreases sharply with decreasing temperature, which likewise accompanies upward motion. Therefore, even though an ascending parcel's pressure decreases exponentially with altitude, the temperature dependence in (5.24) prevails, so its saturation mixing ratio decreases monotonically with altitude.

Consider a moist air parcel ascending in thermal convection. Under unsaturated conditions, the parcel's mixing ratio and saturation mixing ratio satisfy

$$r < r_c.$$

As it rises, the parcel performs work at the expense of its internal energy, which decreases its temperature at the dry adiabatic lapse rate Γ_d. From (5.24), the decrease of temperature is attended by a reduction of saturation mixing ratio r_c (Fig. 5.3). By contrast, the parcel's actual mixing ratio r and potential temperature θ remain constant under the foregoing conditions. Sufficient upward displacement will reduce the saturation mixing ratio to the actual mixing ratio:

$$r_c = r,$$

at which point the parcel is saturated. The elevation where this first occurs is referred to as the *lifting condensation level* (LCL). Because convective clouds

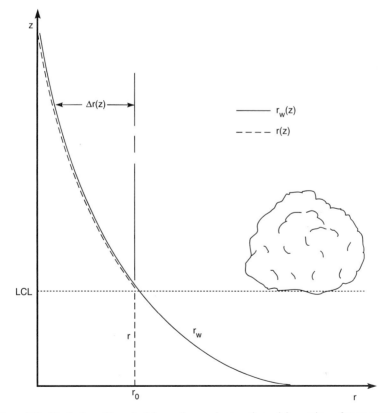

Figure 5.3   Vertical profiles of mixing ratio r and saturation mixing ratio r_w for an ascending air parcel below and above the *lifting condensation level* (LCL).

form through the process just described, the LCL defines the base of cumulus clouds that are fueled by air originating at the surface.

Below the LCL, the parcel's thermodynamic behavior can be regarded as adiabatic because the characteristic timescale for vertical displacement (e.g., from minutes in cumulus convection to 1 day in sloping convection) is small compared to the characteristic timescale for heat transfer. Therefore, the parcel evolves in state space along a dry adiabat, which is described by Poisson's equations (2.30) and characterized by a constant value of θ.

In physical space, the parcel is actually part of a layer that is displaced vertically (Fig. 5.4). *Material contours*, which define a fixed collection of fluid elements, buckle to form a plume of moist air that rises through positive buoyancy. Folds

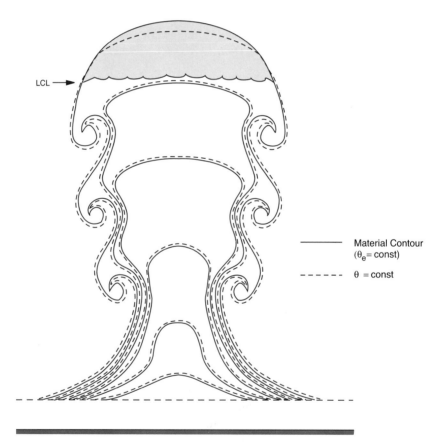

Figure 5.4 An ascending plume of moist air that develops from surface air that has become positively buoyant. Material lines (solid), which are defined by a fixed collection of air parcels, remain coincident with isentropes (θ=const) below the LCL, but they drift to higher θ and higher altitude above the LCL, where air warms through latent heat release. Although they depart from isentropes, material lines remain coincident with pseudo-isentropes (θ_e = const) above the LCL.

in the wall of the plume reflect turbulent entrainment with surrounding air, which dilutes the plume's buoyancy with that of the environment. Associated mixing introduces diabatic effects that are ignored in the present analysis, but play an essential role in dissipating cumulus convection (Chapter 9). Under unsaturated adiabatic conditions, θ and r are both conserved for individual fluid elements. Therefore, a material line that coincides initially with a certain isentrope $\theta = \theta_0$ and mixing ratio contour $r = r_0$ remains coincident with those isopleths—as long as they reside beneath the LCL.

Above the LCL, the foregoing behavior breaks down. Continued ascent and expansion work reduces the saturation mixing ratio r_c below the mixing ratio r of vapor present. To restore chemical equilibrium, some of the vapor must condense—just enough to maintain saturation:

$$r = r_c.$$

Thus, above the LCL, r and r_c both decrease with height (Fig. 5.3), the decrease of vapor being reflected in an increase of condensate (4.6.2).

In this fashion, ascent above the LCL wrings vapor out of solution with dry air and produces condensate (e.g., cloud droplets). The production of condensate is attended by a release of latent heat to the gas phase of the system. Internal to the parcel, that heating alters the potential temperature of the gas phase through (2.36) and adds positive buoyancy, which in turn promotes continued ascent. The change of water vapor mixing ratio that results from a displacement Δz above the LCL is given by

$$\Delta r(z) = r_c\big[p(z), T(z)\big] - r_0,$$

where z, p, and T are understood to refer to the displaced parcel. Because r_c decreases monotonically with altitude, the greater the displacement above the LCL, the less vapor that remains in the gas phase of the parcel and the greater the abundance of condensate and the liberation of latent heat. Cloud droplets produced in this fashion grow (through mechanisms described in Chapter 9) until they can no longer be supported by the updraft, at which point they precipitate out of the parcel.

Since θ and r are no longer conserved, the material contour coincident initially with the isopleths $\theta = \theta_0$ and $r = r_0$ deviates from those isopleths (Fig. 5.4). The release of latent heat increases θ (2.36), so the material contour advances to isentropes of greater potential temperature. For reasons developed in Chapter 7, these lie at a higher altitude than the original isentrope. Similarly, condensation reduces r, so the material line moves to isopleths of smaller mixing ratio, which likewise lie at higher altitude.

Saturated air that is descending undergoes just the reverse behavior. Adiabatic compression then increases the internal energy and temperature of the gas phase, which increases the saturation mixing ratio r_c over the actual mixing ratio r. Condensate can then evaporate to restore chemical equilibrium: $r = r_c$. Evaporation of condensate must be attended by absorption of latent heat from the gas phase, which cools the parcel, introduces negative buoyancy, and thus promotes continued descent. Any condensate that previously precipitated out of the parcel is not available to reabsorb latent heat that it released during ascent, which therefore remains in the gas phase.

The foregoing behavior is responsible for confining water vapor near the earth's surface. Introduced over warm oceans, water vapor is extracted from air that is displaced upward. Moisture is lost altogether when condensate precipitates back to the earth's surface, after cloud particles have become sufficiently large. In this fashion, thermodynamics, in combination with hydrostatic stratification, maintains upper levels of the atmosphere in a very dry state. Even inside a convective tower, mixing ratio decreases vertically because chemical equilibrium requires r to equal r_c, which decreases steadily with altitude. By limiting its vertical transport, thermodynamics prevents water vapor from reaching great altitudes, where it would be photodissociated by energetic radiation and ultimately destroyed when the free hydrogen produced is lost to space (Sec. 1.2.2).

Due to exchange of latent heat with the condensed phase, the gas phase of an air parcel is not adiabatic above the LCL—even though the entire system may still be. Consequently, the potential temperature of the gas phase is no longer conserved. Since the parcel's mass is dominated by dry air, the transformation of mass has only a minor effect on the energetics of the parcel. But the transfer of latent heat that attends the phase transformation has a major effect by serving as an internal heat source for the system. Latent heat released to the gas phase during condensation offsets cooling due to adiabatic expansion work that is performed by the parcel during ascent. Conversely, latent heat absorbed from the gas phase during vaporization offsets warming due to adiabatic compression work that is performed on the parcel during descent. Owing to the transfer of latent heat, the parcel's temperature no longer changes with altitude at the dry adiabatic lapse rate, but rather varies more slowly under saturated conditions.

5.4.2   The Pseudo-Adiabatic Process

If expansion work occurs fast enough for heat transfer with the environment to remain negligible and if none of its moisture precipitates out, the parcel is closed and its behavior above the LCL is described by a reversible saturated adiabatic process. That process depends weakly on the abundance of condensate present (e.g., on how much of the system's enthalpy is represented by condensate) and therefore on the LCL of the parcel. However, because it is present only in trace abundance, the variation of condensate unnecessarily complicates the parcel's description under saturated conditions. This complication is averted by describing the parcel's behavior in terms of a *pseudo-adiabatic process*, in which the system is treated as open and condensate is removed (added) immediately after (before) it is produced (destroyed). Because the water component accounts for only a small fraction of the system's mass, the pseudo-adiabatic process is nearly identical to a reversible saturated adiabatic process.

A pseudo-adiabatic change of state may be constructed in two legs:

1. Reversible saturated adiabatic expansion (compression), which results in the production (destruction) of condensate of mass dm_c and a commensurate release (absorption) of latent heat to (from) the gas phase
2. Removal (addition) of condensate of mass dm_c

Since the phase transformation occurs adiabatically and reversibly, this process is isentropic:

$$ds = 0.$$

Also, since just enough vapor condenses to maintain chemical equilibrium ($r = r_c$), the change of water vapor mixing ratio is given by

$$dr = dr_c.$$

Then (5.22) implies

$$d \ln \theta = -d \left(\frac{l r_c}{c_p T} \right). \tag{5.25}$$

With the removal of condensate, (5.25) describes the change of potential temperature of the gas phase in terms of the transformation of the water component. A decrease of r_c, such as accompanies ascent and expansional cooling, increases θ, whereas an increase of r_c, such as accompanies descent and compressional warming, decreases θ.

Integrating (5.25) obtains

$$\theta \exp \left(\frac{l r_c}{c_p T} \right) = \text{const}, \tag{5.26}$$

which describes a family of paths in the state space of moist air, one analogous to the family of adiabats described by Poisson's equation (2.30) for dry air. Just as the latter was used to introduce the potential temperature, which is preserved for an adiabatic process, (5.26) may be used to introduce another state variable, which is preserved for a pseudo-adiabatic process. Evaluating (5.26) at a reference state of zero pressure (toward which r_c approaches zero faster than does the parcel's temperature) yields

$$\frac{\theta_e}{\theta} = \exp \left(\frac{l r_c}{c_p T} \right), \tag{5.27}$$

which defines the *equivalent potential temperature* θ_e. According to (5.26), θ_e is constant during a pseudo-adiabatic process. Condensation is accompanied by a reduction of r_c and an increase of θ, but the two vary in such proportion as to preserve θ_e. The same holds for vaporization. In physical space, the material line in Fig. 5.4 advances to isentropes of higher θ, due to the release of latent heat, but it remains coincident with the isopleth of θ_e with which it coincided initially. The equivalent potential temperature reflects the maximum temperature a moist air parcel can assume through adiabatic compression and the release of latent heat: namely, if it was displaced to the top of the atmosphere, where all of the moisture condensed and released its latent heat, if the condensate produced subsequently precipitated out, and if the parcel was then brought down adiabatically to the surface.

Like θ, $\theta_e = \theta_e(p, T, r)$ is a state variable. Just as θ is conserved along an adiabat in state space and below the LCL, θ_e is conserved along a *pseudo* or *saturated adiabat* in state space and above the LCL. Because an adiabatic process involving no transformation of phase is also pseudo-adiabatic, θ_e is conserved under unsaturated conditions as well. However, the definition (5.27) can be applied only under saturated conditions because only then does $r = r_c$, as is implicit in the derivation of θ_e. Alternatively, because it is conserved, θ_e can be calculated with r in place of r_c if T is replaced by the parcel's temperature at the LCL, where it just becomes saturated.

5.4.3   The Saturated Adiabatic Lapse Rate

The temperature of a dry air parcel decreases with its altitude at the dry adiabatic lapse rate Γ_d. To a good approximation, the same holds for a moist air parcel under unsaturated conditions because the trace abundance of water vapor modifies thermal properties of air only slightly. Under saturated conditions, the adiabatic description of air breaks down due to the release of latent heat that accompanies transformation of water from one phase to another. Latent heat exchanged with the gas phase then offsets cooling and warming that accompanies adiabatic expansion and compression.

An approximate description of how the temperature of a saturated parcel changes with altitude can be derived from the first law with the aid of (5.19). From (2.35) the first law for the gas phase can be expressed

$$c_p d \ln T - R d \ln p = \frac{\delta q}{T}. \tag{5.28}$$

If the parcel is unsaturated, (5.28) recovers the dry adiabatic lapse rate (2.33). If it is saturated, the heat transferred to the gas phase is given by

$$\delta q = -l d r_c. \tag{5.29}$$

Then (5.28) becomes

$$c_p d \ln T - R d \ln p = -\frac{l}{T} d r_c,$$

where l is treated as constant. Hydrostatic equilibrium (1.16) reduces this to

$$c_p dT + g dz = -l d r_c. \tag{5.30}$$

Strictly, the saturation mixing ratio depends on both pressure and temperature, so

$$dr_c = \left(\frac{\partial r_c}{\partial T}\right)_p dT + \left(\frac{\partial r_c}{\partial p}\right)_T dp.$$

However, the strong dependence on temperature passed on from the Clausius–Clapeyron equation through (5.7) is the dominant influence on r_c. If its dependence on pressure is ignored, the change of saturation mixing ratio can be written

$$dr_c = \frac{dr_c}{dT} dT,$$

where T and z are understood to refer to the displaced parcel. Then (5.30) reduces to

$$c_p dT + g dz = -l\frac{dr_c}{dT} dT$$

or

$$\left(c_p + l\frac{dr_c}{dT}\right) dT + g dz = 0.$$

In terms of the dry adiabatic lapse rate, this can be expressed

$$\left(1 + \frac{l}{c_p}\frac{dr_c}{dT}\right) dT + \Gamma_d\, dz = 0. \tag{5.31}$$

Then the saturated parcel's temperature decreases with altitude according to

$$-\frac{dT}{dz} = \frac{\Gamma_d}{1 + \frac{l}{c_p}\frac{dr_c}{dT}}$$

$$= \Gamma_s, \tag{5.32}$$

which defines the *saturated adiabatic lapse rate*.

Unlike Γ_d, Γ_s varies with the parcel's altitude due to the nonlinear dependence on T of r_c. However, since $(dr_c/dT) > 0$, (5.32) implies

$$\Gamma_s < \Gamma_d, \tag{5.33}$$

so a parcel's temperature decreases with altitude slower under saturated conditions than under unsaturated conditions. This property of saturated vertical motion follows from the release of latent heat to the gas phase, which offsets cooling associated with adiabatic expansion. Although variable, the saturated adiabatic lapse rate has a value of $\Gamma_s \cong 6.5$ K km^{-1} for conditions representative of the troposphere. This is close to the global-mean lapse rate of the troposphere (Fig. 1.2). No accident, this correspondence follows from dynamical processes that are developed in Chapters 6 and 7.

5.5 The Pseudo-Adiabatic Chart

Thermodynamic processes associated with vertical motion are represented conveniently on a diagram of the state space of moist air. Figure 5.5 shows the *pseudo-adiabatic chart*, which displays as functions of T and $-p^\kappa$

1. Adiabats: $\theta = $ const (solid lines)
2. Pseudo-adiabats: $\theta_e = $ const (dashed lines)
3. Isopleths of saturation mixing ratio[3]: $r_w = $ const (thin solid lines)

The ordinate $-p^\kappa$ can be interpreted in terms of altitude (1.17), which is indicated at the right for the U.S. Standard Atmosphere. Adiabats, along which a parcel evolves under unsaturated conditions and its temperature varies

[3]If ice is present, moisture properties actually depend on r_i, but the isopleths of r_i differ only moderately from those of r_w according to (4.39) and (4.40).

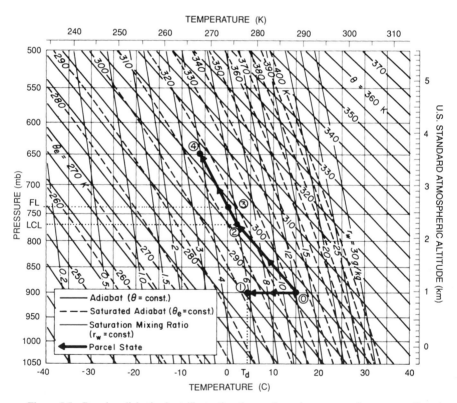

Figure 5.5 Pseudo-adiabatic chart illustrating thermodynamic processes for an ascending air parcel. Shown as functions of temperature and elevation are (1) adiabats (solid), which are labeled by the constant values of θ characterizing those lines, (2) pseudo-adiabats (dashed), which are labeled by the constant values of θ_e characterizing those curves, and (3) isopleths of saturation mixing ratio (thin solid), which are labeled by the constant values of r_w defining those curves.

with height at the dry adiabatic lapse rate Γ_d, appear as straight lines in this representation (2.34). Pseudo-adiabats, along which a parcel evolves under saturated conditions and its temperature varies with height at the saturated adiabatic lapse rate Γ_s, are curved—but only weakly. Consistent with (5.33), pseudo-adiabats have a slope with respect to height that is everywhere smaller than that of adiabats. Both adiabats and pseudo-adiabats are labeled in Kelvin, which correspond to the constant values of θ and θ_e characterizing those paths.

The use of the pseudo-adiabatic chart is best illustrated with an example. Consider conditions leading to the formation of a cumulus cloud. The cloud is fed by air ascending in a moist thermal that is driven by surface heating (Fig. 5.4). At the surface, which is located at 900 mb, air has a temperature of $T_0 = 15°C$ and a mixing ratio $r_0 = 6.0 \ g \ kg^{-1}$. Because air inside the thermal originates at the surface, thermodynamic properties inside the cloud, as well as beneath the cloud, can be inferred from the evolution of an air parcel that is initially at the ground. Once the parcel's initial state (state 0) has been located on the pseudo-adiabatic chart, individual properties follow by allowing the system to evolve along certain paths in the state space of the parcel, which are indicated in Fig. 5.5.

SURFACE RELATIVE HUMIDITY

The saturation mixing ratio at the surface follows from the isopleth of r_w passing through state 0, which gives

$$r_{w0} = 12 \ g \ kg^{-1}.$$

Thus, the initial relative humidity of surface air is

$$RH_0 = \frac{6}{12} = 50\%.$$

SURFACE POTENTIAL TEMPERATURE

The parcel's initial potential temperature is determined by the adiabat passing through state 0, which is characterized by the constant value

$$\theta = \theta_0 = 297 \ K.$$

SURFACE DEW POINT

The dew point of surface air follows from isobaric cooling out of state 0. During that process, the parcel's state evolves along the isobar $p = 900$ mb, which cuts across isopleths of saturation mixing ratio toward lower values of r_w. Eventually, state 1 is reached, at which $r_w = r = 6 \ g \ kg^{-1}$ and the parcel is saturated. The temperature at which this occurs defines the dew point temperature of surface air

$$T_{d0} = 4°C.$$

The foregoing process is responsible for the formation of ground fog. The dew point spread of surface air, which is 11°C in this example, provides an indirect measure of the base of convective clouds, since it reflects the amount of adiabatic cooling necessary to achieve saturation (Problem 5.25).

CUMULUS CLOUD BASE

The base of convective clouds corresponds to the LCL of surface air. The latter may be determined by displacing the parcel upward adiabatically. During that process, the parcel's state evolves along the adiabat passing through state 0 ($\theta = 297$ K), which also cuts across isopleths of saturation mixing ratio toward lower values of r_w. Eventually, state 2 is reached, at which $r_w = r$ and the parcel is again saturated. The level at which this occurs corresponds to the LCL

$$p_{\text{LCL}} = 770 \text{ mb.}$$

The parcel's temperature at this level, which is 13 K colder than its initial temperature, is only 2 K colder than the dew point temperature, reflecting the weak pressure dependence of r_c.

EQUIVALENT POTENTIAL TEMPERATURE AT THE SURFACE

The equivalent potential temperature is determined once the LCL is located. Through state 2 and along subsequent states, passes a saturated adiabat that defines θ_e for the parcel. Because θ_e is conserved under both saturated and unsaturated conditions, that value is also the equivalent potential temperature of the parcel below the LCL, so

$$\theta_{e0} = 315 \text{ K}$$

at 900 mb. Note, even though it is conserved throughout, θ_e must be inferred at and above the LCL, for reasons discussed in Sec. 5.4.2.

FREEZING LEVEL OF SURFACE AIR

At the LCL, the parcel's temperature is greater than 0°C, so the freezing level (FL) lies inside the cloud. Hence, lower portions of the cloud contain water droplets, whereas higher portions contain ice particles.[4] Conditions inside the cloud can be determined by displacing the parcel above the LCL. The parcel's state then evolves along the saturated adiabat passing through state 2. Cutting across isopleths of saturation mixing ratio toward lower r_w and across adiabats toward higher θ, the $\theta_e = 315$ K saturated adiabat eventually reaches state 3, where the temperature is 0°C. That condition is achieved at the level

$$p_{\text{FL}} = 740 \text{ mb.}$$

---

[4] Ice forms only in the presence of a special type of aerosol particle, referred to as a *freezing nucleus* (Chapter 9). Since freezing nuclei are comparatively rare, many cloud droplets are actually "supercooled," that is, they remain liquid at temperatures below 0°C.

LIQUID WATER CONTENT AT THE FREEZING LEVEL

Because the parcel is saturated, its mixing ratio must equal the saturation mixing ratio at the freezing level. The isopleth of saturation mixing ratio passing through state 3 gives for the mixing ratio there

$$r_{FL} = 5.5 \text{ g kg}^{-1}.$$

If no precipitation occurs, the total water content of the parcel is preserved. Therefore, the liquid water content at the freezing level is given by

$$r_l = 0.5 \text{ g kg}^{-1}.$$

Approximately 10% of the parcel's moisture has condensed by this altitude.

TEMPERATURE INSIDE CLOUD AT 650 MB

At 650 mb, the parcel has evolved along the $\theta_e = 315$ K saturated adiabat to state 4, where its temperature is

$$T_{650} = -6°\text{C}.$$

Were the parcel perfectly dry, it would have continued to evolve above 770 mb along the 315 K adiabat. Without the release of latent heat to offset adiabatic cooling, the parcel's temperature would then have decreased more rapidly, resulting in a temperature at 650 mb of $-12°$C.

MIXING RATIO INSIDE CLOUD AT 650 MB

Because the parcel remains saturated, the isopleth of saturation mixing ratio passing through state 4 gives for the mixing ratio at 650 mb

$$r_{650} = 4.0 \text{ g kg}^{-1}.$$

By this level, one-third of the parcel's moisture has transformed into condensate.

The foregoing example illustrates the strong constraint on water vapor imposed by thermodynamics and hydrostatic stratification. Inside convective towers, which transport moisture upward from its source at the earth's surface, the abundance of vapor that can be supported decreases with altitude (Fig. 5.3). By 500 mb, less than 30% of the surface mixing ratio of water inside the parcel described above remains as vapor. According to (1.24), the absolute humidity ρ_v decreases even faster. Thus, less than 15% of the absolute concentration of water vapor remains after the parcel has been displaced to the middle troposphere. It is for this reason that deep convection increases the column abundance of water vapor in Fig. 1.16 only modestly. Much of Σ_{H_2O} resides in the lowest 1 to 2 km, where water vapor is distributed more uniformly.

Through this mechanism, water vapor transported vertically inside convective towers is systematically extracted. Condensation leads to a decrease with altitude of water vapor mixing ratio and, through precipitation, a similar decrease of total water content. Outside convective towers, air is even drier because it has not recently been in contact with the reservoir of moisture at the Earth's surface or it has been dehydrated inside convective towers in the aforementioned manner. The same mechanism operates on large scales inside sloping convection, like that involved in synoptic weather systems. By limiting the vertical transport of water vapor, thermodynamics in combination with hydrostatic stratification confines moisture to a shallow neighborhood of the earth's surface and has preserved water on the planet.

Suggested Reading

A complete discussion of the Gibbs–Dalton law, along with its implications for multi-component systems, is given in *Thermodynamics* (1970) by Keenan.

Atmospheric Thermodynamics (1981) by Iribarne and Godson provides a detailed treatment of moisture-dependent air properties, including those under saturated conditions, and a complete survey of thermodynamic charts.

The Principles of Chemical Equilibrium (1971) by Denbigh discusses the thermodynamics of condensed phases.

Problems

The pseudo-adiabatic chart in Appendix F is to be used only for those problems in which it is explicitly indicated.

5.1. A downslope wind in North America is called a *chinook*, which is an Indian term meaning "snow eater." A chinook often occurs with the mountains blanketed in clouds but with clear skies leeward. Consider the following synoptic situation: Moist air originating in the eastern Pacific is advected over the western United States where it is forced over the continental divide. On the windward side, the surface lies at 800 mb, where the temperature and mixing ratio are 20°C and 15 g kg^{-1}, respectively. If the summit lies at 600 mb and if any condensate that forms precipitates out, determine (a) the surface air temperature at Denver, which lies at 830 mb, (b) the mixing ratio at Denver, and (c) the relative humidity at Denver.

5.2. A refrigerator having an interior volume of 2.0 m^3 is sealed and switched on. If the air initially has a temperature of 30°C and a relative humidity of 0.50, determine (a) the temperature at which condensation forms on the walls, (b) how much moisture will have condensed when the temperature reaches 2°C, and (c) how much heat must be rejected to the surroundings to achieve the final state in part (b).

5.3. A cold front described by the surface

$$z = h(x - ct)$$

moves eastward with velocity c and undercuts warmer air ahead of it. Far ahead of the front, undisturbed air is characterized by the temperature and mixing ratio profiles $T_\infty(z)$ and $r_\infty(z)$, respectively. If air is lifted in unison over the frontal surface, for a given location x, derive an expression for (a) the variation of temperature with altitude above the frontal surface but beneath the LCL as a function of time, (b) the variation of mixing ratio with altitude above the frontal surface but beneath the LCL as a function of time, and (c) an expression for the altitude of the LCL if, far from the frontal zone, pressure decreases with altitude approximately as $e^{-z/H}$, with $H = \text{const}$.

5.4. The development in Sec. 5.1 for a mixture of dry air and water considers the presence of only one condensed phase. Why and under what conditions is this permissible?

5.5. State the number of thermodynamic degrees of freedom for (a) moist unsaturated air and (b) moist saturated air. (c) Explain the numbers in parts (a) and (b) and why they differ.

5.6. Compare θ and θ_v for saturated conditions at 1000 mb and a temperature of (a) 10°C, (b) 20°C, and (c) 30°C.

5.7. Warm moist air leaves an array of cooling towers at a power plant situated at 825 mb. A cloud forms directly overhead. If the ambient temperature profile is isothermal, with $T = 5°C$ and if the initial temperature and mixing ratio of air leaving the towers are 30°C and 25 g kg^{-1}, respectively, determine (a) the relative humidity immediately above the towers, (b) the dew point immediately above the towers, (c) the virtual potential temperature immediately above the towers; compare this value to the potential temperature and discuss it in relation to typical differences between θ and θ_v, (d) the pressure at the cloud base, (e) the equivalent potential temperature at 800 mb, (f) the mixing ratio at 700 mb, and (g) the pressure at cloud top. A pseudo-adiabatic chart is provided in Appendix F.

5.8. Room temperature on a given day is 22°C, while the outside temperature at 1000 mb is 2°C. Calculate the maximum relative humidity that can be accommodated inside without room windows fogging, if the windows can be treated as having a uniform temperature.

5.9. (a) Use the same conditions as in Problem 5.8, but for an aircraft cabin that is pressurized to 800 mb. (b) As in part (a), but noting that the interior pane is thermally isolated from the exterior pane.

5.10. Demonstrate that the pressure dependence in (5.15) is negligible.

5.11. Evaporation is an efficient means of cooling air and, through contact with other media, of rejecting heat to the environment. An evaporative cooler processes ambient air at 35°C and 900 mb to produce room air with a relative humidity of 0.60. Calculate the temperature inside if the relative humidity outside is (a) 0.10 and (b) 0.50.

5.12. A cooling tower situated at 1000 mb processes ambient air to release saturated air at a temperature of 35°C and at a rate of $10 \text{ m}^3 \text{ s}^{-1}$. Calculate the rate at which heat is rejected if ambient air has a temperature of 20°C and a relative humidity of (a) 0.20 and (b) 0.80.

5.13. Show that the exponent in (5.26) vanishes for an air parcel that is displaced to the top of the atmosphere.

5.14. Show that (5.19) holds exactly when referenced to the mass of dry air.

5.15. A morning temperature sounding is plotted in Fig. 5.6. Through absorption of shortwave radiation at the ground, the surface inversion ($\Gamma < 0$) that developed during the night is replaced during the day by adiabatic thermal structure in the lowest half kilometer. A cumulus cloud then forms over an asphalt parking lot at 1000 mb, where the mixing ratio is 10 g kg^{-1} and the relative humidity is 50%. For the air column above the parking lot, determine the (a) surface air temperature, (b) pressure at the cloud base, (c) potential temperature at 900 mb, (d) potential temperature at 700 mb, (e) equivalent potential temperature at the surface, (f) mixing ratio at 700 mb, (g) pressure at the cloud top, and (h) mixing ratio at the cloud top. A pseudo-adiabatic chart is provided in Appendix F.

5.16. Moist air moves inland from a maritime region, where it has a temperature of 16°C and a relative humidity of 66%. Through contact with the ground, the air cools. Calculate the temperature at which fog forms.

5.17. Revisit Problem 2.15, recognizing now that the bubble is surrounded by water. If the bubble is saturated throughout its ascent, yet remains adiabatic, calculate the bubble's temperature upon reaching the surface.

5.18. Use the pseudo-adiabatic chart in Appendix F to determine the LCL above terrain at 1000 mb for (a) moist surface conditions representative of the eastern United States: $T = 30°C$ and $RH = 70\%$, and (b) arid surface conditions representative of the southwestern United States: $T = 30°C$ and $RH = 10\%$. (c) Contrast the heights of cumulus cloud bases under these conditions in relation to the likelihood of precipitation reaching the surface.

5.19. An altitude chamber is used to simulate a sudden decompression from normal aircraft cabin pressure: 800 mb, to ambient pressure at 18,000 ft. If the initial temperature is 22°C, how small must the relative humidity be to avoid spontaneous cloud formation during the decompression? A pseudo-adiabatic chart is provided in Appendix F.

5.20. Outside air has a temperature of −10°C and a relative humidity of 0.50. (a) What is the relative humidity indoors if the room temperature is 22°C and if air is simply heated, without humidification? (b) What mass

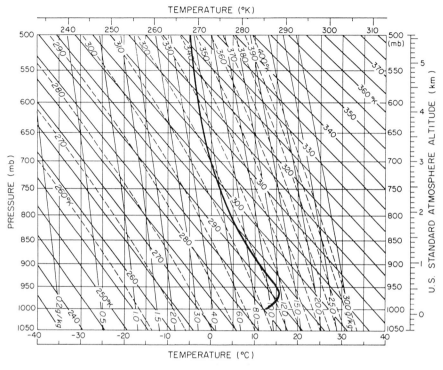

TEMPERATURE (°K)

PRESSURE (mb)

U.S. STANDARD ATMOSPHERE ALTITUDE (km)

TEMPERATURE (°C)

Figure 5.6

of water vapor must be added to a room volume of 75 m³ to elevate its relative humidity to 40%? (c) How much energy is required to achieve the state in part (b)?

5.21. On a given day, the lapse rate and relative humidity are constant, with $\Gamma = 8$ K km^{-1} and $RH = 0.80$. If the surface temperature at 1000 mb is 20°C, (a) estimate the total precipitable water vapor. (b) Below what height is 90% of the water vapor column represented? (c) Now calculate the precipitable water vapor inside a column comprised of ascending air and a cumulonimbus cloud that extends to 100 mb. A pseudo-adiabatic chart is provided in Appendix F.

5.22. A morning temperature sounding over Florida reveals the profile

$$T = \begin{cases} 10 - 6(z - 1) \ (C) & z \geq 1 \ \text{km} \\ 20 - 10z \ (C) & z < 1 \ \text{km}. \end{cases}$$

That afternoon, surface conditions are characterized by a temperature of 30°C and a mixing ratio of 20 g kg^{-1}. (a) To what height will cumulus clouds develop if air motion is nearly adiabatic? (b) Plot the vertical profile of mixing ratio from the surface to the top of a cumulus cloud. (c) Over the same altitude range, plot the vertical profile of mixing ratio but neglect the explicit pressure dependence of the saturation mixing ratio. (d) Discuss how the results in parts (a) and (b) would be modified if air inside the ascending plume is not adiabatic but, rather, mixes with surrounding air in roughly equal proportions. A pseudo-adiabatic chart is provided in Appendix F.

5.23. For the temperature sounding and surface conditions in Problem 5.22, determine the height of convection and the vertical profile of temperature inside an ascending plume if the air were perfectly dry. A pseudo-adiabatic chart is provided in Appendix F.

5.24. An air parcel drawn into a cumulonimbus cloud becomes saturated at 900 mb, where its temperature is 20°C. Calculate the fractional internal energy represented by condensate when the parcel has been displaced to 100 mb. A pseudo-adiabatic chart is provided in Appendix F.

5.25. The base of convective clouds may be estimated from the dew point spread at the surface. (a) Derive an expression for the dew point temperature T_d of an ascending parcel as a function of its height in terms of the surface mixing ratio r_0, presuming environmental conditions to be isothermal. (b) Show that T_d for the parcel varies with height approximately linearly in the lowest few kilometers. (c) Use the lapse rate of T_d in part (b) to derive an expression for the height of the LCL in terms of the dew point spread at the surface: $(T - T_d)_0$. (d) Use the result of part (c) to estimate the cloud base under the conditions given in Problem 5.22.

5.26. Graph the dew point temperature as a function of relative humidity for 1000 mb and 20°C.

5.27. For reasons developed in Chapter 9, many clouds are supercooled (e.g., they contain liquid droplets at temperatures below 0°C). For a cloud that is saturated with respect to water, calculate the relative humidity with respect to ice at a temperature of (a) −5°C and (b) −20°C.

Chapter 6 | Hydrostatic Equilibrium

Changes of thermodynamic state accompanying vertical motion follow from the distribution of atmospheric mass, which is determined ultimately by gravity. In the absence of motion, Newton's second law applied to the vertical reduces to a statement of *hydrostatic equilibrium* (1.16), wherein gravity is balanced by the vertical pressure gradient. This simple form of mechanical equilibrium is accurate even in the presence of motion because the acceleration of gravity is much greater than vertical accelerations of individual air parcels for nearly all motions of dimensions greater than a few tens of kilometers. Only inside deep convective towers and other small-scale phenomena are vertical accelerations large enough to invalidate hydrostatic equilibrium.

Because it is such a strong body force, gravity must be treated with some care. Complications arise from the fact that the gravitational acceleration experienced by an air parcel does not act purely in the vertical and varies with location. According to the discussion above, gravity is large enough to overwhelm other contributions in the balance of vertical forces. The same is true for the horizontal force balance. Horizontal components of gravity introduced by the earth's rotation and additional sources must be balanced by other horizontal forces, which unnecessarily complicate the description of atmospheric motion and can overshadow those forces actually controlling the motion of an air parcel.

6.1 Effective Gravity

The acceleration of gravity appearing in (1.16) is not constant. Nor does it act purely in the vertical. In the reference frame of the earth, the gravitational acceleration experienced by an air parcel follows from three basic contributions:

1. Radial gravitation by the planet's mass
2. Centrifugal acceleration due to rotation of the reference frame
3. Anisotropic contributions,

as illustrated in Fig. 6.1. Radial gravitation by the planet's mass is the dominant contribution. Under idealized circumstances, that component of g acts

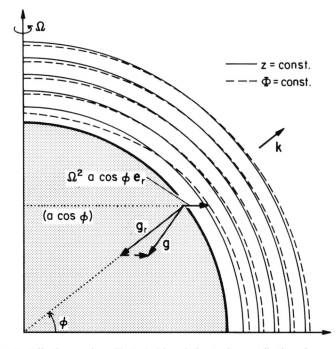

Figure 6.1 Effective gravity **g** illustrated in relation to its contributions from radial gravitation and centrifugal acceleration. Also shown are surfaces of constant geometric altitude z and corresponding surfaces of constant geopotential Φ, to which **g** is orthogonal.

perpendicular to surfaces of constant geometric altitude z, which are concentric spheres. The rotation of the earth introduces another contribution. Because it is noninertial, the reference frame of the earth includes a centrifugal acceleration that acts perpendicular to and away from the axis of rotation. Last, departures of the planet from sphericity and homogeneity (e.g., through variations of surface topography) introduce anisotropic contributions to gravity that must be determined empirically.

Collectively, these contributions determine the *effective gravity*, which can be expressed as follows:

$$g(\lambda, \phi, z) = \frac{a^2}{(a + z)^2} g_0 k + \Omega^2(a + z) \cos \phi e_r + \epsilon(\lambda, \phi, z), \qquad (6.1)$$

where a is the mean radius of the earth, g_0 is the radial gravitation at mean sea level, k is the local upward unit normal, Ω is the planet's angular velocity, λ and ϕ are longitude and latitude, respectively, e_r is a unit vector directed outward from the axis of rotation, and $\epsilon(\lambda, \phi, z)$ represents all anisotropic contributions. Even if the planet were perfectly spherical, effective gravity would not act uniformly across the surface of the earth, nor entirely in the vertical. According to (6.1), the centrifugal acceleration deflects g into the horizontal, which introduces a component of gravity along surfaces of constant z. The centrifugal acceleration also causes the vertical component of g to vary from a minimum of 9.78 m s^{-2} at the equator to a maximum of 9.83 m s^{-2} at the poles. Although more complicated, anisotropic contributions have a similar effect.

These contributions introduce horizontal components of gravity that must be balanced by other forces acting along surfaces of constant geometric altitude. Thus, (1.16) represents but one of several component equations that are required to describe hydrostatic balance. Horizontal forces introduced by gravity unnecessarily complicate the description of atmospheric behavior because they can overshadow the forces actually controlling the motion of an air parcel. For instance, rotation introduces a horizontal component that must be balanced by a variation of pressure along surfaces of constant geometric altitude. That horizontal pressure gradient exists even under static conditions, for which an observer in the reference frame of the earth would anticipate Newton's second law to reduce to a simple hydrostatic balance in the vertical. To avoid such complications, it is convenient to introduce a coordinate system that consolidates horizontal components of gravity into the vertical coordinate.

6.2   Geopotential Coordinates

Because gravity is conservative, the specific work performed during a cyclic displacement of mass through the earth's gravitational field vanishes:

$$\Delta w_g = - \oint g(\lambda, \phi, z) \cdot d\xi = 0, \qquad (6.2)$$

where $d\xi$ denotes the incremental displacement of an air parcel. According to Sec. 2.1.4, (6.2) is a necessary and sufficient condition for the existence of an exact differential

$$d\Phi = \delta w_g$$
$$= -g \cdot d\xi, \qquad (6.3)$$

which defines the gravitational potential or *geopotential* Φ. In (6.3), $d\Phi$ equals the specific work performed against gravity to complete the displacement $d\xi$. Then, per the aforementioned discussion, gravity is an irrotational vector field that can be expressed in terms of its potential function

$$g = -\nabla\Phi. \qquad (6.4)$$

Surfaces of constant geopotential are not spherical like surfaces of constant geometric altitude. Centrifugal acceleration in the rotating reference frame of the earth distorts geopotential surfaces into oblate spheroids (Fig. 6.1). The component of $\nabla\Phi$ along surfaces of constant geometric altitude introduces a horizontal component of gravity that must be balanced by a horizontal pressure gradient force.

Representing gravity is simplified by transforming from pure spherical coordinates, in which elevation is fixed along surfaces of constant geometric altitude, to geopotential coordinates, in which elevation is fixed along surfaces of constant geopotential. Geopotential surfaces can be used for coordinate surfaces because Φ increases monotonically with altitude (6.3), which ensures a one-to-one relationship between those variables. Introducing the aforementioned transformation is tantamount to measuring elevation z along the line of effective gravity, which will be termed *height* to distinguish it from geometric altitude. Defining z in this manner agrees with the usual notion of local vertical being plumb with the line of gravity and will be adopted as the convention hereafter.

In geopotential coordinates, gravity has no horizontal component. Consequently, in terms of height, (6.4) reduces to simply

$$d\Phi = gdz, \tag{6.5}$$

where g denotes the magnitude of g, which is understood to act in the direction of decreasing height—not geometric altitude. Using mean sea level as a reference elevation[1] yields an expression for the absolute geopotential

$$\Phi(z) = \int_0^z gdz. \tag{6.6}$$

Having dimensions of specific energy, the geopotential represents the work that must be performed against gravity to raise a unit mass from mean sea level to a height z. Evaluating Φ from (6.6) requires measured values of effective gravity throughout. However, since it is a point function, Φ is independent of path by (6.2) and the exact differential theorem (2.7). Therefore, Φ depends only on height z and surfaces of constant geopotential coincide with surfaces of constant height.

Using surfaces of constant geopotential for coordinate surfaces consolidates gravity into the vertical coordinate, height. However, g still varies with z through (6.5). To account for that variation, it is convenient to introduce yet another vertical coordinate in which the dependence on height is absorbed. *Geopotential height* is defined as

$$\begin{aligned} Z &= \frac{1}{g_0} \int_0^z gdz \\ &= \frac{1}{g_0} \Phi(z), \end{aligned} \tag{6.7}$$

---

[1] Mean sea level may be used to define a reference value of Φ because it nearly coincides with a surface of constant geopotential.

where $g_0 = 9.8$ m s^{-2} is a reference value reflecting the average over the surface of the earth. Then

$$dZ = \frac{1}{g_0} d\Phi$$

$$= \frac{g}{g_0} dz. \tag{6.8}$$

Since it accounts for the variation of gravity, geopotential height simplifies the expression for hydrostatic balance. In terms of Z, (1.16) reduces to

$$dp = -\rho g dz$$

$$= -\rho g_0 dZ. \tag{6.9}$$

When elevation is represented in terms of geopotential height, gravity can be described by the constant value g_0, which simplifies the description of atmospheric motion.

Although it has these formal advantages, geopotential height is nearly identical to height because $g \cong g_0$ throughout the homosphere. In fact, the two measures of elevation differ by less that 1% below 60 km and only 3% below 100 km. Therefore, Z can be used interchangeably with z, which will be our convention hereafter. Then g is understood to refer to the constant reference value g_0.

6.3 Hydrostatic Balance

In terms of geopotential height, hydrostatic balance can be expressed

$$g dz = -v dp. \tag{6.10}$$

The ideal gas law transforms this into

$$dz = -\frac{RT}{g} d\ln p$$

$$= -H d\ln p, \tag{6.11}$$

where the scale height H is defined in (1.17). According to (6.11), the change of height is proportional to the local temperature and to the change of $-\ln p$. Since changes of temperature and thus H are small compared to changes of pressure, the quantity $-\ln p$ can be regarded as a dimensionless measure of height.

Consider a layer bounded by two isobaric surfaces $p = p_1(x, y, z)$ and $p = p_2(x, y, z)$, as illustrated in Fig. 6.2. By integrating (6.11) between those

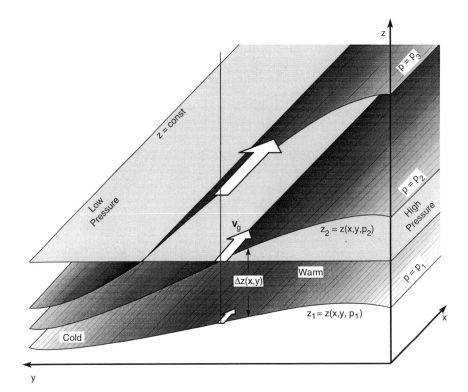

Figure 6.2 Isobaric surfaces (dark shading) in the presence of a horizontal temperature gradient. Vertical spacing of isobaric surfaces is compressed in cold air and expanded in warm air, which makes the height of an individual isobaric surface low in the former and high in the latter. Because pressure decreases upward monotonically, a surface of constant height (light shading) then has low pressure where isobaric height is low and high pressure where isobaric height is high. Therefore, contours of height for an individual isobaric surface can be interpreted similarly to isobars on a constant height surface. Also indicated is the geostrophic velocity $\boldsymbol{v}_g$ (Sec. 12.1), which is directed parallel to contours of isobaric height and has magnitude proportional to its horizontal gradient.

surfaces, we obtain

$$\Delta z = z_2 - z_1 = -\langle H \rangle \ln \left(\frac{p_2}{p_1} \right), \tag{6.12.1}$$

where

$$\langle H \rangle = \frac{R \langle T \rangle}{g} \tag{6.12.2}$$

and

$$\langle T \rangle = \frac{\int_{p_1}^{p_2} T d \ln p}{\int_{p_1}^{p_2} d \ln p}$$

$$= \frac{\int_{p_1}^{p_2} T d \ln p}{\ln \left(\frac{p_2}{p_1} \right)}, \tag{6.12.3}$$

define the layer–mean scale height and temperature, respectively, and $z_1 = z(x, y, p_1)$ and $z_2 = z(x, y, p_2)$ denote the heights of the two isobaric surfaces. Known as the *hypsometric equation*, (6.12) asserts that the *thickness* of a layer bounded by two isobaric surfaces is proportional to the mean temperature of that layer and the pressure difference across it. In regions of cold air, the *e*-folding scale H is small, so pressure in (1.17) decreases with height sharply and the vertical spacing of isobaric surfaces is compressed. By comparison, H is large in regions of warm air, so pressure decreases with height more slowly, and the spacing of isobaric surfaces is expanded.

If $z_1 = 0$, the hypsometric equation provides the height of the upper surface $z(x, y, p_2)$. For example, the height of the 500-mb surface in Fig. 1.9 can then be expressed as a vertical integral of temperature. By (6.12), the height of an isobaric surface can be determined from measurements of temperature between that surface and the ground or, alternatively, between that surface and a reference isobaric surface, the height of which is known. Such temperature measurements are made routinely by ground-launched rawinsondes, which also measure air motion and humidity and ascend via balloon to as high as 10 mb. Launched twice a day, rawinsondes provide dense coverage over populated landmasses, but only sparse coverage over oceans and in the tropics. Temperature measurements are also made remotely by satellite. Satellite measurements of temperature provide continuous global coverage of the atmosphere, albeit asynoptically.[2] Satellite retrievals also have coarser vertical resolution ($\sim$ 4 to 8 km) than that provided by rawinsondes ($\sim$1 km). Both are assimilated operationally in *synoptic analyses* of the instantaneous circulation, like those shown in Figs. 1.9 and 1.10.

According to Fig. 1.9, the height of the 500-mb surface slopes downward toward the pole at midlatitudes, where z_{500} decreases from 5900 to 5000 m. That decrease of 500-mb height mirrors the poleward decrease of layer–mean temperature between the surface and 500 mb, $\langle T \rangle_{500}$, which is shown in Fig. 6.3 for the same time as that in Fig. 1.9a. The wavy region where $\langle T \rangle_{500}$ and z_{500} change abruptly delineates the *polar front*, which separates cold polar air from warmer air at lower latitudes. As is apparent from Fig. 1.9a, that region

---

[2]The term *synoptic* refers to the simultaneous distribution of some property, for example, a snapshot over the earth of the instantaneous height and motion on the 500-mb surface (Fig. 1.9a). However, a polar-orbiting satellite necessarily observes different regions at different instants. Termed *asynoptic*, such measurements sample the entire globe only after 12 to 24 hr.

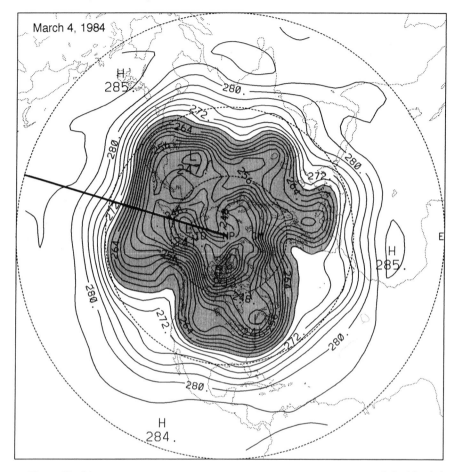

Figure 6.3   Mean temperature (Kelvin) of the layer between 1000 mb and 500 mb for March 4, 1984. The bold solid line marks a meridional cross section displayed in Fig. 6.4.

of sharp temperature change coincides with the strong circumpolar flow of the jet stream, which is tangential to contours of isobaric height. Meridional gradients of temperature and height are particularly steep east of Asia and North America, where they mark the North Pacific and North Atlantic storm tracks that are manifested in the time-mean pattern (Fig. 1.9b) as locally intensified jets. Anomalies of cold air that punctuate the polar front in Fig. 6.3 introduce synoptic-scale depressions of the 500-mb surface in Fig. 1.9a. By deflecting contours of isobaric height, those features disrupt the motion of the jet stream, which, in their absence, is nearly circumpolar.

The hypsometric equation provides a theoretical basis for using pressure to describe elevation. Because H is positive, (6.12) implies a one-to-one relationship between height and pressure: a given height corresponds to a single pressure. Therefore, pressure can be used as an alternative to height for the vertical coordinate. Isobaric coordinates, which are developed in Chapter 11, use surfaces of constant pressure for coordinate surfaces. Although they evolve with the circulation, isobaric coordinates afford several advantages.

In isobaric coordinates, pressure becomes the independent variable and the height of an isobaric surface becomes the dependent variable, for example, $z_2 = z(x, y, p_2)$. Because pressure decreases monotonically with height, low height of an isobaric surface corresponds to low pressure on a surface of constant height and just the reverse for high values (Fig. 6.2). For the same reason, contours of isobaric height resemble isobars on a surface of constant height. Consequently, the horizontal distribution of isobaric height may be interpreted analogously to the distribution of pressure on a surface of constant height. From hydrostatic equilibrium, the latter represents the weight of the atmospheric column above a given height. Therefore, the distribution of isobaric height reflects the horizontal distribution of atmospheric mass above the mean elevation of that isobaric surface.

Contours of 500-mb height that delineate the polar front in Fig. 1.9a correspond to isobars on the constant height surface: $z \sim 5.5$ km, with pressure decreasing toward the pole. Thus, less atmospheric column lies above that elevation poleward of the front than equatorward of the front. Figure 6.4 shows a meridional cross section of the thermal structure and motion in Figs. 6.3 and 1.9a. The horizontal distribution of mass just noted results from compression of isobaric surfaces poleward of the front, which is centered near 40 N, and expansion equatorward of the front. This introduces a tilt to isobaric surfaces at midlatitudes, one that steepens with height. The jet stream is found in the same region and likewise intensifies vertically, to a maximum near the tropopause. Similar behavior at high latitudes marks the base of the polar-night jet.

6.4 Stratification

Expressions (2.33) and (5.32) for the dry adiabatic and saturated adiabatic lapse rates, which describe the evolution of a displaced air parcel under unsaturated and saturated conditions, respectively, hold formally with z representing geopotential height and g constant. Neither Γ_d nor Γ_s has a direct relationship to the temperature of the surroundings because a displaced parcel is thermally isolated under adiabatic conditions. Thermal properties of the environment are dictated by the history of air residing at a given location, for example, by where that air has been and what thermodynamic influences have acted on it. The *environmental lapse rate* is defined as

$$\Gamma = -\frac{dT}{dz}, \tag{6.13}$$

where T refers to the ambient temperature. Like parcel lapse rates, $\Gamma > 0$ corresponds to temperature decreasing with height. Conversely, $\Gamma < 0$ cor-

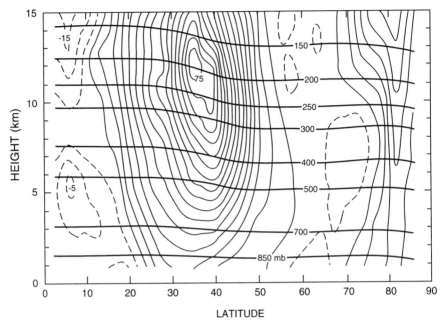

Figure 6.4   Meridional cross section of thermal structure and motion in Figs. 1.9a and 6.3 at the position indicated in the latter. Vertical spacing of isobaric surfaces (bold solid curves) is expanded at low latitude and compressed at high latitude, introducing a slope at midlatitudes that steepens with height. Zonal wind speed (contoured in m s^{-1}) delineates the subtropical jet, which coincides with the steep slope of isobaric surfaces, both intensifying upward to a maximum at the tropopause. Contour increment: 5 m s^{-1}.

responds to temperature increasing with height, in which case the profile of environmental temperature is said to be *inverted*.

The compressibility of air leads to atmospheric mass being stratified, as is reflected in the vertical distributions of density and pressure. Since the environment is in hydrostatic equilibrium, the distribution of pressure may be related to the thermal structure through the hypsometric relation. Incorporating (6.11) transforms (6.13) into

$$-\frac{1}{H}\frac{dT}{d\ln p} = \Gamma$$

or

$$\frac{d\ln T}{d\ln p} = \frac{R}{g}\Gamma$$

$$= \frac{\Gamma}{\Gamma_d}\kappa. \tag{6.14}$$

6.4.1 Idealized Stratification

For certain classes of thermal structure, (6.14) can be integrated analytically to obtain the distributions of pressure and other thermal properties inside an atmospheric layer.

LAYER OF CONSTANT LAPSE RATE

If $\Gamma = \text{const} \neq 0$, (6.14) yields

$$\frac{T}{T_s} = \left(\frac{p}{p_s}\right)^{\left(\frac{\Gamma}{\Gamma_d}\right)\kappa}, \tag{6.15}$$

where the subscript s refers to the base of the layer. Because

$$T(z) = T_s - \Gamma z, \tag{6.16}$$

(6.15) can be used to obtain the vertical distribution of pressure:

$$\frac{p}{p_s} = \left[1 - \kappa\frac{\Gamma}{\Gamma_d}\left(\frac{z}{H_s}\right)\right]^{\left(\frac{\Gamma_d}{\kappa\Gamma}\right)}. \tag{6.17}$$

Then (2.31) implies the vertical profile of potential temperature:

$$\theta(z) = (T_s - \Gamma z)\left[1 - \kappa\frac{\Gamma}{\Gamma_d}\left(\frac{z}{H_s}\right)\right]^{-\frac{\Gamma_d}{\Gamma}}, \tag{6.18}$$

where $p_s = p_0 = 1000$ mb has been presumed for the base of the layer.

As shown in Fig. 6.5, the compressibility of air makes pressure decrease with height for all Γ. The same is true of density. However, potential temperature decreases with height only for $\Gamma > \Gamma_d$, whereas it increases with height for $\Gamma < \Gamma_d$. In both cases, θ varies sharply with height because the pressure term dominates over the temperature term in determining the height dependence in (6.18). As a result, environmental potential temperature can change by several hundred degrees in just a few scale heights, as is typical in the stratosphere (refer to Fig. 17.19). By contrast, the potential temperature of a displaced air parcel is independent of its height under adiabatic conditions.

The thermal structure described by (6.16) through (6.18) can apply locally (e.g., inside a certain layer), even if temperature does not vary linearly throughout the atmosphere. Should it apply for the entire range of height, (6.16) has one further implication. For $\Gamma > 0$, positive temperature requires the atmosphere to have a finite upper bound

$$z_{\text{top}} = \frac{T_s}{\Gamma}, \tag{6.19}$$

where p vanishes. For $\Gamma \leq 0$, no finite upper bound exists.

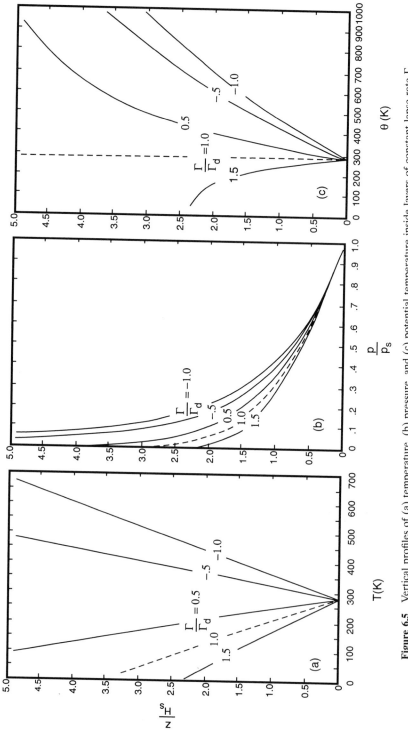

Figure 6.5 Vertical profiles of (a) temperature, (b) pressure, and (c) potential temperature inside layers of constant lapse rate Γ.

ISOTHERMAL LAYER

For the special case $\Gamma = 0$, (6.17) is indeterminate. Reverting to the hydrostatic relation (6.11) yields

$$z = -H \ln \left(\frac{p}{p_s} \right)$$

or

$$\frac{p}{p_s} = e^{-\frac{z}{H}}, \tag{6.20}$$

which is identical to (1.17) under these circumstances. Then the vertical profile of potential temperature is simply

$$\theta(z) = T_s e^{\kappa \frac{z}{H}}$$
$$= \theta_s e^{\kappa \frac{z}{H}}. \tag{6.21}$$

Under isothermal conditions, environmental potential temperature increases by a factor of e every $\kappa^{-1} \cong 3$ scale heights.

ADIABATIC LAYER

For the special case $\Gamma = \Gamma_d$, (6.15) reduces to

$$\frac{T}{T_s} = \left(\frac{p}{p_s} \right)^{\kappa} \tag{6.22}$$

and the distribution of pressure (6.17) becomes

$$\frac{p}{p_s} = \left(1 - \kappa \frac{z}{H_s} \right)^{\frac{1}{\kappa}}. \tag{6.23}$$

Equation (6.22) is the same relationship between temperature and pressure implied by Poisson's equation (2.30.2), which defines potential temperature (2.31). Accordingly, the vertical profile of potential temperature (6.18) reduces to

$$\theta(z) = \theta_s = \text{const} \tag{6.24}$$

inside this layer.

6.5   Lagrangian Interpretation of Stratification

As noted earlier, hydrostatic equilibrium applies in the presence of motion as well as under static conditions. Therefore, each of the stratifications in

Sec. 6.4 is valid even if a circulation is present, as is invariably the case. Under those circumstances, vertical profiles of temperature, pressure, and potential temperature [(6.16) through (6.18)] correspond to the horizontal-mean thermal structure and hence to averages over many ascending and descending air parcels. Interpreting thermal structure in terms of the behavior of individual air parcels provides some insight into the mechanisms controlling mean stratification.

For a layer of constant lapse rate, the relationship between temperature and pressure (6.15) resembles one implied by Poisson's equation (2.30.2), but for a polytropic process (Sec. 2.5.1) with $(\Gamma/\Gamma_d)\kappa$ in place of κ. Consequently, we can associate the thermal structure in (6.16) through (6.18) with a vertical rearrangement of air in which individual parcels evolve diabatically according to a polytropic process. In that description, air parcels moving vertically exchange heat with their surroundings in such proportion for their temperatures to vary linearly with height. For an individual air parcel, the foregoing process has a polytropic specific heat that satisfies

$$\frac{R}{(c_p - c)} = \frac{\Gamma}{\Gamma_d}\kappa$$

or

$$c = c_p \left(1 - \frac{\Gamma_d}{\Gamma}\right). \tag{6.25}$$

The corresponding heat transfer for the parcel is then given by (2.37) and its change of potential temperature follows from (2.40).

6.5.1 Adiabatic Stratification

Consider a layer characterized by

$$\Gamma = \Gamma_d, \tag{6.26.1}$$

the stratification of which is termed *adiabatic*. Then (6.25) implies

$$c = 0,$$
$$\delta q = d\theta = 0, \tag{6.26.2}$$

so individual air parcels evolve adiabatically. Under these circumstances, air parcels move vertically without interacting thermally with their surroundings. Such behavior can be regarded as the limiting situation when vertical motion occurs on a timescale that is short compared to diabatic effects, as is typical of cumulus and sloping convection. Although their vertical motion is adiabatic, individual parcels may still exchange heat at the boundaries of the layer. In fact, such heat transfer is necessary to maintain vertical motion in the absence of mechanical forcing.

The evolution of an individual parcel can be envisaged in terms of a circuit that it follows between the upper and lower boundaries of the layer, where heat transfer is concentrated. Such a circuit is depicted in Fig. 6.6 for a layer representative of the troposphere. At the lower boundary, an individual parcel moves horizontally long enough for heat transfer to occur. Isobaric heat absorption [e.g., through absorption of longwave (LW) radiation and transfer of sensible heat from the surface] increases the parcel's potential temperature (Fig. 6.7a). The accompanying increase of temperature (Fig. 6.7b) causes the parcel to become positively buoyant and rise. If the timescale of vertical motion is short compared to that of heat transfer, the parcel ascends along an adiabat in state space until it reaches the upper boundary, where it again moves horizontally. Isobaric heat rejection (e.g., through LW emission to space) then decreases the potential temperature and temperature of the parcel, which therefore becomes negatively buoyant. The parcel then sinks along a different adiabat until it has returned to the surface and completed a thermodynamic cycle.

If it remains unsaturated [e.g., below its lifting condensation level (LCL)], each air parcel comprising the layer preserves its potential temperature away from the boundaries. Consequently, each parcel traces out a uniform profile of θ as it traverses the layer. The horizontal–mean potential temperature $\overline{\overline{\theta}}(z)$ is equivalent to an average over all such parcels at a given elevation, so it is likewise independent of height, that is,

$$\overline{\overline{\theta}}(z) = \text{const.} \tag{6.27.1}$$

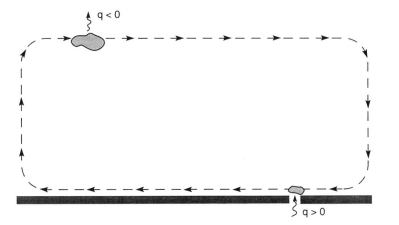

Figure 6.6 Idealized circuit followed by an air parcel during which it absorbs heat at the base of a layer and rejects heat at its top, with adiabatic vertical motion between.

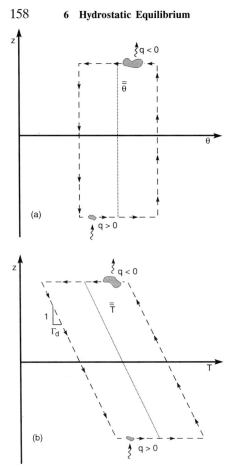

Figure 6.7 Thermodynamic cycle followed by the air parcel in Fig. 6.6 in terms of (a) potential temperature and (b) temperature. Horizontally averaged behavior for a layer composed of many such parcels is also indicated by the dotted lines.

As is true for an individual parcel (Sec. 2.4.2), a uniform distribution of potential temperature corresponds to a constant environmental lapse rate

$$\overline{\overline{\Gamma}} = \Gamma_d. \qquad (6.27.2)$$

Consequently, the thermal structure associated with (6.24) can be interpreted as the horizontal mean for a layer in which air is being actively and adiabatically rearranged in the vertical.

This interpretation can be extended to saturated conditions (e.g., inside cloud). An air parcel's evolution is then pseudo-adiabatic, so its equivalent potential temperature is conserved. Active vertical rearrangement of air will then make the horizontal–mean equivalent potential temperature $\overline{\overline{\theta}}_e(z)$ independent of height

$$\overline{\overline{\theta}}_e(z) = \text{const.} \qquad (6.28.1)$$

From Sec. 5.4.3, this uniform distribution of $\overline{\overline{\theta}}_e(z)$ corresponds to a mean environmental lapse rate

$$\overline{\overline{\Gamma}} = \Gamma_s, \qquad (6.28.2)$$

the corresponding stratification being termed *saturated adiabatic*.

Even though all parcels inside the layer evolve in like fashion, local behavior will differ from the horizontal mean because conserved properties may still vary from one parcel to another (e.g., due to different histories experienced by those parcels at the layer's boundaries). An ascending parcel will have values of θ and θ_e, which are greater than the corresponding horizontal-mean values because that parcel will have recently absorbed heat at the lower boundary. Conversely, a descending parcel will have values of θ and θ_e, which are less than the corresponding horizontal-mean values because that parcel will have recently rejected heat at the upper boundary. Therefore, the actual stratification of the layer will vary with horizontal position.

If heat transfer at the boundaries were to be eliminated, convectively driven motions would then "spin down" through turbulent and molecular diffusion toward a limiting state of no motion. Since diffusion destroys gradients between individual parcels, this limiting state is characterized by homogeneous distributions of θ and θ_e and thus, everywhere, by the environmental lapse rate $\Gamma = \Gamma_d$ or Γ_s. Hence, this limiting homogeneous state has stratification identical to the horizontal-mean stratification of the layer being convectively overturned, which may therefore be regarded as "statistically well mixed."

The stratification (6.28) is characteristic of the troposphere. Close to the saturated adiabatic lapse rate (Fig. 1.2), the mean thermal structure in the troposphere follows from efficient vertical exchange of moist air inside cumulus and sloping convection. Air rearranged by convection is continually replenished with moisture through contact with warm ocean surfaces. Because convective motions operate on timescales of a day or shorter, they make the troposphere statistically well mixed and, in the mean, described approximately by (6.28).

In traversing the circuit, the parcel in Fig. 6.7 absorbs heat at high temperature and rejects heat at low temperature. By the second law (Sec. 3.2), net heat is absorbed over a cycle. Then the first law (2.13) implies that the parcel performs net work during its traversal of the circuit. It follows that an individual parcel in the above circulation behaves as a heat engine. In fact, the thermodynamic cycle in Fig. 6.7 is analogous to the Carnot cycle pictured in Fig. 3.2, except that heat transfer occurs (approximately) isobarically instead of isothermally. More expansion work is performed by the parcel during ascent than is performed on the parcel during descent. The area circumscribed by the parcel's evolution in Fig. 6.7b reflects the net work it performs during the cycle and is proportional to the change of the parcel's potential temperature at the upper and lower boundaries (i.e., to the heat transfer there).

Net heat absorption and work performed by individual air parcels make the general circulation of the troposphere behave as a heat engine, one driven thermally by heat transfer at its lower and upper boundaries. Work performed by individual parcels is associated with a redistribution of mass when air that is effectively warmer and lighter (namely, when compressibility is taken into account) at the lower boundary is exchanged with air that is effectively cooler and heavier air at the upper boundary. The latter represents a conversion of potential energy into kinetic energy, which maintains the general circulation against frictional dissipation.

6.5.2 Diabatic Stratification

The idealized behavior just described relies on heat transfer being confined to the lower and upper boundaries of the layer, where an air parcel resides long enough for diabatic effects to become important. On longer timescales, as are typical of vertical motions in the stratosphere, the evolution of an individual air parcel is not adiabatic. Radiative transfer, which is the primary diabatic influence outside the boundary layer and clouds, is characterized by cooling rates of order 1 K day^{-1} in the troposphere (see Fig. 8.24). Cooling rates as large as 10 K day^{-1} occur in the stratosphere and near cloud (Fig. 9.33).

By comparison, the cooling rate associated with adiabatic expansion follows from (2.33) as

$$\left(\frac{dT}{dt}\right)_{ad} = -\Gamma'\left(\frac{dz}{dt}\right)$$
$$= -\Gamma'w, \tag{6.29.1}$$

where T and z refer to an individual parcel, Γ' denotes its lapse rate, and

$$w = \frac{dz}{dt} \tag{6.29.2}$$

is the parcel's vertical velocity. A vertical velocity of 0.1 m s^{-1}, which is representative of cumulus and sloping convection, gives $(dT/dt)_{ad} = 84$ K day^{-1} under unsaturated conditions and of order 50 K day^{-1} under saturated conditions. Both are much larger than radiative cooling rates, so heat transfer in those motions can be ignored to a good approximation. On the other hand, a velocity of 1 mm s^{-1}, which is representative of vertical motion in the stratosphere, is associated with an adiabatic cooling rate of only $(dT/dt)_{ad} = 0.84$ K day^{-1}. Heat transfer in such motions cannot be ignored and may even dominate temperature changes associated with expansion and compression of individual air parcels.

Under those circumstances, an air parcel moving vertically interacts thermally with its environment. Because θ is no longer conserved, diabatic motion implies a different thermal structure. Consider a layer characterized by

$$\Gamma_d > \Gamma = \text{const}, \tag{6.30.1}$$

the stratification of which is termed *subadiabatic*: Temperature decreases with height slower than dry adiabatic. By (6.18), θ increases with height (compare Fig. 6.5). Then (6.25) together with (2.37) gives

$$c > 0$$
$$\delta q > 0 \qquad dT > 0 \tag{6.30.2}$$
$$\delta q < 0 \qquad dT < 0$$

for $\Gamma < 0$ (temperature increasing with height) and c and δq with reversed inequalities for $\Gamma > 0$. Both imply that an air parcel absorbs heat during ascent and rejects heat during descent, so its potential temperature increases with height. As illustrated in Fig. 6.8, heat is rejected at high temperature in upper levels and absorbed at low temperature in lower levels. By the second law, net heat is rejected over a cycle. Then the first law implies that net work must be performed on the parcel for it to traverse the circuit. More work is performed on the parcel during descent than is performed by the parcel during ascent.[3]

Opposite to the troposphere, this behavior is characteristic of the stratosphere. Net heat rejection and work performed on individual air parcels make the general circulation of the stratosphere behave as a refrigerator, one driven mechanically by waves that propagate upward from the troposphere. By rearranging air, planetary waves exert an influence on the stratosphere analogous to paddle work. When dissipated, they cause air at middle and high latitudes to cool radiatively, so individual parcels sink across isentropic surfaces to lower θ (see Figs. 8.27 and 17.11). Conversely, air at low latitudes warms radiatively, so parcels then rise to higher θ.

For both, vertical motion is slow enough to make diabatic effects important and to allow the temperature of individual parcels to increase with height. This feature of the stratospheric circulation follows from buoyancy. Contrary to the troposphere, buoyancy in the stratosphere strongly restrains vertical motion because, on average, warmer (high θ) air overlies cooler (low θ) air. Work performed on the stratosphere against the opposition of buoyancy leads to a redistribution of mass, when heavier air at lower levels is exchanged with lighter air at upper levels. The latter represents a conversion of kinetic energy

---

[3] In practice, this paradigm can be pushed only so far because stratospheric air must eventually enter the troposphere to complete the thermodynamic cycle (refer to Fig. 17.9), which makes the stratosphere an open system.

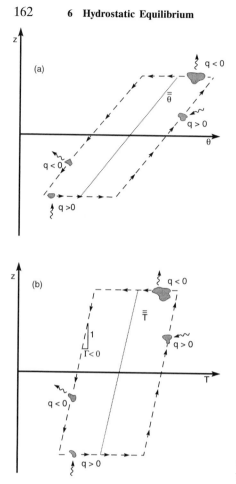

Figure 6.8 As in Fig. 6.7, but for an air parcel whose vertical motion is diabatic and whose temperature increases with height.

into potential energy, which is dissipated thermally through LW emission to space.

Stratification with

$$\Gamma > \Gamma_d, \tag{6.31}$$

which is termed *superadiabatic*, can also be interpreted in terms of diabatic motion of individual air parcels. However, such stratification will be seen to be mechanically unstable. Buoyancy then reinforces vertical motion, so it occurs on timescales too short for diabatic effects to be important. Through convective overturning, a superadiabatic layer quickly adjusts to adiabatic stratification, for reasons developed in Chapter 7.

Suggested Reading

Atmospheric Thermodynamics (1981) by Iribarne and Godson contains a thorough discussion of atmospheric statics.

Problems

6.1. Calculate the discrepancy between geometric and geopotential height, exclusive of anisotropic contributions to gravity, up to 100 km.

6.2. Show that the polytropic potential temperature in Problem 2.9 is conserved under the diabatic conditions described in Sec. 6.5.2.

6.3. Use (2.27) and replace z by $-\ln p$ to relate the area circumscribed in Fig. 6.7 to the work performed by the air parcel during one thermodynamic cycle.

6.4. Approximate the thermodynamic cycle in Fig. 6.7 by one comprised of two isobaric legs and two adiabatic legs to show that (a) net work is performed during a complete cycle and (b) the work performed is proportional to the change of temperature associated with isobaric heat transfer.

6.5. Show that an atmosphere of uniform negative lapse rate need not have an upper bound.

6.6. For reasons developed in Chapter 12, large-scale horizontal motion is nearly tangential to contours of geopotential height and given approximately by the geostrophic velocity

$$v_g = \frac{1}{f}\left(-\frac{\partial \Phi}{\partial y}, \frac{\partial \Phi}{\partial x}\right),$$

where $v = (u, v)$ denotes the horizontal component of motion in the eastward (x) and northward (y) directions and Φ is the geopotential along an isobaric surface. Suppose that satellite measurements provide the three-dimensional distributions of temperature and of the mixing ratio of a long-lived chemical species. From those measurements, construct an algorithm to determine 3-dimensional air motion on large scales at elevations great enough to ignore variations in surface pressure.

6.7. An upper level depression, typical of extratropical cyclones during early stages of development, affects isobaric surfaces aloft but leaves the surface pressure distribution undisturbed. Consider the depression of isobaric height associated with the temperature distribution

$$T(\lambda, \phi, \xi) = \overline{\overline{T}} + \overline{T}\cos(\phi) + T'\exp\left[-\frac{\lambda^2 + (\phi - \phi_0)^2}{L^2}\right] \cdot f(\xi),$$

where $\overline{\overline{T}}$, $\overline{T}$, and T' reflect global–mean, zonal–mean, and disturbance contributions to temperature, respectively, $\xi = -\ln(p/p_0)$, with $p_0 =$

1000 mb, is a measure of elevation,

$$f(\xi) = \begin{cases} -\cos\left(\pi\frac{\xi}{\xi_T}\right) & \xi \le \xi_T \\ 0 & \xi > \xi_T, \end{cases}$$

with $\xi_T = -\ln(100/1000)$ reflecting the tropopause elevation. The surface pressure $p_s = p_0$ remains constant, $\phi_0 = \pi/4$, $L = 0.25$ rads, and $\overline{\overline{T}} = 250$ K, $\overline{T} = 30$ K, and $T' = 5$ K. (a) Derive an expression for the 3-dimensional distribution of geopotential height. (b) Plot the vertical profiles of disturbance temperature and height in the center of the anomaly. (c) Plot the horizontal distribution of geopotential height for the midtropospheric level $\xi = \xi_T/2$, as a function of λ and $\phi > 0$. (d) Plot a vertical section at latitude ϕ_0, showing the elevations ξ of isobaric surfaces as functions of λ.

6.8. Under the conditions of Problem 6.7, how large must the anomalous temperature T' be for the disturbance to form a cut-off low at the midtropospheric level $\xi = \xi_T/2$?

6.9. Unlike midlatitude depressions, hurricanes are warm-core cyclones that are invariably accompanied by a trough in surface pressure. Consider a hurricane associated with the temperature distribution

$$T(\lambda, \phi, \xi) = \overline{T} + T'(\lambda, \phi)e^{-\alpha\xi},$$

where $\overline{T}$ and T' reflect zonal–mean and disturbance contributions to temperature, $\xi = -\ln(p/p_0)$, with $p_0 = 1000$ mb, is a measure of elevation,

$$T'(\lambda, \phi) = \overline{T}\alpha\xi_s(\lambda, \phi)\exp[\alpha\xi_s(\lambda, \phi)],$$

in which

$$\xi_s(\lambda, \phi) = 0.1 \, \exp\left(-\frac{\lambda^2 + \phi^2}{L^2}\right),$$

with $\alpha^{-1} = \xi_T/2$, and $\xi_T = \ln(100/1000)$ reflects the surface and tropopause pressures, respectively, and $L = 0.1$ rads. (a) Derive an expression for the 3-dimensional distribution of geopotential height. (b) Plot the vertical profiles of disturbance temperature and height in the center of the anomaly. (c) Plot the horizontal distribution of geopotential height for the midtropospheric level $\xi = \xi_T/2$. (d) Plot a vertical section at the equator, showing the elevations ξ of isobaric surfaces as functions of λ.

6.10. Show that subadiabatic stratification requires an individual parcel moving vertically to, in general, absorb (reject) heat as it ascends (descends).

6.11. Surface analyses produced by the weather service plot pressure adjusted to the common reference level, mean sea level (MSL), so that different locations can be compared meaningfully. (a) Derive an approximate expression for the pressure at MSL in terms of the local surface height z_s, surface pressure p_s, and the global–mean surface temperature $\overline{\overline{T}} =$ 288 K. (b) A surface analysis shows a ridge with a maximum of 1040 mb over Denver, which lies at an elevation of 5300 ft above MSL. What surface pressure was actually recorded there?

6.12. Deep convection inside a region of the tropics releases latent heat to achieve a column heating rate of 250 W m^{-2}. (a) If, uncompensated, that heating would produce a temperature perturbation that is invariant with height up to the tropopause, calculate the implied change of thickness for the 1000- to 100-mb layer after 1 day (compare Fig. 1.30). (b) What process, in reality, compensates for the tendency of tropospheric thickness implied in part (a)?

Chapter 7 | Hydrostatic Stability

The property responsible for the distinctly different character of the troposphere and stratosphere is vertical stability. Although forces are never far from hydrostatic equilibrium, vertical motions are introduced by forced lifting over elevated terrain and through buoyancy. Buoyantly driven motion is related closely to the stability of the atmospheric mass distribution, which is shaped by transfers of energy between the earth's surface, the atmosphere, and deep space. By promoting convection in some regions and suppressing it in others, vertical stability controls a wide range of properties.

7.1 Reaction to Vertical Displacement

Hydrostatic equilibrium and compressibility lead to density decreasing with height regardless of temperature structure (Fig. 6.5). Thus, mean stratification invariably has lighter air configured over heavier air, which suggests stability with respect to vertical displacements. Were air incompressible, this would indeed be the case.

Consider an air parcel inside the layer shown in Fig. 7.1. In a linear stability analysis, we will use this parcel to establish how the layer reacts to small disturbances from equilibrium. Although the analysis focuses on an individual parcel, bear in mind that stability refers to the layer as a whole. Suppose the parcel is subjected to a virtual displacement $\delta z' = z'$, where primes will distinguish properties of the parcel from those of its environment. Through expansion or compression, the parcel automatically adjusts to the environmental pressure to preserve mechanical equilibrium. Thus,

$$p' = p$$

during the displacement. Further, so long as the displacement occurs on a timescale short compared to that of heat transfer with the environment, the parcel's evolution may be regarded as adiabatic.

Per unit volume, Newton's second law for the parcel is

$$\rho' \frac{d^2 z'}{dt^2} = -\rho' g - \frac{\partial p'}{\partial z'}, \qquad (7.1.1)$$

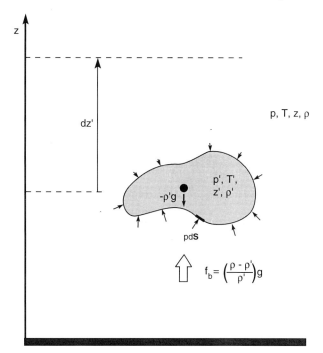

Figure 7.1 Schematic of an air parcel, the properties of which are distinguished by primes, inside a designated layer. The specific buoyancy force f_b results from an imbalance between the parcel's weight and the net pressure force acting over its surface.

which follows from a development analogous to the one leading to (1.16). The momentum balance for the environment is simply hydrostatic equilibrium:

$$0 = \rho g - \frac{\partial p}{\partial z}. \tag{7.1.2}$$

Subtracting (7.1) and incorporating mechanical equilibrium yields for the parcel

$$\rho' \frac{d^2 z'}{dt^2} = (\rho - \rho')g, \tag{7.2.1}$$

which may also be argued from Archimedes' principle (Problem 7.3). Per unit mass, the parcel's momentum balance is then described by

$$\frac{d^2 z'}{dt^2} = \left(\frac{\rho - \rho'}{\rho'}\right)g$$

$$= f_b, \tag{7.2.2}$$

where f_b is the (specific) net buoyancy force acting on the parcel. With (7.1.2), the buoyancy force is recognized as an imbalance between the parcel's weight and the vertical component of pressure force acting over its surface.

Consider how the displaced parcel's temperature varies in relation to that of the environment. With the gas law, (7.2.2) may be expressed

$$\frac{d^2 z'}{dt^2} = g\left(\frac{T' - T}{T}\right). \tag{7.3}$$

For small displacements from its undisturbed elevation, the parcel's temperature decreases with height at the parcel lapse rate: $\Gamma' = \Gamma_d$ (Γ_s) under unsaturated (saturated) conditions. Thus,

$$T' = T_0 - \Gamma' z' \tag{7.4.1}$$

describes the first-order variation of the parcel's temperature from its undisturbed value T_0, which equals the environmental temperature at the parcel's undisturbed elevation. On the other hand, the temperature of the surroundings decreases with height at the local environmental lapse rate Γ, so the corresponding variation of environmental temperature is

$$T = T_0 - \Gamma z'. \tag{7.4.2}$$

The parcel's lapse rate is determined by its thermodynamics. Under adiabatic conditions, Γ' is independent of the environmental lapse rate Γ because the parcel is then thermally isolated. The environmental lapse rate, on the other hand, is determined by the history of air residing in the layer (e.g., by where that air has been and what thermodynamic influences have acted on it). Incorporating (7.4) transforms the vertical momentum balance (7.3) into

$$\frac{d^2 z'}{dt^2} = \frac{g}{T}(\Gamma - \Gamma') z'. \tag{7.5}$$

Hence, the specific buoyancy force experienced by the parcel is proportional to its displacement z' and to the difference between the parcel and environmental lapse rates.

7.2 Stability Categories

If the parcel is unsaturated,

$$\Gamma' = \Gamma_d. \tag{7.6.1}$$

Since Γ' is then constant, (7.4.1) is also valid for finite displacements. If the parcel is saturated,

$$\Gamma' = \Gamma_s, \tag{7.6.2}$$

which implies the local saturated lapse rate at the undisturbed elevation.

7.2.1 Stability in Terms of Temperature

For each of the preceding conditions, three possibilities exist (Fig. 7.2):

1. $\underline{\Gamma < \Gamma'}$: Environmental temperature decreases with height slower than the displaced parcel's temperature. Then (7.5) implies

$$\frac{1}{z'}\frac{d^2z'}{dt^2} < 0.$$

The parcel experiences a buoyancy force f_b that opposes the displacement z'. If it is displaced upward, the parcel becomes heavier than its surroundings and thus negatively buoyant, whereas it becomes positively buoyant and lighter than its surroundings if the parcel is displaced downward. This buoyancy reaction constitutes a *positive restoring force*, one that acts to restore the system to its undisturbed elevation—irrespective of the sense of the displacement. The atmospheric layer is then said to be *hydrostatically stable* and the environmental temperature profile is said to be of *positive stability*.

2. $\underline{\Gamma = \Gamma'}$: Environmental temperature decreases with height at the same rate as the parcel's temperature. Under these circumstances,

$$\frac{1}{z'}\frac{d^2z'}{dt^2} = 0,$$

so the buoyancy force f_b vanishes and the displaced parcel experiences no restoring force. The atmospheric layer is then said to be *hydrostatically neutral* and the environmental temperature profile is said to be of *zero stability*.

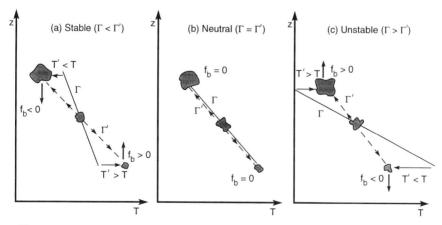

Figure 7.2 Buoyancy reaction experienced by a displaced air parcel, in terms of the environmental lapse rate Γ and the parcel lapse rate Γ', which equals Γ_d (Γ_s) under unsaturated (saturated) conditions. Stability categories: (a) $\Gamma < \Gamma'$, (b) $\Gamma = \Gamma'$, and (c) $\Gamma > \Gamma'$.

3. $\Gamma > \Gamma'$: Environmental temperature decreases with height faster than the parcel's temperature. Then

$$\frac{1}{z'} \frac{d^2 z'}{dt^2} > 0.$$

The parcel experiences a buoyancy force f_b that reinforces the displacement z'. If it is displaced upward, the parcel becomes lighter than its surroundings and thus positively buoyant, whereas it becomes negatively buoyant and heavier than its surroundings if the parcel is displaced downward. This buoyancy reaction constitutes a *negative restoring force*, one that drives the system away from its undisturbed elevation—irrespective of the sense of the displacement. The atmospheric layer is then said to be *hydrostatically unstable* and the temperature profile is said to be of *negative stability*.

Even though density decreases with height in all three cases, each leads to a very different response through buoyancy. The layer's stability is not determined by its vertical profile of density because the density of a displaced air parcel changes through expansion and compression. Depending on whether its density decreases with height faster or slower than environmental density, a displaced parcel will find itself heavier or lighter than its surroundings and thus experience either a positive or a negative restoring force.

The degree of stability or instability is reflected in the magnitude of the restoring force, which is proportional to the difference between the environmental and parcel lapse rates. According to (7.5), the parcel lapse rate Γ' serves as a reference against which the local environmental lapse rate Γ is measured to determine stability (Fig. 7.3). Under unsaturated conditions, $\Gamma' = \Gamma_d$. If environmental temperature decreases with height slower than Γ_d (i.e., the lapse rate is subadiabatic), the layer is stable. If environmental temperature decreases with height at a rate equal to Γ_d (i.e., the lapse rate is adiabatic), the layer is neutral. If its temperature decreases with height faster than Γ_d (i.e., the lapse rate is superadiabatic), the layer is unstable.

Under saturated conditions (e.g., inside a layer of stratiform cloud), the same criteria hold with Γ_s in place of Γ_d. Since

$$\tilde{\Gamma}_s < \Gamma_d,$$

Fig. 7.3 implies that hydrostatic stability is violated more easily under saturated conditions than under unsaturated conditions. That is, the range of stable lapse rate is narrower under saturated conditions than under unsaturated conditions. Reduced stability of saturated air follows from the release (absorption) of latent heat, which warms (cools) an ascending (descending) parcel. By adding positive (negative) buoyancy, heat transfer between the gas and condensed phases of saturated air reinforces vertical displacements and thus destabilizes the layer.

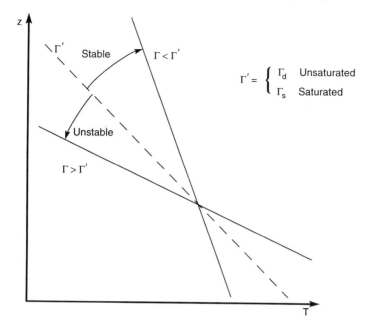

Figure 7.3 Vertical stability in terms of temperature and the environmental lapse rate Γ.

7.2.2 Stability in Terms of Potential Temperature

Stability criteria can also be expressed in terms of potential temperature. For unsaturated conditions, logarithmic differentiation of (2.31) gives

$$\frac{1}{\theta}\frac{d\theta}{dz} = \frac{1}{T}\frac{dT}{dz} - \frac{\kappa}{p}\frac{dp}{dz}.$$

Hydrostatic equilibrium and the gas law transform this into

$$\frac{1}{\theta}\frac{d\theta}{dz} = \frac{1}{T}\frac{dT}{dz} + \frac{\kappa}{H}.$$

Then with (1.17), the vertical gradient of potential temperature can be expressed in terms of the lapse rate

$$\frac{d\theta}{dz} = \frac{\theta}{T}(\Gamma_d - \Gamma). \tag{7.7}$$

Since it follows from the definitions of potential temperature and lapse rate, (7.7) applies to the parcel as well as to the environment. For the latter, an environmental lapse rate $\Gamma < \Gamma_d$ (subadiabatic) corresponds to $d\theta/dz > 0$ and potential temperature increasing with height. An environmental lapse rate $\Gamma = \Gamma_d$ (adiabatic) corresponds to $d\theta/dz = 0$ and a uniform profile of potential temperature. Last, an environmental lapse rate $\Gamma > \Gamma_d$ (superadiabatic) corresponds to $d\theta/dz < 0$ and potential temperature decreasing with height.

Thus, for unsaturated conditions, the stability criteria established previously in terms of temperature translate into the simpler criteria in terms of potential temperature:

$$\frac{d\theta}{dz} > 0 \quad \text{(stable)},$$

$$\frac{d\theta}{dz} = 0 \quad \text{(neutral)}, \qquad (7.8)$$

$$\frac{d\theta}{dz} < 0 \quad \text{(unstable)}.$$

Illustrated in Fig. 7.4, these criteria have the same form as those expressed in terms of temperature, except that the reference profile corresponding to neutral stability is just a uniform distribution of θ. This has a parallel in the vertical stability of an incompressible fluid, like water. If potential temperature increases with height, effectively lighter air (when compressibility is taken into account) overlies effectively heavier air, so the layer is stable. If potential temperature is uniform with height, the layer is neutral, whereas if θ decreases with height, effectively heavier air overlies effectively lighter air, so the layer is unstable.

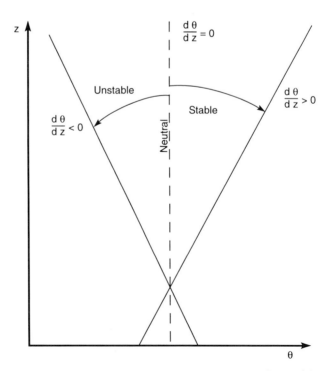

Figure 7.4 Vertical stability, under unsaturated conditions, in terms of potential temperature.

The simplified stability criteria (7.8) reflect the close relationship between θ and atmospheric motion. Recall that potential temperature is conserved for individual air parcels because it includes the effect of expansion work. Therefore, a displaced parcel traces out a uniform profile of θ, which defines the reference thermal structure corresponding to neutral stability. The same is true of temperature for an incompressible fluid. It is for this reason that potential temperature plays a role for air analogous to the one that temperature plays for water. Hence, the vertical distribution of θ describes the effective stratification of mass when compressibility is taken into account.

Under saturated conditions, θ is not conserved. However, θ_e is conserved under those conditions and bears a relationship to Γ_s analogous to the one that θ bears to Γ_d under unsaturated conditions. Using arguments similar to that above, it can be shown that the criteria for stability under saturated conditions are also given by (7.8), but with θ_e in place of θ (Problem 7.15). Thus, the vertical distribution of θ_e describes the effective stratification of mass, when both compressibility and latent heat release are taken into account.

7.2.3 Moisture Dependence

Moisture complicates the description of vertical stability by introducing the need for two categories: One valid under unsaturated conditions and another valid under saturated conditions, for which the release of latent heat must be accounted. Since $\Gamma_s < \Gamma_d$, three possibilities exist (Fig. 7.5). If

$$\Gamma < \Gamma_s, \tag{7.9.1}$$

a layer is *absolutely stable*. A small rearrangement of air will be met by a positive restoring force, regardless of whether the air is saturated, because (7.9.1) ensures that displaced parcels will cool (warm) faster than their surroundings. If

$$\Gamma > \Gamma_d, \tag{7.9.2}$$

the layer is *absolutely unstable*. A small rearrangement of air will result in a negative restoring force, regardless of whether the air is saturated, because (7.9.2) ensures that displaced parcels will cool (warm) slower than their surroundings. If

$$\Gamma_s < \Gamma < \Gamma_d, \tag{7.9.3}$$

the layer is *conditionally unstable*. A small rearrangement of air will then result in a positive restoring force if the layer is unsaturated, but a negative restoring force if it is saturated. For infinitesimal displacements, these three possibilities are mutually exclusive because a parcel that is unsaturated remains so and likewise for one that is saturated. However, for finite displacements, a moist air parcel may change from unsaturated to saturated conditions or vice versa.

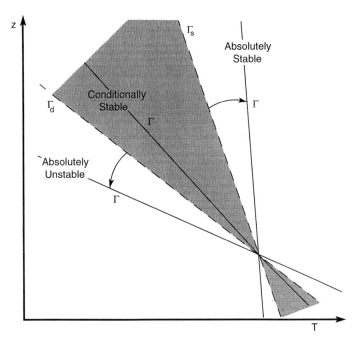

Figure 7.5   Vertical stability of moist air in terms of temperature and the environmental lapse rate Γ.

The restoring force experienced by the parcel can then change sign if the layer is conditionally unstable, a possibility that is treated in Sec. 7.4.

7.3 Implications for Vertical Motion

The stability of a layer determines its ability to support vertical motion and thus to support transfers of heat, momentum, and constituents. Since vertical motion must be compensated by horizontal motion to conserve mass (compare Fig. 5.4), hydrostatic stability also influences horizontal transport. Three-dimensional (3-D) turbulence that disperses atmospheric constituents involves both vertical and horizontal motion. Therefore, suppressing vertical motion also suppresses the horizontal component of 3-D eddy motion and thus turbulent dispersion.

A layer that is stably stratified inhibits vertical motion. Small vertical displacements introduced mechanically by flow over elevated terrain or thermally through isolated heating are then opposed by the positive restoring force of buoyancy (Problem 7.5). Conversely, a layer that is unstably stratified promotes vertical motion through the negative restoring force of buoyancy. Work performed by or against buoyancy reflects a conversion between potential and kinetic energy, which controls the layer's evolution.

The vertical momentum balance for an unsaturated air parcel can be expressed as

$$\frac{d^2 z'}{dt^2} + N^2 z' = 0,$$ (7.10.1)

where

$$N^2 = \frac{g}{T}(\Gamma_d - \Gamma)$$

$$= -\frac{f_b}{z'}$$ (7.10.2)

defines the *buoyancy* or *Brunt–Väisäillä frequency* N. Equation (7.10.1) describes a simple harmonic oscillator, with N^2 reflecting the "stiffness" of the buoyancy spring. From (7.7), the Brunt–Väisäillä frequency can be expressed simply

$$N^2 = g\frac{d \ln \theta}{dz}.$$ (7.10.3)

Since it is proportional to the restoring force of buoyancy, N^2 is a measure of stability.

If a layer is stable, $N^2 > 0$. Then (7.10.1) has solutions of the form

$$z'(t) = Ae^{iNt} + Be^{-iNt}.$$ (7.11.1)

The parcel oscillates about its undisturbed elevation, with kinetic energy repeatedly exchanged with potential energy. The stronger the stability (i.e., the stiffer the spring of buoyancy), the more rapid the parcel's oscillation and the smaller the displacements. Owing to the positive restoring force of buoyancy, small imposed displacements remain small. By limiting vertical motion, positive stability also imposes a constraint on horizontal motion (e.g., to suppress eddy motions associated with 3-D turbulence).

Under neutral conditions, $N^2 = 0$ and the restoring force vanishes. Imposed displacements are then met with no opposition. In that event, the solution of (7.10.1) grows linearly with time, which corresponds to a displaced parcel moving inertially with no conversion between potential and kinetic energy. Under these circumstances, small disturbances ultimately evolve into finite displacements of air. That long-term behavior violates the linear analysis leading to (7.10.1), which is predicated on infinitesimal displacements. However, in the presence of friction, displacements remain bounded (Problem 7.17).

Under positive and neutral stability, air displacements can remain small enough for the mean stratification of a layer to be preserved. A layer of negative stability evolves very differently. In that event, $N^2 = -\hat{N}^2 < 0$ and solutions of (7.10.1) have the form

$$z'(t) = Ae^{\hat{N}t} + Be^{-\hat{N}t}.$$ (7.11.2)

A parcel's displacement then grows exponentially with time through reinforcement by the negative restoring force of buoyancy. The first term in (7.11.2), which dominates the long-term behavior, violates the linear analysis leading to (7.10.1). Unlike behavior under positive and neutral stability, air displacements need not remain bounded. Except for small $\hat{N}$, displacements amplify exponentially—even in the presence of friction (Problem 7.16). Small initial disturbances then evolve into fully developed convection, in which nonlinear effects limit subsequent amplification by modifying the stratification of the layer. By rearranging mass, convective cells alter N^2 and hence the buoyancy force experienced by individual air parcels. Because it is based on mean properties of the layer remaining constant, the simple linear description (7.10) then breaks down.

Amplifying motion described by (7.11.2) is fueled by a conversion of potential energy, which is associated with the vertical distribution of mass, into kinetic energy, which is associated with convective motions. By feeding off of the layer's potential energy, air motions modify its stratification. Fully developed convection, which results in efficient vertical mixing, rearranges the conserved property θ (θ_e) into a distribution that is statistically homogeneous (Sec. 6.5). According to (7.8), this limiting distribution corresponds to a state of neutral stability. Thus, small disturbances to an unstable layer amplify and eventually evolve into fully developed convection, which neutralizes the instability by mixing θ (θ_e) into a uniform distribution. In that limiting state, no more potential energy is available for conversion to kinetic energy, so convective motions then decay through frictional dissipation.

7.4   Finite Displacements

Stability criteria established in Sec. 7.2 hold for infinitesimal displacements. Then a parcel that is initially unsaturated remains so and likewise for one that is initially saturated. For finite displacements, a disturbed air parcel can evolve between saturated and unsaturated conditions. Such displacements are relevant to conditional instability (7.9.3), because a disturbed parcel may then experience a restoring force of one sign during part of its displacement and of opposite sign during the remainder of its displacement. If the layer itself is displaced vertically (e.g., in sloping convection or through forced lifting over elevated terrain), finite displacements allow the layer's stratification and hence its stability to change.

7.4.1   Conditional Instability

Consider an unsaturated parcel that is displaced upward inside the conditionally unstable layer shown in Fig. 7.6a. Depicted there is a simple representation

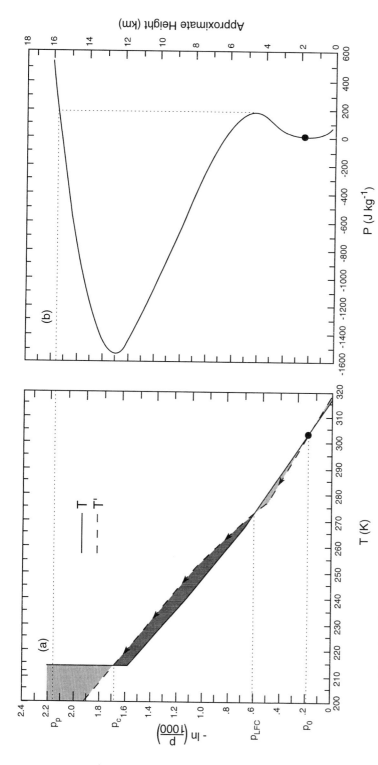

Figure 7.6 (a) Environmental temperature (solid line), under conditionally unstable conditions representative of the tropical troposphere ($\Gamma \cong 8.5 \text{ K km}^{-1}$) and stratosphere, and parcel temperature (dashed line) displaced from the undisturbed level $p_0 = 800$ mb. Parcel temperature decreases with height along a dry adiabat below the *lifting condensation level* (LCL) and along a saturated adiabat above the LCL. It crosses the profile of environmental temperature at the *level of free convection* p_{LFC} and again at the crossing level p_c above the tropopause. (b) Potential energy of the displaced parcel, which is proportional to minus the cumulative area $A(p)$ under the parcel's temperature profile and above the environmental temperature profile that is shaded in (a).

of the tropical troposphere, which is assigned a constant and conditionally unstable lapse rate, and the lower stratosphere, which is assigned a lapse rate of zero. Below its LCL, the displaced parcel cools at the dry adiabatic lapse rate Γ_d and therefore more rapidly than its surroundings (7.9.3). Consequently, the parcel experiences a positive restoring force, one which increases with upward displacement (7.5) and which must be offset for that displacement to be completed. Once it passes its LCL, the parcel cools slower at the saturated adiabatic lapse rate Γ_s, due to the release of latent heat. Since the layer is conditionally unstable, the parcel then cools slower than its surroundings (7.9.3). As a result, the temperature difference between the parcel and its environment and hence the positive restoring force of buoyancy diminish with height. After sufficient displacement, the temperature profile of the parcel crosses the profile of environmental temperature. The parcel then becomes warmer than its surroundings, so it experiences a negative restoring force and thus can ascend of its own accord. The height where this occurs is the *level of free convection* (LFC). Reinforcement by buoyancy then accelerates the parcel upward until its temperature again crosses the profile of environmental temperature at the crossing level p_c. Above that level, the parcel is again cooler than its surroundings, so buoyancy opposes further ascent.

Energetics of the displaced parcel provide some insight into deep convective motions in the tropical troposphere. Under adiabatic conditions, the buoyancy force is conservative (e.g., the work performed along a cyclic path vanishes). Therefore, the potential energy P of the displaced parcel can be introduced in a manner similar to that used in Sec. 6.2 to introduce the geopotential:

$$dP = \delta w_b$$
$$= -f_b dz', \tag{7.12}$$

where δw_b is the incremental work performed against buoyancy. By defining a reference value of zero potential energy at the undisturbed height z_0 and incorporating (7.2), we obtain

$$P = \int_{z_0}^{z} \left(\frac{v - v'}{v} \right) g \, dz'$$
$$= \int_{p_0}^{p} (v' - v) \, dp. \tag{7.13}$$

Then, with the gas law, the parcel's potential energy becomes

$$P(p) = -R \int_{p_0}^{p} (T' - T)(-d \ln p)$$
$$= -RA(p), \tag{7.14}$$

where $A(p)$ is the cumulative area between the temperature profile of the parcel T' and that of the environment T to the level p (Fig. 7.6a).

Above the undisturbed level but below the LFC, $T' < T$, so P increases upward (Fig. 7.6b). Beneath the undisturbed level, T' and T are reversed, so P increases downward as well. Increasing P away from the equilibrium level describes a "potential well" in which the parcel is bound. Work must be performed against the positive restoring force of buoyancy to liberate the parcel from this potential well.

Once the parcel reaches its LFC, where P is a maximum, that work is available for conversion to kinetic energy K. Above the LFC, $T' > T$, so P decreases upward. Under conservative conditions,

$$\Delta K = -\Delta P. \tag{7.15.1}$$

An equivalent expression follows from (7.2), the derivation of which is left as an exercise. By (7.15.1), decreasing P above the LFC represents a conversion of potential energy to kinetic energy, which drives deep convection through buoyancy work.

The total potential energy available for conversion to kinetic energy is termed the *convective available potential energy* (CAPE). Represented by the darkly shaded area in Fig. 7.6a, that energy reflects the total work performed by buoyancy above the LFC and the large drop of potential energy in Fig. 7.6b. Since the parcel's temperature eventually crosses the environmental temperature profile a second time (at the crossing level p_c), CAPE is necessarily finite, as is the kinetic energy that can be acquired by the parcel. An upper bound on the parcel's kinetic energy follows from (7.15.1) as

$$\frac{w'^2}{2} = P(p_{\mathrm{LFC}}) - P(p_c)$$

$$= R \int_{p_{\mathrm{LFC}}}^{p_c} (T' - T)(-d \ln p) = \mathrm{CAPE}, \tag{7.15.2}$$

which corresponds to adiabatic ascent under conservative conditions. However, in practice, mixing with its surroundings makes the behavior inside convection inherently nonconservative (Sec. 7.4.2), so the upward velocity in (7.15.2) is seldom observed.

Above the crossing level, where the parcel is neutrally buoyant, T' and T are again reversed, so P increases upward above p_c. The parcel then becomes negatively buoyant and is bound in another potential well. Despite opposition by buoyancy, the parcel overshoots its new equilibrium level p_c due to the kinetic energy it acquired above the LFC. However, P increases sharply above p_c because environmental temperature above the tropopause diverges from the parcel's temperature. Consequently, penetration into the stable layer aloft is shallow compared to the depth traversed through the conditionally unstable layer below. When the parcel reaches the penetration level p_p, where P equals the previous maximum at the LFC, all of the kinetic energy it acquired above the LFC has been reconverted into potential energy, so the parcel is again bound.

The penetration level in this analysis provides an estimate of the height of "convective overshoots," which typify deep cumulus towers. However, like the maximum updraft in (7.15.2), this estimate is only an upper bound. In practice, some of the kinetic energy driving convective overshoots is siphoned off by mixing with the surroundings. Figure 7.7 shows an overshooting cumulonim-

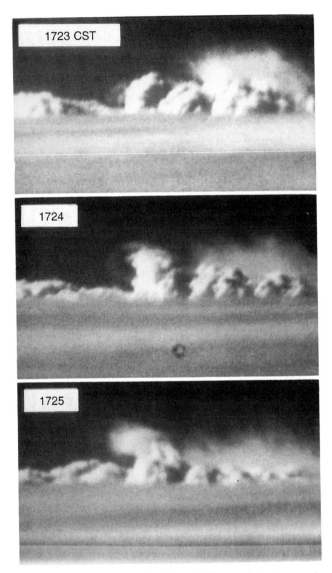

Figure 7.7   Convective domes overshooting the anvil of a cumulonimbus complex, at 2-min intervals. Fibrous features visible in the second plate mark ice particles that nucleate spontaneously when stratospheric air is displaced by convection. After Fujita (1992).

bus complex at 2-min intervals. Several convective domes have penetrated above the anvil, where air eventually fans out horizontally. Negatively buoyant, overshooting cloud splashes back into the anvil, but not before it has been sheared and mixed with stratospheric air. Thus, convective overshoots lead to an irreversible rearrangement and eventually mixing of tropospheric and stratospheric air.

This feature makes penetrative convection instrumental in the competition between vertical overturning in the troposphere (which drives the stratification toward neutral stability) and radiative transfer in the stratosphere (which drives the stratification toward positive stability) that controls the mean elevation of the tropopause. Figure 7.8 displays anomalous temperature, moisture, and cloud cover for an area in the tropics, all as functions of time. Episodes of towering moisture (light shading in Fig. 7.8b) and cold high cloud (Fig. 7.8a), which mark deep cumulus convection, are contemporaneous with anomalously high tropopause (dashed line). Tropopause temperature during those periods is also colder than normal (contours) because the positive lapse of temperature extends upward to heights that are normally stable.

7.4.2   Entrainment

In practice, the vertical velocity and penetration height of a cumulus tower are reduced from adiabatic values by nonconservative effects. Cooler and drier environmental air that is entrained into and mixed with a moist thermal (Fig. 5.4) depletes ascending parcels of positive buoyancy and kinetic energy. Only in the cores of broad convective towers are adiabatic values ever approached. Mixing also modifies the surroundings of a cumulus tower, which mediates gradients of temperature and moisture that control buoyancy.

For reasons established in Chapter 9, entrainment of environmental air is an intrinsic feature of cumulus convection, one that makes the behavior of ascending parcels diabatic. The impact this process has on convection can be illustrated by regarding entrained environmental air to become uniformly mixed with ascending air. Following Holton (1992), consider a parcel of mass m inside the moist thermal pictured in Fig. 5.4. Through mixing, additional mass is incorporated into that parcel, which may then be identified with a fixed percentage of the overall mass crossing a given level. If μ is a conserved property (e.g., $\mu = r$ beneath the LCL), $m\mu'$ represents how much of that property is associated with the parcel under consideration. As before, primes will distinguish properties of the parcel from those of the environment. After mass dm of environmental air has been mixed into it, the parcel will have mass $m + dm$ and property $\mu' + d\mu'$. Conservation of μ then requires

$$(m + dm)(\mu' + d\mu') = m\mu' + dm\mu, \qquad (7.16.1)$$

where $dm\mu$ represents how much conserved property has been incorporated into the parcel from the environment. If the foregoing process occurs during

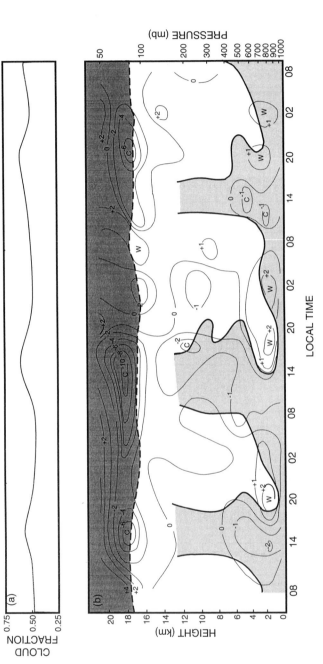

Figure 7.8 Time series of (a) high cloud cover and (b) moisture and temperature derived from radiosondes and satellite IR radiances during the winter MONEX observing campaign. Relative humidities exceeding 80% are lightly shaded. Temperature departures from local time-mean values are contoured. Dark shading indicates stratospheric air above the local tropopause (heavy dashed line). Adapted from Johnson and Kriete (1982).

the time dt, expanding (7.16.1) and neglecting terms higher than first order leads to

$$\frac{d\mu'}{dt} = \frac{d(\ln m)}{dt}(\mu - \mu'), \tag{7.16.2}$$

which governs the conserved property inside the entraining thermal. More generally, if μ is not conserved but, rather, is produced per unit mass at the rate S_μ, its evolution inside the aforementioned parcel is governed by

$$\frac{d\mu'}{dt} = \frac{d(\ln m)}{dt}(\mu - \mu') + S_\mu. \tag{7.17}$$

The time rate of change following the parcel is related to its vertical velocity $w' = dz'/dt$ through the chain rule:

$$\frac{d}{dt} = w\frac{d}{dz}, \tag{7.18}$$

where d/dz refers to a change with respect to the parcel's height. Then (7.17) becomes

$$w\frac{d\mu'}{dz} = \frac{w}{H_e}(\mu - \mu') + S_\mu, \tag{7.19}$$

in which $H_e^{-1} = d(\ln m)/dz$ defines the mass *entrainment height*. Representing the height for upwelling mass to increase by a factor of e, H_e reflects the rate that the entraining thermal expands due to incorporation of environmental air.[1]

If $\mu = \ln \theta_e$, the source term $S_{\theta_e} = 0$ because θ_e is conserved. Then, with (5.27), (7.19) reduces to

$$\frac{d(\ln \theta'_e)}{dz} = \frac{1}{H_e}\left[\ln\left(\frac{\theta}{\theta'}\right) + \frac{l}{c_p}\left(\frac{r_c}{T} - \frac{r'_c}{T'}\right)\right]. \tag{7.20}$$

Replacing saturation values by the actual mixing ratios and temperatures by those corresponding to the LCL (Sec. 5.4.2) and noting that differences of temperature are small compared to differences of moisture leads to

$$\frac{d(\ln \theta'_e)}{dz} \cong \frac{1}{H_e}\left[\ln\left(\frac{\theta}{\theta'}\right) + \frac{l}{c_pT}(r - r')\right]$$
$$= \frac{1}{H_e}\left[\ln\left(\frac{T}{T'}\right) + \frac{l}{c_pT}(r - r')\right]. \tag{7.21}$$

Because $r' > r$ and $T' > T$, θ'_e decreases with the ascending parcel's height. This is to be contrasted with the parcel's behavior under conservative conditions (e.g., in the absence of entrainment), in which case its mass is fixed and

---

[1]A cumulus cloud expands with height slower that the moist thermal supporting it because mixing with drier environmental air leads to evaporation of condensate and dissolution of the cloud along its periphery (Sec. 9.3.3).

θ'_e = const. The two sinks of θ'_e on the right-hand side of (7.21) reflect transfers of latent and sensible heat to the environment, which are mixed across the parcel's control surface (compare Fig. 2.1).

Suppose now μ equals the specific momentum w. Then $w = 0$ for the environment, so (7.19) reduces to

$$w\frac{dw}{dz} = -\frac{w^2}{H_e} + S_w$$

or

$$\frac{dK'}{dz} = -\frac{2}{H_e}K' + \frac{S_K}{w}, \tag{7.22}$$

where $K' = w^2/2$ is the specific kinetic energy of the parcel and production of momentum and kinetic energy are related as $S_K = wS_w$. Unlike the source term for θ_e, $S_K \neq 0$. Even in the absence of entrainment, K is not conserved because the parcel's kinetic energy changes through work performed on it by buoyancy. Differentiating (7.12) along with (7.15) implies

$$\frac{dK'}{dz} = f_b \tag{7.23.1}$$

under conservative conditions (e.g., for a parcel of fixed mass, wherein $H_e \to \infty$). Hence, we identify

$$S_K = wf_b$$
$$= wg\left(\frac{T' - T}{T}\right). \tag{7.23.2}$$

Then (7.22) becomes

$$\frac{dK'}{dz} = g\left(\frac{T' - T}{T}\right) - \frac{2}{H_e}K'. \tag{7.24}$$

The sink of K' on the right-hand side represents turbulent drag that is exerted on the ascending parcel due to incorporation of momentum from the environment. By siphoning off kinetic energy, entrainment reduces the parcel's acceleration from what it would be under the action of buoyancy alone and limits its penetration into a stable layer to about an entrainment height (Problem 7.20).

7.4.3   Potential Instability

Until now, we have considered displacements of individual air parcels within a fixed layer. Suppose the layer itself is displaced vertically (e.g., in sloping convection or forced lifting over elevated terrain). Then changes of the layer's thermodynamic properties alter its stability. Stratification of moisture plays a key role in this process because different levels need not achieve saturation simultaneously.

Consider an unsaturated layer in which potential temperature increases with height,

$$\frac{d\theta}{dz} > 0, \tag{7.25.1}$$

but in which equivalent potential temperature decreases with height,

$$\frac{d\theta_e}{dz} < 0. \tag{7.25.2}$$

By (7.8), the layer is stable. The difference $\theta_e - \theta$ for an individual parcel reflects the total latent heat available for release. Therefore, (7.25.2) reflects a decrease with height of mixing ratio (e.g., as would develop over a warm ocean surface).

Suppose this layer is displaced vertically and changes in its thickness can be ignored (Fig. 7.9). Since the layer is unsaturated, air parcels along a vertical section all cool at the dry adiabatic lapse rate Γ_d (e.g., between positions

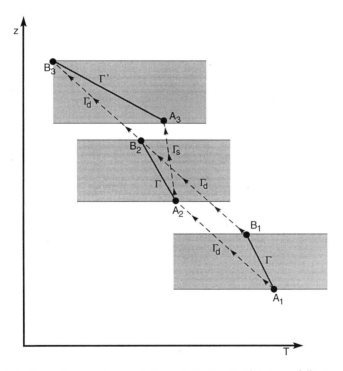

Figure 7.9 Successive positions and thermal structures of a potentially unstable layer (shaded), in which moisture decreases sharply with height, that is displaced vertically. Air is initially unsaturated, so all levels cool at the dry adiabatic lapse rate. However displacement to position 2 leads to saturation of air at the layer's base (A_2). Thereafter, that air cools at the saturated adiabatic lapse rate, while air at the layer's top (B_2) continues to cool at the dry adiabatic lapse rate.

1 and 2). Hence, the layer's profile of temperature is preserved through the displacement, as are the layer's lapse rate and stability. Once some of the layer becomes saturated, this is no longer true. Because they have greater mixing ratios, lower levels achieve saturation sooner than upper levels. Above their LCL, lower levels (e.g., at A_2) cool slower at the saturated adiabatic lapse rate Γ_s, whereas upper levels (e.g., at B_2) continue to cool at the dry adiabatic lapse rate Γ_d. Differential cooling between lower and upper levels then swings the temperature profile counterclockwise and hence destabilizes the layer. This destabilization follows from the release of latent heat at lower levels, much as would occur were the layer heated from below. Sufficient vertical displacement will make the layer unstable with respect to saturated conditions (e.g., at position 3), even though initially it may have been absolutely stable. Such a layer is *potentially unstable*, the criterion for which is given by (7.25.2).

A layer characterized by (7.25.2), but with the reversed inequality, is *potentially stable*. Reflecting an increase with height of mixing ratio, this condition accounts for the release of latent heat in upper levels, which stabilizes the layer when it becomes saturated. Once a potentially stable or unstable layer becomes fully saturated, it satisfies the usual criteria for instability because θ_e is conserved and hence (7.25.2) obeys (7.8) with θ_e in place of θ.

7.4.4   Modification of Stability under Unsaturated Conditions

A layer's stability can also change under unsaturated conditions through vertical compression and expansion. Consider the layer in Fig. 7.10, which is displaced to a new elevation and thermodynamic state, as denoted by primed variables, from an initial elevation and state denoted by unprimed variables. Because θ is conserved under unsaturated conditions, the change of θ across the layer is preserved:

$$d\theta' = d\theta.$$

Then the vertical gradient of potential temperature satisfies

$$\left(\frac{d\theta}{dz}\right)' dz' = \left(\frac{d\theta}{dz}\right) dz. \tag{7.26.1}$$

A horizontal section of layer having initial area dA must also satisfy conservation of mass:

$$\rho \, dA \, dz = \rho' \, dA' \, dz'. \tag{7.26.2}$$

By combining (7.26), we obtain

$$\left(\frac{d\theta}{dz}\right)' = \left(\frac{d\theta}{dz}\right)\left(\frac{\rho'}{\rho}\right)\left(\frac{dA'}{dA}\right) \tag{7.27}$$

for the new vertical gradient of potential temperature inside the layer.

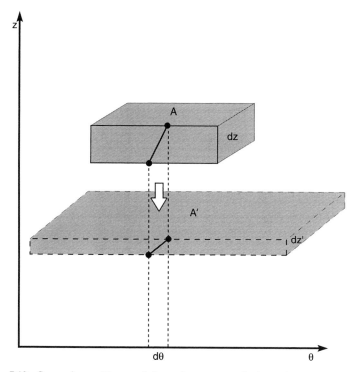

Figure 7.10   Successive positions and thermal structures of a layer that remains unsaturated and therefore preserves the potential temperature difference across it.

Sinking motion and horizontal divergence of air correspond to

$$\frac{\rho'}{\rho} > 1,$$

$$\frac{dA'}{dA} > 1,$$

(7.28)

respectively. Each implies vertical compression of the layer: the former through a reduction of the layer's overall volume and the latter to compensate horizontal expansion of the layer. By (7.27), these motions steepen the vertical gradient of potential temperature. Consequently, they increase the stability (instability) of a layer that is initially stable (unstable). In anticyclones, sinking or "subsidence" of stable air and horizontal divergence that compensates it (Sec. 12.3) lead to the formation of a *subsidence inversion*, which caps pollutants near the ground.

Ascending motion and horizontal convergence imply (7.28) with the reversed inequalities. Vertical expansion then weakens the vertical gradient of potential temperature and therefore drives a stable layer toward neutral stability. Ascending motion in cyclones and horizontal convergence compensating it can virtually eliminate the stability of a layer, providing conditions favorable for convection.

7.5   Stabilizing and Destabilizing Influences

The preceding development illustrates that stability can be altered by internal motions that rearrange air. More important to the overall stability of the atmosphere is external heat transfer, which shapes thermal structure. The vertical distribution of θ provides insight into how stability is established because it reflects the effective stratification of mass.

In light of the first law (2.36), the stability criteria (7.8) imply that a layer is destabilized by heating from below or cooling from aloft (Fig. 7.11a). Each rotates the vertical profile of θ counterclockwise and thus toward instability. These are precisely the influences exerted on the troposphere by exchanges of energy with the earth's surface and deep space. Absorption of longwave (LW) radiation and transfers of latent and sensible heat from the Earth's surface warm the troposphere from below, whereas LW emission at upper levels cools the troposphere from above. Both drive the vertical profile of θ counterclockwise and the lapse rate toward superadiabatic values. By destabilizing the stratification, these forms of heat transfer sustain convective overturning, which in turn maintains the troposphere close to moist neutral stability and makes its general circulation behave as a heat engine (Fig. 6.7).

Conversely, cooling from below or heating from above stabilize a layer (Fig. 7.11b). Each rotates the profile of θ clockwise and thus toward increased

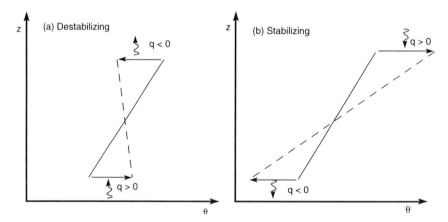

Figure 7.11   (a) Destabilizing influences: Heating from below and cooling from above act to swing the profile of potential temperature counterclockwise. (b) Stabilizing influences: Cooling from below and heating from above act to swing the profile of potential temperature clockwise.

stability. These forms of heat transfer are symbolic of the stratosphere, where ozone heating increases with height to stabilize the thermal structure. By inhibiting vertical motion, strong stability suppresses turbulent dispersion and allows constituents to become highly stratified. Work must then be performed against buoyancy to drive vertical motion, which makes the general circulation of the stratosphere behave as a refrigerator (Fig. 6.8).

According to Secs. 7.2 and 7.4, moisture also figures in the stability of a layer. If the lapse rate is conditionally unstable (7.9.3), increasing a layer's mixing ratio lowers the LCL and hence the LFC of individual parcels, which in turn increase CAPE to drive deep convection (compare Fig. 7.6). Likewise, increasing r in lower portions of the layer (e.g., through absorption of water vapor from a warm ocean surface) increases the equivalent potential temperature there, which swings the profile of θ_e counterclockwise and drives $d\theta_e/dz$ toward negative values. Thus, introducing moisture from below, such as occurs over warm ocean surfaces and reflects a transfer of latent heat across the layer's lower boundary, drives the tropical troposphere toward potential instability.

For conditional and potential instability, finite displacements of air—whether they be of an individual air parcel or of a layer *in toto*, eventually result in buoyantly driven convection. However, for each, a finite potential well must be overcome before the instability can be released. Due to the source of water vapor at its base, lower portions of the tropical troposphere are potentially unstable. Figure 7.12 shows mean distributions of θ_e and N^2 over the equatorial western Pacific. In the lowest 3 km, θ_e decreases with height, whereas $N^2 > 0$ implies that θ increases with height. Despite such instability, deep convection breaks out in preferred regions, where forced lifting is prevalent (Fig. 1.25). Over the Indian Ocean and western Pacific, equatorward-moving air driven by thermal contrasts between sea and neighboring continents converges horizontally to produce forced ascent. Similarly, diurnal heating of tropical landmasses imparts enough buoyancy for surface air over the maritime continent, South America, and tropical Africa to overcome the potential well.

7.6 Turbulent Dispersion

7.6.1 Convective Mixing

Hydrostatic stability controls a layer's ability to support vertical motion and hence mixing by 3-D turbulence. Regions of weak or negative stability favor convective overturning, which results in thorough mixing of air. This situation is common near the ground because absorption of shortwave (SW) radiation destabilizes the surface layer. Introducing moisture from below has a similar effect by increasing the equivalent potential temperature of surface air, which

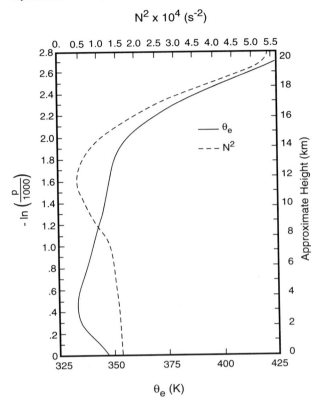

Figure 7.12 Mean vertical profiles of equivalent potential temperature (solid line) and buoyancy frequency squared (dashed line) over Indonesia, obtained by averaging radiosonde observations between longitudes of 90°E and 105°E and latitudes of 10°S and 10°N during March 1984 (refer to Fig. 1.25).

makes the overlying layer potentially unstable. When conditions are favorable, small disturbances evolve into fully developed convection, which in turn neutralizes the instability by rearranging mass vertically.

As noted earlier, the troposphere has a global–mean lapse rate of about 6.5 K km^{-1}, which is close to neutral stability with respect to saturated conditions. This mean stratification reflects two salient features of the troposphere: (1) Tropospheric air is moisture laden and (2) tropospheric air is efficiently overturned in cumulus and sloping convection. Efficient vertical exchange leads to a characteristic time for air to move between the surface and the tropopause of only a couple of days. Because it operates on a timescale that is short compared to diabatic effects, vertical mixing drives the thermal structure of the troposphere toward that of saturated adiabatic motion (Sec. 6.5) and hence toward a mean state of moist neutral stability.

Cumulus convection is favored in the tropical troposphere by high sea surface temperature (SST) and strong insolation. Both support large transfers of radiative, latent, and sensible heat from the earth's surface, which destabilize the overlying atmosphere. Organized convection, which is reinforced by the release of latent heat, opposes those destabilizing influences to maintain much of the tropical troposphere close to moist neutral stability. Cumulus convection also operates outside the tropics, but it is organized in continental monsoons (e.g., over India and Australia) and synoptic disturbances. Frontal motions associated with cyclones, like the one in the eastern Atlantic in Fig. 1.23, cause organized lifting of air through sloping convection. By displacing moist and often conditionally or potentially unstable air that originated in the tropics, those motions favor convection, which likewise drives mean stratification toward moist neutral stability.

7.6.2 Inversions

Contrary to the troposphere, the stratosphere is characterized by strong positive stability. Ozone heating makes temperature increase and potential temperature increase sharply with height (Fig. 6.5). The potential temperature increases from about 350 K near the tropopause to more than 1000 K near 10 mb. The sharp vertical gradient of θ is associated with a strong positive restoring force, one that suppresses vertical motions and 3-D turbulence and allows chemical constituents to become highly stratified.

Unlike the troposphere, where air is efficiently rearranged in the vertical, the stratosphere has a characteristic timescale for vertical exchange of months to years. The layered nature of the stratosphere is occasionally demonstrated by volcanic eruptions, which loft debris above the tropopause. During the 1950s and 1960s, nuclear detonations introduced radioactive debris into the stratosphere. Because they have very slow sedimentation rates, aerosol particles introduced in this manner behave as tracers. Volcanic and nuclear debris are observed to undergo little or no vertical motion, instead fanning out horizontally into clouds of global dimension that remain intact for many months (see Fig. 9.6). The great eruption of Krakatoa, which destroyed the Indonesian island of the same name in 1883, altered radiative properties of the stratosphere and the color of sunsets for years afterward. The long residence time of stratospheric aerosol allows it to alter SW absorption at the earth's surface for long durations, which makes it an important consideration in the global energy budget. In sharp contrast, tropospheric aerosol (e.g., from deforestation and the Kuwaiti oil fires) is eliminated by convective processes on a timescale of only days (Chapter 9).

Convective cells do not penetrate appreciably above the tropopause, where N^2 and the restoring force of buoyancy increase abruptly (Fig. 7.12). Upon encountering this sharp increase of stability, an ascending plume is quickly drained of positive buoyancy and fans out horizontally to form a cloud anvil, as typifies the mature stage of cumulonimbus thunderstorms. Similar behavior is occasionally observed at intermediate levels of the troposphere, when developing towers encounter a shallow inversion through which the buoyant plume is able to penetrate (see Fig. 9.18).

Temperature inversions frequently accompany anticyclones. Often less than a kilometer above the surface, a subsidence inversion (Sec. 7.4) confines pollutants to a shallow layer near the ground, where they become airborne. Because winds accompanying an anticyclone also tend to be light, pollutants are neither dispersed nor advected away from source regions, allowing concentrations to increase during periods of high production. The same mechanisms inhibit cloud formation by confining moisture near the surface. Consequently, anticyclones are typified by cloud-free but often polluted conditions.

7.6.3   Life Cycle of the Nocturnal Inversion

The competition between heat transfer and convection for control of the stratification of a layer is illustrated by the formation and breakup of the *nocturnal inversion*, which often develops at night. Under arid and cloud-free conditions, which favor efficient LW cooling to space, the surface temperature decreases sharply after sunset. To preserve thermal equilibrium, heat is transferred to the ground radiatively and conductively from the overlying air, which cools the surface layer from below and stabilizes it (Fig. 7.13). Sufficient cooling will swing the temperature structure toward an inverted profile, which corresponds to the potential temperature increasing sharply with height in a shallow layer adjacent to the ground.

Strongly stable, the nocturnal inversion suppresses vertical motion, as well as 3-D turbulence. Without turbulent transport to disperse them, pollutants introduced at the surface are frozen in the stable air, so their concentrations increase steadily near pollution sources. Such behavior is particularly noticeable in urban areas during early morning periods of heavy traffic.

After sunrise, absorption of SW radiation warms the ground, which influences the surface layer in just the reverse manner. To preserve thermal equilibrium, heat is then transferred to the overlying air, initially through conduction and radiative transfer. These forms of heat transfer warm the surface layer from below, which increases the potential temperature and swings its vertical profile counterclockwise near the ground. Inside the shallow surface layer that has been destabilized, thermal convection then develops spontaneously to neutralize the instability. By rearranging air vertically, heat is transferred from the surface (in high values of θ) and mixed across the shallow layer of convective overturning. The same is true for other conserved properties, such as moisture and pollutants.

Absorption of SW radiation and vertical heat transfer continue to destabilize the surface layer and thus maintain convective overturning. To preserve neutral stability, additional heat transferred to the atmosphere must be mixed over an ever deeper layer. In this fashion, the layer driven toward neutral stability expands upward, as do pollutants that were frozen near the surface in

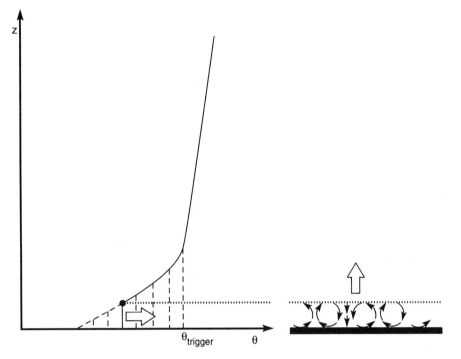

Figure 7.13 Schematic illustrating the breakdown of the nocturnal inversion, which forms during night. Following sunrise, surface heating and heat transfer to the overlying air destabilize a shallow layer adjacent to the ground. Convection then develops spontaneously and, through vertical mixing, restores the thermal structure inside that layer to neutral stability. Continued heat transfer from the ground requires θ to be mixed over a progressively deeper layer, which erodes the nocturnal inversion from below. When convection has advanced through the entire layer of strong stability, the nocturnal inversion has been destroyed. Deep convection can then penetrate into weakly stable air overhead, to disperse pollutants that were previously trapped near the surface in stable air.

stable air. Convective mixing erodes the nocturnal inversion from below and reduces pollution concentrations by diluting them over an ever deeper layer.

After sufficient surface heating, convection will have advanced through the entire nocturnal inversion, at which point it reaches air of weaker stability overhead (e.g., air that was unaffected by surface cooling during the night). The surface temperature when this condition is reached is termed the *break* or *trigger temperature* because deep convection can then develop from the source of positive buoyancy at the ground. Pollutants are then dispersed vertically over a deep volume, as well as horizontally by 3-D turbulence, resulting in a sharp reduction of pollution concentrations. By permitting vertical transport of water vapor, this condition also sets the stage for cumulus cloud formation, which appears soon thereafter if sufficient moisture is present.

Suggested Reading

Vertical stability is treated in detail, along with other considerations relevant to cloud formation, in *Atmospheric Thermodynamics* (1981) by Iribarne and Godson.

Environmental Aerodynamics (1978) by Scorer contains an advanced treatment of thermal entrainment, including laboratory simulations.

Problems

A pseudo-adiabatic chart is provided in Appendix F.

7.1. Use Newton's second law to derive the momentum balance (7.1.1) for an individual air parcel.

7.2. In early morning, the profile of potential temperature in a layer adjacent to the surface is described by

$$\frac{\theta}{\theta_0} = e^{\frac{z}{h_1}} - \frac{z}{h_2},$$

with $h_1 > h_2$. If condensation can be ignored, (a) characterize the static stability of this layer, (b) determine the profile of buoyancy frequency, and (c) determine the limiting distribution of θ for $t \to \infty$, after sufficient surface heating for convection to develop.

7.3. Use Archimedes' principle to establish (7.2.2).

7.4. A simple harmonic oscillator comprised of a 1-kg weight and an elastic spring with constant k provides an analog of buoyancy oscillations. Calculate the effective spring constant for unsaturated conditions representative of (a) the troposphere: $\Gamma = 6.5 \text{ K km}^{-1}$ and $T = 260$ K and (b) the stratosphere: $\Gamma = -4 \text{ K km}^{-1}$ and $T = 250$ K.

7.5. An impulsive disturbance imposes a vertical velocity w_0 on an individual air parcel. If the layer containing that parcel has nonnegative stability N, (a) determine the parcel's maximum vertical displacement, as a function of N, under linear conditions and (b) describe the layer's evolution under nonlinear conditions, in relation to N.

7.6. Upstream of isolated rough terrain, the surface layer has thermal structure

$$T(z) = T_0 \frac{a}{a+z},$$

where $a > 0$ and $T_0 > a\Gamma_d$. (a) Determine the upstream profile of potential temperature. (b) Characterize the upstream vertical stability. (c) Describe the mean stratification far downstream of the rough terrain.

7.7. Derive the identity

$$\frac{1}{\theta}\frac{d\theta}{dz} = \frac{1}{T}\frac{dT}{dz} + \frac{\kappa}{H}$$

relating the vertical gradients of temperature and potential temperature.

7.8. An early morning sounding reveals the temperature profile

$$T(z) = \begin{cases} T_0 + \left(\frac{\Gamma_d}{2}\right)z & z < 1 \text{ km} \\ T_0 + \left(\frac{\Gamma_d}{2}\right) - \left(\frac{\Gamma_d}{4}\right)(z - 1) & z \geq 1 \text{ km}. \end{cases}$$

(a) Determine the *trigger temperature* for deep convection to develop.
(b) If, following sunrise, the ground warms at 3°C per hour, estimate the time when cumulus clouds will appear.

7.9. For the circumstances in Problem 7.8, describe the transformations between radiative, internal and potential, and kinetic energies involved in the development of convection.

7.10. The potential temperature inside a layer varies as

$$\frac{\theta}{\theta_0} = -(z - a)^2 + a^2 + 1 \qquad z > 0.$$

(a) Characterize the stability of this layer. (b) How large a vertical displacement would be required to liberate a parcel originally at $z = 0$ from this layer?

7.11. Radiative heating leads to temperature in the stratosphere increasing up to the stratopause, above which it decreases in the mesosphere. If the temperature profile at a certain station is approximated by

$$T(z) = T_0 \frac{z}{b + \left(\frac{z}{h}\right)^2},$$

determine (a) the profile of potential temperature, (b) the profile of Brunt–Väisäillä frequency squared N^2. (c) For reasons developed in Chapter 14, gravity waves can propagate through regions of positive static stability, which provides the positive restoring force for their oscillations. In terms of the variable b, how small must h be to block propagation through the mesosphere? (d) By noting that gravity waves impose vertical displacements, discuss the behavior of air in the mesosphere under the conditions given for part (c).

7.12. Analyze the stability and energetics for the stratification and parcel in Fig. 7.6, but under saturated conditions.

7.13. To what height will the parcel in Problem 7.12 penetrate under purely adiabatic conditions? (See Problem 7.20.)

7.14. Estimate the maximum height of convective towers under the conditions in Fig. 7.6, but with the isothermal stratosphere replaced by one with a lapse rate of -3 K km^{-1}.

7.15. Establish stability criteria analogous to (7.8) under saturated conditions.

7.16. Rayleigh friction approximates turbulent drag in proportion to an air parcel's velocity:

$$D = Kw,$$

where D is the specific drag experienced by a parcel, w is its velocity, and the Rayleigh friction coefficient K has dimensions of inverse time. For given Rayleigh friction coefficient K, show that vertical displacements remain bounded for small instability but become unbounded for sufficiently large instability.

7.17. Show that, in the presence of Rayleigh friction, vertical displacements remain bounded under conditions of neutral stratification.

7.18. The lapse rate at a certain tropical station is constant and equal to 7 K km^{-1} from 1000 to 200 mb, above which conditions are isothermal. At the surface, the temperature is 30°C and the relative humidity is sufficiently great for vertical displacements to bring about almost immediate saturation. Approximate the saturated adiabat by the constant lapse rate, $\Gamma_s \cong 6.5$ K km^{-1}, to estimate the maximum vertical velocity attainable inside convective towers. (See Problem 7.20.)

7.19. The weather service forecasts the intensity of convection in terms of a *thermal index* (TI), which is defined as the local ambient temperature minus the temperature anticipated inside a thermal under adiabatic conditions. The more negative TI, the stronger are updrafts inside thermals. A morning sounding over an arid region reveals a lapse rate of -4 K km^{-1} in the lowest kilometer and a constant lapse rate of 6 K km^{-1} above, with the surface temperature being 10°C. If air is sufficiently dry to remain unsaturated and if the surface temperature is forecast to reach a maximum of 40°C, (a) determine the thermal index then as a function of height for $z > 1$ km. (b) To what height will strong updrafts be expected? (c) Calculate the maximum updraft and maximum height reached by updrafts anticipated under adiabatic conditions if, at the time of maximum temperature, mean thermal structure has been driven adiabatic in the lowest kilometer.

7.20. In practice, vertical velocities inside a thermal differ significantly from adiabatic values, entrainment limiting the maximum updraft to that near the level of maximum buoyancy. Under the conditions in Problem 7.19, but accounting for entrainment, calculate the profile of kinetic energy and note the height of maximum updraft inside a thermal for an entrainment height of (a) infinity, (b) 1 km, and (c) 100 m.

7.21. Establish relationship (7.21) governing behavior inside an entraining thermal.

7.22. According to Fig. 1.27, the troposphere is heated below by transfers of radiative, latent, and sensible heat, while it is cooled aloft by LW cooling. Discuss why, on average, the troposphere remains close to moist neutral stability in the presence of these destabilizing influences.

7.23. The approach of a warm front is often heralded by deteriorating visibility and air quality. Consider a frontal surface that slopes upward and to the right, separates warm air on the left from cold air on the right, and moves to the right. The intersection of this surface with the ground marks the front. (a) Characterize the vertical stability at a station ahead of the front. (b) Discuss the vertical dispersion of pollutants that are introduced at the ground. (c) Describe how the features in (a) and (b) and pollution concentrations vary as the front approaches the station.

7.24. Consider the conditions in Problem 7.18, but in light of entrainment. If the ambient mixing ratio varies as $r = r_0 e^{-z/h}$, with $h = 2$ km, and the entrainment height is 1 km, calculate the profile of (a) equivalent potential temperature inside a moist thermal (b) specific cooling rate for air entrained into the thermal.

7.25. Consider a constant tropospheric lapse rate of $\Gamma = 7$ K km^{-1} and mixing ratio that varies as $r = r_0 e^{-z/h}$, with $h = 2$ km. If the troposphere can be regarded as being close to saturation, characterize the stability of the troposphere to large-scale lifting (e.g., in sloping convection) in the presence of moisture that is representative of (a) midlatitudes, $r_0 = 5$ g kg^{-1}, and (b) the tropics, $r_0 = 25$ g kg^{-1}.

7.26. A morning temperature sounding over Florida reveals the profile

$$T = \begin{cases} 20 - 8(z - 1) \ [°C] & z \geq 1 \text{ km} \\ 20 \ [°C] & z < 1 \text{ km}. \end{cases}$$

If the mixing ratio at the surface is 10 g kg^{-1}, estimate the LFC.

7.27. See Problem 13.19.

Chapter 8 | Atmospheric Radiation

The thermal structure and stratification discussed in Chapters 6 and 7 are shaped in large part by radiative transfer. According to the global–mean energy budget (Fig. 1.27), of some 542 W m^{-2} that is absorbed by the atmosphere, more than 80% is supplied through radiative transfer: 368 W m^{-2} through absorption of longwave (LW) radiation from the earth's surface and another 68 W m^{-2} through direct absorption of incoming shortwave (SW) radiation. Similarly, the atmosphere loses energy through LW emission to the earth's surface and to space.

To maintain thermal equilibrium, components of the atmospheric energy budget must, on average, balance. Vertical transfers of radiant energy involved in this balance are instrumental in determining thermal structure and motion, which characterize individual layers. Likewise, the dependence of optical properties on temperature, variable constituents, and clouds makes radiative transfer horizontally nonuniform. Geographical variations of radiative heating (Fig. 1.29c) must be compensated by horizontal transfers of energy that maintain thermal equilibrium locally. That energy transfer is accomplished by the general circulation, which, in turn, is driven by the nonuniform distribution of heating. Thus, understanding these and other essential features requires an understanding of how radiation interacts with the atmosphere.

8.1 Shortwave and Longwave Radiation

Energy transfer in the atmosphere involves radiation in two distinct bands of wavelength. SW radiation emitted by the sun and LW radiation emitted by the earth's surface and atmosphere are concentrated in wavelengths λ that are widely separated due to the disparate temperatures of the emitters. Figure 8.1a shows blackbody emission spectra for temperatures of 6000 and 288 K, which correspond to the equivalent blackbody temperature of the sun and the mean surface temperature of the earth. The representation in Fig. 8.1a, logarithmic in wavelength versus wavelength times energy flux, is area preserving. Equal areas under the curves in different bands of wavelength represent equal power.

The spectrum of SW radiation, which is concentrated at wavelengths shorter than 4 μm, peaks in the visible near $\lambda = 0.5$ μm. Tails of the SW spectrum

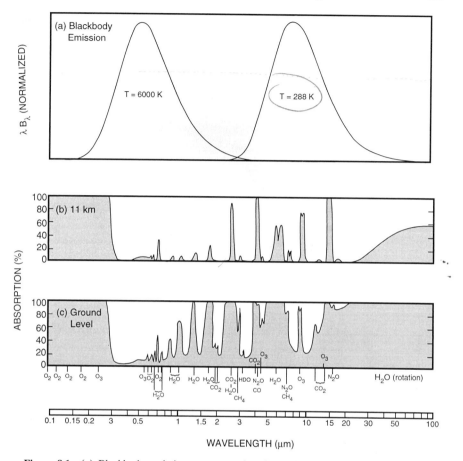

Figure 8.1 (a) Blackbody emission spectra as functions of wavelength λ for temperatures corresponding to the sun and the earth's global–mean surface temperature. The representation: λ times energy flux versus $\log(\lambda)$, is area preserving—equal areas under a spectral curve represent equal power. (b) Absorption as a function of λ along a vertical path from the top of the atmosphere to a height of 11 km. (c) As in (b), but to the surface. Adapted from Goody and Yung (1989).

extend into the UV ($\lambda < 0.3$ μm) and into the near-infrared region ($\lambda > 0.7$ μm). On the other hand, the LW spectrum peaks well into the IR at a wavelength of about 10 μm. Tails of the LW spectrum extend down to wavelengths of about 5 μm and out to the microwave region ($\lambda > 100$ μm). Overlap between the spectra of SW and LW radiation is negligible, so they may be treated independently by considering wavelengths shorter and longer than 4 μm.

Also shown as a function of wavelength is the fractional absorption of radiation passing from the top of the atmosphere to a height of 11 km (Fig. 8.1b) and to the ground (Fig. 8.1c). Energetic radiation, $\lambda < 0.3$ μm, is absorbed at high levels in connection with photodissociation and photoionization of O_2 and O_3. As a result, very little of the UV incident on the top of the atmosphere reaches the tropopause. By contrast, the bulk of the spectrum in the visible and near-IR regions arrives at the tropopause unattenuated, with the exception of narrow absorption bands of water vapor and carbon dioxide. However, in passing from 11 km to the ground, the remaining SW spectrum is substantially absorbed in the IR by water vapor and carbon dioxide, absolute concentrations of which are large in the troposphere. Consequently, the spectrum of SW radiation reaching the surface is concentrated at visible wavelengths, for which the atmosphere is mostly transparent.

In contrast to solar radiation, LW radiation emitted by the earth's surface is almost completely absorbed—most notably by H_2O in a wide band centered at 6.3 μm and another in the far infrared, by CO_2 in a band centered at 15 μm near the peak of the LW emission spectrum, and by a variety of trace gases including O_3, CH_4, and N_2O. Consequently, most of the LW energy emitted at the ground is captured in the overlying air. That energy must be reemitted, half upward and half back downward. The upward component then undergoes repeated absorption and reemission in adjacent layers until it is eventually rejected to space. This sequence of radiative exchanges traps LW radiation and, through mechanisms developed later in this chapter, elevates the surface temperature of the earth.

A comparison of Figs. 8.1b and 8.1c indicates that most of the LW energy emitted by the surface is absorbed in the troposphere. Only in the *atmospheric window* at wavelengths of 8 to 12 μm is absorption weak enough for much of the LW radiation emitted by the surface to pass freely through the atmosphere. The 9.6-μm band of ozone, which is positioned inside this window, is the only strong absorber at those wavelengths. Most of that absorption takes place in the stratosphere, where ozone concentrations are large.

8.1.1 Spectra of Observed SW and LW Radiation

The most important property of the SW spectrum is the *solar constant*, which represents the flux of radiant energy integrated over wavelength reaching the top of the atmosphere at the mean earth–sun distance. The solar constant is weakly variable, having a value of about 1370 W m^{-2} according to recent satellite measurements. When distributed over the globe, this SW flux leads to the daily *insolation* shown in Fig. 1.28, which follows primarily from the length of day and solar inclination for particular latitudes and times of year.

Spectra of observed SW and LW radiation are shaped by the absorption characteristics described above. At the top of the atmosphere (Fig. 8.2), the solar spectrum resembles the emission spectrum of a blackbody at about 6000 K. That temperature is characteristic of the sun's *photosphere*, where most of the radiation is emitted. Wavelengths shorter than 300 nm deviate from this simple picture, instead reflecting temperatures of 4500 to 5000 K, which are

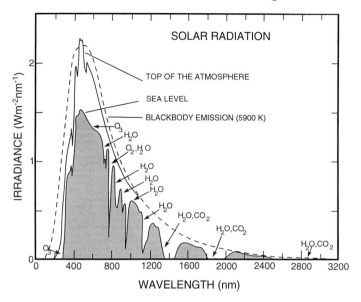

Figure 8.2 Spectrum of SW radiation at the top of the atmosphere (solid) and at the earth's surface (shaded), compared against the emission spectrum of a blackbody at 6000 K (dashed). Individual absorbing species indicated. Adapted from Coulson (1975).

found somewhat higher near the base of the sun's *chromosphere*. At the earth's surface, the wings of the solar spectrum have been mostly stripped off, in the UV by ozone and molecular oxygen and in the IR by absorption bands of water vapor and carbon dioxide, leaving wavelengths of 0.3 to 2.4 μm.

Wavelengths shorter than 300 nm, which constitute a small fraction of the solar energy flux, are absorbed fairly high in the earth's atmosphere through photodissociation and photoionization. As shown in Fig. 8.3, these energetic components penetrate to altitudes in inverse relation to their wavelengths. Wavelengths shorter than 200 nm do not penetrate appreciably below 50 km, whereas $\lambda < 150$ nm tend to be confined to the thermosphere. Exceptional is the discrete *Lyman–α emission* at 121 nm, which corresponds to transitions from the gravest excited state of hydrogen. The Lyman–α line happens to co-incide with an optical window in the banded absorption structure at these wavelengths that allows it to reach the upper mesosphere and lower thermo-sphere, where it is involved in photoionization and photodissociation.

The wavelength of SW radiation is also inversely related to the altitude in the solar atmosphere where it is emitted. As indicated in Fig. 8.4, λ of 300 nm and longer originate in the photosphere, whereas $\lambda < 200$ nm are emitted by the chromosphere, and very energetic components with $\lambda < 50$ nm originate higher yet in the corona. Solar variability increases with altitude in the solar atmosphere. Consequently, temporal variations in the SW spectrum are most

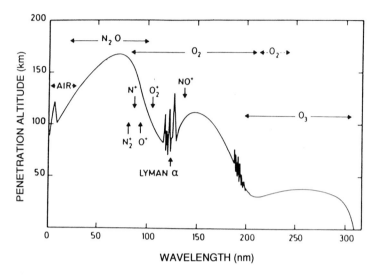

Figure 8.3 Penetration altitude, where SW flux at wavelength λ is attentuated by a factor of e^{-1}. Absorbing species and thresholds for photoionization indicated. After Herzberg (1965).

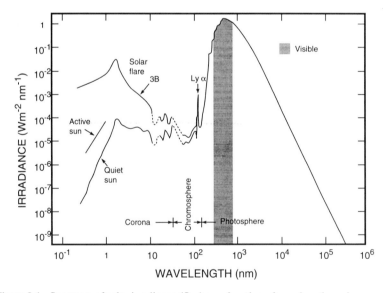

Figure 8.4 Spectrum of solar irradiance (flux) as a function of wavelength, under normal and disturbed conditions. Visible band and emission levels in the solar atmosphere indicated. Adapted from Smith and Gottlieb (1974) with permission of Kluwer Academic Publishers.

apparent in the far-UV region, which makes only a small contribution to the solar constant.

Two phenomena dominate solar variability. The 27-day rotation cycle of the sun leads to variations of several percent in the UV and fall to under 1% for wavelengths longer than 250 nm (WMO, 1986). Sunspot activity also evolves through an 11-year cycle that modulates solar emissions. Between its extremes, *solar-max* and *solar-min*, activity changes noticeably in the outer reaches of the sun's atmosphere. Owing to the inverse relationship between solar altitude and the emitted wavelength, SW variability decreases sharply with increasing λ (Fig. 8.4), which, in turn, strongly limits the altitudes in the earth's atmosphere affected by such variations. At $\lambda = 160$ nm, the peak-to-peak variation in flux between solar-max and solar-min is about 10% and it drops to less than 1% by 300 nm. In terms of the solar constant and most of the energy reaching the earth's surface, the variation between solar-max and solar-min is of order 0.1% (Willson *et al.*, 1986).

The spectrum of LW radiation emitted to space (Fig. 8.5) also resembles that of a blackbody. However, it corresponds to a temperature of 288 K only in the atmospheric window: wavenumbers λ^{-1} of 800 to 1200 cm^{-1}, where radiation emitted by the surface passes to space relatively freely. Other wavenumbers are absorbed and reemitted in the overlying atmosphere by water vapor and clouds, which emit radiation at colder temperatures.

The strong absorption bands of CO_2 and O_3 sharply reduce emission to space at 15 and 9.6 μm, respectively. Each corresponds to a blackbody temperature distinctly colder than that within the atmospheric window, where outgoing radiation emanates from the earth's surface. The 15-μm emission of CO_2, which is positioned near the center of the LW spectrum, corresponds to a blackbody temperature of about 220 K and an altitude in the upper troposphere. The 9.6-μm emission of ozone corresponds to a temperature of about 270 K and an altitude in the upper stratosphere. Along with water vapor, these minor constituents trap LW radiation emitted by the earth's surface and elevate its temperature over what it would be in the absence of an atmosphere (e.g., over the effective blackbody temperature of 255 K). Known as the *greenhouse effect*, this phenomenon follows from the atmosphere's transparency to visible radiation and opacity to infrared radiation, which are controlled by optically active species.

8.2 Description of Radiative Transfer

8.2.1 Radiometric Quantities

Understanding how radiation influences atmospheric properties requires a quantitative description of radiative transfer, which is complicated by its three-

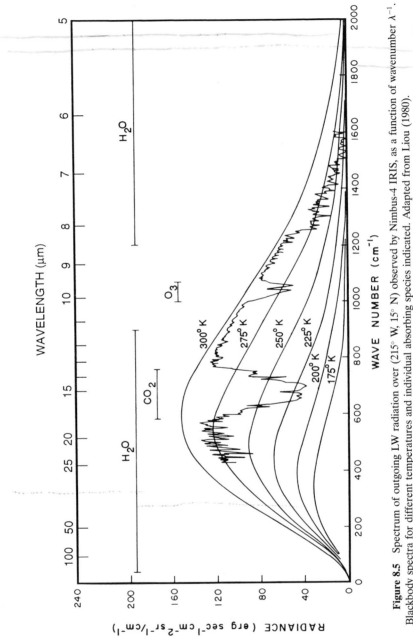

Figure 8.5 Spectrum of outgoing LW radiation over (215° W, 15° N) observed by Nimbus-4 IRIS, as a function of wavenumber λ^{-1}. Blackbody spectra for different temperatures and individual absorbing species indicated. Adapted from Liou (1980).

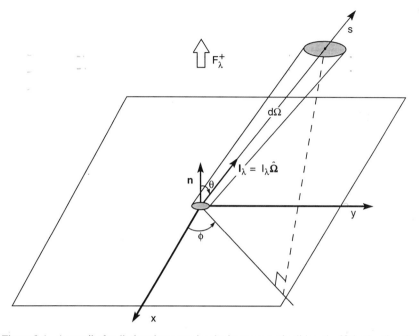

Figure 8.6 A pencil of radiation that occupies the increment of solid angle $d\Omega$ in the direction $\hat{\boldsymbol{\Omega}}$ and traverses a surface with unit normal $\boldsymbol{n}$. The monochromatic *intensity* or *radiance* passing through the pencil: $\boldsymbol{I}_\lambda = I_\lambda \hat{\boldsymbol{\Omega}}$, describes the rate at which energy inside the pencil crosses the surface per unit area, steradian, and wavelength. Integrating the component normal to the surface: $I_\lambda(\hat{\boldsymbol{\Omega}} \cdot \boldsymbol{n}) = I_\lambda \cos \theta$, over the half-space of 2π steradians in the positive $\boldsymbol{n}$ direction yields the monochromatic forward *flux* or *irradiance* F_λ^+.

dimensional nature and its dependence on wavelength and directionality. As illustrated in Fig. 8.6, radiant energy traversing a surface with unit normal $\boldsymbol{n}$ can have contributions from all directions, which are characterized by the unit vector $\hat{\boldsymbol{\Omega}}$. Such multidimensional radiation is termed *diffuse*, whereas, if it arrives along a single direction (like incoming solar), it is termed *parallel-beam radiation*. The monochromatic *intensity* or *radiance* $\boldsymbol{I}_\lambda = I_\lambda \hat{\boldsymbol{\Omega}}$ represents the rate that energy of wavelengths λ to $\lambda + d\lambda$ flows per unit area through an increment of solid angle $d\Omega$ in the direction $\hat{\boldsymbol{\Omega}}$. Having dimensions of power/(wavelength $\cdot$ area $\cdot$ steradian), $I_\lambda(x, \hat{\boldsymbol{\Omega}})$ is a function of position x and direction $\hat{\boldsymbol{\Omega}}$ and characterizes a pencil of radiation that traverses the surface at a *zenith angle* θ from its normal. Then the rate energy flows in the direction $\hat{\boldsymbol{\Omega}}$ per unit cross-sectional area and wavelength interval is $I d\Omega$.

Integrating the component of intensity normal to the surface: $\boldsymbol{I}_\lambda \cdot \boldsymbol{n} = I_\lambda(\hat{\boldsymbol{\Omega}} \cdot \boldsymbol{n}) = I_\lambda \cos \theta$, over the half-space of solid angle in the positive $\boldsymbol{n}$ direction defines the monochromatic *flux* or *irradiance* crossing the surface

$$F_\lambda^+ = \int_{2\pi} I_\lambda \cdot n d\Omega^+$$

$$= \int_{2\pi} I_\lambda \cos\theta d\Omega^+$$

$$= \int_0^{2\pi} \int_0^{\pi/2} I_\lambda(\phi, \theta) \cos\theta \sin\theta d\theta d\phi, \tag{8.1}$$

which has dimensions of power/(wavelength · area). The *total flux* then follows by integrating over the electromagnetic spectrum

$$F^+ = \int_0^\infty F_\lambda^+ d\lambda, \tag{8.2}$$

which has the dimensions of energy flux: power/area. It follows that the monochromatic flux F_λ^+ represents the spectral density at wavelength λ of the total flux F^+.

The forward energy flux in the n direction is described by F_λ^+. The backward energy flux F_λ^- may be obtained by integrating I in the negative n direction over the opposite half-space. Both are nonnegative quantities. The monochromatic *net flux* in the n direction is then the difference between the forward and backward components:

$$F_\lambda = F_\lambda^+ - F_\lambda^-. \tag{8.3}$$

By applying (8.1) and (8.3) to the three coordinate directions, we obtain the net flux vector F_λ. In the special case of *isotropic* radiation, for which the intensity is independent of direction: $I_\lambda \neq I_\lambda(\hat{\Omega})$, (8.1) reduces to

$$F_\lambda^+ = \pi I_\lambda. \tag{8.4}$$

Since the forward and backward components are then equal, the net flux in any direction vanishes.

Electromagnetic radiation can interact with matter through three basic mechanisms: absorption, scattering, and emission. A pencil of radiation occupying the solid angle $d\Omega$ is attenuated in proportion to the density and absorbing characteristics of the medium through which it passes. Energy can also be scattered out of the solid angle, which likewise attenuates energy flowing through $d\Omega$. Together, absorption and scattering constitute the net *extinction* of energy passing through the pencil of radiation. Conversely, energy emitted into the solid angle or scattered into $d\Omega$ from other directions intensifies the flow of energy through the pencil of radiation. These interactions are governed by a series of laws that describe the transfer of radiation through matter.

8.2.2   Absorption

In the absence of scattering, absorption of energy from a pencil of radiation is expressed by the following principle.

LAMBERT'S LAW

The fractional energy absorbed from a pencil of radiation is proportional to the mass traversed by the radiation. For an incremental distance ds, this principle may be written

$$\frac{dI_\lambda}{I_\lambda} = -\rho\sigma_{a\lambda}ds, \tag{8.5.1}$$

where the constant of proportionality $\sigma_{a\lambda}$ is the specific *absorption cross section* (also referred to as the *mass absorption coefficient*), which has units of area/mass. The cross section symbolizes the area of the pencil lost by passing through an increment of mass $\rho dAds$ or, equivalently, the effective absorbing area of that mass—each for the wavelength λ. The foregoing principle can also be expressed in terms of the particle number density n (e.g., of molecules or aerosol) through which the radiation passes

$$\frac{dI_\lambda}{I_\lambda} = -n\hat{\sigma}_{a\lambda}ds, \tag{8.5.2}$$

where the absorption cross section $\hat{\sigma}_{a\lambda}$ is referenced to an individual particle and has dimensions of area. Thus, $\hat{\sigma}_{a\lambda}$ symbolizes the effective absorbing area posed by a particle in the path of the radiation. In either representation, Lambert's law for attenuation of energy is linear in the intensity I_λ.

The *absorption coefficient*

$$\beta_{a\lambda} = \rho\sigma_{a\lambda}$$

$$= n\hat{\sigma}_{a\lambda}, \tag{8.6.1}$$

which has dimensions of inverse length, measures the characteristic distance over which energy is attenuated. The density in (8.5) corresponds to the mass of the absorber, which, in a mixture, may follow from several constituent gases. Collecting contributions from different constituents then gives

$$\beta_{a\lambda} = \sum_i r_i\rho\sigma_{a\lambda i} \tag{8.6.2}$$

$$= \sum_i r_i n\hat{\sigma}_{a\lambda i}, \tag{8.6.3}$$

where r_i denotes the mass mixing ratio of the ith absorbing species.

By integrating Lambert's law along the path of radiation, we obtain

$$I_\lambda(s) = I_\lambda(0) \exp\left(-\int_0^s \rho\sigma_{a\lambda}ds'\right). \tag{8.7}$$

In the absence of scattering and emission, the intensity inside a pencil of radiation decreases exponentially with the *optical path length*

$$u(s) = \int_0^s \rho \sigma_{a\lambda} ds'$$
$$= \int_0^s \beta_{a\lambda} ds', \tag{8.8}$$

which is the dimensionless distance traversed by radiation weighted according to the density and absorption cross section of the medium. Because Lambert's law involves no directionality, it applies to flux as well as intensity.

The monochromatic *transmissivity*, which describes the fraction of incident radiation remaining in the pencil at a given distance, is then

$$\mathcal{T}_\lambda(s) = \frac{I_\lambda(s)}{I_\lambda(0)}$$
$$= e^{-u(s)}. \tag{8.9}$$

Transmissivity decreases exponentially with optical path length through an absorbing medium. Conversely, the monochromatic *absorptivity* a_λ represents the fraction of incident radiation that has been absorbed from the pencil during the same traversal. Since

$$\mathcal{T}_\lambda + a_\lambda = 1 \tag{8.10}$$

in the absence of scattering and emission, the absorptivity follows as

$$a_\lambda(s) = 1 - e^{-u(s)}. \tag{8.11}$$

Absorptivity exponentially approaches unity with path length through an absorbing medium, which, in that limit, is said to be "optically thick."

8.2.3 Emission

To maintain thermal equilibrium, a substance that absorbs radiant energy must also emit it. Like absorption, the emission of energy into a pencil of radiation is proportional to the mass involved. The basis for describing thermal emission is the theory of blackbody radiation, which was developed by Planck and contemporaries near the turn of the century.

Blackbody radiation is an idealization that corresponds to the energy emitted by an isolated cavity in thermal equilibrium. Such radiation is characterized by the following properties:

- The radiation is uniquely determined by the temperature of the emitter.
- For a given temperature, the radiant energy emitted is the maximum possible at all wavelengths.
- The radiation is isotropic.

In addition to being a perfect emitter, a blackbody is also a perfect absorber: Radiation incident on a blackbody is completely absorbed at all wavelengths.

PLANCK'S LAW

To explain blackbody radiation, Planck postulated that the energy E of molecules is quantized and can undergo only discrete transitions that satisfy

$$\Delta E = \Delta n \cdot h\nu, \tag{8.12}$$

where n is integer, h is Planck's constant (Appendix C), and $\nu = c/\lambda$ is the frequency of electromagnetic radiation emitted or absorbed to accomplish the energy transition ΔE. Hence, the radiation emitted or absorbed by individual molecules is quantized in *photons* that carry energy in integral multiples of $h\nu$. On this basis, Planck derived the following relationship between the spectrum of intensity B_λ emitted by a population of molecules and their absolute temperature:

$$B_\lambda(T) = \frac{2hc^2}{\lambda^5(e^{\frac{hc}{K\lambda T}} - 1)}, \tag{8.13}$$

where K is the Boltzmann constant. Plotted in Fig. 8.7, the *Planck* or *blackbody spectrum* (8.13) possesses a single maximum at wavelength λ_m and increases with temperature. At wavelengths shorter than λ_m, B_λ decreases sharply, whereas a comparatively wide band is involved at longer wavelengths. Because blackbody radiation is isotropic, the form of the Planck spectrum applies to flux (8.4) as well as to intensity.

WIEN'S DISPLACEMENT LAW

The wavelength of maximum intensity follows from (8.13) as

$$\lambda_m = \frac{2897}{T} \quad [\mu\text{m}], \tag{8.14}$$

which decreases with increasing temperature of the emitter. This feature of the emission spectrum allows the *brightness temperature* of a body to be inferred from radiation emitted by it. For instance, SW radiation incident on the top of the atmosphere is concentrated in the visible and peaks near a wavelength of 0.480 μm (Fig. 8.2), which corresponds to blue light. Wien's displacement law then implies an effective solar temperature of about 6000 K. On the other hand, the 288 K mean surface temperature of earth corresponds to maximum emission at a wavelength of about 10 μm (Fig. 8.1). Similar reasoning can be applied to individual wavelengths via (8.13). For instance, features in the observed spectrum of outgoing LW radiation (Fig. 8.5) at 15 and 9.6 μm can be identified with emission by CO_2 in the upper troposphere and by O_3 in the upper stratosphere. Each is significantly colder than the surface temperature, which is manifested in the outgoing spectrum only in the atmospheric window at 8 to 12 μm.

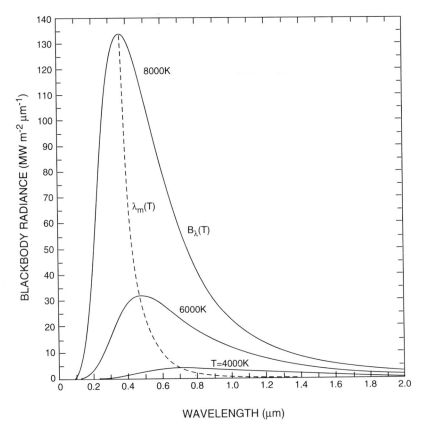

Figure 8.7   Spectra of emitted intensity $B_\lambda(T)$ for blackbodies at several temperatures, with wavelength of maximum emission $\lambda_m(T)$ indicated.

THE STEFAN–BOLTZMANN LAW

The total flux emitted by a blackbody follows by integrating over the electromagnetic spectrum, which, in combination with (8.4), yields

$$F = \pi B(T)$$
$$= \sigma T^4, \tag{8.15}$$

where σ is the Stefan–Boltzmann constant (Appendix C).[1]

The quartic temperature dependence in (8.15) implies LW emissions that differ sharply between cold and warm objects. This feature of LW radiation

[1] Chronologically, the Stefan–Boltzmann law and Wien's displacement law predate Planck's postulate of the quantization of energy and his derivation of (8.13). While accounting for these earlier results, Planck's law resolved the failure of classical mechanics to explain the thermal emission spectrum at short wavelengths. See Eisberg (1967) for an overview.

is inherent to IR imagery like that shown in Figs. 1.23 and 1.25. The Sahara desert in North Africa at 1200 GMT (Fig. 1.25a), which is very dark and has temperatures in excess of 315 K, corresponds to emitted fluxes exceeding 550 W m^{-2}. (This strong surface emission is mediated by the atmosphere, which traps LW radiation and limits outgoing fluxes at the top of the atmosphere to 350 W m^{-2} or less.) At the same time, LW emission from deep convective towers over Indonesia, which have temperatures as cold as 180 K, is only of order 60 W m^{-2}. These large differences of LW emission symbolize geographical variations in the earth's energy budget. They are also symbolic of diurnal variations, which follow from changes during the day of surface temperature and cloud cover.

KIRCHHOFF'S LAW

The preceding laws determine the emission into a pencil of radiation during its passage through a medium that behaves as a blackbody. Real substances absorb and emit radiation at rates smaller than those of a perfect absorber and emitter. A *graybody* absorbs and emits with efficiencies that are independent of wavelength. The *absorptivity* a_λ of a substance is defined as the ratio of the intensity it absorbs to $B_\lambda(T)$. According to (8.11), the absorptivity of a layer approaches unity with increasing optical path length, at which point it behaves as a blackbody. Similarly, the *emissivity* ϵ_λ is defined as the ratio of intensity emitted by a substance to $B_\lambda(T)$, which, according to the properties of blackbody radiation, must likewise approach unity with increasing optical path length.

Consider two infinite plates that communicate with one another only radiatively: one a blackbody and the other a graybody with constant absorptivity a and emissivity ϵ. Suppose the plates are in thermal equilibrium, so they absorb and emit energy in equal proportion. Suppose further that this state is characterized by the plates having different temperatures. Introducing a conducting medium between the plates will then drive the system out of thermal equilibrium by permitting heat to flow from the warmer to the colder plate. To restore thermal equilibrium, radiation must then transfer energy from the colder plate back to the warmer plate—which violates the second law of thermodynamics. It follows that the *radiative equilibrium temperature* of the gray plate, that temperature at which it emits energy at the same rate as it absorbs energy from its surroundings, must be identical to that of the black plate.

For the gray plate to be in equilibrium, the flux of energy it emits, $\epsilon \sigma T^4$, must equal the flux of energy it absorbs from the black plate: $a\sigma T^4$. Applying this reasoning to substances of arbitrary monochromatic absorptivity and emissivity leads to the conclusion

$$\epsilon_\lambda = a_\lambda. \qquad (8.16)$$

Known as *Kirchhoff's law*, (8.16) asserts that a substance emits radiation at each wavelength as efficiently as it absorbs it.

In general, the radiative efficiency of a substance varies with wavelength. As illustrated in Problem 8.12, an illuminated surface with an absorptivity of a_{SW} in the SW and a_{LW} in the LW has a radiative–equilibrium temperature that varies according to the ratio a_{SW}/a_{LW}. Thus, a surface like snow, which absorbs weakly in the visible but strongly in the IR, has a lower radiative–equilibrium temperature than a gray surface, for which the absorptivity is independent of wavelength. Likewise, the selective transmission characteristics of the atmosphere elevate the radiative–equilibrium temperature of the surface over what it would be in the absence of an atmosphere. SW radiation that passes freely through the atmosphere is absorbed at the surface. But LW radiation reemitted by the surface to preserve thermal equilibrium is trapped by the overlying atmosphere, which absorbs at those wavelengths (Problem 8.16). To offset reduced transmission, the surface must emit more LW radiation, which requires a higher temperature (8.15), until the LW radiation rejected at the top of the atmosphere just balances the SW radiation absorbed. The increased surface temperature (e.g., the greenhouse effect) is related directly to the difference in opacity between the visible and IR, the demonstration of which is left as an exercise.

Kirchhoff's law is predicated on a state of *local thermodynamic equilibrium* (LTE), wherein temperature is uniform and radiation is isotropic within some small volume. Those conditions are satisfied when energy transitions are dominated by molecular collisions, which is true for the most important radiatively active gases at pressures greater than 0.01 mb (e.g., at altitudes below 60 km). At higher altitudes, LTE breaks down because the interval between collisions is no longer short compared to the lifetime of excited states associated with absorption and emission. Kirchhoff's law is then invalid.

With (8.16), the fractional energy emitted into a pencil of radiation along an incremental distance ds can be expressed

$$\frac{dI_\lambda}{B_\lambda} = d\epsilon_\lambda = da_\lambda$$

$$= \rho\sigma_{a\lambda}ds$$

by Lambert's law. Then

$$dI_\lambda = \rho\sigma_{a\lambda}J_\lambda ds, \tag{8.17}$$

where

$$J_\lambda = B_\lambda(T) \tag{8.18}$$

denotes the monochromatic *source function* in the absence of scattering. Equation (8.18) holds under the conditions of Kirchhoff's law, namely, under LTE. Otherwise, the contribution to the source function from emission is more complex.

8.2.4   Scattering

Beyond absorption and emission, energy passing through a pencil of radiation is also modified by scattering, which refers to the extraction and subsequent reemission of energy by matter. A population of molecules possesses, in addition to the translational energy associated with temperature, electronic, vibrational, and rotational energies. Those forms of internal energy can be excited by absorbing a photon, which can then be released in several ways. The simplest is when the excited internal energy is converted into translational energy or "thermalized" through molecular collision, which corresponds to thermal absorption of radiation. Thermal emission occurs through the reverse process. The excited energy can also be reemitted—in wavelengths and directions different from those of the incident radiation (Fig. 8.8), which then constitutes scattering. Analogous interactions occur between radiation and atmospheric aerosol.

The foregoing process is described in terms of the monochromatic *scattering cross section* $\sigma_{s\lambda}$, which symbolizes the fractional area removed from a pencil of radiation through scattering. Effects of absorption and scattering are additive since both are linear. Then the monochromatic *extinction cross section* is defined by

$$k_\lambda = \sigma_{a\lambda} + \sigma_{s\lambda}, \tag{8.19.1}$$

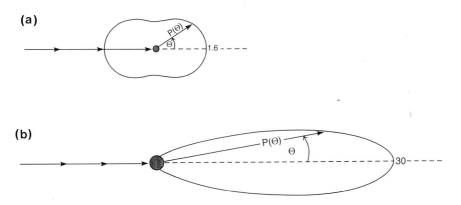

Figure 8.8   Angular distribution of radiation scattered from (a) small particles (of radius $a \ll \lambda$), which is representative of *Rayleigh scattering* of SW radiation by air molecules (Sec. 9.4.1), and (b) large particles ($a \gg \lambda$), which is representative of *Mie scattering* of SW radiation by cloud droplets (Sec. 9.4.1). Phase function P is plotted in terms of the scattering angle Θ and in (b) for a scattering population with the refractive index of water and an effective size parameter $x_e = 2\pi(a_e/\lambda) = 5$. *Note:* The compressed scale in (b) implies that energy redirected by large particles is dominated by forward scattering. Larger particles produce even stronger forward scattering (compare Fig. 9.27). Data in (b) courtesy of F. Evans (U. Colorado).

from which the *extinction coefficient* follows as

$$\beta_{e\lambda} = \rho k_\lambda, \qquad (8.19.2)$$

where ρ is understood to refer to the optically active species. Lambert's law (8.5) and its consequences for attenuation then hold with k_λ in place of $\sigma_{a\lambda}$.

Scattering also modifies how energy is introduced into a pencil of radiation, which makes the source function more complex than the contribution from emission alone (8.18). While scattering removes radiation from one direction, it introduces it into other directions. Alternatively, photons can be scattered into a particular solid angle $d\Omega$ from all directions. If photons are introduced into the pencil of radiation through only one encounter with a particle, the process is termed *single scattering*. If more encounters are involved, the process is termed *multiple scattering*.

The *single scattering albedo*

$$\omega_\lambda = \frac{\sigma_{s\lambda}}{k_\lambda} \qquad (8.20)$$

represents the fraction of radiation lost through extinction that is scattered out of a pencil of radiation. Then

$$1 - \omega_\lambda = \frac{\sigma_{a\lambda}}{k_\lambda} \qquad (8.21)$$

represents the fraction lost through extinction that is absorbed from the pencil of radiation. The directionality of the scattered component is described in the *phase function* $P_\lambda(\hat{\boldsymbol{\Omega}}, \hat{\boldsymbol{\Omega}}')$, which corresponds to the fraction of radiation scattered by an individual particle from the direction $\hat{\boldsymbol{\Omega}}'$ into the direction $\hat{\boldsymbol{\Omega}}$. If the phase function is normalized according to

$$\frac{1}{4\pi} \int_{4\pi} P_\lambda(\hat{\boldsymbol{\Omega}}, \hat{\boldsymbol{\Omega}}') d\Omega' = 1, \qquad (8.22)$$

then

$$\omega_\lambda \frac{P_\lambda(\hat{\boldsymbol{\Omega}}, \hat{\boldsymbol{\Omega}}')}{4\pi} d\Omega'$$

represents the fraction of radiation lost through extinction from the pencil in the direction $\hat{\boldsymbol{\Omega}}'$ that is scattered into the pencil in the direction $\hat{\boldsymbol{\Omega}}$. The scattered contribution to the source function is then

$$J_{s\lambda} = \frac{\omega_\lambda}{4\pi} \int_{4\pi} I_\lambda(\hat{\boldsymbol{\Omega}}') P_\lambda(\hat{\boldsymbol{\Omega}}, \hat{\boldsymbol{\Omega}}') d\Omega'. \qquad (8.23)$$

Combining the contributions from emission and scattering yields the total source function

$$J_\lambda = (1 - \omega_\lambda) B_\lambda(T) + \frac{\omega_\lambda}{4\pi} \int_{4\pi} I_\lambda(\hat{\boldsymbol{\Omega}}') P_\lambda(\hat{\boldsymbol{\Omega}}, \hat{\boldsymbol{\Omega}}') d\Omega'. \qquad (8.24)$$

8.2.5 The Equation of Radiative Transfer

Collecting the extinction of energy (8.19) and the introduction of energy (8.17) yields the net rate at which the intensity inside a pencil of radiation changes with distance s,

$$\frac{dI_\lambda}{\rho k_\lambda ds} = -I_\lambda + J_\lambda, \tag{8.25}$$

which is the *radiative transfer equation* in general form. In the absence of scattering, (8.25) reduces to

$$\frac{dI_\lambda}{\rho k_\lambda ds} = -I_\lambda + B_\lambda(T), \tag{8.26}$$

where k_λ then equals the absorption cross section $\sigma_{a\lambda}$. Introducing the *optical thickness*

$$\chi_\lambda(s) = \int_s^0 \rho k_\lambda ds'$$
$$= -u(s), \tag{8.27.1}$$

which increases in the negative s direction (i.e., opposite to $\hat{\boldsymbol{\Omega}}$):

$$d\chi_\lambda = -\rho k_\lambda ds, \tag{8.27.2}$$

casts the energy budget for a pencil of radiation into the canonical form

$$\frac{dI_\lambda}{d\chi_\lambda} = I_\lambda - B_\lambda(T), \tag{8.28}$$

which is known as *Schwartzchild's equation*.

The solution of Schwartzchild's equation can be expressed formally with the aid of an integrating factor. Multiplying by $e^{-\chi_\lambda}$ allows (8.28) to be written

$$\frac{d}{d\chi_\lambda}(e^{-\chi_\lambda}I_\lambda) = -e^{-\chi_\lambda}B_\lambda(T),$$

which upon integrating from 0 to $\chi_\lambda(s)$ yields

$$I_\lambda(s) = I_\lambda(0)e^{\chi_\lambda(s)} - \int_0^{\chi_\lambda(s)} B_\lambda[T(\chi_\lambda')]e^{\chi_\lambda(s)-\chi_\lambda'}d\chi_\lambda'. \tag{8.29}$$

The first term in (8.29) describes an exponential decrease with optical path length of the incident intensity $I_\lambda(0)$, as embodied in Lambert's law (8.7). The second term describes the cumulative emission and absorption between 0 and $\chi_\lambda(s)$. If the properties T, ρ, and k_λ are known as functions of s, (8.29) uniquely determines the intensity along the path of radiation.

8.3   Absorption Characteristics of Gases

8.3.1   Interactions between Radiation and Molecules

The absorptivity of a medium approaches unity with increasing optical path length (Sec. 8.2.1), irrespective of wavelength. Along finite path lengths (e.g., through an atmosphere of bounded mass), absorption may be large at some wavelengths and small at others, according to the optical properties of the medium. For a gas, the absorption spectrum a_λ is concentrated in a complex array of lines that correspond to transitions between the discrete electronic, vibrational, and rotational energy levels of molecules. Electronic transitions are stimulated by radiation at UV and visible wavelengths, whereas vibrational and rotational transitions occur at IR wavelengths. At pressures greater than about 0.1 mb (e.g., below 60 km), the internal energy acquired by absorbing a photon is quickly thermalized. In addition to discrete absorption characteristics, a continuum of absorption occurs at shorter wavelengths in connection with the photodissociation and photoionization of molecules. The latter occur at wavelengths of X-ray, UV, and, to a lesser degree, visible radiation and are possible for all λ shorter than the threshold to break molecular and electronic bonds.

Figure 8.9 shows absorption spectra in the IR for optically active gases corresponding to a vertical path through the atmosphere. All of the species are minor constituents of air, the most important being water vapor, carbon dioxide, and ozone. Each of these triatomic molecules is capable of undergoing simultaneous vibration–rotation transitions, which produce a clustering of absorption lines. At coarse resolution, like spectra in Figs. 8.1 and 8.9, these clusters appear as continuous bands of absorption. Water vapor absorbs strongly in a broad band centered near 6.3 μm and in another at 2.7 μm. Both bands correspond to transitions from vibrationally excited states: 6.3-μm absorption to the ν_2 vibrational mode (Fig. 8.10a), whereas the band at 2.7 μm involves both ν_1 and ν_3 modes of vibration. Rotational transitions of H_2O, which are less energetic, lead to absorption at wavelengths longer than 12 μm (compare Fig. 8.1). Carbon dioxide is excited vibrationally by wavelengths near 15 μm and also near 4.3 μm. The former corresponds to the transverse mode ν_2 (Fig. 8.10b), whereas the latter corresponds to the longitudinal vibration ν_3. Ozone absorbs strongly at wavelengths near 9.6 μm in connection with vibrational transitions. Because it coincides with the atmospheric window at 8 to 12 μm (Fig. 8.1), this absorption band allows stratospheric ozone to interact radiatively with the troposphere and the earth's surface. Except for ozone, most of the absorption by these species takes place in the troposphere, where absolute concentrations are large (compare Figs. 8.1b and c). Nitrous oxide, methane, carbon monoxide, and chlorofluorocarbon (CFC) -11 and -12 also have absorption lines within the range of wavelength shown in Fig. 8.9, but are of secondary importance.

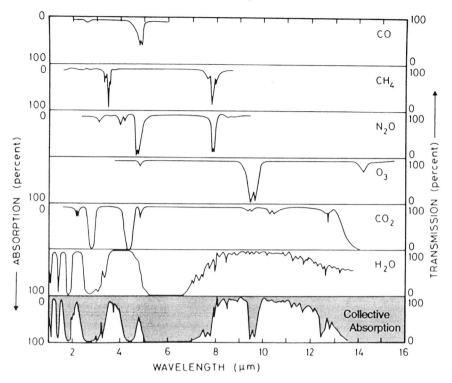

Figure 8.9 Absorption spectra for LW radiation passing vertically through the atmosphere as contributed by strong absorbing gases. Adapted from *Handbook of Geophysics and Space Environment* (1965).

At higher resolution, IR absorption features in Fig. 8.9 are actually comprised of a complex array of lines. Figure 8.11 illustrates this in high-resolution spectra for the 14-μm band of CO_2 (Fig. 8.11a) and for the rotational band of H_2O at wavelengths of 27 to 31 μm (Fig. 8.11b). Absorption lines in the 15-μm band of CO_2 are spaced regularly in wavenumber. By contrast, the rotational band of water vapor is characterized by a random distribution of absorption lines. In principle, all such lines must be accounted for in radiative calculations. However, their number and complexity make *line-by-line calculations* impractical for most applications. Instead, radiative calculations rely on *band models* to represent the gross characteristics of absorption spectra in particular ranges of wavelength.

Similar features characterize the absorption spectrum at visible and UV wavelengths, where radiation interacts with molecular oxygen and ozone. In addition to discrete features, the more energetic radiation at these wavelengths produces continuous bands of absorption in connection with photodissociation and photoionization of molecules. Figure 8.12 shows the absorption

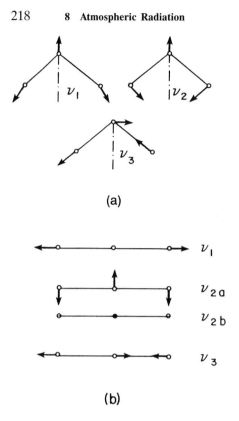

(a)

ν_1

ν_{2a}

ν_{2b}

ν_3

(b)

Figure 8.10 Normal modes of vibration for triatomic molecules corresponding to (a) H_2O and O_3, in which atoms are configured triangularly, and (b) CO_2, in which atoms are configured longitudinally. After Herzberg (1945).

cross section $\hat{\sigma}_{a\lambda}$ for O_2 in the UV. The *Herzberg continuum* lies at wavelengths of 242 to 200 nm, where O_2 is dissociated into two ground-state oxygen atoms. At shorter wavelengths, the absorption spectrum is marked by the discrete *Schumann–Runge bands*, where O_2 is vibrationally excited. These excited bound states are unstable and eventually produce two ground-state oxygen atoms. Each of the vibrational bands in the Schumann–Runge system is actually comprised of many rotational lines, which are shown at high resolution in Fig. 8.13. The rotational line structure is fairly regular at long wavelength, but gradually degenerates to an almost random distribution at short wavelength. Wavelengths shorter than 175 nm are absorbed in the *Schumann–Runge continuum* (Fig. 8.12), in which O_2 is dissociated into two oxygen atoms, one electronically excited. Absorption at still shorter wavelengths exhibits an irregular banded structure that eventually merges into the ionization continuum at wavelengths shorter than 102 nm.

Actual absorption by individual bands in Fig. 8.12 varies strongly with altitude. Even though it has the smallest values of $\hat{\sigma}_{a\lambda}$, the Herzberg continuum at 200 to 242 nm dominates absorption by O_2 up to 60 km because shorter wavelengths have already been removed at higher altitude (Fig. 8.3). Above 60 km,

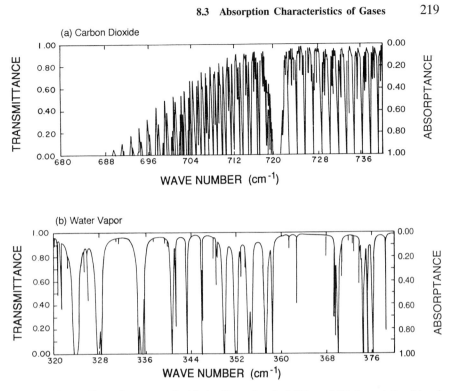

Figure 8.11 Absorption spectra in (a) the 14-μm band of CO_2 and (b) the rotational band of H_2O at 27 to 31 μm. Adapted from McClatchey and Selby (1972).

the Schumann–Runge bands are dominant. The Schumann–Runge continuum prevails only in the thermosphere, where very energetic UV is present.

Ozone absorbs at longer wavelengths, which penetrate to lower altitudes. The primary absorption of UV by ozone is in the *Hartley band* at wavelengths of 200 to 310 nm (Fig. 8.14). At long wavelength, the Hartley band merges with the *Huggins bands* at 310 to 400 nm. Ozone also absorbs in the visible in the *Chappuis band* at wavelengths of 400 to 850 nm, which is important at altitudes below 25 km. In all of these bands, absorption follows from photodissociation of ozone. The Hartley and Chappuis bands are continuous, whereas the Huggins bands contain a spectrum of diffuse lines. The Chappuis band is weaker than the Hartley and Huggins bands, but its importance is underscored by its overlap with the peak of the SW spectrum, which leads to rapid photodissociation of O_3 in sunlight—even at the surface. This property couples ozone in the stratosphere to snow and cloud cover in the troposphere, which sharply increase the underlying reflectivity and SW flux.

Absorption and therefore radiative heating follow from the cross sections of individual species weighted by their respective mixing ratios (Fig. 8.15). Absorption at wavelengths longer than 200 nm is dominated by photodissociation

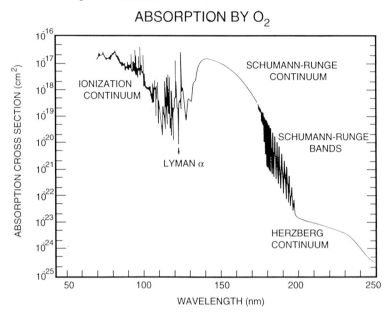

Figure 8.12   Absorption cross section as a function of wavelength for molecular oxygen. After Brasseur and Solomon (1986). Reprinted by permission of Kluwer Academic Publishers.

of O_3, whereas photodissociation of O_2 dominates at shorter wavelengths. Owing to the inverse relationship between wavelength and penetration altitude (Fig. 8.3), these bands influence different levels. Absorption by O_3 in the Hartley and Huggins bands at $\lambda > 200$ nm dominates in the stratosphere and mesosphere, where it provides the primary source of heating. Ozone absorption in the Chappuis band at $\lambda > 400$ nm becomes important below 25 km. On the other hand, absorption by O_2 at $\lambda < 200$ nm prevails only only above 60 km. Despite its secondary contribution to absorption, photodissociation of O_2 plays a key role in the energetics of the stratosphere and mesosphere because it produces atomic oxygen, which supports ozone formation (1.27). In addition to molecular oxygen and ozone, water vapor, methane, nitrous oxide, and CFCs also absorb in the UV through photodissociation.

8.3.2   Line Broadening

Absorption lines in Figs. 8.11 and 8.13 are not truly discrete; rather, they occupy finite bands of wavelength due to practical considerations surrounding molecular absorption and emission. The spectral width of an absorbing line is described in terms of a shape factor. The absorption cross section at frequency ν is expressed

$$\sigma_{a\nu} = Sf(\nu - \nu_0), \tag{8.30.1}$$

SCHUMANN-RUNGE BANDS

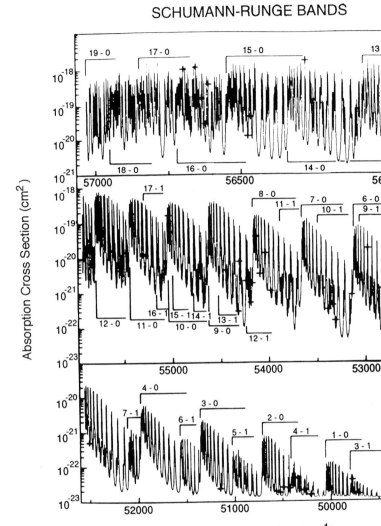

Figure 8.13 Absorption cross section in the Schumann–Runge absorption bands of molecular oxygen. Shown at high spectral resolution. After Kocharts (1971). Reprinted by permission of Kluwer Academic Publishers.

where

$$S = \int \sigma_\nu d\nu \qquad (8.30.2)$$

is the *line strength*, ν_0 is the line center, and the shape factor f accounts for line broadening.

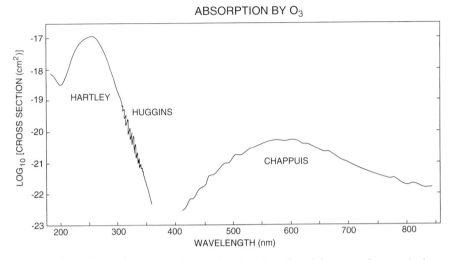

Figure 8.14   Absorption cross section as a function of wavelength for ozone. Sources: Andrews *et al.* (1987) and WMO (1986).

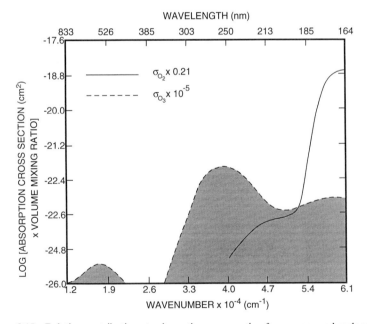

Figure 8.15   Relative contributions to absorption cross section from ozone and molecular oxygen. Adapted from Brasseur and Solomon (1986). Reprinted by permission of Kluwer Academic Publishers.

A fundamental source of line width is *natural broadening*, which results from the finite lifetime of excited states. Perturbations to the radiation stream then introduce a natural width that has the *Lorentz line shape*

$$f_L(\nu - \nu_0) = \frac{\alpha_L}{\pi(\nu - \nu_0)^2 + \alpha_L^2}, \qquad (8.31.1)$$

where

$$\alpha_L = (2\pi \bar{t})^{-1} \qquad (8.31.2)$$

is the half-width at half-power of the line (Fig. 8.16) and $\bar{t}$ is the mean lifetime of the excited state. For vibrational and rotational transitions in the IR, natural broadening is insignificant compared to other sources of line width. The two most important stem from the random motion of molecules and collisions between them.

Collisional or *pressure broadening* results from perturbations to the absorbing or emitting molecules, which are introduced through encounters with other molecules and which destroy the phase coherence of radiation. The spectral width introduced by collision is modeled in the Lorentz line shape (8.31.1), but with the collisional half-width

$$\alpha_c = \alpha_0 \left(\frac{p}{p_0}\right) \left(\frac{T_0}{T}\right)^{\frac{1}{2}}, \qquad (8.32)$$

which, from kinetic theory, is inversely proportional to the mean time between collisions and where $\alpha_0 \cong 0.1 \text{ cm}^{-1}$ is the half-width at standard temperature and pressure T_0, p_0.

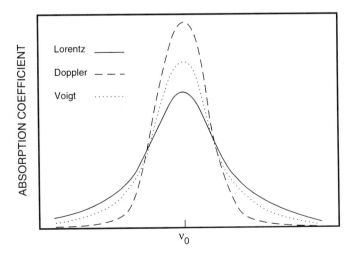

Figure 8.16 Lorentz, Doppler, and Voigt shape factors describing finite spectral line width. Reconstructed from Andrews *et al.* (1987).

Molecular motion v along the line of sight introduces another source of line width. *Doppler broadening* follows from the frequency shift

$$\nu = \nu_0 \left(1 \pm \frac{v}{c}\right) \tag{8.33.1}$$

of individual molecules undergoing emission or absorption. In combination with the Boltzmann probability distribution

$$p(v) = \sqrt{\frac{m}{2\pi KT}} \exp\left(-\frac{mv^2}{2KT}\right), \tag{8.33.2}$$

where m is the molecular mass, (8.33.1) leads to the shape factor

$$f_D(\nu - \nu_0) = \frac{1}{\alpha_D\sqrt{\pi}} \exp\left[-\left(\frac{\nu - \nu_0}{\alpha_D}\right)^2\right], \tag{8.34.2}$$

where

$$\alpha_D = \frac{\nu_0}{c}\sqrt{\frac{2KT}{m}} \tag{8.34.2}$$

is the Doppler half-width divided by a factor of $\sqrt{\ln 2}$. The Doppler line shape is illustrated in Fig. 8.16 along with the collisional line shape for $\alpha_D = \alpha_c$. Unlike the collisional half-width, α_D does not depend on pressure.

Below 30 km, the pressure dependence in α_c makes collisional broadening dominant for IR bands of CO_2 and H_2O. At higher altitudes, Doppler and natural broadening become significant at wavelengths in the visible and UV. Natural broadening of the Schumann–Runge bands of O_2 becomes comparable to collisional broadening in the upper stratosphere and mesosphere. Those two sources of line width are treated jointly in the *Voigt line shape*, which is superposed in Fig. 8.16. When they are comparable, the Doppler profile is dominant near the line center, whereas the Lorentz profile prevails in the wings of the line. The Voigt profile gives a line shape between the two.

8.4   Radiative Transfer in a Plane Parallel Atmosphere

In a stratified atmosphere, properties vary sharply with height. It is therefore convenient to treat radiative transfer within the framework of a *plane parallel atmosphere*, in which

- Curvature associated with sphericity of the Earth is ignored.
- The medium is regarded as horizontally homogeneous and the radiation field horizontally isotropic.

Then, along a slant path ds (Fig. 8.17), a pencil of radiation inclined from the vertical at a zenith angle θ traverses an atmospheric slab of thickness

$$dz = \mu ds, \tag{8.35.1}$$

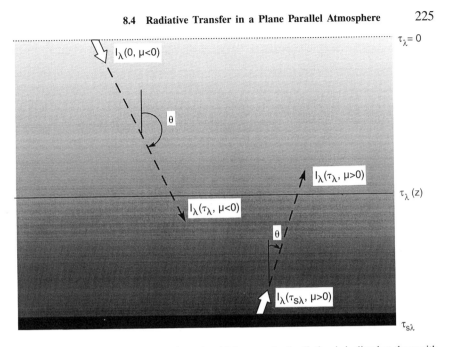

Figure 8.17   Plane parallel atmosphere, in which a pencil of radiation is inclined at the zenith angle $\theta = \cos^{-1} \mu$. Elevation is measured by the optical depth for a given wavelength τ_λ, which increases downward from zero at the top of the atmosphere to a surface value of $\tau_{s\lambda}$.

where

$$\mu = \cos \theta. \tag{8.35.2}$$

Introducing the *optical depth*

$$\tau_\lambda = \int_z^\infty \rho k_\lambda dz'$$

$$= \frac{1}{g} \int_0^p k_\lambda dp' = \mu \chi_\lambda, \tag{8.36}$$

which is measured downward from the top of the atmosphere, transforms the budget for a pencil of radiation (8.28) into

$$\mu \frac{dI_\lambda}{d\tau_\lambda} = I_\lambda - J_\lambda, \tag{8.37}$$

where $I_\lambda = I_\lambda(\tau_\lambda, \mu)$ and $J_\lambda = J_\lambda(\tau_\lambda, \mu)$. Proceeding as in the treatment of (8.28) leads to a formal expression for upwelling radiation $(0 < \mu \leq 1)$ in terms of that at the surface ($\tau_\lambda = \tau_{s\lambda}$),

$$I_\lambda(z, \mu) = I_\lambda(\tau_{s\lambda}, \mu) \exp\left[\frac{\tau_\lambda(z) - \tau_{s\lambda}}{\mu}\right]$$

$$+ \int_{\tau_\lambda(z)}^{\tau_{s\lambda}} J_\lambda(\tau'_\lambda, \mu) \exp\left[\frac{\tau_\lambda(z) - \tau'_\lambda}{\mu}\right] \frac{d\tau'_\lambda}{\mu}$$

$$0 < \mu \le 1, \tag{8.38.1}$$

and for downwelling radiation ($-1 \le \mu < 0$) in terms of that at the top of the atmosphere ($\tau_\lambda = 0$),

$$I_\lambda(z, \mu) = I_\lambda(0, \mu) \exp\left[\frac{\tau_\lambda(z)}{\mu}\right]$$

$$- \int_0^{\tau_\lambda(z)} J_\lambda(\tau'_\lambda, \mu) \exp\left[\frac{\tau_\lambda(z) - \tau'_\lambda}{\mu}\right] \frac{d\tau'_\lambda}{\mu}$$

$$-1 \le \mu < 0. \tag{8.38.2}$$

As before, the first terms on the right-hand sides of (8.38) describe the extinction of radiation incident at the boundaries, whereas the second terms account for the cumulative emission, absorption, and scattering into the pencil between the boundaries and the height z. If I_λ is known at the upper and lower boundaries and if J_λ can be specified in terms of known properties, (8.38) determines the radiation field throughout the atmosphere. Alternatively, if J_λ, ρ, and k can all be specified in terms of temperature [e.g., in the absence of scattering (8.18) and under hydrostatic equilibrium (1.17)], then (8.38) determines the thermal structure of the atmosphere. The expressions for the radiation field then provide integral equations for the temperature distribution $T(z)$. Their solution defines the *radiative–equilibrium thermal structure*, which corresponds to a balance between radiative components of the energy budget.

Horizontal homogeneity and isotropy make the upwelling and downwelling fluxes

$$F_\lambda^\uparrow(\tau_\lambda) = 2\pi \int_0^1 I_\lambda(\tau_\lambda, \mu) \mu \, d\mu$$

$$\tag{8.39}$$

$$F_\lambda^\downarrow(\tau_\lambda) = -2\pi \int_{-1}^0 I_\lambda(\tau_\lambda, \mu) \mu \, d\mu$$

in (8.1) and (8.3) the essential descriptors of radiative transfer in a plane parallel atmosphere. By integrating (8.38) over zenith angle and incorporating

(8.4), we obtain

$$F_\lambda^\uparrow(\tau_\lambda) = 2\pi \int_0^1 I_\lambda(\tau_{s\lambda}, \mu) \exp\left(\frac{\tau_\lambda - \tau_{s\lambda}}{\mu}\right) \mu d\mu$$

$$+ 2\pi \int_{\tau_\lambda}^{\tau_{s\lambda}} \int_0^1 J_\lambda(\tau_\lambda', \mu) \exp\left(\frac{\tau_\lambda - \tau_\lambda'}{\mu}\right) d\mu d\tau_\lambda', \quad (8.40.1)$$

$$F_\lambda^\downarrow(\tau_\lambda) = -2\pi \int_{-1}^0 I_\lambda(0, \mu) \exp\left(\frac{\tau_\lambda}{\mu}\right) \mu d\mu$$

$$+ 2\pi \int_0^{\tau_\lambda} \int_{-1}^0 J_\lambda(\tau_\lambda', \mu) \exp\left(\frac{\tau_\lambda - \tau_\lambda'}{\mu}\right) d\mu d\tau_\lambda'. \quad (8.40.2)$$

For LW radiation and in the absence of scattering, the source function is just $B_\lambda(T)$. Upwelling radiation at the surface, which is taken to be black, is then given by

$$I_\lambda(\tau_{s\lambda}, \mu) = B_\lambda(T_s) \qquad 0 < \mu \le 1. \quad (8.41.1)$$

Likewise, downwelling LW radiation at the top of the atmosphere must vanish, so

$$I_\lambda(0, \mu) = 0 \qquad -1 \le \mu < 0. \quad (8.41.2)$$

The upwelling and downwelling fluxes of LW radiation can then be written

$$F_\lambda^\uparrow(\tau_\lambda) = 2\pi B_\lambda(T_s) \int_0^1 \exp\left(\frac{\tau_\lambda - \tau_{s\lambda}}{\mu}\right) \mu d\mu$$

$$+ 2 \int_{\tau_\lambda}^{\tau_{s\lambda}} \int_0^1 \pi B_\lambda[T(\tau_\lambda')] \exp\left(\frac{\tau_\lambda - \tau_\lambda'}{\mu}\right) d\mu d\tau_\lambda', \quad (8.42.1)$$

$$F_\lambda^\downarrow(\tau_\lambda) = 2 \int_0^{\tau_\lambda} \int_0^1 \pi B_\lambda[T(\tau_\lambda')] \exp\left(\frac{\tau_\lambda' - \tau_\lambda}{\mu}\right) d\mu d\tau_\lambda'. \quad (8.42.2)$$

The integrations over zenith angle may be expressed in terms of the exponential integral (Abramowitz and Stegun, 1972)

$$E_n(\tau) = \int_1^\infty \frac{e^{-x\tau}}{x^n} dx, \quad (8.43.1)$$

which satisfies

$$\frac{dE_n}{d\tau} = -E_{n-1}(\tau). \quad (8.43.2)$$

Letting $x = \mu^{-1}$ transforms (8.42) into

$$F_\lambda^\uparrow(\tau_\lambda) = 2\pi B_\lambda(T_s)E_3(\tau_{s\lambda} - \tau_\lambda)$$
$$+ 2\int_{\tau_\lambda}^{\tau_{s\lambda}} \pi B_\lambda[T(\tau_\lambda')]E_2(\tau_\lambda' - \tau_\lambda)d\tau_\lambda', \qquad (8.44.1)$$

$$F_\lambda^\downarrow(\tau_\lambda) = 2\int_0^{\tau_\lambda} \pi B_\lambda[T(\tau_\lambda')]E_2(\tau_\lambda - \tau_\lambda')d\tau_\lambda'. \qquad (8.44.2)$$

Then integrating over wavelength recovers the total upward and downward fluxes

$$F^\uparrow(\tau) = 2\int_0^\infty \pi B_\lambda(T_s)E_3(\tau_{s\lambda} - \tau_\lambda)d\lambda$$
$$+ 2\int_\tau^{\tau_s}\int_0^\infty \pi B_\lambda[T(\tau_\lambda')]E_2(\tau_\lambda' - \tau_\lambda)d\lambda d\tau', \qquad (8.45.1)$$

$$F^\downarrow(\tau) = 2\int_0^\tau \int_0^\infty \pi B_\lambda[T(\tau_\lambda')]E_2(\tau_\lambda - \tau_\lambda')d\lambda d\tau', \qquad (8.45.2)$$

which constitute the formal description of LW radiative transfer in a nonscattering atmosphere.

8.4.1 Transmission Function

Implementing (8.45) is complicated by the rapid variation with wavenumber of absorption cross section, which involves thousands of vibrational and rotational lines in the IR. In place of line-by-line calculations, radiative transfer is calculated with the aid of band-averaged transmission functions that embody the general characteristics of the absorption spectrum in designated ranges of wavelength.

Integrating (8.44) over a frequency interval $\Delta\nu$ yields the band-averaged fluxes

$$F_{\bar\nu}^\uparrow(\tau_{\bar\nu}) = 2\pi B_{\bar\nu}(T_s)\int_{\Delta\nu} E_3(\tau_{s\nu} - \tau_\nu)\frac{d\nu}{\Delta\nu}$$
$$+ 2\int_{\tau_{\bar\nu}}^{\tau_{s\nu}} \pi B_{\bar\nu}[T(\tau_{\bar\nu}')]\int_{\Delta\nu} E_2(\tau_\nu' - \tau_\nu)\frac{d\nu}{\Delta\nu}d\tau', \qquad (8.46.1)$$

$$F_{\bar\nu}^\downarrow(\tau_{\bar\nu}) = 2\int_0^{\tau_{\bar\nu}} \pi B_{\bar\nu}[T(\tau_{\bar\nu}')]\int_{\Delta\nu} E_2(\tau_\nu - \tau_\nu')\frac{d\nu}{\Delta\nu}d\tau_\nu', \qquad (8.46.2)$$

where $B_{\bar\nu}$ can be used for the band-averaged emission because the Planck function varies smoothly with frequency. The *band transmissivity* or *transmission function* is defined as

$$\mathcal{T}_{\bar\nu}(\tau_{\bar\nu}, \mu) = \frac{1}{\Delta\nu}\int_{\Delta\nu} e^{-\frac{\tau_\nu}{\mu}}d\nu. \qquad (8.47)$$

Decreasing downward, $\mathscr{T}_{\bar{\nu}}$ represents the fractional intensity in the band and in direction μ that reaches optical depth $\tau_{\bar{\nu}}$. The *band absorptivity* is then

$$a_{\bar{\nu}} = 1 - \mathscr{T}_{\bar{\nu}}, \qquad (8.48)$$

analogous to (8.11).

Averaging (8.47) over zenith angle defines the *diffuse flux transmission function* for the band

$$\mathscr{T}_{\bar{\nu}}(\tau_{\bar{\nu}}) = \frac{\int_0^1 \mathscr{T}_{\bar{\nu}}(\tau_{\bar{\nu}}, \mu)\mu\,d\mu}{\int_0^1 \mu\,d\mu}$$

$$= 2\int_0^1 \mathscr{T}_{\bar{\nu}}(\tau_{\bar{\nu}}, \mu)\mu\,d\mu, \qquad (8.49)$$

which can be written in terms of the exponential integral as

$$\mathscr{T}_{\bar{\nu}}(\tau_{\bar{\nu}}) = 2\int_{\Delta\nu} E_3(\tau_\nu)\frac{d\nu}{\Delta\nu}. \qquad (8.50)$$

With the aid of (8.43.2), the band-averaged fluxes can then be expressed in terms of the flux transmission function

$$F_{\bar{\nu}}^{\uparrow}(\tau_{\bar{\nu}}) = \pi B_{\bar{\nu}}(T_s)\mathscr{T}_{\bar{\nu}}(\tau_{s\bar{\nu}} - \tau_{\bar{\nu}})$$
$$- \int_{\tau_{\bar{\nu}}}^{\tau_{s\bar{\nu}}} \pi B_{\bar{\nu}}[T(\tau_\nu')]\frac{d\mathscr{T}_{\bar{\nu}}(\tau_\nu' - \tau_{\bar{\nu}})}{d\tau_\nu'}d\tau_\nu', \qquad (8.51.1)$$

$$F_{\bar{\nu}}^{\downarrow}(\tau_{\bar{\nu}}) = +\int_0^{\tau_{\bar{\nu}}} \pi B_{\bar{\nu}}[T(\tau_\nu')]\frac{d\mathscr{T}_{\bar{\nu}}(\tau_{\bar{\nu}} - \tau_\nu')}{d\tau_\nu'}d\tau_\nu'. \qquad (8.51.2)$$

In principle, absorption characteristics of the band $\Delta\nu$ determine the corresponding transmission function and the vertical fluxes through (8.51). Collecting contributions from all absorbing bands then obtains the total upwelling and downwelling fluxes. However, in practice, even individual absorption bands are so complex as to make direct calculations impractical (refer to Figs. 8.11, and 8.13). Instead, the transmission function is evaluated with the aid of band models that capture the salient features of the absorption spectrum in terms of properties like mean line strength, line spacing, and line width. The regular band model of Elsasser (1938) treats a series of evenly spaced Lorentz lines, like those comprising the CO_2 14-μm band (Fig. 8.11a). The Goody (1952) random model treats a spectrum of lines that are randomly spaced, as is typical of the 6.3-μm band of water vapor (Fig. 8.11b). See Liou (1980) for a development of these models.

The transmission function, although more complex than the exponential attenuation of monochromatic radiation (8.9), is far simpler than a line-by-line calculation over the band. It is instructive to consider how the transmission function, or alternatively the absorptivity (8.48), behaves with optical depth for an individual absorption line. Below a certain height, the absorptivity of a line

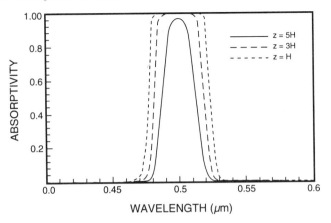

Figure 8.18 Absorptivity contributed by an individual absorption line at 0.5 μm for an absorber with a uniform mixing ratio and for the layer extending from the top of the atmosphere to a height z of 5 scale heights, 3 scale heights, and 1 scale height. As the layer extends downward, where absorber concentration increases exponentially, the line becomes saturated, first at its center but eventually over a widening range of wavelength.

may be represented in terms of the Lorentz profile (8.31.1). As τ_ν increases, so does a_ν (Fig. 8.18). Eventually, the absorptivity reaches unity near the line center, at which point the corresponding frequencies are fully absorbed. The line is then said to be *saturated*, which corresponds to the limit of *strong absorption*.

Subsequent absorption can occur only in the wings of the line, which are pushed out by the ever-widening region where the line is saturated. In a band containing many lines, saturation leads to the absorption spectrum filling in between discrete features. This occurs first for the narrowly separated vibration–rotation transitions (e.g., Figs. 8.11, and 8.13). Over sufficient optical depth, those clusters merge to produce continuous bands of absorption, such as those appearing in Figs. 8.1 and 8.9. The strong absorption limit applies to bands that dominate vertical energy transfer in the atmosphere. It is the basis for a powerful approximation that reduces the three-dimensional description of diffuse LW radiation to a one-dimensional description.

8.4.2 Two-Stream Approximation

Embodied in the flux transmission function (8.50) is an integral over zenith angle, which collects contributions from diffuse radiation, and another integral over frequency. Because $E_3(\tau_\nu)$ depends on frequency, the two integrations are related. In the strong absorption limit, the integral over zenith angle can be approximated fairly accurately by incorporating the spectral character of $\mathscr{T}_\nu$.

Integrals on the right-hand sides of (8.51) involve the term

$$\frac{d\mathcal{F}_{\bar{\nu}}}{d\tau'_{\nu}}(\Delta\tau_{\nu})d\tau'_{\nu} = 2\int_{\Delta\nu} E_2(\Delta\tau_{\nu})\rho k_{\nu}dz'\frac{d\nu}{\Delta\nu}, \qquad (8.52.1)$$

where

$$\Delta\tau_{\nu} = |\tau'_{\nu} - \tau_{\nu}|. \qquad (8.52.2)$$

If the band described by $\mathcal{F}_{\bar{\nu}}$ is saturated, the factors k_{ν} and $E_2(\Delta\tau_{\nu})$ have reciprocal behavior in frequency, which sharply limits their product. As depicted in Fig. 8.19, $E_2(\Delta\tau_{\nu})$ vanishes near the center of the band, for which the medium is optically thick, and approaches unity far from the band center, for which the medium is optically thin. A transition between these extremes occurs at frequencies corresponding to an optical thickness $\Delta\tau_{\nu}$ of unity. On the other hand, k_{ν} is large near the center of the band and vanishes in the wings.

Due to the reciprocal nature of $E_2(\Delta\tau_{\nu})$ and k_{ν}, contributions to (8.52.1) are concentrated in a fairly narrow spectral interval that corresponds to an optical thickness of about 1. Consequently, most of the energy exchange between two levels in (8.51) occurs along paths having an optical thickness of unity. Along paths having optical thicknesses much greater than 1, the medium is optically thick. Radiation is then absorbed before it can traverse the levels, which therefore experience little interaction. Along paths having optical thicknesses much smaller than 1, the medium is optically thin. Radiation then traverses the levels with little attenuation, so the levels again experience little interaction.

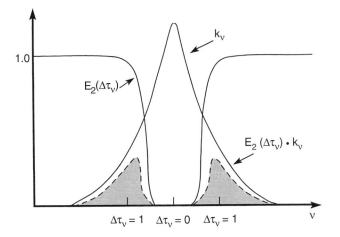

Figure 8.19   Factors $E_2(\Delta\tau_{\nu})$ and k_{ν} appearing in the integral over zenith angle that is implicit to the band-averaged upwelling and downwelling fluxes (8.51). Due to their reciprocal nature, the factors' product is concentrated at frequencies corresponding to an optical depth of $\Delta\tau_{\nu} \cong 1$, which is associated with an effective zenith angle of $\bar{\mu}^{-1} \cong \frac{5}{3}$. Adapted from Andrews *et al.* (1987).

For any specified $\Delta\tau_\nu$, an effective inclination $\bar{\mu}$ may be found such that the exponential integral over zenith angle is given by

$$2E_3(\Delta\tau_\nu) = \exp\left(-\frac{\Delta\tau_\nu}{\bar{\mu}}\right). \tag{8.53.1}$$

Because contributions to (8.51) are restricted to $\Delta\tau_\nu \cong 1$, the parameter $\bar{\mu}^{-1}$, which is referred to as the *diffusivity factor*, is likewise restricted to a narrow range of values. For $\Delta\tau_\nu = 1$, the diffusivity factor assumes the value

$$\bar{\mu}^{-1} \cong \frac{5}{3}. \tag{8.53.2}$$

The implication of this analysis is that the zenith angle dependence in (8.51) can be eliminated in favor of an effective inclination $\bar{\mu}$, which reduces the multidimensional description to integrals over optical depth alone. Diffuse transmission which is embodied in $\mathcal{T}_{\bar{\nu}}(\tau_{\bar{\nu}})$ is then equivalent to that of a collimated beam inclined at a zenith angle of 53°

$$\mathcal{T}_{\bar{\nu}}(\tau_{\bar{\nu}}) \cong \mathcal{T}_{\bar{\nu}}(\tau_{\bar{\nu}}, \bar{\mu}) \tag{8.53.3}$$

or, alternatively, inclined at zero zenith angle but through an optical thickness expanded by a factor of $\frac{5}{3}$.

Known as the *exponential kernel approximation*, this simplification follows from the fact that radiative exchange in strong bands is dominated by spectral intervals in which $\Delta\tau_\nu = 1$. For this reason, most of the LW radiation emitted by the earth's surface is captured in the lower troposphere, which is made optically thick by strong absorption bands of water vapor and carbon dioxide. Likewise, incident SW radiation in particular wavelengths of UV is absorbed in the stratosphere over a limited range of altitude (Fig. 8.3), which is made optically thick by photodissociation at those wavelengths.[2] The development leading to (8.53) is one of several so-called *two-stream approximations* that eliminate the zenith angle dependence in $F^\uparrow$ and $F^\downarrow$. Applicable under fairly wide circumstances (e.g., in the presence of scattering), this formalism leads to diffusivity factors in the range $\frac{3}{2} \leq \bar{\mu}^{-1} \leq 2$, depending on the particular approximation adopted.

Because μ enters the equations jointly with τ_ν, the full description of radiative transfer in a plane parallel atmosphere can be reduced to a vertical description by taking $\mu = \pm\bar{\mu}$ for upwelling and downwelling radiation and introducing the transformation

$$\tau_\lambda^* = \bar{\mu}^{-1}\tau_\lambda, \tag{8.54.1}$$

$$F_\lambda^{\uparrow\downarrow}(\tau_\lambda^*) = \pi I_\lambda(\tau_\lambda, \pm\bar{\mu}), \tag{8.54.2}$$

$$J_\lambda^*(\tau_\lambda^*) = \pi J_\lambda(\tau_\lambda, \pm\bar{\mu}), \tag{8.54.3}$$

---

[2] For weak bands that are unsaturated near their centers, energy exchange may arrive from smaller values of $\Delta\tau_\nu$, which correspond to larger values of $\bar{\mu}^{-1}$.

which relies on hemispheric isotropy. Then integrating the radiative transfer equation (8.37) over upward and downward half-spaces yields for the budgets of upwelling and downwelling radiation

$$\frac{dF_\lambda^\uparrow}{d\tau_\lambda^*} = F_\lambda^\uparrow - J_\lambda^*, \tag{8.55.1}$$

$$-\frac{dF_\lambda^\downarrow}{d\tau_\lambda^*} = F_\lambda^\downarrow - J_\lambda^*. \tag{8.55.2}$$

The foregoing development applies to LW radiation, which is inherently diffuse. SW radiation, which is parallel beam in the absence of scattering, involves only a single direction μ.

8.5 Thermal Equilibrium

8.5.1 Radiative Equilibrium in a Gray Atmosphere

We are now in a position to evaluate the thermal structure toward which radiative transfer drives an atmosphere. To do so, we consider a simple model of the earth's atmosphere: a *gray atmosphere* that is transparent to SW radiation but absorbs LW radiation with a constant absorption cross section which is independent of wavelength, temperature, and pressure. Consider a plane parallel gray atmosphere that is nonscattering, motionless, and in thermal equilibrium with incoming SW radiation and with a black underlying surface. Individual layers of the atmosphere then interact only through LW radiation.

For an incremental slab of thickness dz, the first law (2.22.2) implies

$$\rho c_p \frac{dT}{dt} - \frac{dp}{dt} = \rho \dot{q}$$

$$= -\frac{dF}{dz}, \tag{8.56.1}$$

where

$$F = F^\uparrow - F^\downarrow \tag{8.56.2}$$

is the net LW flux integrated over wavelength and $\dot{q}$ is the local rate at which heat is absorbed per unit mass or the *specific heating rate*. The convergence of LW flux on the right-hand side of (8.56.1) represents the local radiative heating rate per unit volume, which forces the atmosphere's thermal structure. In equilibrium, the left-hand side vanishes. Then the net flux must be independent of height

$$F = \text{const}, \tag{8.57}$$

which implies that radiative energy does not accumulate within individual layers.

The upwelling and downwelling components of F are governed by (8.55). Adding and subtracting yields the equivalent system

$$\frac{dF}{d\tau^*} = \bar{F} - 2B^*,\tag{8.58.1}$$

$$\frac{d\bar{F}}{d\tau^*} = F,\tag{8.58.2}$$

where

$$\bar{F} = F^\uparrow + F^\downarrow\tag{8.58.3}$$

represents the total flux emanating from an incremental slab and all quantities have been integrated over wavelength. With (8.57), (8.58.2) gives

$$\bar{F} = F\tau^* + c\tag{8.59.1}$$

and (8.58.1) reduces to

$$\bar{F} = 2B^*.\tag{8.59.2}$$

Thus, the emission

$$B^*(\tau^*) = \frac{F}{2}\tau^* + B_0^*\tag{8.60}$$

increases linearly with optical depth from its value B_0^* at the top of the atmosphere (Fig. 8.20).

For thermal equilibrium of the atmosphere and surface collectively, the incident SW flux

$$F_0 = (1 - \mathscr{A})\bar{F}_s\tag{8.61}$$

(Sec. 1.4.1) must be balanced by the outgoing LW flux at the top of the atmosphere

$$F^\uparrow(0) = F_0.\tag{8.62}$$

Because the downwelling LW flux at the top of the atmosphere vanishes, $F(0)$ and $\bar{F}(0)$ each reduce to $F^\uparrow(0)$, so by (8.57)

$$F = F_0.\tag{8.63}$$

The net LW flux at any level equals the SW flux at the top of the atmosphere. Then incorporating (8.59.2) yields

$$B^*(\tau^*) = \frac{F_0}{2}(\tau^* + 1),\tag{8.64}$$

which is shown in Fig. 8.20 along with $F^\uparrow$ and $F^\downarrow$. Under radiative equilibrium, the upwelling and downwelling fluxes differ by a constant, so the net heating at any level vanishes.

Consider next thermal equilibrium of the surface. Since it is black, the surface absorbs the incident SW flux in addition to the downwelling LW flux

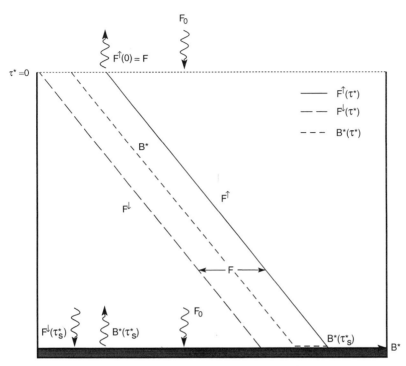

Figure 8.20   Upwelling and downwelling fluxes and emission in a gray atmosphere that is in radiative equilibrium with an incident SW flux F_0 and a black underlying surface. *Note:* the emission profile is discontinuous at the surface.

emitted by the atmosphere $F^{\downarrow}(\tau_s^*)$. Together, these must balance the upwelling flux emitted by the surface

$$B^*(T_s) = F_0 + F^{\downarrow}(\tau_s^*). \tag{8.65}$$

Subtracting (8.59.2) and (8.63) gives the downwelling LW flux emitted by the atmosphere

$$F^{\downarrow}(\tau_s^*) = B^*(\tau_s^*) - \frac{F_0}{2}, \tag{8.66}$$

which, when combined with (8.65), implies

$$B^*(T_s) = B^*(\tau_s^*) + \frac{F_0}{2}. \tag{8.67}$$

According to (8.67), the temperature predicted by radiative equilibrium is discontinuous at the surface, the ground being warmer than the overlying air.

The emission profile (8.64) determines the radiative equilibrium temperature through

$$B^*(\tau^*) = \pi B[T(\tau^*)]$$
$$= \sigma T^4. \tag{8.68}$$

Because the atmosphere is hydrostatic, density decreases with height approximately exponentially (1.17) and by (8.36) so does τ^*. Further, most of the atmosphere's opacity follows from water vapor and clouds (Fig. 8.1), which are concentrated in the lowest levels. Therefore, the linear variation with optical depth of B^* translates into a steep decrease with height of temperature. At the surface, the lapse rate is infinite because the radiative equilibrium temperature is discontinuous there.

Taking the optical depth to vary with height exponentially

$$\tau = \tau_s e^{-\frac{z}{h}}, \tag{8.69}$$

where $h = 2$ km is symbolic of the absolute concentration of water vapor (Fig. 1.14), and $F_0 = 240$ W m^{-2} results in the temperature profiles (solid lines) shown in Fig. 8.21. For given $\tau_s > 0$, the surface temperature exceeds that in the absence of an atmosphere ($\tau_s = 0$) and increases with the atmosphere's optical depth. These features are manifestations of the greenhouse effect. Radiative equilibrium temperature is maximum at the ground, where it varies sharply with height and jumps discontinuously to the surface temperature T_s. As $z \to \infty$, the radiative equilibrium temperature approaches a finite limiting value $\left(\frac{F_0}{2\sigma}\right)^{\frac{1}{4}} \cong 215$ K, which is called the *skin temperature*.

In a neighborhood of the surface, temperature decreases with height more rapidly than the saturated adiabatic lapse rate $\Gamma_s \cong 6.5$ K km^{-1} (dotted lines), which characterizes moist neutral stability. Thus, each of the radiative equilibrium temperature profiles in Fig. 8.21 is hydrostatically unstable in its lowest levels. For an optical depth of $\tau_s = 4$, the radiative equilibrium profile is conditionally unstable below 6 km, absolutely unstable below about 5 km, and has a lapse rate of some 36 K km^{-1} at the ground.

8.5.2 Radiative–Convective Equilibrium

If air motion is accounted for, convection develops spontaneously to neutralize the unstable stratification introduced by radiative transfer. The *radiative–convective equilibrium* that ensues, which is superposed in Fig. 8.21 for $\tau_s = 4$ (dashed line), produces two layers of distinctly different character. In the layer below a height z_T, which constitutes the troposphere in this framework, thermal structure is controlled by convective overturning that is driven by radiative heating and its continual destabilization of the stratification. Since it operates on a timescale much shorter than radiative transfer, convective overturning

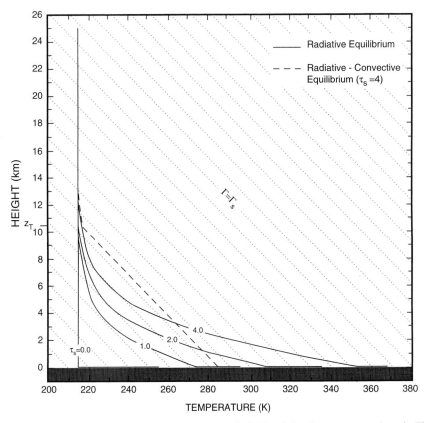

Figure 8.21 Radiative equilibrium temperature (solid lines) for the gray atmosphere in Fig. 8.20, with a profile of optical depth representative of water vapor (8.69), presented for several atmospheric optical depths τ_s. Saturated adiabatic lapse rate (dotted lines) and radiative–convective equilibrium temperature for $\tau_s = 4$ (dashed line) superposed.

drives the stratification below z_T to neutral stability (Sec. 7.6.1), which, under saturated conditions, corresponds to a lapse rate of Γ_s. Heat supplied at the ground is carried upward and mixed vertically over the convective layer to maintain a uniform profile of equivalent potential temperature θ_e. In the layer above z_T, which constitutes the stratosphere in this framework, the thermal structure remains close to radiative equilibrium because radiative transfer stabilizes the stratification at those levels. Since tropospheric air displaced vertically by convection cools along the saturated adiabat, it cannot penetrate appreciably above the tropopause height z_T, where the saturated adiabatic and radiative equilibrium profiles cross.

The particular form of radiative–convective equilibrium assumed depends on the parameterization of ingredients like clouds and convective heat flux at the surface, which dictate the height of the tropopause. For instance, taking the height of the convective layer equal to the maximum height of instability under radiative equilibrium predicts one tropopause height. On the other hand, taking the air temperature at $z = 0$ equal to the surface temperature under radiative equilibrium leads to a different tropopause height.

An alternative that circumvents ambiguities surrounding convection and is consistent with the radiative framework behind (8.64) requires the convective layer to supply the same upward radiation flux at the tropopause as that under radiative equilibrium (Goody and Yung, 1989). Tantamount to presuming the stratosphere is unaffected by vertical motions below, this formalism provides a self-consistent (albeit limited) rationale for determining radiative–convective equilibrium. Integrating (8.55) as before leads to expressions valid under arbitrary conditions for the upward and downward radiation fluxes

$$F^\uparrow(\tau^*) = B(T_s)\exp(\tau^* - \tau_s^*) - \int_{\tau_s^*}^{\tau^*} \exp(\tau^* - \tau')B^*[T(\tau')]d\tau' \quad (8.70.1)$$

$$F^\downarrow(\tau^*) = \int_0^{\tau^*} \exp(\tau' - \tau^*)B(\tau')d\tau', \quad (8.70.2)$$

where convection has been presumed efficient enough to maintain the surface at the same temperature as the overlying air. Then equating (8.70.1) to the upward flux under radiative equilibrium

$$F^\uparrow(\tau^*) = \frac{F_0}{2}(\tau^* + 2), \quad (8.71)$$

which follows in the same manner as (8.66), obtains a surface temperature of $T_s \cong 285$ K for $\tau_s = 4$. Superposed in Fig. 8.21, the radiative–convective equilibrium temperature that follows (dashed line) is cooler in the lower troposphere and warmer in the upper troposphere than the radiative equilibrium profile and has a tropopause at about 10.5 km, in qualitative agreement with observed thermal structure. For a characteristic extinction scale of $h = 2$ km, z_T lies several optical depths above the ground, so the stratosphere is close to the atmosphere's skin temperature.

The corresponding upward and downward fluxes (dashed lines) are shown as functions of height in Fig. 8.22a along with those under radiative equilibrium (solid lines). Reduced temperature in the lower troposphere leads to smaller upward and downward fluxes that merge with the radiative equilibrium profiles above z_T. To satisfy overall equilibrium, $F^\downarrow$ approaches 0 and $F^\uparrow$ approaches 240 W m^{-2} as $z \to \infty$. Whereas the radiative equilibrium fluxes preserve a constant difference, the net flux under radiative–convective equilibrium varies with altitude below z_T, which implies nonzero radiative heating in the troposphere. The specific heating rate $\dot{q}/c_p$ (Fig. 8.22b), which has dimensions of degrees per day, is negative throughout the troposphere and

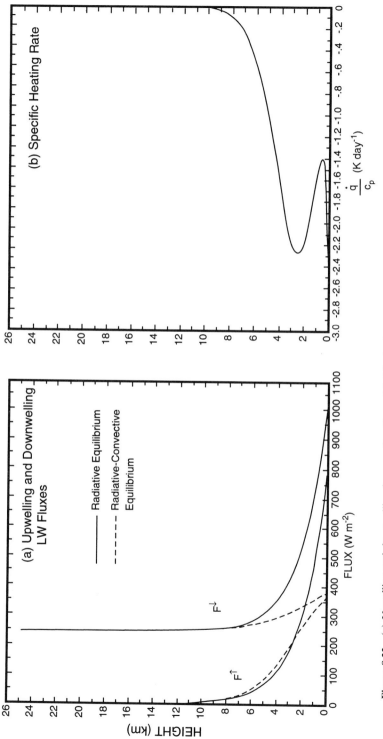

Figure 8.22 (a) Upwelling and downwelling fluxes as functions of height in the gray atmosphere in Fig. 8.20 for radiative equilibrium (solid lines) and radiative–convective equilibrium (dashed lines). (b) Specific heating rate under radiative–convective equilibrium.

approaches zero inside the region under radiative equilibrium. Thus, radiative transfer cools the troposphere, which is heated by convective transfer from the surface. Reaching a maximum of about 2 K day^{-1}, the cooling rate in Fig. 8.22b is in qualitative agreement with detailed calculations of radiative cooling based on observed behavior (compare Fig. 8.24a).

More sophisticated calculations of radiative-convective equilibrium incorporate distributions of optically active species, mean cloud cover, and direct absorption of SW radiation. Figure 8.23 shows the results of a calculation that accounts for radiative transfer by water vapor, carbon dioxide, and ozone. Resembling the idealized behavior in Fig. 8.21, the radiative equilibrium temperature (solid line) drops sharply from about 340 K at the surface to a tropopause value of about 185 K near 10 km. Most of the troposphere is absolutely unstable, whereas temperature increases in the stratosphere due to absorption of SW radiation by ozone. Radiative–convective equilibrium (dashed line) produces a thermal structure which also resembles that of the gray atmosphere, with temperatures colder in the lower troposphere and warmer in the upper troposphere than those under pure radiative equilibrium.

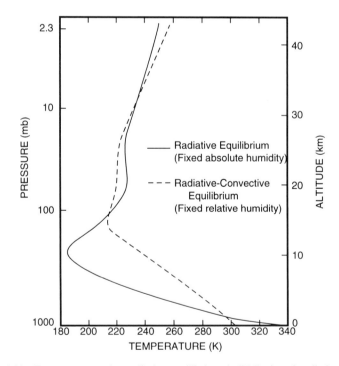

Figure 8.23 Temperature under radiative equilibrium (solid line) and radiative–convective equilibrium (dashed line) from calculations that include mean distributions of water vapor, carbon dioxide, and ozone. Adapted from Manabe and Wetherald (1967).

8.5.3 Radiative Heating

Knowledge of the mean distributions of temperature and optically active constituents allows the LW heating to be inferred via (8.56) from the upwelling and downwelling fluxes. For an individual band, the specific heating rate can be expressed

$$
\begin{aligned}
\frac{\dot{q}}{c_p} &= -\frac{1}{\rho c_p}\frac{dF}{dz^*} \\
&= \frac{k}{c_p}\frac{dF}{d\tau^*},
\end{aligned}
\tag{8.72}
$$

where (8.53) has been incorporated into height and the frequency dependence is implicit. Differentiating and subtracting (8.51) gives

$$
\frac{dF}{d\tau^*} = +B^*(T_s)\frac{d\mathcal{T}(\tau_s^* - \tau^*)}{d\tau^*} - \int_{\tau^*}^{\tau_s^*} B^*[T(\tau')]\frac{d^2\mathcal{T}(\tau' - \tau^*)}{d\tau'd\tau^*}\,d\tau'
$$
$$
- \int_0^{\tau^*} B^*[T(\tau')]\frac{d^2\mathcal{T}(\tau^* - \tau')}{d\tau'd\tau^*}\,d\tau' + 2B^*[T(\tau^*)]\frac{d\mathcal{T}(0)}{d\tau^*},
$$

from which the specific heating rate can be arranged into the form

$$
\begin{aligned}
\frac{\dot{q}(\tau^*)}{c_p} = \frac{k(\tau^*)}{c_p}\Bigg\{ &B^*[T(\tau^*)]\frac{d\mathcal{T}(\tau^* - 0)}{d\tau^*} + \Big[B^*(T_s) - B^*(\tau^*)\Big]\frac{d\mathcal{T}(\tau_s^* - \tau^*)}{d\tau^*} \\
&- \int_{\tau^*}^{\tau_s^*}\Big[B^*(\tau') - B^*(\tau^*)\Big]\frac{d^2\mathcal{T}(\tau' - \tau^*)}{d\tau^* d\tau'}\,d\tau' \\
&- \int_0^{\tau^*}\Big[B^*(\tau') - B^*(\tau^*)\Big]\frac{d^2\mathcal{T}(\tau^* - \tau')}{d\tau^* d\tau'}\,d\tau'\Bigg\}.
\end{aligned}
\tag{8.73}
$$

Equation (8.73) collects contributions to $\dot{q}/c_p$ from LW interactions between the level τ^* and its environment, with $d\mathcal{T}(\tau^* - \tau')/d\tau'$ serving as an influence function between that level and another τ'. The first term on the right-hand side describes interaction with space. Since $d\mathcal{T}/d\tau < 0$ (8.47), this term is always negative and represents "cooling to space." The second term, which describes interaction with the surface, is negative when the level τ^* is warmer than the ground. The last two terms are *exchange integrals* that collect contributions from levels τ' above and below τ^*. Because $d^2\mathcal{T}/d\tau d\tau' < 0$, those contributions represent cooling when the temperature at level τ^* exceeds that at level τ'.

Figure 8.24a illustrates the dominant contributions to LW heating. Except for ozone, the primary LW absorbers cool the atmosphere. Water vapor dominates LW cooling in the troposphere, where it leads to a globally averaged

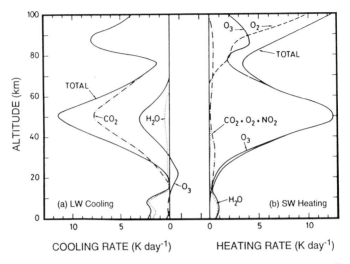

Figure 8.24   Global–mean profiles of (a) LW cooling and (b) SW heating. Contributions from individual radiatively active constituents also shown. After London (1980).

cooling rate of about $2\ \mathrm{K\,day}^{-1}$. In the stratosphere, the 15-μm band of CO_2 dominates LW cooling, which, together with the 9.6-μm band of O_3, produces a maximum cooling rate near the stratopause of some $12\ \mathrm{K\,day}^{-1}$. Because it lies within the atmospheric window, the 9.6-μm band of ozone also produces heating in the lower stratosphere by absorbing upwelling LW radiation from below.[3]

In contrast to LW radiation, SW radiation produces only heating because the atmosphere does not emit at those wavelengths. For a particular frequency, the specific heating rate is given by

$$\dot{q} = -k\mu_s \frac{dF}{d\tau},$$

where $\mu_s = -\mu > 0$ refers to the *solar zenith angle* and, as above, the frequency dependence is implicit. Incorporating (8.7) transforms this into

$$\dot{q} = -k\mu_s \frac{d}{d\tau}\left[I_0 \exp\left(-\frac{\tau}{\mu_s} \right) \right]$$

$$= kI_0 \exp\left(-\frac{\tau}{\mu_s} \right). \tag{8.74}$$

---

[3] Even though specific heating rates in the stratosphere are larger than those in the troposphere, volume heating rates $\rho\dot{q}$ are smaller by two orders of magnitude.

Then the volume heating rate follows as

$$\rho\dot{q} = \rho k I_0 \exp\left(-\frac{\tau}{\mu_s}\right)$$
$$= n\hat{k} I_0 \exp\left(-\frac{\tau}{\mu_s}\right), \qquad (8.75)$$

(8.6), where ρ and n refer to absorbers at the frequency under consideration.

Heating varies strongly with the absolute concentration n of absorber. Dominated by the variation of air density, the absorber concentration can be modeled as

$$\frac{n}{n_0} = e^{-\frac{z}{H}}.$$

Then the optical depth also varies exponentially and (8.75) becomes

$$\rho\dot{q} = n_0\hat{k} I_0 \exp\left[-\left(\tau_0 e^{-\frac{z}{H}} + \frac{z}{H}\right)\right], \qquad (8.76.1)$$

where

$$\tau_0 = \frac{n_0\hat{k}H}{\mu_s}. \qquad (8.76.2)$$

The volume heating rate (8.76) possesses a single maximum at the height

$$\frac{z_0}{H} = \ln\tau_0, \qquad (8.77)$$

which corresponds to a slant optical path from the top of the atmosphere of unity. Above that level, heating decreases exponentially with the air density, whereas it decreases even faster below due to attenuation from the first exponential term in (8.76.1). Therefore, SW heating is concentrated within about a scale height of the level z_0, which defines the *Chapman layer* for this frequency. Consequently, the penetration altitude in Fig. 8.3 for a particular wavelength also describes where most of the SW absorption and heating occur. As shown in Fig. 8.25, increased solar zenith angle reduces SW heating, because the vertical flux is diminished according to μ_s, and increases the altitude z_0, because it elongates the optical path length traversed by solar radiation.

Because the extinction cross section k for different wavelengths results from different species, which, in turn, have different vertical profiles, the full spectrum of SW radiation produces a series of Chapman layers that follow from the vertical distributions of optically active species. According to Fig. 8.3, energetic radiation at wavelengths shorter than 200 nm is absorbed in the upper mesosphere and thermosphere by O_2 in the Schumann–Runge bands and continuum. Wavelengths of 200 to 300 nm are absorbed in the mesosphere and stratosphere by O_3 in the Hartley band. Longer wavelengths penetrate to

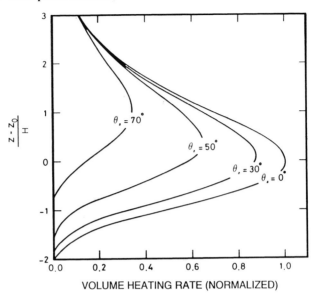

Figure 8.25 Volume heating rate due to absorption of SW radiation at solar zenith angle θ_s. Concentration of heating within a depth of order H defines the *Chapman layer* for a particular wavelength. After Banks and Kocharts (1973).

the surface, with partial absorption in the visible by ozone and in the near IR by water vapor and carbon dioxide.

Ozone is responsible for most of the SW absorption in the stratosphere and mesosphere (Fig. 8.24b). It produces maximum heating near the stratopause, even though the absolute concentration of ozone maximizes considerably lower (Fig. 1.17). In the global–mean, SW heating and LW cooling at these heights nearly cancel, indicating that the middle atmosphere is close to radiative equilibrium. In the troposphere, water vapor is the dominant SW absorber. Achieving a global–mean heating rate of about 1 K day^{-1} , SW absorption does not completely offset LW cooling. Thus, radiative transfer leads to net cooling of the troposphere, which must be compensated by mechanical energy transfer from the surface.

Heating rates depicted in Fig. 8.24 apply to a clear atmosphere under global–mean conditions. Clouds and aerosol influence the budgets of SW and LW radiation dramatically. Figure 8.26 shows that SW heating rates can be substantially larger under conditions typical of the tropical troposphere (see also Reynolds *et al.*, 1975). Likewise, the radiative balance in the stratosphere suggested by Fig. 8.24 does not apply locally. As shown in Fig. 8.27, the winter stratosphere experiences strong net cooling—more than 8 K day^{-1} in the polar night, which implies that region is warmer than radiative equilibrium. As in the troposphere, net radiative cooling must be balanced by mechanical energy

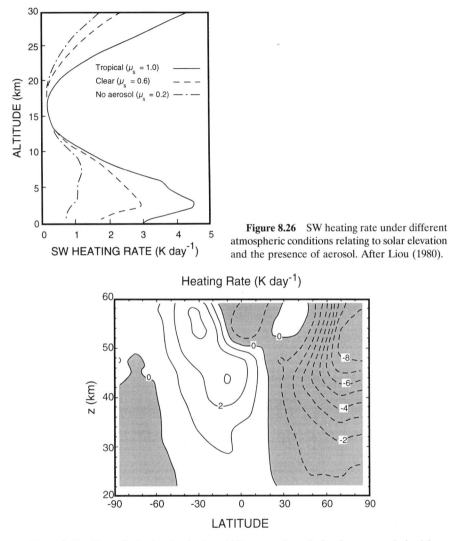

Figure 8.26 SW heating rate under different atmospheric conditions relating to solar elevation and the presence of aerosol. After Liou (1980).

Figure 8.27 Net radiative heating in the middle atmosphere during January, as derived from Nimbus-7 LIMS. After Kiehl and Solomon (1986).

transfer. In the stratosphere, that energy transfer occurs through planetary waves, which drive the circulation out of radiative equilibrium (refer to Fig. 1.10).

8.6 Thermal Relaxation

The foregoing discussion illustrates that a layer will cool or warm according to the temperature of its surroundings. If a layer is in radiative equilibrium with its surroundings, LW and SW contributions to its radiation budget balance, so net heating vanishes. If disturbed from radiative-equilibrium (e.g., by motions that displace air into different radiative environments), a layer will experience net heating or cooling (8.73), which tends to restore the layer to radiative equilibrium. For example, polar air drawn equatorward by a cyclone will find itself colder than the radiative-equilibrium temperature of its new environment (Fig. 1.15). Net radiative heating then acts to destroy the temperature anomaly by driving that air toward the local radiative equilibrium temperature. Since the circulation is related to the thermal structure (Figs. 1.9a and 10a), dissipating anomalous temperature also dissipates anomalous motion.

In the absence of other forms of heat transfer, the temperature field is described by (8.56), with the heating rate given by (8.73). The first term in (8.73), which describes cooling to space, often dominates radiative heating. Figure 8.28 compares the total heating for water vapor, carbon dioxide, and ozone against cooling to space alone. Over much of the troposphere and stratosphere, cooling to space provides an accurate approximation to net radiative heating. Exceptional is ozone in the lower stratosphere, which absorbs upwelling 9.6-μm radiation from below.

The agreement in Fig. 8.28 under widely varying circumstances is the basis for the *cool-to-space approximation*, which, for a particular frequency band, can be expressed

$$\frac{dT}{dt} \cong -\frac{1}{\rho c_p} B^*[T(z^*)] \frac{d\mathcal{T}_{\bar{\nu}}(z^* - \infty)}{dz^*}, \tag{8.78}$$

where vertical motion (e.g., dp/dt) is ignored, z^* includes the diffusivity factor (8.54.1), and summation over different absorbers is understood. Away from the surface and in the absence of strong curvature of the emission profile, (8.78) accurately represents cooling due to both water vapor and carbon dioxide. These conditions hold for temperature disturbances of large vertical scale, for which interactions with underlying and overlying layers is negligible. Emission to space of monochromatic radiation then takes place from a Chapman layer, analogous to absorption of SW radiation, so emission over a finite spectral band can be interpreted in terms of a series of Chapman layers.

Consider a disturbance $T'(z, t)$ to an equilibrium temperature profile $T_0(z)$.[4] If $T'/T_0 \ll 1$ and if the cool-to-space approximation applies, the emission in (8.78) can be linearized in T' to obtain

$$\frac{\partial T'}{\partial t} = -\frac{1}{\rho_0 c_p} \frac{dB^*}{dT}[T_0(z^*)] \frac{d\mathcal{T}_{\bar{\nu}}(z^* - \infty)}{dz^*} \cdot T'$$

$$= -\alpha(z^*)T' \tag{8.79}$$

---

[4]The equilibrium thermal structure can be maintained by heat transfer other than radiation alone, but we will take T_0 to represent the radiative–equilibrium temperature.

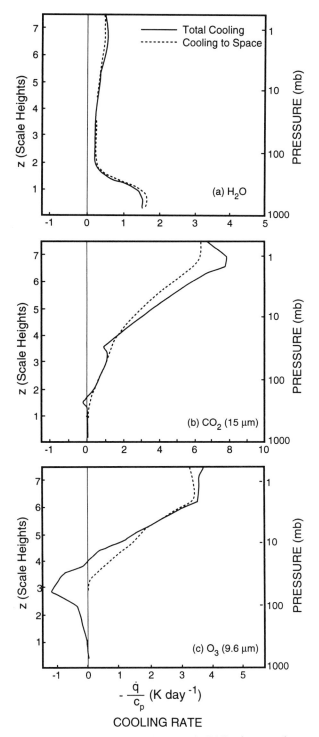

Figure 8.28 Comparison of the total cooling rate (solid lines) versus the contribution from cooling to space (dashed lines) for (a) water vapor, (b) carbon dioxide, and (c) ozone. Adapted from Rodgers and Walshaw (1966).

Known as the *Newtonian cooling approximation*, (8.79) governs the evolution of anomalous temperature. The Newtonian cooling coefficient $\alpha(z^*)$ varies spatially through the equilibrium thermal structure $T_0(z^*)$. This approximation is of great practical importance because it eliminates the rather cumbersome interactions in (8.73) in favor of a simple expression that depends only on the local temperature and is linear in the disturbance. Even though it breaks down for CO_2 in the mesosphere (Andrews *et al.*, 1987), the Newtonian cooling approximation is widely adopted.

Under the influence of Newtonian cooling, a temperature disturbance (e.g., introduced through motion) relaxes exponentially toward radiative equilibrium. Owing to its linearity, (8.79) can be cast into the form of an e-folding time

$$-\frac{1}{T'}\frac{\partial T'}{\partial t} = \frac{4\sigma T_0^3}{\rho_0 c_p}\frac{d\mathcal{T}_{\nu}(z^* - \infty)}{dz^*}$$

$$= t_{rad}^{-1}, \tag{8.80}$$

which measures the efficiency of radiative transfer in relation to other factors influencing thermal structure. Taking values representative of the troposphere (Problem 8.32) yields a radiative timescale of order 10 days.

The timescale $t_{rad} = \alpha^{-1}$ is an order of magnitude longer than the characteristic timescale of air motion, which makes radiative transfer inefficient compared to dynamical influences. For this reason, adiabatic behavior of individual air parcels is a good approximation on timescales comparable to advection. Because it operates on a timescale much shorter than radiative transfer, air motion controls tropospheric properties like thermal structure and vertical stability. In the stratosphere, where water vapor and clouds are virtually absent, t_{rad} decreases to only 3 to 5 days. Although shorter, this is still long enough to permit air motions to influence many stratospheric properties.

More accurate formulations of heat transfer apply to temperature disturbances whose vertical scales are not large enough to neglect exchange with neighboring layers. Figure 8.29 shows thermal damping rates for sinusoidal vertical structure with several vertical wavelengths λ_z. Minimum dissipation occurs in the limit $\lambda_z \to \infty$, which corresponds to deep temperature structure described by Newtonian cooling (8.79). Under those circumstances, t_{rad} decreases from 10 to 15 days in the troposphere to about 5 days near the stratopause. Smaller vertical scales experience significantly faster dissipation due to exchange with neighboring layers, which produces relaxation times of only a couple of days in the upper stratosphere and mesosphere.

8.7   The Greenhouse Effect

The radiative–equilibrium surface temperature T_s is significantly warmer than that in the absence of an atmosphere. Combining (8.67) and (8.64) for a gray

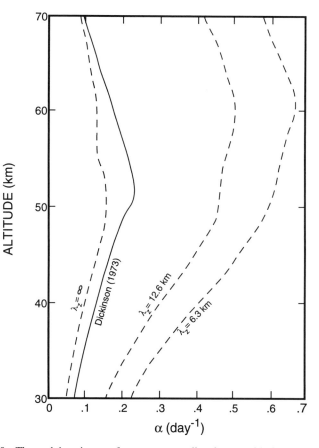

Figure 8.29 Thermal damping rate for temperature disturbances with sinusoidal vertical structure of vertical wavelength λ_z (dashed lines). After Fels (1982). The Newtonian cooling rate (solid line), from a calculation by Dickinson (1973), is superposed.

atmosphere yields

$$B^*(T_s) = \frac{F_0}{2}(\tau_s^* + 2), \qquad (8.81)$$

which determines T_s through (8.68). Surface temperature increases with the optical depth of the atmosphere, which is a statement of the greenhouse effect. Although smaller, the surface temperature under radiative–convective equilibrium is likewise controlled by τ_s, for example, through (8.70) and (8.71).

The atmosphere's optical depth depends on its mass and on its composition through optically active species. Such constituents are produced and destroyed by surface processes. For example, water vapor is introduced at ocean surfaces through evaporation. Carbon dioxide is produced by oxidation of organic matter, whereas ozone is destroyed at the Earth's surface through oxidation. These and other processes controlling optically active species are temperature dependent, which introduces the possibility of feedback between the two sides of (8.81).

Noteworthy is the mechanism controlling water vapor, which is the primary absorber in the IR. Temperature dependence in the Clausius–Clapeyron relation (4.38) makes the water vapor abundance vary sharply with T_s. The exponential dependence of saturation vapor pressure implies that an increase of T_s can sharply increase the water vapor content of the atmosphere. Enhanced cloud cover, which is likewise implied, further increases τ_s. By (8.81), the increased optical depth results in a higher equilibrium surface temperature, which, in turn, increases the saturation vapor pressure and H_2O content of the atmosphere, and so forth.

Positive feedbacks like the one between temperature and water vapor support potentially large changes of T_s and other properties that characterize climate. A paradigm of such changes is the so-called "runaway greenhouse effect," which is used to explain the present state of the Venusian atmosphere. The evolution of a planet's atmosphere is thought to occur through slow discharge of gases from the planet's surface. Atmospheric uptake of those gases is limited by their saturation vapor pressures, which depend only on temperature (e.g., on T_s). Because those gases are responsible for atmospheric opacity, their saturation vapor pressures translate into the optical depth τ_s. Plotting T_s against $\log(e_w)$, which is symbolic of $\log(\tau_s)$, produces the saturation curve (dotted line) in Fig. 8.30 separating heterogeneous states of water from vapor phase alone.

The optical depth τ_s and incident solar flux F_0 determine the surface temperature of the planet, for example, via (8.81) under radiative equilibrium or its counterpart (8.70) under radiative–convective equilibrium. Then the surface temperature T_s defines a family of radiative–convective equilibrium curves, two of which are superposed in Fig. 8.30. The curve for a particular F_0 can be thought to represent the evolution of a given planet. For small τ_s, $T_s \cong (F_0/\sigma)^{\frac{1}{4}}$, which reflects the initial state of the planet when all of its water resided at the surface in condensed phase. The situation for Earth is represented in the curve for planet 1 (solid line). Initially unsaturated, the atmosphere incorporates water vapor from the surface, with concomitant increases of its opacity τ_s and surface temperature T_s. Positive feedback between temperature and water vapor then drives the state of the atmosphere to the right along its radiative–convective equilibrium curve. Eventually, the atmosphere's state encounters the saturation curve, beyond which further increases of τ_s and T_s are prevented. For the conditions of planet 1, saturation is achieved at a surface temperature of about 287 K. Subsequent increases of water vapor are offset by condensation, which returns H_2O to the planet's surface. Those conditions describe a saturated system that contains multiple phases of water, reflecting the present state of Earth's atmosphere. By preventing H_2O from

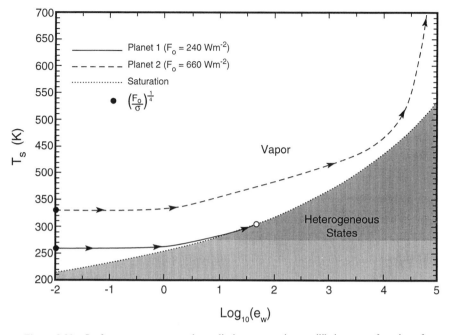

Figure 8.30   Surface temperature under radiative–convective equilibrium, as a function of saturation vapor pressure for water (which is symbolic of atmospheric optical depth to LW radiation τ_s). Radiative–convective equilibrium curves shown for conditions representative of Earth (planet 1) and Venus (planet 2), which characterize atmospheric evolution from an initial state when all water resided at the surface in condensed phase. Positive feedback between temperature and water vapor drives the state of a planet's atmosphere to greater temperature and humidity. For planet 1 (solid line), the radiative–equilibrium curve eventually encounters the saturation curve (dotted line), where no more water is absorbed by the atmosphere and positive feedback ceases. However, the radiative equilibrium curve for planet 2 (dashed line) never encounters the saturation curve, so it continues to evolve through positive feedback until all water has been incorporated into the atmosphere. The calculation of radiative–convective equilibrium presumes a constant albedo and relates optical depth to saturation vapor pressure as $\tau = 4[e_w(T)/e_w(288\ K)]$.

reaching great heights, where it would be photodissociated and the resulting atomic hydrogen would be lost to space (Sec. 1.2.2), saturation is responsible for maintaining water on the planet.

The curve for planet 2, which corresponds to Venus, describes a very different evolution. Closer to the sun, Venus has a larger F_0 and hence a higher initial surface temperature. As for planet 1, the atmosphere's state evolves to the right along the radiative–convective equilibrium curve. However, planet 2 never encounters the saturation curve. Instead, water continues to boil off the surface, which reinforces the surface temperature, until all H_2O has been incorporated into the atmosphere. Unchecked, feedback between T_s and τ_s leads to temperatures that are greatly elevated over the initial radiative–convective equilibrium temperature. This is symbolic of the 750 K mean surface temperature of Venus—which is some 400 K warmer than would result in the absence of its atmosphere.

Atmospheric composition suggested by the evolution of planet 2 also differs fundamentally from that of planet 1. For initial water abundance comparable to that of Earth, the Venusian atmosphere would reach a state in which H_2O is its primary constituent. Water vapor would then be vulnerable to photodissociation by UV, allowing atomic hydrogen to escape to space and free oxygen to recombine at the surface through oxidation processes. In this fashion, atmospheric water vapor would be systematically depleted and replaced by carbon dioxide, which is consistent with the predominance of CO_2 in the present-day Venusian atmosphere.

For Earth, gases other than water vapor also contribute to the greenhouse effect. Absorption by carbon dioxide is situated near the maximum of the LW emission spectrum (Fig. 8.5). In tandem with the steadily increasing concentration of CO_2 (Fig. 1.11), this feature suggests that increased surface temperatures are inevitable if the present trend continues. CFCs, methane, and nitrous oxide follow carbon dioxide in importance. All have anthropogenic sources that have been associated with increasing concentrations.

Were the preceding feedback to operate in isolation, small changes of temperature (e.g., stimulated by occasional volcanic eruptions or anthropogenic activities) could result in large shifts of the Earth's climate. In fact, the temperature—water vapor feedback is but one of several believed to shape climate. The involvement of properties like surface reflectivity (e.g., snow and ice cover), clouds, and the distributions of optically active species is strongly suggested by models of radiative–convective equilibrium, which exhibit marked sensitivity to how such ingredients are prescribed. Clouds and aerosol, which are treated in the next chapter, are especially important because they sharply alter the absorption and scattering characteristics of the atmosphere.

Suggested Reading

An Introduction to Atmospheric Radiation (1980) by Liou is a very readable treatment of radiative transfer. It includes discussions of solar variability and band models.

Fundamentals of Modern Physics (1967) by Eisberg provides an excellent overview of Planck's theory and the emergence of modern physics. It includes a basic treatment of line broadening.

Aeronomy of the Middle Atmosphere (1986) by Brasseur and Solomon and *Middle Atmosphere Dynamics* (1987) by Andrews *et al.* contain excellent treatments of radiative transfer in the stratosphere and mesosphere. They also include descriptions of solar variability and its implications to UV radiation.

An advanced treatment of radiative transfer can be found in *Atmospheric Radiation* (1989) by Goody and Yung, which contains an enlightening comparison of two-stream approximations.

Atmospheric Ozone: Assessment of Our Understanding of Processes Controlling Its Present Distribution and Change (WMO, 1986) includes a detailed overview of solar variability based on satellite measurements.

Special Functions and Their Applications (1972) by N. Lebedev contains a proper development of the exponential integral.

Hartmann (1994) provides an excellent overview of the earth's energy budget, including a treatment of horizontal energy transport by the general circulation.

Ramanathan *et al.* (1989) provides an overview of recent satellite measurements of the earth's radiation budget and observations of the solar constant.

Manabe and Strickler (1964) is a classic calculation of radiative-convective equilibrium, which contains an estimate of thermal damping time inside the troposphere.

Climate Change: The IPCC Scientific Assessment (Houghton *et al.*, 1990) contains an overview of climate feedback mechanisms and assessments of how well they are understood.

Problems

8.1. Calculate the brightness temperature of the earth based on (a) an observed broadband OLR of 235 W m^{-2}, (b) the outgoing narrowband flux (Fig. 8.5) at $\lambda = 11$ μm, (c) the outgoing narrowband flux at $\lambda = 9.6$ μm, and (d) the outgoing narrowband flux at $\lambda = 15$ μm.

8.2. Derive the Stefan–Boltzmann law (8.15).

8.3. As the sun cools, its spectrum will shift toward longer wavelengths. Estimate the change in the earth's equivalent blackbody temperature if the peak in the SW spectrum is displaced from its current position of 0.48 μm to a yellower wavelength of 0.55 μm.

8.4. Orbital ellipticity brings the earth some 3.5% nearer the sun during January than during July, which is responsible for the slight asymmetry of insolation apparent in Fig. 1.28. (a) Calculate the corresponding change in the earth's equivalent blackbody temperature. (b) Discuss this change in relation to other changes that are introduced by hemispheric asymmetries of the earth's surface and atmospheric circulation.

8.5. Consider a 250 K isothermal atmosphere that contains a single absorbing gas of uniform mixing ratio 0.05. The specific absorption cross section of that gas is given as a function of wavelength by

$$\sigma_{a\lambda} = 1.0 \exp\left[-\left(\frac{\lambda - \lambda_0}{\alpha}\right)^2\right] \quad [\text{m}^2\text{kg}^{-1}],$$

where $\lambda_0 = 0.5$ μm and $\alpha = 0.01$ μm. (a) Calculate, as a function of λ, the optical path length for downward-traveling radiation from the top of the atmosphere to the height z. (b) Plot the absorptivity of this layer, as a function of λ, for $z = 5$, 3, and 1 scale heights.

8.6. Show that a layer having optical thickness $\Delta u << 1$ has absorptivity $\Delta a = \Delta u$.

8.7. Prove Kirchhoff's law (8.16) in generality.

8.8. The top of the photosphere has a radius of 7.0×10^8 m. If the mean earth–sun distance is 1.5×10^{11} m, calculate the equivalent blackbody temperature of the photosphere.

8.9. Venus has an albedo of 0.77 and a mean distance from the sun of 1.1×10^{11} m. (a) Calculate the equivalent blackbody temperature of Venus. (b) Discuss this value in relation to the observed surface temperature of 750 K.

8.10. Consider an isothermal atmosphere that is optically thick to LW radiation. (a) How much of the LW radiance emitted to space at zero zenith angle originates in its uppermost three optical depths? (b) Determine the altitude range, in scale heights, occupied by the layer between optical depths of 0.1 and 3.0 under isothermal conditions.

8.11. A flat plate sensor on board a satellite behaves as a graybody with constant absorptivity a. Calculate its radiative–equilibrium temperature when the sensor faces the sun if (a) $a = 0.8$ and (b) $a = 0.2$.

8.12. As for Problem 8.11, but if the sensor has different absorptivities to SW and LW radiation of (a) a_{SW} and a_{LW}, respectively, and (b) $a_{SW} = 0.2$ and $a_{LW} = 0.8$.

8.13. Under the conditions of Problem 8.12, calculate the sensor's temperature as a function of time after the satellite has been rotated to face the sensor away from the sun, if it has an area of 0.1 m^2 and a specific heat of 2 J K^{-1}.

8.14. The hood of an automobile can be modeled as a gray flat plate with LW absorptivity 0.9 and the atmosphere as a gray layer with LW absorptivity 0.8. In this framework, estimate the hood's temperature for a solar zenith angle of 25° if (a) the automobile's color is light with a SW absorptivity of 0.2 and (b) the automobile's color is dark with a SW absorptivity of 0.6. (c) What physical process would, in practice, mediate the values in parts (a) and (b)?

8.15. A simple model approximates the earth's surface as a graybody and the atmosphere as a gray layer with LW absorptivities of 1.0 and 0.8, respectively. Use this model to estimate the radiative–equilibrium temperature of the ground in the presence of 50% cloud cover (fully reflective) and under (a) perpetual summertime conditions over vegetated terrain: a solar zenith angle of 25° and the surface having a SW absorptivity of 0.6, and (b) perpetual wintertime conditions over snow-covered terrain: a solar zenith angle of 60° and the surface having a SW absorptivity of 0.2.

8.16. Consider a simple model of the earth comprised of an isothermal gray layer with SW and LW absorptivities a_{SW} and a_{LW}, respectively, and a black underlying surface, both of which are illuminated by a SW flux F_0 and are in thermal equilibrium. The energy budget can be formulated by tracing energy through repeated absorptions and reemissions by the surface and atmosphere. (a) Develop an arithmetic progression for the fractional energy absorbed by the atmosphere during successive trans-missions from the surface to construct a series representation for the net energy absorbed by the atmosphere. (b) Use the series representation to calculate the radiative-equilibrium temperature of the atmosphere for $F_0 = (1 - \mathscr{A})(F_s/4) = 240$ W m^{-2} (Sec. 1.3), $a_{SW} = 0.20$, and $a_{LW} = 0.94$, which are representative of values in Fig. 1.27.

8.17. Use the results of Problem 8.16 to determine the radiative–equilibrium temperature of the surface (a) for the conditions given, (b) in the absence of an atmosphere, and (c) as in part (a), but with the atmosphere's LW absorptivity increased to $a_{LW} = 0.98$.

8.18. Estimate the level where the collisional half-width equals the Doppler half-width for (a) the water vapor absorption line (Fig. 8.11) at $\lambda^{-1} = 352$ cm^{-1} and (b) the CO$_2$ absorption line at $\lambda^{-1} = 712$ cm^{-1}.

8.19. Consider a discretely stratified atmosphere comprised of N isothermal layers, each of which is transparent to SW radiation but has a LW ab-sorptivity a and which, collectively, are in radiative equilibrium with an incident SW flux F_0 and a black underlying surface. (a) Derive expres-sions for the flux F_s emitted by the surface for $N = 1$, $N = 2$, and then generalize those expressions to arbitrary N. (b) Let $a = \tau_s/N$ and take the limit $N \to \infty$ to recover expression (8.81) for a continuously stratified atmosphere. (c) Verify that, even though they have different optical characteristics, the atmospheres for different N all have the same equivalent blackbody temperature. Explain why.

8.20. Derive expressions (8.38) for the upwelling and downwelling intensities in a plane parallel atmosphere as functions of optical depth.

8.21. Show that the upwelling and downwelling fluxes for horizontally isotropic radiation are described in terms of the exponential integral E_2 by (8.45).

8.22. An 80% solar eclipse occurs during morning, when the tempera-ture would normally be 20° C. Use a gray atmosphere in radiative–equilibrium with a black surface to estimate the change of surface temperature that would result were the eclipse to persist indefinitely (compare Problem 8.33).

8.23. Show that the two-stream transformation (8.54) casts the budgets of upwelling and downwelling radiation into the form of (8.55).

8.24. Consider a gray atmosphere that is moisture laden, in hydrostatic equilibrium, and has a LW specific absorption coefficient k. The atmosphere is in radiative equilibrium with a black surface, which, in turn, is in thermal equilibrium with a downwelling SW flux F_0. (a) Provide an expression relating the radiative–equilibrium temperature T_{RE} to height z. Discuss the form of this relationship and how it can be solved for $T_{RE}(z)$. (b) Determine $T_{RE}(z)$ in a neighborhood of the surface if the atmosphere is optically thick (e.g., in the limit $k \to \infty$) and if the relative change of temperature there can be ignored. (c) Under the conditions of part (b) but in light of hydrostatic stability and if air near the surface is almost saturated, derive an expression involving the height of the tropopause. Discuss how the tropopause height depends on surface temperature and the physical processes controlling it.

8.25. Consider a gray atmosphere with a LW specific absorption coefficient k and an underlying surface with SW and LW absorptivities a_{SW} and a_{LW}, respectively. (a) Determine the distribution of radiative–equilibrium temperature as a function of optical depth, a_{SW}, and a_{LW}. (b) For $a_{LW} = 1$, discuss the limiting behavior: $a_{SW} \to 0$. (c) For $a_{SW} = 1$, discuss the limiting behavior: $a_{LW} \to 0$.

8.26. Discuss how the competing influences of radiative transfer and convective mixing control stratification, in light of the diurnal variation of insolation.

8.27. Show that the upwelling flux inside a gray atmosphere that is in radiative equilibrium with a black underlying surface is given by (8.71).

8.28. Revisit the radiative–convective equilibrium in Figs. 8.21 and 8.22. (a) Determine the temperature distribution under the same conditions, but now using the CAPE (Sec. 7.4.1) of the radiative equilibrium atmosphere to infer the tropopause height z_T and presuming the air to be virtually saturated. (b) Plot the profiles of upwelling and downwelling fluxes under this radiative-convective equilibrium. (c) Plot the profile of radiative heating.

8.29. The Venusian atmosphere is composed chiefly of carbon dioxide, with $g = 8.8$ m s^{-2}, $c_p = 8.44 \times 10^2$ J kg^{-1} K^{-1} for CO_2, a mean distance from the sun of 0.70 that of Earth, and an albedo of $\mathscr{A} = 0.77$. (a) Use radiative equilibrium and the observed surface temperature $T_s = 750$ K to estimate the optical depth of the Venusian atmosphere. (b) Calculate the temperature distribution under radiative–convective equilibrium based on (1) the profile of optical depth (8.69), with $h = RT_s/g$, (2) the presumption that temperatures inside the convective layer are hot enough to ignore condensation, and (3) in place of the constraint used in Sec. 8.5.2 to determine the tropopause height, requiring the temperature at the midlevel of the hydrostatically unstable layer to equal that under radiative equilibrium. (c) Plot the profiles of upwelling radiation, downwelling radiation, and radiative heating.

8.30. The CO_2 absorption band is already saturated in the earth's atmosphere (Fig. 8.1). In light of this feature, explain how increased levels of CO_2 could result in global warming.

8.31. Derive the expression for Newtonian cooling (8.79) from the full representation of radiative heating.

8.32. (a) Calculate the characteristic timescale of thermal damping in the troposphere based on $\rho_0 = 1.2$ kg m^{-3}, the equivalent blackbody temperature $T_e = 255$ K (which is symbolic of where most OLR is emitted), and a characteristic depth of 5 km (which corresponds to the preceding temperature under radiative–convective equilibrium: Fig. 8.21). (b) What is the heat capacity of the surface under this approximation?

8.33. Use Newtonian cooling in Problem 8.32 to estimate the amplitude ΔT_s of the diurnal cycle of surface temperature, where ΔT_s is approximated by half the nocturnal depression of temperature from its daytime maximum. Estimate ΔT_s under equinoctial conditions, for a surface temperature maximum of 300 K, and for an optical depth (a) $\tau_s = 4$, which is representative of the global–mean conditions, and (b) $\tau_s = 1$, which is representative of arid conditions.

8.34. Construct a counterpart to Fig. 8.30 for the runaway greenhouse effect, using the same initial values but based on radiative equilibrium. At what temperature does the earth's atmosphere become saturated? Compare this value with the one obtained under radiative-convective equilibrium.

Chapter 9 | Aerosol and Clouds

Radiative transfer is modified importantly by clouds. Owing to their high reflectivity in the visible, clouds shield the earth-atmosphere system from solar radiation and thus represent cooling in the shortwave (SW) energy budget. Conversely, the strong infrared (IR) absorptivity of water and ice particles sharply increases the optical depth of the atmosphere, which magnifies the greenhouse effect and represents warming in the longwave (LW) energy budget. Atmospheric aerosol has optical properties similar to clouds.

We develop cloud processes from a morphological description of atmospheric aerosol, without which clouds would not form in the earth's atmosphere. We then examine the microphysics controlling cloud formation. Macrophysical characteristics of clouds and accompanying microphysical properties are developed in terms of environmental conditions controlling the formation of particular cloud types. These elemental considerations culminate in descriptions of radiative and chemical processes that operate inside clouds and figure in climate.

9.1 Morphology of Atmospheric Aerosol

Small particulates are produced and removed through a variety of processes, which make the composition, size, and distribution of atmospheric aerosol widely variable (Table 9.1). Aerosol concentrations are smallest over the oceans (Fig. 1.22), where a particle number density of $n = 10^3$ cm^{-3} is representative, and greatest over industrial areas, where $n > 10^5$ cm^{-3} is observed. These and other distinctions lead to two broad classes of tropospheric aerosol: *continental* and *marine*.

9.1.1 Continental Aerosol

Continental aerosol includes (1) crustal species that are produced by erosion of the earth's surface, (2) combustion and secondary components related to anthropogenic activities, and (3) carbonaceous components consisting of hydrocarbons and elemental carbon. Crustal aerosol is produced in subtropical deserts like the Sahara, the southwestern United States, and southern

Table 9.1 Properties of Atmospheric Aerosols and Clouds

Type of Particulate	Altitude (km)	Horizontal Scale (km)	Frequency of Occurrence	Composition	Mass Loading (mg/m³)	Optical Depth (at 0.55 μm)	Mean Particle Radius (μm)	Principal Size Range (μm)
Stratus, cumulus, nimbus clouds	1–18	10–1,000	0.5	Water, ice	1,000–10,000	~1–100	10–1,000	Variable
Cirrus clouds	7–16	10–1,000	0.3	Ice	10–100	~1	~10–100	Variable
Fog	0–1	10–100	Sporadic	Water	10–100	1–10	~10	10–50
Tropospheric aerosols	0–10	1,000–10,0000 (ubiquitous)	1	Sulfate, nitrate, minerals	0.01–0.1	~0.1	0.1–1	~0.3
Ocean haze	0–1	100–1,000	0.3	Sea salt sulfate	0.1–1	0.1–1	0.5	~0.3
Dust storms	0–3	10–1,000	Sporadic	Silicates, clays	< 1 – > 100	1–10	1–10	10–100
Volcanic clouds	5–35	100–10,000	Sporadic	Mineral ash, sulfates	< 1 – > 1,000	0.1–10	0.1–10	1–10
Smoke	0–10	1–100	Sporadic (from fires)	Soot, ash, tars	0.1–1	~0.1–10	0.1–1	~0.3
Stratospheric aerosols	10–30	1,000–10,000 (ubiquitous)	1	Sulfate	0.001–0.01	~0.01	0.1	~0.1
Polar stratospheric clouds	15–25	10–1,000	0.1 (winter only)	HNO_3/H_2O ice	0.001–0.01	~0.01–1	1–10	~1
Polar mesospheric clouds	80–85	~200 (polar regions above 50°)	0.1 (summer only)	Ice	~0.0001	~0.0001–0.01	~0.05	~0.02
Meteoric dust	50–90	10–1,000	0.5–1	Minerals, carbon	~0.00001	~0.00001	≲ 0.01	Wide range incl. micro-meteors

After WMO (1988).

Asia (see Fig. 9.40). Combustion and secondary aerosols originate chiefly in the industrialized regions of Europe and North America. These continental source regions give the Northern Hemisphere greater number densities than the Southern Hemisphere. Carbonaceous aerosol has large sources in tropical regions due to agricultural burning and, through reaction with other species, to vegetative emissions.

Aerosol number density decreases vertically away from the surface (Fig. 9.1), where particles and gaseous precursors originate. In the troposphere, n has a characteristic height of only 2 km. This shallow depth reflects the absolute concentration of water vapor (Fig. 1.14) and removal by precipitation processes. In the stratosphere, n decreases more slowly.

Aerosol is characterized by its *size distribution, n(a)*, which denotes the number density of particles with radii smaller than a [a monotonically increasing function of a, with $n(a)$ approaching n as $a \to \infty$]. The *size spectrum, dn/da* (Fig. 9.2), represents the contribution to $n = \int (dn/da)da$ from particles with

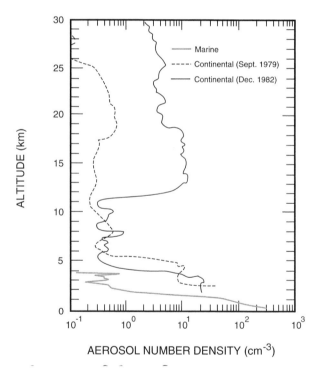

AEROSOL NUMBER DENSITY (cm⁻³)

Figure 9.1 Aerosol number density n as a function of height in continental and marine environments. Profile of marine aerosol corresponds to particle radii $a > 0.5$ μm. The steep decrease of n above the marine boundary layer mirrors a similar decrease in dew point temperature (not shown). Profiles of continental aerosol correspond to $a > 0.15$ μm and are shown before and after the eruption of El Chichon. Sources: Patterson (1982) and WMO (1988).

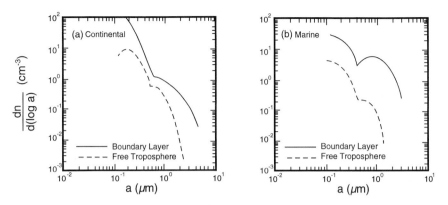

Figure 9.2 Aerosol size spectra, as functions of particle radius a, inside the planetary boundary layer and aloft for (a) continental environment and (b) marine environment. From Patterson (1982). Copyright by Spectrum Press (A. Deepak Publishing).

radii between a and $a + da$. As before, plotting $dn/d(\log a)$ versus $\log a$ is area preserving: Equal areas under the spectrum represent equal contributions to n from different size ranges. Crustal aerosol, which is comprised chiefly of wind-blown silicates, involves particles with radii $a > 1$ μm. Secondary and combustion aerosols, which form through condensation, contain submicron-scale particles. These different production mecnanisms give continental aerosol a bimodal size spectrum (Fig. 9.2a), in which particles are concentrated in two distinct ranges of a. The size spectrum diminishes sharply with increasing a, the greatest contribution to n arriving from particles with $a < 0.1$ μm—the so-called *Aitken nuclei*. Despite their relative abundance, Aitken nuclei constitute less than half of the surface area of aerosol (Fig. 1.22), which figures centrally in thermodynamic and chemical processes, and only about 10% of its mass.

Combustion and secondary aerosols form through several mechanisms: (1) gas-to-particle conversion, wherein gaseous precursors undergo physical and chemical changes that result in particle formation, (2) direct condensation of combustion products, and (3) direct emission of liquid and solid particles. Gas-to-particle conversion is responsible for much of the sulfate and nitrate that dominate urban aerosol. That process is exemplified by a series of reactions which begins with the gaseous pollutant sulfur dioxide

$$OH + SO_2 \rightarrow HSO_3, \tag{9.1}$$

followed by oxidation of HSO_3 to form H_2SO_4 vapor. Owing to its low saturation pressure, sulfuric acid vapor readily condenses, or *nucleates*, into liquid particles—especially in the presence of water vapor. Once formed, nuclei grow rapidly to radii of 0.01 to 0.1 μm. Larger aerosol can then develop through coagulation of individual nuclei to form *accumulation particles*. Sulfate aerosol is also produced through reaction of sulfur dioxide with ammonia inside cloud droplets, which, upon evaporating, leave behind sulfate particles. In the troposphere, $(NH_4)_2SO_4$, NH_4HSO_4, and H_2SO_4 are all present, whereas supercooled H_2SO_4 droplets are the prevalent form of sulfate in the stratosphere. Nitrogen emissions from combustion and natural sources lead to analogous products. Sulfate and nitrate are both important in the troposphere, but nitrate is especially prevalent in urban areas because it is an inevitable by-product of combustion.

Carbonaceous aerosols like soot consist of submicron-size particles. Natural sources include pollen and spores, as well as complex hydrocarbon vapors, like isoprene. Emitted via plant transportation, those vapors react with oxides of nitrogen and subsequently nucleate.

Once formed, aerosol nuclei interact strongly with water vapor. Through condensation (Fig. 9.3), liquid particles like H_2SO_4 enlarge steadily with increasing relative humidity (RH). Conversely, hygroscopic solids like NaCl and $(NH_4)_2SO_4$ remain dry below a threshold of $RH \cong 80\%$. Particles then dissolve, undergo a discontinuous enlargement, and thereafter exhibit dependence on RH similar to liquid aerosol.

Anthropogenic sources account for about 30% of aerosol production. The anthropogenic component of sulfate, which plays a key role in cloud formation, exceeds 60% over urban areas. The distribution of continental aerosol is strongly affected by sedimentation, especially for large crustal species. Fall speeds range from 50 cm s^{-1} for large silicates ($a \cong 50$ μm) to 0.03 cm s^{-1} for smaller ($a \cong 1$ μm) particles. As a result, larger particles tend to be confined to a neighborhood of their source regions.

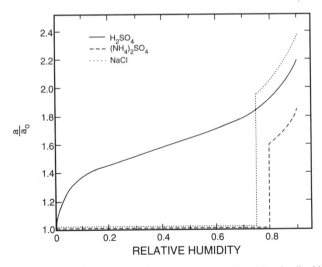

Figure 9.3 Aerosol particle radius as a function of relative humidity for liquid droplets and hygroscopic solids. Adapted from Patterson (1982). Copyright by Spectrum Press (A. Deepak Publishing).

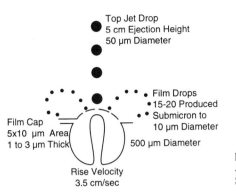

Top Jet Drop
5 cm Ejection Height
50 µm Diameter

Film Drops
15-20 Produced
Submicron to
10 µm Diameter

Film Cap
5x10 µm Area
1 to 3 µm Thick

500 µm Diameter

Rise Velocity
3.5 cm/sec

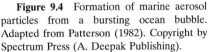

Figure 9.4   Formation of marine aerosol particles from a bursting ocean bubble. Adapted from Patterson (1982). Copyright by Spectrum Press (A. Deepak Publishing).

9.1.2   Marine Aerosol

Composed primarily of sea salt, marine aerosol has a smaller overall concentration. Its number density drops sharply above the boundary layer (Fig. 9.1), resembling the distribution of moisture. Like continental aerosol, marine aerosol has a size spectrum that is bimodal (Fig. 9.2b), which reflects two classes of droplets that form when ocean bubbles burst (Fig. 9.4). Water entrained into a bubble is ejected vertically in a stream of drops ($a \cong 25$ μm). Upon evaporating, those drops leave behind large ($a > 1$ μm) particles of sea salt. The thin film comprising the bubble's surface shatters to release droplets of 1 μm and smaller, which evaporate and produce smaller particles of sea salt.

9.1.3   Stratospheric Aerosol

Aerosol is introduced into the stratosphere through penetrative convection and volcanic eruptions. It also forms *in situ* through gas-to-particle conversion from precursors like SO_2, which maintains a background level of H_2SO_4 droplets. Its long residence time makes stratospheric aerosol fundamentally different from tropospheric aerosol. Small particulates with slow fall speeds are dynamically isolated by strong static stability. Sequestered from removal processes associated with precipitation, aerosol can survive in the stratosphere long after it is introduced. This feature enables stratospheric aerosol to alter SW absorption at the earth's surface for long durations. The great eruption of Krakatoa in 1883, which destroyed the Indonesian island of the same name, altered SW transmission and sunsets for 3 years afterward (see, e.g., Humphreys, 1964).

Aerosol number density decreases vertically in the stratosphere more slowly than in the troposphere (Fig. 9.1). Following major volcanic eruptions, aerosol number density actually increases into the stratosphere, as was evident following the eruption of El Chichon in 1982. A clear signature of El Chichon was registered in SW transmission (Fig. 9.5), which decreased sharply in 1982. In-

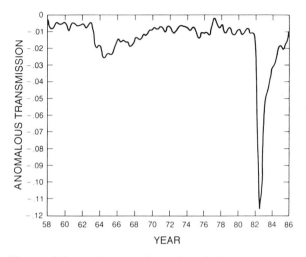

Figure 9.5   Change of direct solar transmission through the atmosphere, as observed over Hawaii by ground-based instruments. *Note:* Reduction of direct transmission is offset by an increase of diffuse transmission (see, e.g., Hoffman, 1988). After WMO (1988).

creased extinction in volcanic debris can increase radiative heating, which, in turn, can produce ascent of air in which the debris is suspended.

A characteristic vertical scale of 7 km for n in the stratosphere resembles the scale height of air density, which suggests that the relative concentration of aerosol (e.g., n/n_d) is approximately conserved for individual air parcels. Owing to the absence of removal processes, stratospheric aerosol behaves as a tracer. Figure 9.6 shows the cloud of volcanic debris introduced by the eruption of Mount St. Helens. Two months after the eruption, the cloud has dispersed zonally but not meridionally. Its confinement to high latitudes of the summer hemisphere follows from certain characteristics of the stratospheric circulation (Chapter 17).

9.2   Microphysics of Clouds

Large particles ($a > 1$ μm) are vulnerable to sedimentation, which accounts for 10 to 20% of the overall removal of atmospheric aerosol. A more efficient removal mechanism is *washout*, which occurs when cloud droplets that form on aerosol particles carry them back to the earth's surface. This removal mechanism is closely related to cloud formation.

9.2.1   Droplet Growth by Condensation

The simplest means of forming cloud is through *homogeneous nucleation*, wherein pure vapor condenses to form droplets. Suppose a small embryonic

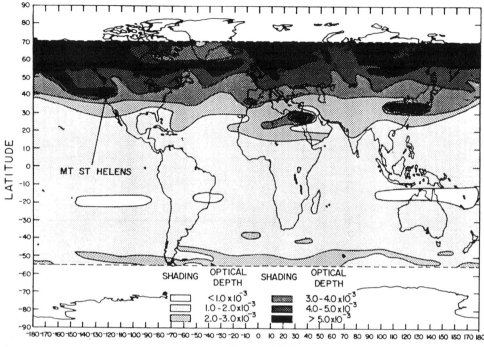

Figure 9.6 Mean aerosol optical depth at 1 μm two months after the eruption of Mt. St. Helens. Constructed by averaging satellite measurements from the Stratospheric Aerosol and Gas Experiment (SAGE) between July and August 1980. After Kent and McCormick (1984). Copyright by the American Geophysical Union.

droplet forms through chance collisions of vapor molecules. The survival of that droplet is determined by a balance between condensation and evaporation. The equilibrium between the droplet and surrounding vapor is described by the Gibbs free energy (3.16.2), but with one modification. In addition to expansion work, a spherical droplet performs work in association with its surface tension σ, which has dimensions of energy/area. Thus, σdA represents the work to form the incremental area dA of interface between vapor and liquid.

For a heterogeneous system comprised of the droplet and surrounding vapor, the fundamental relation for the Gibbs free energy (4.13) becomes

$$dG = -SdT + Vdp + (\mu_v - \mu_w)dm_v + \sigma dA, \qquad (9.2.1)$$

where

$$dm_v = -dm_w$$
$$= -n_w dV_w, \qquad (9.2.2)$$

and n_w and V_w denote the number density and volume of the droplet, respectively.[1]

The difference of chemical potential between the vapor and liquid phases can be expressed in terms of the vapor and saturation vapor pressures by appealing to the fundamental relation (3.16) for the individual phases.[2] Under an isothermal and reversible change of pressure de,

$$d\mu_v = v_v de \tag{9.3.1}$$

for the vapor and

$$d\mu_w = v_w de \tag{9.3.2}$$

for the liquid. Subtracting gives

$$d(\mu_v - \mu_w) = (v_v - v_w)de$$
$$\cong v_v de. \tag{9.4}$$

Then applying the gas law for an individual molecule of vapor

$$ev_v = KT \tag{9.5}$$

(e.g., Lee $et\ al.$, 1973), where K is the Boltzmann constant, gives

$$d(\mu_v - \mu_w) = KTd\ln e,$$

which upon integrating from the saturation pressure (at which $\mu_v = \mu_w$) obtains

$$\mu_v - \mu_w = KT\ln\left(\frac{e}{e_w}\right). \tag{9.6}$$

Incorporating (9.6) into (9.2) and integrating from a reference state of pure vapor (with the vapor pressure maintained) yields the change of Gibbs free energy

$$\Delta G = -V_w n_w KT\ln\left(\frac{e}{e_w}\right) + \sigma A.$$

Thus, forming a spherical droplet of radius a corresponds to the change of Gibbs free energy

$$\Delta G = -4\pi a^2\sigma - \frac{4}{3}\pi a^3 n_w KT\ln\left(\frac{e}{e_w}\right). \tag{9.7}$$

---

[1] In phase transformations away from saturation, the system is out of chemical equilibrium, so the chemical potentials in (9.2) need not be equal.

[2] Surface tension introduces a pressure difference between the droplet and surrounding vapor, which makes the pressure during a phase transformation variable. Strictly, that process should be treated as isothermal and isochoric, for which F is the appropriate free energy. However, discrepancies with G under isothermal and isobaric conditions are small enough to be ignored (see, e.g., Pruppacher and Klett, 1978).

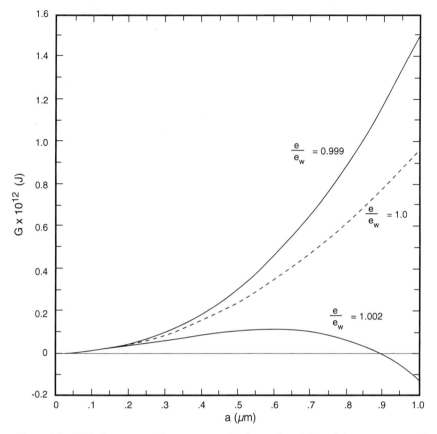

Figure 9.7 Gibbs free energy of a pure-water droplet produced through homogeneous nucleation, as a function of droplet radius. Shown for subsaturation ($e/e_w < 1$), saturation ($e/e_w = 1$), and supersaturation ($e/e_w > 1$).

Figure 9.7 shows the free energy as a function of droplet radius for subsaturation ($e/e_w < 1$), saturation ($e/e_w = 1$), and supersaturation ($e/e_w > 1$). For subsaturation as well as for saturation, G increases monotonically with droplet radius a. Under isothermal isobaric conditions, a system approaches thermodynamic equilibrium by reducing its free energy (3.23), so droplet formation is not favored for $e/e_w \leq 1$. Those conditions characterize stable equilibrium and $G(a)$ an unbounded potential well. A droplet formed through chance collision of vapor molecules will spontaneously evaporate (e.g., in association with latent heat released during its formation). Under supersaturated conditions, $G(a)$ possesses a maximum at a critical radius a_c, beyond which the free energy decreases and condensation is more efficient than evaporation. $G(a)$ then characterizes conditional equilibrium and a bounded potential well. A droplet formed with $a < a_c$ evaporates back to the initial state ($G = 0$). But one formed with $a > a_c$ grows spontaneously through condensation of vapor and is said to be *activated*.

The critical radius for a given temperature and vapor pressure follows by differentiating (9.7):

$$a_c = \frac{2\sigma}{n_w K T \ln\left(\frac{e}{e_w}\right)}. \qquad (9.8)$$

Known as *Kelvin's formula*, (9.8) can be rearranged for the *critical* or *equilibrium supersaturation*

$$\epsilon_c = \left(\frac{e}{e_w}\right)_c - 1, \qquad (9.9)$$

above which a droplet of radius a grows spontaneously through condensation. As shown in Fig. 9.8, small droplets have higher ϵ_c than large droplets, with $\epsilon_c \to 0$ as $a \to \infty$. This behavior reflects the so-called *curvature effect* of surface tension (9.2), which makes the equilibrium vapor pressure over a spherical droplet higher than that over a plane surface.

The curve for ϵ_c describes an unstable equilibrium, which therefore cannot be maintained. Adding a vapor molecule drives a droplet of radius a above the equilibrium curve. The actual supersaturation ϵ then lies above the equilibrium value ϵ_c, so the droplet continues to grow through condensation. Removing a water molecule has the reverse effect, wherein the droplet continues to evaporate. In either case, a perturbed droplet evolves away from the equilibrium state. More importantly, values of ϵ_c implied by Kelvin's formula pose a practical barrier to droplet formation via chance collection of vapor molecules. An embryonic droplet as large as 0.01 μm still requires a supersaturation of 12% to be sustained. Yet, supersaturations exceeding 1% are rarely observed, $\epsilon = 0.1\%$ being typical. Therefore, cloud formation cannot be explained by homogeneous nucleation.

Instead, cloud droplets form through *heterogeneous nucleation*, when water vapor condenses onto existing particles of atmospheric aerosol. Termed *cloud condensation nuclei* (CCN), such particles support condensation at supersaturations well below those required for homogeneous nucleation. Particles that are *wettable* allow water to spread over their surfaces. Such particles provide ideal sites for condensation because they then resemble a droplet of pure water, which, for observed supersaturations, could not attain that size through homogeneous nucleation. According to Fig. 9.8, the larger such a nucleus, the lower its equilibrium supersaturation and the more it favors droplet growth through condensation. Other conditions equal, large nuclei then activate sooner than small nuclei.

Even more effective are hygroscopic particles, like sodium chloride and ammonium sulfate. In the presence of moisture, NaCl and $(NH_4)_2SO_4$ absorb vapor and readily dissolve (Fig. 9.3). The resulting solution has a saturation

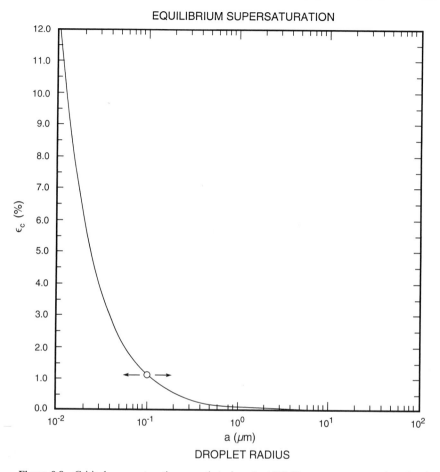

Figure 9.8 Critical supersaturation $\epsilon_c = (e/e_w)_c - 1$ at 278 K necessary to sustain a droplet of radius a. The critical value of ϵ describes an unstable equilibrium, wherein perturbations in a are reinforced. A perturbed droplet then either evaporates or grows through condensation away from the equilibrium state.

vapor pressure below that of pure water—because e_w is proportional to the absolute concentration of water molecules on the surface of the droplet. Consequently, a droplet containing dissolved salt favors condensation more than would a pure-water droplet of the same size. The so-called *Köhler curves* (Fig. 9.9) describe the equilibrium supersaturations for solutions containing specified amounts of solute. Hence, a droplet which develops on a soluble nucleus evolves along the curve corresponding to the fixed mass of that solute. The presence of NaCl sharply reduces the equilibrium supersaturation below that for a pure-water droplet (homogeneous nucleation), which is superposed on Fig. 9.9. For fixed a, ϵ_c decreases with increasing solute and eventually

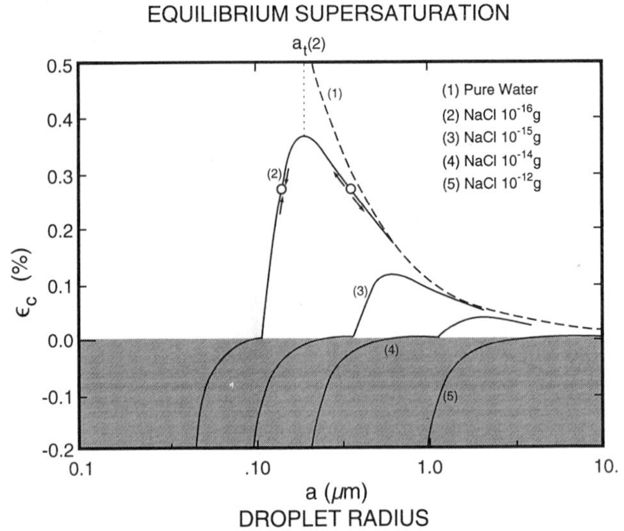

EQUILIBRIUM SUPERSATURATION

Figure 9.9   Critical supersaturation necessary to sustain a droplet of radius a and containing specified amounts of solute (solid lines). Critical supersaturation for a pure-water droplet (dashed line) is superposed. For a given amount of solute, the corresponding *Köhler curve* describes different forms of equilibrium, depending on whether a is smaller or larger than the threshold radius at the maximum of the curve, a_t. For $a < a_t$, a droplet is in stable equilibrium: Perturbations in a are opposed by condensation and evaporation, which restore the droplet to its original state. For $a > a_t$, a droplet is in unstable equilibrium: Perturbations in a are reinforced, so the droplet either evaporates toward a_t or grows through condensation away from its original state.

becomes negative (i.e., $RH < 100\%$). Hence, nuclei that are more soluble activate sooner and are more favorable to droplet growth.

The equilibrium described by a given Köhler curve differs according to whether the droplet radius is larger or smaller than a at the maximum of the curve, which is hereafter referred to as the threshold radius a_t. A droplet with $a < a_t$ is in stable equilibrium: Adding a vapor molecule drives the droplet along its Köhler curve to higher ϵ_c. The actual supersaturation ϵ then lies beneath the equilibrium value, so the droplet evaporates back to its original size. Removing a water molecule has the reverse effect. In either case, a perturbed droplet is restored to its original state. Droplets with $a < a_t$ cannot grow through condensation and comprise *haze*. Conversely, a droplet with $a > a_t$ is in unstable equilibrium: Adding a vapor molecule drives that droplet along its Köhler curve to lower ϵ_c. The actual supersaturation then exceeds the equilibrium value, so the droplet continues to grow through condensation. Removing a water molecule has the reverse effect. In either case, a perturbed droplet evolves away from its initial state, analogous to the behavior of a pure-water droplet (Fig. 9.8). Droplets with $a > a_t$ can enlarge into cloud drops.

Only a small fraction of tropospheric aerosol actually serves as CCN because many particles are too small, not wettable, or insoluble. Continental aerosol is the dominant source of CCN because of its greater number density and high concentration of soluble sulfate, which is found in many cloud droplets. By contrast, sea salt is not common in cloud droplets—even over maritime regions. Instead, CCN in maritime clouds are comprised chiefly of sulfate particles, which, through anthropogenic production, have nearly doubled globally (Houghton *et al.*, 1990). Sulfate aerosol also forms through condensation of dimethylsulphide (DMS), which is emitted by phytoplankton at the ocean surface.

Cumulus clouds have droplet number densities of order 10^2 cm^{-3} over maritime regions and 10^3 cm^{-3} over continental regions, as is reflected in observed droplet size spectra (Fig. 9.10). The greater number density of continental cloud droplets mirrors the greater number density of continental aerosol. It also implies a smaller mean droplet size because the density of condensate, or *liquid water content* ρ_l (g m^{-3}), does not differ appreciably between these cloud categories. In addition to being more numerous and smaller, droplets inside continental cumulus are more homogeneous in size than droplets inside maritime cumulus. The wider size spectrum of maritime cloud droplets

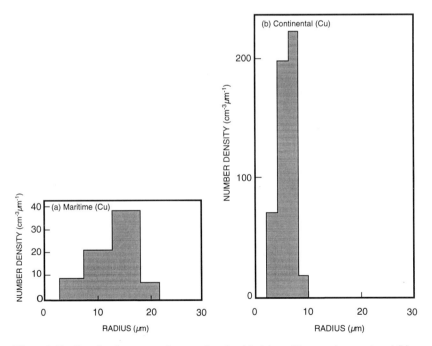

Figure 9.10   Droplet size spectra for cumulus cloud in (a) maritime environment and (b) continental environment. Note the compressed scale in (b). Area under each curve reflects the overall number density *n*, which is an order of magnitude greater in continental cumulus than in maritime cumulus. Source: Pruppacher (1981).

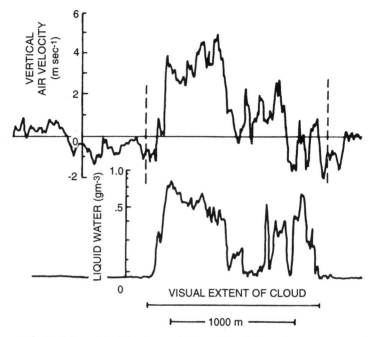

Figure 9.11 Variation of liquid water content ρ_l and vertical motion across a cumulus complex. After Warner (1969).

follows from the fact that they form chiefly on soluble sulfates that activate over a wide range of sizes.

Liquid water content varies strongly over dimensions of a cumulus complex (Fig. 9.11), but remains highly correlated with upward motion. Liquid water content ρ_l increases vertically to a maximum in the upper half of a cloud and then decreases sharply to zero at the cloud top. It is noteworthy that ρ_l observed in nonprecipitating clouds seldom approaches adiabatic values ρ_l^{ad}, which describe air parcels rising in isolation from environmental air. In fact, $\rho_l/\rho_l^{ad} < 0.5$ and vanishes at the cloud top, which reflects entrainment of drier environmental air through the walls of the cloud system (Sec. 7.4.2). For similar reasons, the greatest supersaturations are found in the core of a convective cloud, where moisture is relatively undiluted.

Before precipitation can occur, cloud droplets must enlarge by several orders of magnitude. After passing over the peak of its Köhler curve, a droplet can grow through condensation. The droplet's mass m then increases at the rate vapor diffuses across its surface

$$\frac{dm}{dt} = 4\pi r^2 \nu \frac{d\rho_v}{dr}, \tag{9.10}$$

where r denotes radial distance, ν is the diffusion coefficient for water vapor, and ρ_v is its density. By integrating from the droplet radius $r = a$ to infinity and expressing the droplet's mass in terms of the density of liquid water ρ_w, we obtain

$$\frac{da}{dt} = \frac{\nu}{a\rho_w}[\rho_v(\infty) - \rho_v(a)],$$

which, with the gas law, can be written

$$a\frac{da}{dt} = \nu\left[\frac{\rho_v(\infty)}{\rho_w}\right]\frac{e(\infty) - e(a)}{e(\infty)}. \tag{9.11}$$

For droplets larger than 1 μm, the curvature and solute effects (Figs. 9.8 and 9.9) are small enough to take $e(a) \cong e_w$. Then (9.11) can be expressed

$$a\frac{da}{dt} \cong \nu\left[\frac{\rho_v(\infty)}{\rho_w}\right]\epsilon, \tag{9.12.1}$$

which, if the right-hand side is approximately constant, gives

$$a(t) = \left\{2\nu\left[\frac{\rho_v(\infty)}{\rho_w}\right]\epsilon\right\}^{\frac{1}{2}}. \tag{9.12.2}$$

Droplet growth through condensation (Fig. 9.12) is rapid at small radius, but slows as a droplet enlarges. This implies that a droplet population will become increasingly homogeneous in size as it matures. While explaining the preliminary stage of droplet growth, condensation is too slow at large a to account for observed droplets in precipitating clouds.

9.2.2 Droplet Growth by Collision

Clouds that lie entirely beneath the freezing level are called *warm clouds*. Droplets inside warm clouds also grow through coalescence, when they collide with other droplets. Large droplets have faster fall speeds than small droplets. A large *collector drop* then sweeps through a volume of smaller droplets. The collection efficiency E, which is a dimensionless quantity that reflects the effective collecting area posed to other droplets, is small for collector radii less than 20 μm and increases sharply with collector size. For this reason, the tail of the droplet spectrum at large a is instrumental in forming precipitation.[3]

The rate at which a collector drop grows through collision with a homogeneous population of smaller droplets can be expressed as

$$\frac{dm}{dt} = \pi a^2 E \rho_l \Delta w, \tag{9.13}$$

---

[3] The collection efficiency also increases with the radius of the droplet population encountered by the collector (small droplets being swept aside without impact) up to a maximum. Beyond that droplet radius, coalescence is not favored (see, e.g., Wallace and Hobbs, 1977).

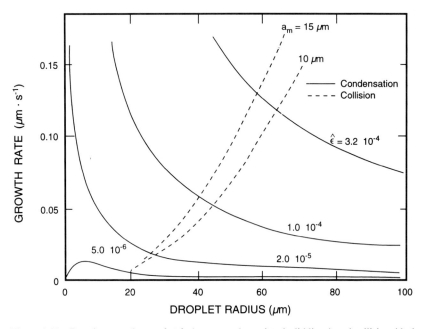

Figure 9.12 Droplet growth rate da/dt due to condensation (solid lines) and collision (dashed lines) inside a cloud with liquid water content of $\rho_l = 1.0$ g m^{-3}, as functions of droplet radius a. Growth rate due to condensation shown for different supersaturations $\hat{e} = \left[\frac{\rho_v(\infty)}{\rho_w} \right] \epsilon$ (see text). Growth rate due to collision shown for a droplet population whose size spectrum has a maximum at radius a_m. Adapted from Matveev (1967).

where a refers to the collector and Δw is the difference in fall speed between the collector and the population of smaller droplets. Rearranging as in the derivation of (9.12) yields

$$\frac{da}{dt} = \frac{Ew}{4} \frac{\rho_l}{\rho_w}, \qquad (9.14)$$

where $\Delta w \cong w$ has been presumed for the collector. Because w and E both increase with the radius of the collector, (9.14) predicts growth that accelerates with the size of the droplet. Therefore, growth due to collision eventually dominates that due to condensation. According to Fig. 9.12, droplets smaller than 20 μm grow primarily through condensation (solid lines), because fall speeds are small enough for such droplets to move in unison with the surrounding air and make collisions infrequent. However, collision (dashed lines) becomes dominant for droplets larger than 20 μm. Since it favors large drops, collision tends to broaden the size spectrum from the homogeneous population favored by condensation. It thus provides a mechanism for a small fraction of the drops to grow much faster than the rest of the population, with $a > 100$ μm dominating precipitation reaching the ground. Marine cumulus, which have larger droplet sizes due to comparatively fewer CCN, are more likely to precipitate than continental cumulus of similar size. Collision also tends to limit the size of cloud droplets to a couple of millimeters, larger drops fragmenting during impact.

9.2.3 Growth of Ice Particles

Clouds extending above the freezing level are called *cold clouds*. Cold clouds are seldom composed entirely of ice. In fact, nearly 50% of clouds warmer than $-10°C$ and virtually all clouds warmer than $-4°C$ contain no ice at all (Fig. 9.13). Droplets in such clouds remain in the metastable state of supercooled water (Problem 3.2). However, 90% of clouds colder than $-20°C$ contain some ice.

The reason cold clouds are not composed entirely of ice, or *glaciated*, is that ice can form only under special circumstances. Homogeneous nucleation of ice is not favored at temperatures above $-36°C$. The free energy of an ice embryo formed when water molecules collect inside a droplet increases for particles smaller than a critical dimension—analogous to homogeneous nucleation of droplets (Fig 9.7). Therefore, such particles tend to disperse. Because the critical dimension is greater than 20 μm for temperatures above $-36°C$, homogeneous nucleation of ice occurs only in very high clouds (see Fig. 7.7).

Heterogeneous nucleation occurs when water molecules collect onto a special type of particle known as a *freezing nucleus*, which has a molecular structure similar to that of ice. Because it assumes the size of the freezing nucleus, an ice embryo formed in this manner can grow at temperatures much warmer than that for homogeneous nucleation. Maritime clouds, whose droplet size spectrum is broad (Fig. 9.10), glaciate more readily than continental clouds. Crustal aerosol composed of clay is a prevalent form of freezing nucleus. But

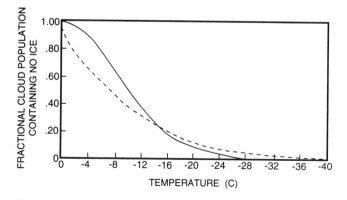

Figure 9.13 Fractional occurrence of unglaciated cloud, as a function of temperature, observed over Germany (solid line) and Minnesota (dashed line). Source: Pruppacher (1981).

even at $-20°$C, less than one out of every hundred million aerosol particles serves as an ice nucleus. This leaves many cloud droplets supercooled, which makes aircraft encountering such clouds prone to icing.

Once formed, ice particles can grow through condensation of vapor, or *deposition*. Since the saturation vapor pressure with respect to ice is lower than that with respect to water (4.40), a cloud that is nearly saturated with respect to water may be supersaturated with respect to ice.[4] A particle's growth is therefore accelerated if it freezes and, in a mixed population, ice particles will grow faster than droplets. Ice particles also grow through collision with supercooled droplets, or *riming*. This produces a layered structure about the original ice particle. Beyond a certain dimension, the original structure is no longer discernible. The resulting particle is then referred to as *grauple*, which is irregularly shaped and often proliferated with cavities. Last, ice particles can grow through coagulation, which is favored at temperatures above $-5°$C. These mechanisms lead to a wide array of shapes and sizes that include bullets, plates, slender needles, and complex configurations of hexagonal crystals (Fig. 9.14).

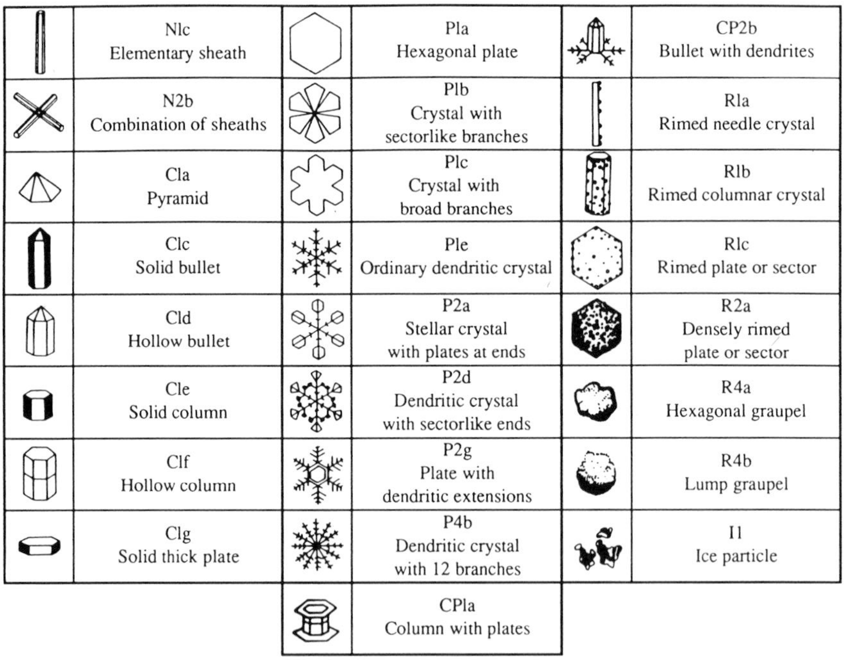

	N1c Elementary sheath		P1a Hexagonal plate		CP2b Bullet with dendrites
	N2b Combination of sheaths		P1b Crystal with sectorlike branches		R1a Rimed needle crystal
	C1a Pyramid		P1c Crystal with broad branches		R1b Rimed columnar crystal
	C1c Solid bullet		P1e Ordinary dendritic crystal		R1c Rimed plate or sector
	C1d Hollow bullet		P2a Stellar crystal with plates at ends		R2a Densely rimed plate or sector
	C1e Solid column		P2d Dendritic crystal with sectorlike ends		R4a Hexagonal graupel
	C1f Hollow column		P2g Plate with dendritic extensions		R4b Lump graupel
	C1g Solid thick plate		P4b Dendritic crystal with 12 branches		I1 Ice particle
			CP1a Column with plates		

Figure 9.14 Ice crystals observed in cold clouds. Adapted from Pruppacher and Klett (1978). Reprinted by permission of Kluwer Academic Publishers.

---

[4] Saturation mixing ratios with respect to water and ice typically differ by 10% or less, but at very cold temperatures representative of the upper troposphere the differences can be substantial (see Problems 5.27 and 17.9).

9.3 Macroscopic Characteristics of Clouds

9.3.1 Formation and Classification of Clouds

Most clouds develop through vertical motion, when moist air is lifted above its lifting condensation level (LCL) and becomes supersaturated. Clouds fall into three broad categories:

1. *Stratiform* (meaning "layered") *clouds* develop from large-scale lifting of a stable layer. Characteristic of sloping convection, this process is exemplified by warm moist air that overrides cold heavier air along a warm front. Vertical motion accompanying stratus cloud development is of order 1 cm s^{-1} and the characteristic lifetime of these clouds is of order 1 day.

2. *Cumuliform* (meaning "piled") *clouds* develop from isolated air plumes that ascend buoyantly. Associated with cellular convection, cumulous clouds grow through positive buoyancy supplied via sensible heat transfer from the surface and latent heat released to the air during condensation, both of which make these clouds dynamic. Updrafts are of order 1 m s^{-1} in developing cumulus, but can be several tens of m s^{-1} in organized mature cells like *cumulus congestus* and *cumulonimbus*. The characteristic lifetime of cumulus clouds ranges from a few minutes to hours.

3. *Cirriform* (meaning "fibrous") *clouds* develop through either of the preceding forms of lifting. Found at high altitudes, cirrus clouds are composed chiefly of ice particles, the larger components of which descend in fallstreaks or *mares' tails* that give these clouds their characteristic wispy appearance (Fig. 9.15).

Additional terms like *nimbo* and *alto*, which denote precipitation bearing and midlevel, respectively, are used to specify particular cloud types (WMO, 1969).

Cumulus clouds develop in association with a plume of rising air known as a *thermal* (Fig. 5.4), which can be treated as a succession of buoyant air parcels. Laboratory simulations reveal the development of a thermal to be closely related to its dissipation (Fig. 9.16a). A mass of buoyant fluid is transformed into a toroidal vortex ring (analogous to a smoke ring), which is characterized by a hole in its center and a complex array of protrusions along its advancing surface. Buoyant fluid penetrates its surroundings by entraining environmental fluid into its center (Fig. 9.16b). This process dilutes its buoyancy and causes the thermal to expand at an angle of about 15° from the vertical. The buoyant mass is turned completely inside out after advancing about 1.5 diameters, during which it has become thoroughly mixed with surrounding fluid and its buoyancy has been destroyed. The direct involvement of vorticity is illustrated in the trajectory of individual air parcels, which may complete several circuits about the advancing vortex ring before falling out of its influence.

Expansion is less evident in cumulus cloud because mixing with drier environmental air leads to evaporation along its periphery. Because observed liquid water

Figure 9.15 Hook-shaped *cirrus uncinus*. After WMO (1969).

contents are well below adiabatic values, mixing occurs throughout a cumulus cloud and some environmental air is communicated into its very core. Consequently, a cumulus can grow to appreciable height only by expanding enough to isolate saturated air in its interior from drier environmental air being mixed across its walls. Rising motion inside the thermal is compensated by gradual sinking motion outside the thermal, which is distributed over a broader area. That subsidence is promoted by evaporative cooling when saturated air mixes with unsaturated environmental air. By inhibiting ascent, subsiding motion organizes new parcels into the existing cloud, which, in turn, favors its growth. Likewise, buoyant parcels are most likely to produce cloud growth if the region

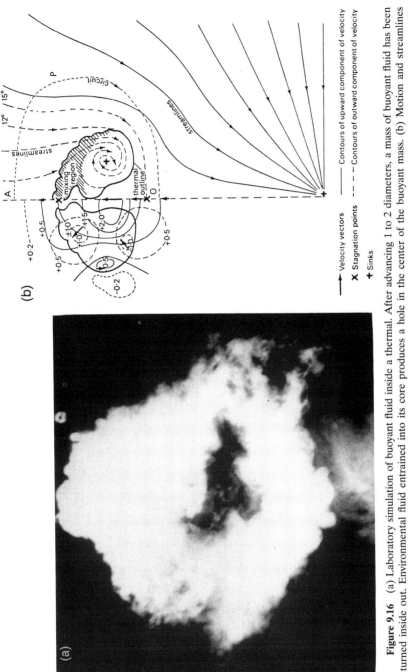

Figure 9.16 (a) Laboratory simulation of buoyant fluid inside a thermal. After advancing 1 to 2 diameters, a mass of buoyant fluid has been turned inside out. Environmental fluid entrained into its core produces a hole in the center of the buoyant mass. (b) Motion and streamlines about the moving body of fluid. Mixing with surrounding fluid causes the advancing mass to expand at an angle from the vertical of about 15°. Adapted from Scorer (1978).

into which they advance has already been moistened and its stability reduced by preceding parcels.

A developing cumulus is marked by well-defined protrusions along its boundary, especially at its top. Those dissolve through evaporation as they penetrate into unsaturated air and are replaced from the rear by new protrusions. The base of a developing cumulus is also well defined, being level or even concave down—because warmer temperature in the thermal's core elevates the LCL (5.24). Conversely, a decaying cumulus is rendered amorphous by turbulent mixing, which leads to evaporation and quickly blurs detailed features created during its development. The cloud's base may then become concave up due to a reversal of vertical motion and the horizontal temperature gradient. Cloud matter descending beneath a cloud's base is termed *virga*, and referred to as precipitation only if it reaches the ground.

Thermal growth is inhibited by strong vertical stability. Figure 9.17 illustrates a thermal that encounters and partially penetrates through a layer of strong stability. After penetrating the stable layer, negatively buoyant fluid continues to advance and eventually diffuses via turbulent mixing with surrounding fluid. Fluid deflected by the stable layer fans out to form an anvil. Such behavior appears occasionally at intermediate levels in the troposphere (Fig. 9.18) and invariably when a cumulonimbus encounters stable air at the tropopause. Turbulent entrainment that accompanies penetrative overshoots at the tropopause (Fig. 7.7) then mixes tropospheric and stratospheric air inside the resulting anvil.

The stratiform anvil of a mature cumulonimbus often evolves into small cellular convection. LW emission at its top and absorption at its base (Sec. 9.4.2) destabilize the anvil layer and introduce convective overturning. Domes of sinking virga then define *mammatus* (meaning "breast-shaped")

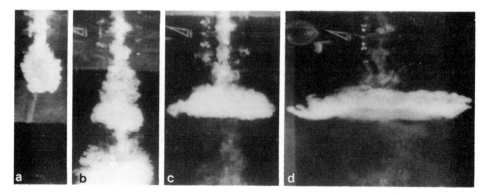

Figure 9.17 Laboratory simulation of a thermal encountering a layer of strong static stability. Upon reaching the stable layer, the negatively buoyant plume is deflected horizontally into an anvil. Fluid penetrating the stable layer continues to advance and eventually diffuses through turbulent mixing. After Scorer (1978).

Figure 9.18 A developing cumulus that encounters and penetrates through a shallow layer of strong static stability. Courtesy of Ronald L. Holle (Holle Photography).

clouds (Fig. 9.19), the descent of which is promoted by evaporative cooling. Spaced quasi-regularly, mammatus features resemble cells in Rayleigh–Bénard convection (Fig. 9.20), which develops in a shallow fluid that is heated differentially at its top and bottom. The horizontal dimension of the cells reflects the depth of the layer involved.[5] Stratiform cloud is often destabilized by absorption of LW radiation from the surface and emission to space, which can transform a continuous cloud layer into an array of cumulus cells having a preferred dimension. *Stratocumulus* and *fair weather cumulus* develop behind a cold front in this fashion (e.g., Fig. 9.21), when cold air that is advected over a warmer surface is heated at its base and cooled radiatively at its top.

Clouds also form in the absence of buoyancy through forced lifting over elevated terrain. *Orographic clouds* do not delineate a particular body of air but, rather, air flows through them. The *lenticular* (meaning "lens-shaped") *cloud* forms inside an organized wave pattern (Fig. 9.22a) that develops when the air stream is deflected over a mountain range. Each lenticular cloud in Fig. 9.22a continually forms along its leading edge, where air is displaced above its LCL, and dissolves along its trailing edge, where air returns below

---

[5] According to Rayleigh-Bénard theory, conversion of potential energy into kinetic energy becomes increasingly efficient with decreasing horizontal dimension of the cells. Energy conversion is eventually offset by frictional dissipation (e.g., mixing of momentum across the walls of the cells), which becomes important at horizontal dimensions comparable to the depth of the fluid. It is no accident that cumulonimbus cells have dimensions comparable to the depth of the troposphere.

Figure 9.19 *Mammatus cloud* that develops from a cumulonimbus anvil. Supplied by the National Center for Atmospheric Research, University Corporation for Atmospheric Research, National Science Foundation.

its LCL. Also known as *mountain wave clouds*, lenticular clouds are often found at great heights and, on occasion, in the stratosphere, where they are termed *nacreous clouds*. Vertical motion accompanying wave clouds has been harnessed by sailplanes to achieve record-breaking altitudes—as high as 50,000 feet. The vertical structure of lenticular clouds is controlled by environmental conditions. When moisture is stratified in layers, lenticular clouds form in a stack (Fig. 9.22b). Such stratification can result from towers of anomalous moisture left upstream by prior convection, which are then sheared by the circulation and folded into a layered structure.

The smooth form of lenticular clouds characterizes air flow that is turbulence free or *laminar*. That steady motion is often replaced just downstream by severe turbulence, which is made partly visible by *fractus* (meaning "fragmented") *clouds*. So-called *rotor clouds*, which are evident at levels beneath the lenticular clouds on Figs. 9.22a and b, are "dynamic." Looping cloud matter in Fig. 9.22c makes visible strong updrafts and downdrafts and severe turbulence, which have been faulted for numerous aircraft disasters over mountainous terrain.

Because they remain fixed, orographic clouds make cloud behavior unreliable as a tracer of air motion. Cumulus clouds more faithfully track the

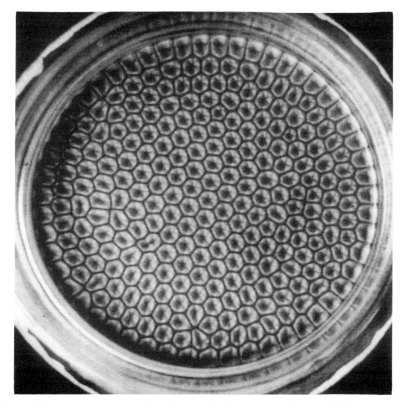

Figure 9.20 Laboratory simulation of Rayleigh–Bénard convection. A fluid layer heated at its base evolves into a regular array of hexagonal cells with a preferred horizontal scale. Fluid ascends in the center of each cell and descends along its edges. After Koschmieder and Pallas (1974).

movement of individual bodies of air, but even they are complicated by non-conservative behavior (e.g., condensation and evaporation), which limits their usefulness for diagnosing air motion.

All of the foregoing clouds develop from upward motion, which reduces the saturation mixing ratio of air through adiabatic cooling. Cloud can also develop isobarically through heat rejection, which cools air below its dew point. Conductive and radiative cooling lead to the formation of fog and marine stratus over cold oceans.

The stratosphere is isolated from convective motions and therefore from the source of water vapor at the earth's surface. Consequently, clouds rarely form above the tropopause. Exceptional are nacreous clouds that develop in connection with the mountain wave. Nacreous clouds are thought to be a subset of more ubiquitous *polar stratospheric clouds* (PSCs). Because the stratosphere

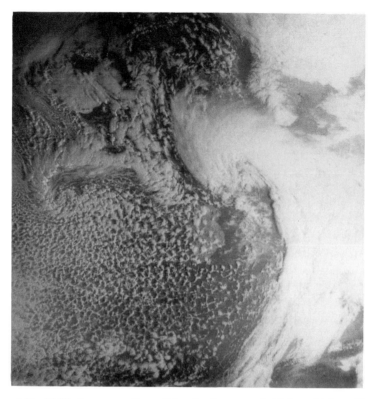

Figure 9.21 Visible image revealing cellular cloud structure behind a cold front. The cloud pattern is organized into a quasi-regular array of cells with a preferred horizontal scale. The cold front is part of a mature cyclone that advances across the British isles. Cloud cover ahead of the cold front marks the warm sector of the cyclone, which is just occluding north of Ireland. After Scorer (1986).

is very dry, PSCs form only under very cold conditions. Zonal–mean temperatures over the Antarctic fall below 190 K during Austral winter, more than 20 K colder than temperatures over the wintertime Arctic (Chapter 17). For this reason, PSCs occur over the Antarctic with much greater frequency and depth than over the Arctic.

Figure 9.22 (a) Schematic of the mountain wave that develops leeward of an elongated range. *Lenticular clouds* are fixed with respect to orography, while air flows through them. The positions of *rotor clouds* are also fixed, but those fragmented features are highly unsteady, with turbulent ascent (descent) along their leading (trailing) edges. Westward tilt with height (e.g., of wave crests) corresponds to upward propagation of energy (Chapter 14), a feature that is often visible in the structure of lenticular clouds. (b) Longitudinal view through the mountain wave. Smooth lenticular clouds form in a stack, separated by clear regions, due to stratification of moisture. These laminar features contrast sharply with turbulent rotor clouds below. (c) Transverse view from the ground. Vorticity and severe turbulence are made visible by looping cloud matter that comprises rotor clouds. Courtesy of V. Haynes and L. Feierabend (Soaring Society of Boulder).

Even over Antarctica, PSCs are tenuous by comparison with tropospheric clouds, almost all of which are optically thick. Some 3 to 5 km in depth, PSCs form at heights of 15 to 25 km and are often layered. Their occurrence is strongly correlated with temperature. Few clouds are sighted at temperatures warmer than 195 K, whereas the likelihood of finding PSC at temperatures below 185 K approaches 100% (WMO, 1988). Two distinct classes of PSCs are observed. Type I PSC is comprised of frozen droplets of nitric acid trihydrate (NAT), which has a low saturation vapor pressure. Submicron in scale, PSC I particles form at temperatures of 195 to 190 K and have the appearance of haze. Type II PSC may be nucleated by Type I particles or by background aerosol, but contains much larger crystals of water ice. PSC II particles, which are several tens of microns and larger, form at temperatures of 190 to 185 K and have features similar to cirrus. Their larger sizes give type II particles significant fall speeds, which makes them important in chemical considerations (Chapter 17).

9.3.2   Microphysical Properties of Clouds

Microphysical properties vary with cloud type (Table 9.2). Stratus have number densities $n \sim 300$ cm^{-3} and droplet radii of $a \sim 4$ μm. Liquid water content is more variable, but of order 0.5 g m^{-3}. Similar numbers apply for cumulus. However, cumulonimbus are distinguished by much lower number density and higher liquid water content, which imply significantly larger droplets. Compared to water clouds, cirrus have much lower number densities that are ob-

Table 9.2

Microphysical Properties of Clouds*

Cloud type	n (cm^{-3})	$\langle a \rangle$ (μm)	ρ_l (g m^{-3})	$\langle \hat{\sigma}_e \rangle$ @ 0.5 μm (μm^2)	$\langle \beta_e \rangle$ @ 0.5 μm (km^{-1})
Stratus (St)	300.	3.	0.15	450.	135.
Stratocumulus (Sc)	250.	5.	0.3	120.	30.
Nimbostratus (Ns)	300.	4.	0.4	400.	120.
Cumulus (Cu)	300.	4.	0.5	200.	60.
Cumulonimbus (Cb)	75.	5.	2.5	500.	38.
Cirrus (Ci)	0.03	250.	0.025[†]	1×10^4	0.5
Tropical cirrus (Cs)	0.10	800.	0.20[†]	4×10^4	0.4
PSC I	≤ 1	0.5	10^{-6} [†]	10	0.005
PSC II	≤ 0.10	50.	0.05[†]	100	0.015

*Representative values of number density n, mean droplet radius $\langle a \rangle$, liquid water content ρ_l, mean extinction cross section $\langle \hat{\sigma}_e \rangle$, and mean extinction coefficient $\langle \beta_e \rangle$ for different cloud types. Typical values are subject to a wide range of variability, especially for cirrus. Sources: Carrier *et al.* (1967), WMO(1988), Dowling *et al.* (1990), Liou (1990), Knollenberg *et al.* (1993), and Turco *et al.* (1993).

[†]Ice water content.

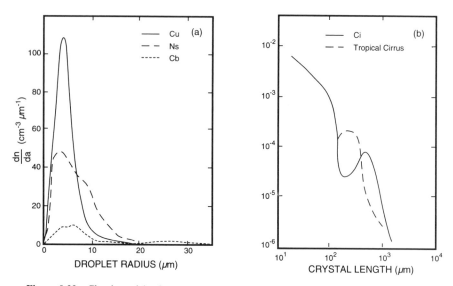

Figure 9.23 Cloud particle size spectra for (a) droplets inside water clouds and (b) crystals inside ice clouds. Adapted from Liou (1990).

served over a wide range: 10^{-7} to 10 cm^{-3} (Dowling *et al.*, 1990; Knollenberg *et al.*, 1993). Their ice water contents ρ_i are likewise smaller (e.g., $\rho_i < 0.1$ g m^{-3}). The characteristic dimension of cirrus ice crystals ranges from several tens of microns to as large as 1 mm. PSCs have number densities similar to cirrus and may form under analogous environmental conditions.

Figure 9.23 presents size spectra representative of several cloud types. Stratus and cumulus have the simplest spectra, which possess a single maximum. Spectra of nimbostratus and cumulonimbus exhibit secondary maxima at radii larger than the mean: $\langle a \rangle \cong 5$ μm. While comparatively unpopulated, the tails of these size spectra ($a > 20$ μm) describe particles that grow most rapidly through collision (Sec. 9.2.2) and consequently favor precipitation. Spectra of cirrus can also possess a secondary maximum, but at much larger size (e.g., at 100 to 1000 μm). These larger particles have rapid fall speeds, which produce the mares' tails characteristic of hook-shaped *cirrus uncinus* (Fig. 9.15).

9.3.3 Cloud Dissipation

Clouds dissolve through evaporation when unsaturated environmental air is entrained into and mixed with saturated air. By diluting moisture inside a thermal with cooler and drier environmental air, entrainment also siphons off positive buoyancy, kinetic energy, and excess water vapor that would otherwise be available to reinforce upward motion through latent heat release (Sec. 7.4.2). Entrainment is the principal destruction mechanism for cumulus cloud and must be offset for such cloud to grow.

When particles grow so large that they can no longer be supported by the updraft (e.g., $a > 100 \ \mu$m), clouds are dissipated by precipitation or, in the case of cirrus, sedimentation. Evaporative cooling and downward drag exerted on air by falling cloud particles oppose the updraft and the resupply of moisture and positive buoyancy necessary to sustain the cloud against entrainment (Problem 9.16). Strong LW cooling at the top of a cumulus tower, which promotes subsidence, can likewise oppose cloud growth.

Entrainment is less effective in stratiform cloud because of its horizontal extent and weaker vertical velocities. However, radiative heating at its base and cooling at its top destabilize stratus, which stimulates entrainment of drier air overhead and acts to dissolve such cloud. Radiative heating can also dissolve cloud by elevating its temperature above the dew point, which breaks up fog and marine stratus. This dissipation mechanism and formation mechanisms associated with surface heating make diurnal variations an important feature of the cloud field (Fig. 9.36).

9.4   Radiative Transfer in Aerosol and Cloud

The presence of liquid and solid particles greatly elongates the path traveled by photons of SW radiation, which then undergo repeated reflection and diffraction. Along with absorption inside particles, this sharply increases the optical depth posed to incident radiation.

9.4.1   Scattering by Molecules and Particles

RAYLEIGH SCATTERING

The simplest treatment of scattering describes the interaction of sunlight with molecules and is due to Rayleigh (1871). *Rayleigh scattering*, which applies to particles much smaller than the wavelength of radiation, considers a molecule exposed to electromagnetic radiation as an oscillating dipole. The strength of this oscillation is measured by the *dipole moment p*, which is related to the electric field E_0 of incident radiation as

$$p = \alpha E_0, \tag{9.15}$$

where α is the *polarizability* of the scatterer. Established through interaction with the wave's electromagnetic field, the dipole in turn radiates a scattered wave with electric field E. At large distances r from the dipole, radiation emitted at an angle γ from p behaves as a plane wave whose amplitude satisfies

$$E = \frac{1}{c^2} \frac{\sin \gamma}{r} \frac{\partial^2 p}{\partial t^2}, \tag{9.16.1}$$

where c is the speed of light (see, e.g., Jackson, 1975). Incorporating (9.15) and plane wave radiation of the form $\exp[ik(r - ct)]$ obtains

$$E = -E_0 k^2 \alpha \frac{\sin \gamma}{r} \exp[ik(r - ct)]. \qquad (9.16.2)$$

Consider sunlight at a *scattering angle* Θ from the path of incident radiation (Fig. 9.24). Sunlight is unpolarized, so its electric field E_0 is distributed isotropically over directions orthogonal to the path of propagation. It can, therefore, be represented in independent components E_{0x} and E_{0y} of equal magnitude. These incident electric fields induce dipole moments p_x and p_y, which, in turn, radiate scattered waves. Radiation scattered at the angle Θ makes an angle $\gamma_x = \pi/2$ from p_x and an angle $\gamma_y = (\pi/2) - \Theta$ from p_y. Then (9.16.2) gives

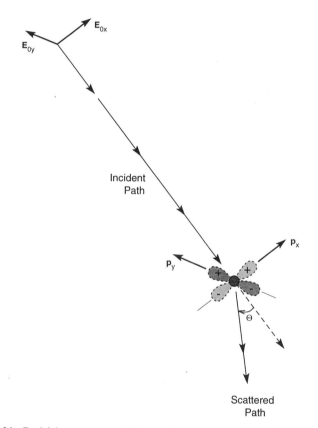

Figure 9.24   Rayleigh scattering of SW radiation by air molecules. Unpolarized sunlight is characterized by radiation with equal and independent electric fields E_{0x} and E_{0y}, which induce dipole moments p_x and p_y at the scatterer. Those in turn emit radiation at angles Θ from the path of incident radiation with electric fields E_x and E_y that are determined by p_x and p_y (see text). The orthogonal electric fields E_{0x} and E_{0y} have been chosen to position the scattering angle Θ in the plane formed by the path of incident radiation and E_{0y}.

the corresponding electric fields of scattered radiation

$$E_x = -E_{0x}k^2\alpha\frac{1}{r}\exp[ik(r-ct)],$$

$$E_y = -E_{0y}k^2\alpha\frac{\cos\Theta}{r}\exp[ik(r-ct)]. \tag{9.17}$$

The intensity of radiation is proportional to the average of $|E|^2$. Incident sunlight of intensity I_0 is then scattered into the angle Θ with intensity

$$I(\Theta, r) = I_0 k^4 \frac{\alpha^2}{r^2}\frac{(1+\cos^2\Theta)}{2}. \tag{9.18}$$

In terms of the phase function (8.22) and wavelength $\lambda = 2\pi/k$, (9.18) may be cast into the canonical form

$$I(\Theta, r) = I_0\frac{\alpha^2}{r^2}\frac{32\pi^4}{3\lambda^4}P(\Theta), \tag{9.19.1}$$

with

$$P(\Theta) = \frac{3}{4}(1+\cos^2\Theta). \tag{9.19.2}$$

The scattered intensity (Fig. 8.8a) has maxima in the forward ($\Theta = 0°$) and backward ($\Theta = 180°$) directions, with equal energy directed into each half-space.

The scattered flux follows as in (9.19), but with the incident flux F_0 in place of I_0. Integrating the scattered flux over a sphere of radius r obtains the scattered power

$$\mathscr{P} = F_0\alpha^2\frac{128\pi^5}{3\lambda^4}, \tag{9.20}$$

which has dimensions of energy/time. Then the scattering cross section for an individual molecule is

$$\hat{\sigma}_s = \frac{\mathscr{P}}{F_0}$$

$$= \alpha^2\frac{128\pi^5}{3\lambda^4}. \tag{9.21}$$

Finally, the scattered intensity at distance r can be expressed

$$I(\Theta, r) = I_0\frac{\hat{\sigma}_s}{4\pi r^2}P(\Theta). \tag{9.22}$$

Electromagnetic field theory relates the polarizability α to the dimensionless refractive index

$$m = m_r - m_i. \tag{9.23}$$

The real and imaginary parts of m refer to the phase speed and absorption of electromagnetic radiation in a medium relative to those in a vacuum. An ensemble of scatterers with number density n has polarizability

$$\alpha = \frac{3}{4\pi n}\left(\frac{m^2 - 1}{m^2 + 2}\right). \tag{9.24}$$

At wavelengths of visible radiation, absorption by air molecules is small enough for m_i to be ignored, while m_r is close to unity. Then (9.26) reduces to

$$\alpha \cong \frac{m_r - 1}{2\pi n}, \tag{9.25}$$

so the scattering cross section for an individual molecule is given by

$$\hat{\sigma}_s = \frac{32\pi^3(m_r - 1)^2}{3n^2\lambda^4}. \tag{9.26}$$

Scattering then yields an optical depth for the atmosphere of

$$\tau = \hat{\sigma}_s(\lambda) \int_0^\infty n(z)dz. \tag{9.27}$$

Rayleigh scattering explains why the sky appears blue. The λ^{-4} dependence of $\hat{\sigma}_s$ implies that shorter wavelengths are scattered by air molecules much more effectively than longer wavelengths. Blue light ($\lambda \sim 0.42 \ \mu$m) is scattered five times more than red light ($\lambda \sim 0.65 \ \mu$m), so it arrives at the earth's surface as diffuse radiation emanating from all directions. When the sun is near the horizon, sunlight passes obliquely through the atmosphere and therefore along an extended path. Shorter wavelengths are then scattered out of the incident beam, leaving longer (red) wavelengths to illuminate the sky. On average, about 40% of the SW flux is scattered out of the incident beam in the near-UV region, whereas less than 1% is removed in the near-IR region. Overall, about 10% of SW radiation incident on the atmosphere is scattered by molecules, half of which is returned to space and contributes to the earth's albedo (Fig. 1.27). Most of that scattering takes place in the lowest 10 km, where $n(z)$ is large. For this reason, increasing elevation witnesses a darkening of the sky and an intensification and whitening of direct sunlight. Rayleigh scattering influences ozone photochemistry and stratospheric heating, for example, by enhancing the SW flux available to the Chappuis bands of O_3 (Sec. 8.3.1).

Owing to its relationship to dipole radiation, Rayleigh scattering has certain polarization properties. In the forward and backward directions ($\Theta = 0$, $180°$), scattered radiation is unpolarized—like incident sunlight. However, scattered sunlight becomes increasingly polarized at intermediate angles and is fully polarized at directions orthogonal to the path of incident radiation ($\Theta = 90°$). It is for this reason that the sky assumes a deeper blue when viewed through a polarizer at directions orthogonal to incident sunlight.

MIE SCATTERING

Scattering from spherical particles of arbitrary dimension was first treated by Mie (1908). *Mie scattering* applies to the interaction of radiation with aerosol and cloud droplets. Figure 9.25 shows the refractive indices for water and ice, as functions of wavelength. Although m_r is of order unity, it varies with λ, which makes diffraction and reflection by spheres of condensate wavelength-dependent. On the other hand, m_i, which is proportional to the absorption coefficient $\beta_{a\lambda}$,

$$m_i = \beta_{a\lambda}\frac{\lambda}{4\pi}, \qquad (9.28)$$

is small in the visible. However, it increases sharply at $\lambda > 1$ μm for both condensed phases, which makes all but tenuous clouds optically thick in the IR.

Application of Mie scattering to atmospheric aerosol relies on particles being sufficiently separated for their interactions with the radiation field to be treated independently. Mie theory solves Maxwell's equations for the electromagnetic field about a dielectric sphere of radius a in terms of an expansion in spherical harmonics and Bessel functions (see Liou, 1980). Properties of the scattered wave field emerge in terms of the dimensionless size parameter

$$x = 2\pi\frac{a}{\lambda}. \qquad (9.29)$$

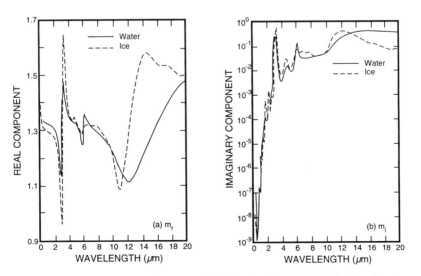

Figure 9.25 Index of refraction for water (solid line) and ice (dashed line) as functions of wavelength. (a) Real component and (b) imaginary component. Adapted from Liou (1990).

The dimensionless *scattering* and *extinction efficiencies*

$$Q_s = \frac{\hat{\sigma}_s}{\pi a^2}$$

$$Q_e = \frac{\hat{k}}{\pi a^2}$$

(9.30)

represent the fractional area of the incident beam removed through interaction with the sphere by scattering and *in toto*. The scattering efficiency is shown in Fig. 9.26 for a sphere with $m_r = 1.33$ and for several values of m_i. For conservative scattering ($m_i = 0$), Q_s increases quartically for $x << 1$ (which corresponds to Rayleigh scattering) and attains a maximum near $x \cong 2\pi$ (i.e., $\lambda \cong a$). The scattering efficiency then passes through a series of oscillations that result from interference between light transmitted through and diffracted about the sphere.

In the limit of large radius, the oscillations in scattering efficiency damp out to give

$$Q_s \sim 2 = \text{const} \qquad x \to \infty$$

(9.31)

for conservative scattering. Hence, a sphere much larger than the wavelength of incident radiation scatters twice the energy it intercepts. Scattered radiation then includes contributions from energy that is diffracted about the sphere as well as energy that is redirected by reflection inside the sphere. This limiting behavior applies to the interaction of visible radiation with cloud droplets

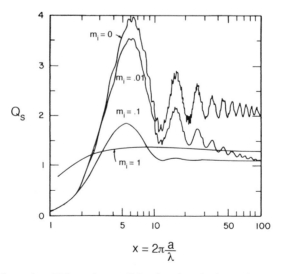

Figure 9.26 Scattering efficiency from a dielectric sphere having real component of refractive index $m_r = 1.33$ for several imaginary components m_i, as functions of the dimensionless size parameter $x = 2\pi(a/\lambda)$. After Hansen and Travis (1974). Reprinted by permission of Kluwer Academic Publishers.

(compare Fig. 9.23). The weak dependence on wavelength and the fact that droplet size spectra span several oscillations in Q_s explain why clouds appear white.

For nonconservative scattering ($m_i > 0$), Q_s exhibits similar behavior, but the oscillations due to interference are damped out with increasing absorption (since radiation emerges from the sphere progressively weaker). By $m_i = 1$, only the maximum near $x \cong 2\pi$ remains. Increased attenuation inside the sphere also reduces the limiting value of Q_s for $x \to \infty$, by absorbing energy that would otherwise contribute to the scattered wave field. Far from the sphere, radiation exhibits strong forward scattering (e.g., Fig. 8.8b), but is complicated for an individual sphere by rapid fluctuations with Θ that result from interference.

Under the conditions of independent scattering, properties of a population of spheres follow by collecting contributions from the individual particles. The scattering and extinction coefficients are then given by

$$\beta_s = \int \hat{\sigma}_s dn(a),$$

$$\beta_e = \int \hat{k} dn(a),$$

(9.32)

where $dn(a) = (dn/da)da$ describes the droplet size spectrum. The single scattering albedo for the population is characterized by

$$\omega = \frac{\beta_s}{\beta_e}.$$

(9.33)

Similarly, the optical depth of a cloud of thickness Δz_c can be estimated as

$$\tau_c = \beta_e \Delta z_c,$$

(9.34)

where β_e is representative of the cloud as a whole.

Figure 9.27 shows, for several wavelengths, the phase function (solid lines) corresponding to a droplet size spectrum representative of cumulus (Fig. 9.23). Compared to Rayleigh scattering (dashed line), Mie scattering possesses much stronger directionality. All λ shown exhibit strong forward scattering, especially shorter wavelengths, which are sharply peaked about $\Theta = 0$ and for which absorption is small (Fig. 9.25b). A weaker maximum appears at backward scattering for wavelengths in the near-IR and visible regions.

The mean extinction cross section for the population: $\langle \sigma_e \rangle = \beta_e/n$ (Fig. 9.28a), is nearly constant across the visible, where cloud droplets are large (9.31). Weak absorption at those wavelengths leads to a single scattering albedo (Fig. 9.28b) of near unity, so $\sigma_e \cong \sigma_s$ and scattering is likewise wavelength independent. Individual components of sunlight are then scattered with equal efficiency, so the cloud appears white. The extinction cross section attains a maximum near 5 μm, at a wavelength comparable to the mean radius of cloud droplets (Fig. 9.23), and falls off at longer wavelength. Because $\omega \sim 1$

DIRECTIONALITY OF SCATTERING

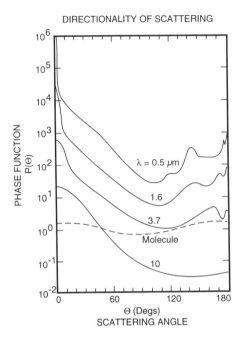

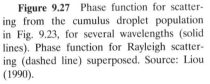

Figure 9.27 Phase function for scattering from the cumulus droplet population in Fig. 9.23, for several wavelengths (solid lines). Phase function for Rayleigh scattering (dashed line) superposed. Source: Liou (1990).

for $\lambda < 10$ μm, most of the extinction results from scattering, which dominates over absorption at wavelengths of SW radiation. Exceptional are narrow bands at $\lambda \cong 3$ and 6 μm, where absorption spectra of water and ice are peaked (Fig. 9.25b). At $\lambda > 10$ μm, ω decreases due to diminishing Q_s and increasing absorption. Similar optical properties follow from a droplet size spectrum representative of stratus.

Extinction by cloud droplets greatly increases the SW opacity of the atmosphere. For cumulus 1 km thick and a droplet number density of 300 cm^{-3}, (9.34) gives an optical depth at $\lambda = 0.5$ μm of $\tau_c = 50$, so the cloud is op-

CLOUD OPTICAL PROPERTIES

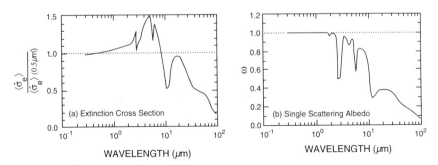

Figure 9.28 (a) Extinction cross section normalized by $\langle \sigma_e \rangle (0.5 \ \mu\text{m}) = 166 \ \mu\text{m}^2$ and (b) single scattering albedo for the cumulus droplet population in Fig. 9.23. Adapted from Liou (1990).

tically thick to visible radiation. Organized cumulus of the same dimensions can have $\tau_c > 100$. Even shallow stratus have optical depths in the visible well in excess of 10. In the IR, τ_c is smaller but still large enough to render all but tenuous clouds optically thick to LW radiation. At 10 μm, the foregoing conditions for cumulus lead to an optical depth of about 20.

Scattering by nonspherical ice particles is complicated by their irregular structure and anisotropy. Even though their rudimentary geometry is hexagonal, ice crystals assume a wide range of shapes (Fig. 9.14), which can produce a myriad of optical effects associated with orientation, directionality, and polarization. The familiar halo of thin cirrus results from prismatic crystals with a preferred orientation.

9.4.2 Radiative Transfer in a Cloudy Atmosphere

Influences of cloud and aerosol on atmospheric thermal structure can now be evaluated with the radiative transfer equation. Consider a homogeneous cloud layer that is illuminated by a solar flux F_s inclined at the zenith angle $\theta_s = \cos^{-1} \mu_s$ (Fig. 9.29). The source function for diffuse SW radiation (8.24) then becomes

$$
J(\phi, \mu, \tau) = \frac{\omega}{4\pi} \int_0^{2\pi} \int_{-1}^1 I(\phi', \mu', \tau)P(\phi, \mu; \phi', \mu')d\phi'd\mu'
$$

$$
+ \frac{\omega}{4\pi} F_s P(\phi, \mu; \phi', -\mu_s)e^{-\frac{\tau}{\mu_s}}, \tag{9.35}
$$

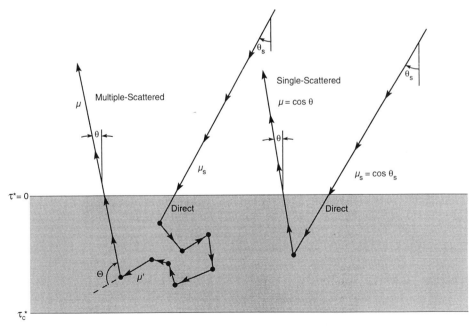

Figure 9.29 Transmission of diffuse SW radiation through a scattering layer.

where the wavelength dependence is implicit and emission is ignored. The first term on the right-hand side represents contributions to diffuse radiation from multiple scattering of the diffuse intensity I, whereas the second term represents the contribution from single scattering of direct solar radiation. Transmission of diffuse SW radiation through a cloudy atmosphere is then governed by the radiative transfer equation.

Scattering transforms (8.25) into an integrodifferential equation. It is then treated by expanding the phase function $P(\cos\Theta)$, where Θ is the three-dimensional scattering angle, in a series of spherical harmonics and the diffuse intensity I in a like manner; see Liou (1980) for a formal treatment. Expressing $P(\cos\Theta)$ in terms of the spherical coordinates (ϕ, θ) and integrating over ϕ then yields the transfer equation governing the azimuthal–mean component $I(\mu, \tau)$ of the diffuse radiation field

$$\mu\frac{dI}{d\tau} = I - \frac{\omega}{2}\int_{-1}^{1} I(\mu', \tau)P(\mu; \mu')d\mu' - \frac{\omega}{4\pi}F_s P(\mu; -\mu_s)e^{-\frac{\tau}{\mu_s}}, \qquad (9.36.1)$$

where

$$P(\mu; \mu') = \sum_j c_j P_j(\mu)P_j(\mu'), \qquad (9.36.2)$$

P_j is the Legendre polynomial of degree j, and the expansion coefficients c_j follow from orthogonality properties of the $P_j(\mu)$. For single scattering, the convolution integral is omitted. The diffuse intensity is then directly proportional to the phase function P. Approximate solutions for multiple scattering can be obtained by numerical techniques. A simple but enlightening solution follows from the two-stream approximation.

Owing to the involvement of Legendre polynomials, the integral in (9.36) can be evaluated efficiently with Gaussian quadrature at a finite number of zenith angles μ_j (see, e.g., Rektorys, 1969), each one describing a different stream of radiation. The two-stream approximation then follows by taking just two angles: $j = \pm1$, which correspond to the diffusivity factor $\bar{\mu}^{-1} = \sqrt{3}$, that is, to the upwelling and downwelling streams $I^{\uparrow\downarrow} = I(\pm\bar{\mu}, \tau)$. Incorporating the orthogonality properties of Legendre polynomials (Abramowitz and Stegun, 1972) and the transformation (8.54) leads to a system of two ordinary differential equations for the downwelling and upwelling fluxes

$$-\frac{dF^{\downarrow}}{d\tau^*} = F^{\downarrow} - \omega\left[fF^{\downarrow} + (1-f)F^{\uparrow}\right] - S^+ \exp\left(-\frac{\bar{\mu}}{\mu_s}\tau^*\right), \qquad (9.37.1)$$

$$\frac{dF^{\uparrow}}{d\tau^*} = F^{\uparrow} - \omega\left[fF^{\uparrow} + (1-f)F^{\downarrow}\right] - S^- \exp\left(-\frac{\bar{\mu}}{\mu_s}\tau^*\right), \qquad (9.37.2)$$

where

$$f = \frac{1+g}{2} \qquad (9.37.3)$$

and $(1 - f)$ represent the fractional energy that is forward and backward scattered, respectively,

$$g = \frac{1}{2} \int_{-1}^{1} P(\cos \Theta) \cos \Theta d(\cos \Theta) \qquad (9.37.4)$$

is an *asymmetry factor*, and

$$S^{\pm} = \omega \frac{F_s}{4}(1 \pm 3g\bar{\mu}\mu_s). \qquad (9.37.5)$$

The asymmetry factor g, which equals the first moment of the phase function, measures the difference in scattering between the forward and backward half-spaces. For isotropic scattering, $g = 0$ and $f = \frac{1}{2}$. The same holds for Rayleigh scattering. However, the strong forward lobe of Mie scattering (Figs. 8.8b and 9.27) leads to $f > \frac{1}{2}$, with $g \cong 0.85$ being representative for water clouds (see, e.g., Liou, 1990).

Subtracting and adding (9.36), introducing the net and total-emitted fluxes

$$F = F^{\uparrow} - F^{\downarrow}, \qquad (9.38.1)$$

$$\bar{F} = F^{\uparrow} + F^{\downarrow}, \qquad (9.38.2)$$

and differentiating with respect to τ^* results in the equivalent second-order system

$$\frac{d^2 F}{d\tau^{*2}} - \gamma^2 F = Z \exp\left(-\frac{\bar{\mu}}{\mu_s}\tau^*\right), \qquad (9.39.1)$$

$$\frac{d^2 \bar{F}}{d\tau^{*2}} - \gamma^2 \bar{F} = \bar{Z} \exp\left(-\frac{\bar{\mu}}{\mu_s}\tau^*\right), \qquad (9.39.2)$$

where

$$\gamma^2 = (1 - \omega)(1 + \omega - 2\omega f) \qquad (9.39.3)$$

$$Z = \frac{\bar{\mu}}{\mu_s}\bar{S} + (1 - \omega)S$$

$$\bar{Z} = -\frac{\bar{\mu}}{\mu_s}S - (1 + \omega - 2\omega f)\bar{S} \qquad (9.39.4)$$

and

$$S = S^+ - S^-$$

$$\bar{S} = S^+ + S^-. \qquad (9.39.5)$$

The general solution of (9.39) can be expressed

$$F^\uparrow = C\alpha_+ \exp(\gamma\tau^*) + D\alpha_- \exp(-\gamma\tau^*) + E_+ \exp\left(-\frac{\bar{\mu}}{\mu_s}\tau^*\right),$$

$$F^\downarrow = C\alpha_- \exp(\gamma\tau^*) + D\alpha_+ \exp(-\gamma\tau^*) + E_- \exp\left(-\frac{\bar{\mu}}{\mu_s}\tau^*\right),$$

$$(9.40.1)$$

where

$$\alpha_\pm = \frac{1 \pm \beta}{2}$$

$$\beta^2 = \frac{1 - \omega}{1 + \omega - 2\omega f}$$

$$E_\pm = \frac{\bar{\delta} \pm \delta}{2}$$

$$(9.40.2)$$

$$\bar{\delta} = \frac{\mu_s^2}{\bar{\mu}^2 - \mu_s^2\gamma^2}\bar{Z} \qquad \delta = \frac{\mu_s^2}{\bar{\mu}^2 - \mu_s^2\gamma^2}Z.$$

The integration constants C and D are determined by boundary conditions. If τ^* is measured from the top of the cloud layer, which has an effective optical depth τ_c^*, and if the underlying surface is black, the downwelling diffuse flux vanishes at the top of the scattering layer and the upwelling diffuse flux vanishes at the bottom

$$F^\downarrow(0) = 0,$$

$$F^\uparrow(\tau_c^*) = 0.$$

$$(9.41)$$

Then

$$C = \frac{E_-\alpha_- \exp(-\gamma\tau_c^*) - E_+\alpha_+ \exp(-\frac{\bar{\mu}}{\mu_s}\tau_c^*)}{\alpha_+^2 \exp(\gamma\tau_c^*) - \alpha_-^2 \exp(-\gamma\tau_c^*)},$$

$$D = \frac{E_+\alpha_- \exp(-\frac{\bar{\mu}}{\mu_s}\tau_c^*) - E_-\alpha_+ \exp(\gamma\tau_c^*)}{\alpha_+^2 \exp(\gamma\tau_c^*) - \alpha_-^2 \exp(-\gamma\tau_c^*)}.$$

$$(9.42)$$

The solution can then be used to evaluate the cloud albedo and transmissivity

$$\mathscr{A}_c = \frac{F^\uparrow(0)}{\mu_s F_s},$$

$$(9.43.1)$$

$$\mathscr{T}_c = \frac{F^\downarrow(\tau_c^*) + \mu_s F_s \exp(-\frac{\bar{\mu}}{\mu_s}\tau_c^*)}{\mu_s F_s},$$

$$(9.43.2)$$

where the direct solar contribution in (9.43.2) is negligible for cloud that is optically thick. From these, the cloud absorptivity follows as $1 - \mathscr{A}_c - \mathscr{T}_c$.

Figure 9.30 shows $\mathscr{A}_c$ and $\mathscr{T}_c$ as functions of optical depth and solar zenith angle for microphysical properties representative of cumulus (Fig. 9.28) and for a wavelength of 0.5 μm. Cloud albedo increases sharply at small optical depth (Fig. 9.30a) and exceeds 50% for $\tau_c > 20$, which is typical of water clouds. Hence, even shallow stratus are highly reflective. Transmissivity decreases with increasing τ_c—nearly complementary to cloud albedo, so most incident SW energy is either reflected or transmitted. The absorptivity is of order 10% across much of the range of optical depth shown, with contributions from droplets and surrounding vapor being comparable (Stephens, 1978a). Because the cloud layer is optically thick, most of the absorption occurs near its top, where heating rates $\dot{q}/c_p$ are 10 K day^{-1} and greater. Cloud albedo also increases with increasing solar zenith angle (Fig. 9.30b), which elongates the

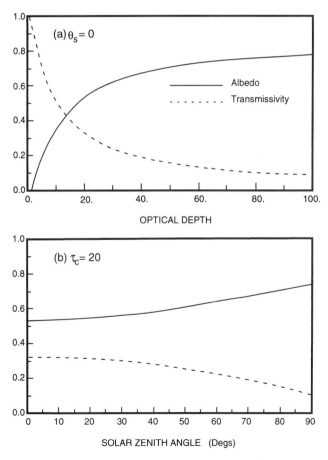

Figure 9.30 Albedo and transmissivity at 0.5 μm for a cloud layer with microphysical properties of cumulus in Fig. 9.23 and nearly conservative scattering (a) for overhead sun, as functions of cloud optical depth, and (b) for a cloud optical depth of 20, as functions of solar zenith angle.

slant optical path. For $\tau_c = 20$, $\mathscr{A}_c$ increases from 53% for overhead sun to 75% when the sun is on the horizon. The behavior of $\mathscr{T}_c$ is nearly complementary, decreasing from 32% for overhead sun to only 10% when the sun is on the horizon. Shallower cloud depends even more strongly on solar zenith angle because the slant optical path then varies between optically thin and optically thick conditions.

The factor most strongly influencing cloud optical properties is liquid water content, which determines τ_c (Stephens, 1978b). The optical depth of a cloud can be expressed in terms of the column abundance of liquid water,

$$\Sigma_l = \int_0^\infty \rho_l dz$$

$$= \Delta z_c \cdot \frac{4\pi\rho_w}{3} \int a^3 dn(a), \tag{9.44}$$

which is called the *liquid water path*. From (9.34) and the fact that $\hat{\sigma}_e = Q_e\pi a^2 \cong 2\pi a^2$ for SW radiation, the optical depth can be expressed as

$$\tau_c = \frac{3\Sigma_l}{2\rho_w a_e}, \tag{9.45.1}$$

where

$$a_e = \frac{\int a(\pi a^2)dn(a)}{\int (\pi a^2)dn(a)} \tag{9.45.2}$$

is a scattering-equivalent mean droplet radius. Because most water clouds have liquid water contents greater than 0.1 $g\,m^{-3}$ (Table 9.2), cloud layers more than a kilometer thick have liquid water paths Σ_l of 100 $g\,m^{-2}$ and greater. Those clouds then have optical depths in the visible of several tens and greater. Deep clouds with $\Sigma_l > 1000$ $g\,m^{-2}$ have albedos exceeding 80% (see the bright features in Fig. 1.24).

For given Σ_l, clouds with smaller droplets have a higher albedo and smaller absorptivity than clouds with larger droplets (Fig. 9.31). The increased reflectivity of clouds containing smaller droplets follows from the increased number densities in such clouds and reduced absorption by smaller droplets. Their smaller droplets and greater number densities make continental clouds more reflective than maritime clouds of the same type and liquid water path (see, e.g., Twomey, 1977). For the same reason, precipitation ($a > 100$ μm) sharply increases the absorptivity of cloud. Ice clouds, while more complex, are likewise highly reflective across the visible, where their dependence on microphysical properties resembles that of water clouds. Thin cirrus with ice water paths of $\Sigma_i < 5$ $g\,m^{-2}$ have albedos of order 10% and absorptivities of only a couple of percent (see, e.g., Paltridge and Platt, 1981).

According to Fig. 9.28b, scattering by cloud particles falls off at $\lambda > 4$ μm, where water absorbs strongly (Fig. 9.25b). Figure 9.32 illustrates that water clouds 3 km thick, which are highly reflective at $\lambda < 1$ μm, become fully absorbing at $\lambda > 4$ μm. The interaction of cloud with LW radiation can be treated

CLOUD OPTICAL PROPERTIES

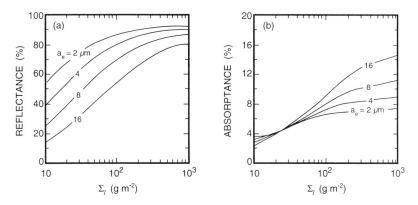

Figure 9.31   SW (a) reflectance and (b) absorptance, as functions of liquid water path, for a solar zenith angle of 60° and for different scattering-equivalent droplet radii a_e. Adapted from Slingo (1989).

as in the foregoing development for SW radiation, but with the scattering of direct solar in (9.37) replaced by emission of terrestrial radiation: $(1-\omega)B^*[T]$. Inside a cloud layer of temperature T_c, upwelling and downwelling LW fluxes are then described by

$$\frac{dF^\uparrow}{d\tau^*} = F^\uparrow - \omega\left[fF^\uparrow + (1-f)F^\downarrow\right] - (1-\omega)B^*[T_c], \qquad (9.46.1)$$

$$-\frac{dF^\downarrow}{d\tau^*} = F^\downarrow - \omega\left[fF^\downarrow + (1-f)F^\uparrow\right] - (1-\omega)B^*[T_c], \qquad (9.46.2)$$

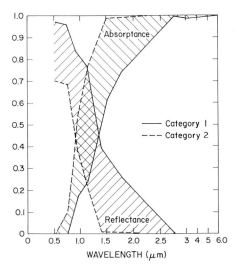

Figure 9.32   Reflectance and absorptance of 3-km-thick stratus, as functions of wavelength, for overhead sun and for a range of microphysical properties bounded by categories 1 and 2. Category 1 is based on a droplet distribution of characteristic radius 6 μm, $n=10^2$ cm^{-3}, and $\rho_l=0.30$ g m^{-3}. Category 2 includes rain with a droplet distribution of characteristic radius 600 μm, $n=0.001$ cm^{-3}, and $\rho_l=2.1$ g m^{-3}. Source: Cox (1981).

where the cloud is sufficiently thin for T_c to be treated as constant. The blackbody spectrum B^* is a particular solution of (9.46). Proceeding as in the treatment of SW radiation leads to the general solution

$$
\begin{aligned}
F^\uparrow &= C\alpha_+ e^{\gamma \tau^*} + D\alpha_- e^{-\gamma \tau^*} + (1 - \omega)B^*[T_c], \\
F^\downarrow &= C\alpha_- e^{\gamma \tau^*} + D\alpha_+ e^{-\gamma \tau^*} + (1 - \omega)B^*[T_c],
\end{aligned}
\tag{9.47}
$$

which applies inside the scattering layer.

To illustrate the influence of clouds on the LW energy budget, we consider a statistically homogeneous scattering layer within the framework of radiative–convective equilibrium (Sec. 8.5.2). Microphysical properties incorporated in the treatment of SW radiation (Fig. 9.28) are used to define a cloud layer of thickness $\Delta z_c = 1$ km and fractional coverage η that is situated at a height z_c in the troposphere—all under the conditions of radiative–convective equilibrium (Fig. 8.21). In the spirit of a gray atmosphere, radiative characteristics at 10 μm are used to define the broadband transmission properties for LW radiation, whereas the fractional cloud cover is adjusted to give an albedo of 30%. The fluxes $F^\uparrow$ and $F^\downarrow$ are described by (8.70) outside the cloud layer and by (9.47) inside the cloud layer. Matching fluxes across the scattering layer then leads to behavior analogous to that under simple radiative–convective equilibrium (Fig. 8.22), except for the addition of the cloud layer.

Figure 9.33a shows upwelling and downwelling fluxes in the presence of a cloud layer of $\tau_c = 20$ that is situated at $z_c = 8$ km. The equilibrium temperature structure (not shown) is identical to that in Fig. 8.21, except that temperatures beneath the cloud layer are uniformly warmer. The surface temperature T_s has increased from its cloud-free value of 285 to 312 K. Below the cloud layer, upwelling and downwelling fluxes resemble those under cloud-free conditions (Fig. 8.22a), except likewise displaced to higher values. Inside the cloud layer, the profiles of $F^\uparrow$ and $F^\downarrow$ undergo a sharp adjustment that drives them nearly into coincidence. This behavior follows from the cloud being optically thick, so, away from its boundaries, the LW flux is approximately isotropic and equal to σT^4. Above the cloud layer, $F^\uparrow$ and $F^\downarrow$ recover the cloud-free forms in Fig. 8.22a to satisfy boundary conditions at the top of the atmosphere (8.63).

Sharp changes of $F^\uparrow$ and $F^\downarrow$ in Fig. 9.33a introduce strong LW warming and cooling near the base and top of the cloud layer. The cloud-induced heating,

$$
\frac{\dot{q}}{c_p} - \left(\frac{\dot{q}}{c_p}\right)^{\mathrm{CS}},
$$

where $(\dot{q}/c_p)^{\mathrm{CS}}$ is the clear-sky heating profile under radiative–convective equilibrium (Fig. 8.22b), is shown in Fig. 9.33b. Absorption (warming) is concentrated within an optical depth of the cloud base, whereas emission (cooling) is concentrated within an optical depth of the cloud top. These regions of net ab-

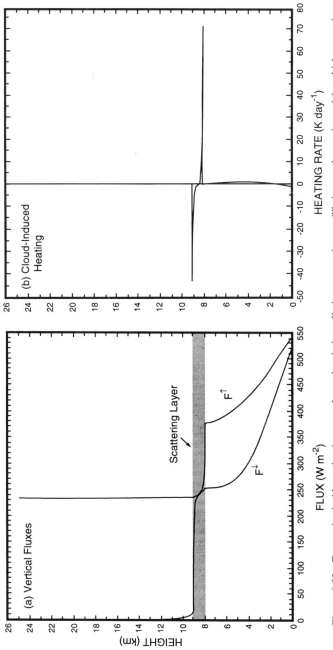

Figure 9.33 Energetics inside a cloudy atmosphere that is in radiative–convective equilibrium and contains a 1-km-thick scattering layer with microphysical properties of cumulus in Fig. 9.23. The fractional cloud cover is adjusted to give an albedo of 0.30. (a) Upwelling and downwelling LW fluxes. (b) Cloud-induced heating: $\frac{q}{c_p} - \left(\frac{q}{c_p}\right)^{cs}$, where the superscript refers to radiative–convective equilibrium under clear-sky conditions (Fig. 8.22).

sorption and emission are shallow, analogous to Chapman layers (Sec. 8.5.3), so they introduce substantial heating and cooling locally (8.72). The LW heating rate approaches 70 K day^{-1} near the cloud's base, whereas the LW cooling rate approaches 50 K day^{-1} near its top. By reemitting energy downward, the cloud also introduces heating beneath its base, but much smaller than the heating and cooling at the cloud's boundaries.

LW cooling in upper portions of a cloud is offset by SW heating, but only partially (see, e.g., Stephens *et al.*, 1978). Anvil cirrus experience net radiative cooling at their tops of several tens of K day^{-1} (see, e.g., Webster and Stephens, 1980) and heating at their bases as large as 100 K day^{-1} (Platt *et al.*, 1984a). Radiative cooling at its top and heating at its base destabilize a cloud layer, introducing convection that entrains drier environmental air and tends to dissolve the cloud (Fig. 9.19). Radiative heating also tends to dissolve cloud by elevating its saturation mixing ratio.

The cloud layer in Fig. 9.33 is optically thick. Therefore, it influences the LW energy budget chiefly through cloud-top emission, which decreases with the cloud's temperature and elevation. Figure 9.34 shows the dependence on cloud height of surface temperature under radiative–convective equilibrium.

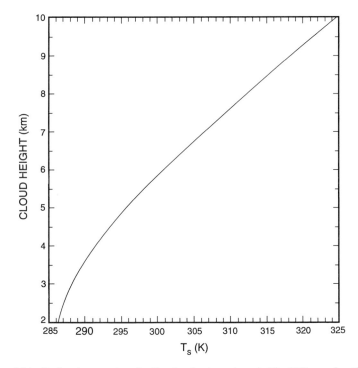

Figure 9.34 Surface temperature for the cloudy atmosphere in Fig. 9.33, as a function of the cloud layer's height.

Situating the cloud below 2 km recovers nearly the same surface temperature as under cloud-free conditions. The scattering layer is then embedded inside the optically thick layer of water vapor, so the cloud does not alter the effective emission temperature of the atmosphere. However, $z_c > 2$ km leads to warmer surface temperature, which increases steadily with the cloud's height. By $z_c = 8$ km, T_s has increased 27 K to maintain the cloud at the equivalent blackbody temperature of the earth (Problem 9.40).

Water clouds become optically thick to LW radiation for liquid water paths Σ_l greater than about 20 g m^{-2} (see, e.g., Stephens, 1984), wherein they behave as blackbodies. The LW characteristics of ice clouds are more complex, at least for those that are optically thin. The emissivity (LW absorptivity) of thin cirrus with ice water paths less than 10 g m^{-2} (which is typical of cirrus 1 km thick) is 0.5 or smaller and may vary diurnally with convection (Platt et al., 1984b). However, even cirrus emits approximately as a blackbody for Δz_c sufficiently large. Figure 9.35 shows spectra of outgoing LW radiation (OLR) in the tropics over a clear-sky region and several neighboring cirrus-covered regions of different optical depths. Cirrus reduces the radiation emitted to space throughout the LW, except near 15 μm, where radiation emitted by the surface has already been absorbed by underlying CO_2. By an optical depth of $\tau_c = 5$, OLR has approached the blackbody spectrum with a temperature equal to the cirrus temperature of 230 K. The effect of cirrus is to remove the layer emitting to space to a higher level and colder temperature. Outgoing radiation is most reduced in the atmospheric window between 8 and 12 μm, where radiation emitted by the surface passes relatively freely through the atmosphere under cloud-free conditions.

Contrary to the preceding treatment, real clouds have only finite horizontal extent. A finite cloud scatters energy out its sides, which allows neighboring clouds to interact. Energy lost through the sides of a cloud exceeds SW absorption for aspect ratios as small as 1/20 (Cox, 1981), so finite extent is important for all but extensive stratiform decks. Optical characteristics are then sensitive to the vertical distribution of microphysical properties. A cloud having the smallest droplets near its base will be less reflective than one having the smallest droplets near its top because, in the former case, much of the SW energy will have already been scattered out the sides before the highly reflective droplets are encountered. Finite cloud can also increase SW absorption by reducing the effective solar zenith angle.

9.5   Roles of Clouds and Aerosol in Climate

Clouds modify the global energy balance by altering the absorption and scattering characteristics of the atmosphere. The largest variations of albedo and OLR in Fig. 1.29 are associated with clouds (see, e.g., Ramanathan et al., 1989). Convection also supports large transfers of sensible and latent heat

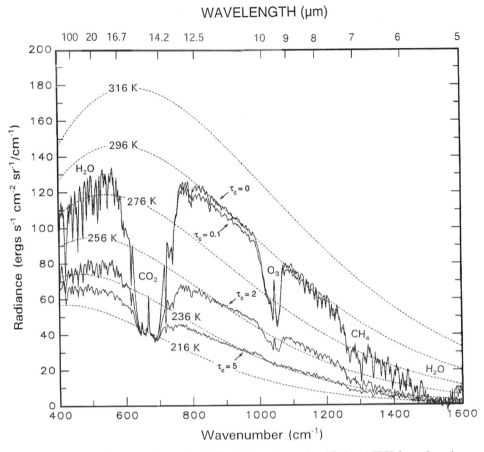

Figure 9.35 Spectrum of outgoing LW radiation observed by Nimbus-4 IRIS for a clear-sky region ($\tau_c = 0$) at (134°E,12°N) and neighboring cirrus-covered regions ($\tau_c > 0$). Courtesy of B. Carlson (NASA/Goddard Institute for Space Studies).

from the earth's surface (Fig. 1.27), which represent another important heat source for the atmosphere. Similar radiative effects are introduced by aerosol, which controls cloud formation through microphysical processes (Sec. 9.2).

9.5.1   Involvement in the Global Energy Budget

INFLUENCE OF CLOUD COVER

The role of clouds in global energetics can be interpreted as a forcing in the radiative energy balance. For LW radiation, the large opacity of clouds increases the optical depth of the atmosphere, which promotes increased surface temperature by enhancing the greenhouse effect (8.81). Thus, clouds introduce warming in the LW energy budget, and this warming is proportional to the cloud-top temperature and height (Fig. 9.34). Cold cirrus also cools the lower stratosphere via exchange in the 9.6-μm band of ozone (Sec. 8.5.3). For SW radiation, the high reflectivity of clouds decreases the incoming solar flux, which favors reduced surface temperature (8.81). Thus, clouds introduce cooling in the SW energy budget. Unlike the LW effect, cloud albedo is insensitive to cloud height, so low clouds in visible imagery (Fig. 1.24) are almost as bright as high clouds. The zenith angle dependence of scattering (Fig. 9.30b) enhances albedo at high latitudes. Clouds also introduce heating by absorbing SW radiation.

Clouds routinely cover about 50% of the earth and account for about half of its albedo. Marine stratocumulus (e.g., over the southeastern Atlantic in Fig. 1.24) are particularly important because they cover a large fraction of the globe and reflect much of the incident SW radiation. However, because their cloud-top temperatures do not differ substantially from those of the surface, marine stratocumulus do not appreciably modify OLR (refer to Fig. 1.23) and hence the LW energy budget. In contrast, deep cumulus over Africa, South America, and the maritime continent have cloud-top temperatures as much as 100 K colder than the underlying surface, so they sharply reduce OLR from that under cloud-free conditions.

The SW and LW effects of clouds are both variable. Owing to the zenith angle dependence of scattering (Fig. 9.30b) and to systematic variations in cloud cover and type, the most important component of cloud variability is diurnal. Figure 9.36 shows the daily variation of cloud properties over the southeastern Pacific, which is populated by marine stratocumulus, and over the Amazon basin, where deep continental convection prevails. Marine stratocumulus (Fig. 9.36a) is present throughout the day, but undergoes a 50% variation in albedo. This resembles the variation of fractional cloud cover, which results from SW heating during daylight hours. The corresponding variation of OLR is small. Continental convection (Fig. 9.36b) undergoes a 25% variation of OLR, which resembles the daily life cycle of cumulonimbus towers and anvil cirrus. The corresponding variation of fractional cloud cover is small. Cirriform shields, which are glaciated and markedly colder than the surface, persist several hours after convection has dissipated (Houze, 1982; Platt *et al.*, 1984a).

A quantitative description of how clouds figure in the global energy budget is complicated by their dependence on microphysical properties and interactions with the surface. These complications are circumvented by comparing radiative fluxes at the top of the atmosphere (TOA) under cloudy versus clear-sky conditions. Over a given region, the column-integrated radiative heating rate must equal the difference between the energy flux absorbed and that emitted

$$\dot{Q} = \int_0^{\infty} \rho \dot{q} \, dz$$
$$= (1 - \mathscr{A})F_s - F_{\text{LW}}, \tag{9.48}$$

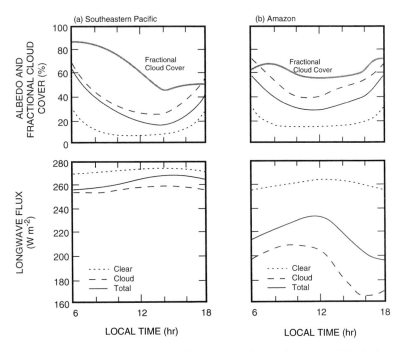

Figure 9.36 Top-of-the-atmosphere radiative properties, as functions of local time, over (a) the southeastern Pacific, which is populated by marine stratocumulus, and (b) the Amazon basin, which is populated by organized cumulonimbus and anvil cirrus. Adapted from Minnis and Harrison (1984).

where $\mathscr{A}$ is the albedo of the region and F_{LW} and F_s denote the downward SW and upward LW fluxes at the TOA. The *cloud radiative forcing*

$$C = \dot{Q} - \dot{Q}^{CS}, \tag{9.49}$$

where $\dot{Q}^{CS}$ is the heating rate under clear-sky conditions, then represents the net radiative influence of clouds on the column energy budget of the region. Incorporating (9.49) into (9.48) yields

$$C = (\mathscr{A}^{CS} - \mathscr{A})F_s + (F_{LW}^{CS} - F_{LW}), \tag{9.50}$$

from which we identify the SW and LW components of cloud forcing

$$C_{SW} = (\mathscr{A}^{CS} - \mathscr{A})F_s$$
$$= F_{SW}^{CS} - F_{SW} \tag{9.51.1}$$

$$C_{LW} = F_{LW}^{CS} - F_{LW}. \tag{9.51.2}$$

If the region has fractional cloud cover η, which refers to overcast subregions, the outgoing SW and LW fluxes are given by

$$F_{SW} = (1 - \eta)F_{SW}^{CS} + \eta F_{SW}^{OC},$$
$$F_{LW} = (1 - \eta)F_{LW}^{CS} + \eta F_{LW}^{OC}, \tag{9.52}$$

where the superscripts refer to clear-sky and overcast subregions. Then the SW and LW components of cloud radiative forcing can be written

$$\begin{aligned} C_{SW} &= \eta(F_{SW}^{CS} - F_{SW}^{OC}) \\ &= \eta F_s(\mathscr{A}^{CS} - \mathscr{A}^{OC}), \end{aligned} \tag{9.53.1}$$

$$C_{LW} = \eta(F_{LW}^{CS} - F_{LW}^{OC}). \tag{9.53.2}$$

The components of cloud forcing (9.53) can be evaluated directly from broadband fluxes of outgoing LW and SW radiation measured by satellite. Figure 9.37 shows time-averaged distributions of C_{SW}, C_{LW}, and C. Longwave forcing (Fig. 9.37a) is large in centers of deep convection over tropical Africa, South America, and the maritime continent, where C_{LW} approaches 100 W m^{-2} (compare Fig. 1.25b). Secondary maxima appear in the maritime Inter Tropical Convergence Zone (ITCZ) and in the North Pacific and North Atlantic storm tracks. Shortwave forcing (Fig. 9.37b) maximizes in the same regions, where $C_{SW} < -100$ W m^{-2}. Negative SW forcing over extensive marine stratocumulus is smaller but significant. Inside the centers of deep tropical convection, SW and LW cloud forcing nearly cancel, leaving small values of C (Fig. 9.37c) throughout the tropics. Negative C_{SW} in the storm tracks and over marine stratocumulus then dominate positive C_{LW} to prevail in the global–mean cloud forcing. Globally averaged values of C_{LW} and C_{SW} are about 30 and -45 W m^{-2}, respectively, which may be compared against the 90 W m^{-2} transfer of latent heat to the atmosphere (Fig. 1.27). The global–mean cloud forcing is then -15 W m^{-2}. This represents cooling about three times as great the warming that would be introduced by a doubling of CO_2. Thus, even a small change of cloud radiative forcing could overshadow the direct effect of increased greenhouse gases.

While circumventing many of the complications surrounding cloud behavior, cloud forcing is limited in several respects. Foremost is the fact that C gives only the column-integrated effect of clouds—it provides no information on the vertical distribution of that heating. Shortwave cloud forcing (which represents cooling) is concentrated near the surface, because the principal effect of increased albedo is to shield the ground from incident solar radiation. Longwave cloud forcing (which represents warming) is manifested in heating near the base of a cloud and cooling near its top (Fig. 9.33b). That

(a) Longwave Cloud Forcing

(b) Shortwave Cloud Forcing

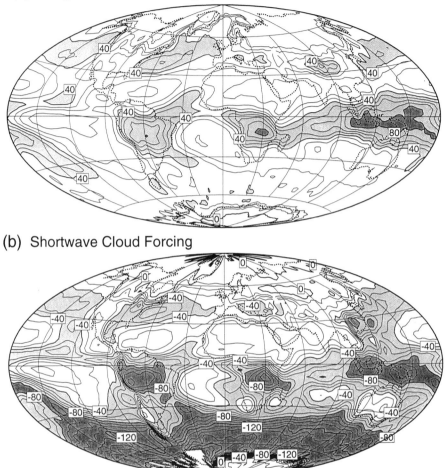

Figure 9.37 Cloud radiative forcing during northern winter derived from ERBE measurements on board the satellites ERBS and NOAA-9 for the (a) LW energy budget, (b) SW energy budget. (*continues*).

radiative forcing depends importantly on the vertical distribution of cloud. For instance, deep cumulonimbus and comparatively shallow cirrus with the same cloud-top temperature yield identical LW forcing, yet imply very different optical depths. The strong correlation between water vapor and cloud cover introduces another source of uncertainty because it biases C_{LW} toward higher values (Hartmann and Doelling, 1991).

Clouds introduce heating through the release of latent heat. Precipitation leads to a net transfer of heat from the oceans to the atmosphere. Latent heating is particularly important for organized deep cumulus clouds, which produce large volumes of precipitation in the tropics. The column-integrated

(c) Net Radiative Cloud Forcing

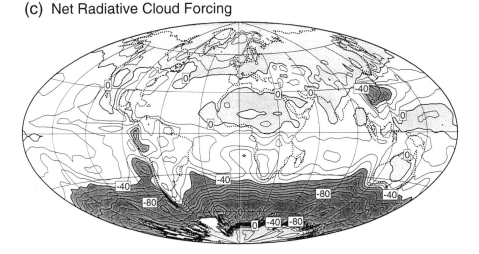

Figure 9.37 (*Continued*). (c) net radiative energy budget. Courtesy of D. Hartmann (U. Washington).

latent heating rate follows from the specific heat l and precipitation rate $\dot{P}$ as

$$\dot{Q} = l\rho_w\dot{P}. \tag{9.54}$$

A counterpart of TOA radiative fluxes in (9.48), (9.54) gives approximately 25 $W\,m^{-2}$ for each $mm\,day^{-1}$ of precipitation.

Figure 9.38 illustrates the geographical distribution of precipitation. Large precipitation is found inside the ITCZ, where $\dot{P} \sim 10\ mm\,day^{-1}$. Thus, latent heating contributes of order 250 $W\,m^{-2}$ to the energy budget of the atmospheric column. These are the same regions implied by cloud forcing to experience strong radiative heating and cooling. But latent heating of the column is three times greater than SW or LW forcing individually and an order of magnitude greater than net cloud radiative forcing in those regions. Further, whereas cloud radiative forcing is sharply concentrated in the vertical, latent heating inside deep cumulus affects much of the tropical troposphere. Latent heat release inside tropical convection leads to a specific heating rate $\dot{q}/c_p$ of order 10 $K\,day^{-1}$ (Fig. 9.39). Radiative heating and cooling can exceed this value near cloud boundaries, but the deep distribution of latent heat release makes it a key source of energy for the tropical troposphere.

Influence of Aerosol

Atmospheric aerosol introduces similar, albeit weaker, effects. Like cloud, aerosol is an efficient scatterer of SW radiation, one that alters optical properties following major volcanic eruptions (Fig. 9.5). Depending on its com-

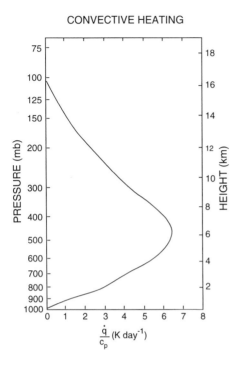

CONVECTIVE HEATING

Figure 9.39   Vertical profile of the heating rate inside tropical convection, normalized by that accompanying a precipitation rate of 10 mm day^{-1}. Source: Webster and Lukas (1992); adapted from data in Yanai *et al.* (1973).

position and the underlying surface, aerosol can either enhance or diminish albedo. Aerosol increases the albedo over otherwise dark oceans to introduce cooling in the SW energy budget. However, arctic haze (which is comprised of anthropogenic pollutants) reduces the albedo over bright polar ice (Ackerman, 1988). Aerosol also contributes significantly to SW absorption in the tropical troposphere (Fig. 8.26). To illustrate how these factors determine the net effect of aerosol, we consider a natural species: desert dust.

During disturbed periods, Saharan dust has optical depths exceeding 2 (Fig. 9.40). Convection carries desert aerosol as high as 500 mb. Dust absorbs SW radiation, so it reduces the albedo over a highly reflective surface, like desert or shallow stratus, to introduce warming in the SW energy budget of the column (9.48). However, dust changes the albedo over a dark surface like ocean only slightly—because both absorb strongly. The LW energy budget is influenced by desert aerosol in a manner similar to clouds. When convected

Figure 9.38   Annual-mean precipitation rate inferred from convective clouds colder than 230 K in global cloud imagery (Fig. 1.25) between August 1983 and July 1984 [see Hendon and Woodberry (1993) for details]. Deep convection in the maritime ITCZ and over tropical landmasses leads to mean precipitation rates exceeding 10 mm day^{-1}. Note also the North Atlantic and North Pacific storm tracks, which lie east of North America and Asia, respectively.

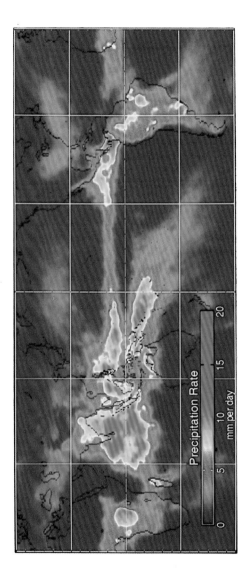

Precipitation Rate

mm per day

0   5   10   15   20

Figure 9.40   Visible image from Meteosat on March 25, 1991, which reveals a tongue of Saharan dust that is drawn northward from the African coast ahead of a cold front. Courtesy of D. Rosenfeld and Y. Levi (Hebrew University).

to great heights, desert aerosol can reduce the effective emission temperature significantly, with a commensurate warming in the column energy budget.[6]

SW effects dominate the energy budget for desert aerosol and produce net warming of the column, at least over surfaces with appreciable albedo. The same is true for the atmosphere alone, but LW and SW components both determine the vertical distribution of heating. Figure 9.41 shows the net heating rate for Saharan dust that is distributed between the surface and 500 mb. Over cloud-free ocean (Fig. 9.41a), absorption of upwelling LW radiation introduces significant heating near the surface, where $\dot{q}/c_p > 2$ K day^{-1}. For optical depths greater than 1, SW absorption introduces a second heating maximum near the middle of the dust layer. Over cloud-free desert (Fig. 9.41b), higher surface albedo backscatters much of the transmitted SW radiation, which is then absorbed inside the dust layer. Net heating is then dominated by SW absorption, which leads to $\dot{q}/c_p$ of order 1 to 2 K day^{-1} across much of the dust layer. Because it heats the atmosphere and cools the underlying surface, the net effect in both cases is to stabilize the stratification.

Aerosols are central to cloud formation, a role that is even more important than their direct radiative effect. Outbreaks of Saharan dust alter both the

---

[6]The reduction of effective emission temperature can exceed that of cloud occupying the same volume because the temperature at the top of the dust layer decreases with height "dry adiabatically."

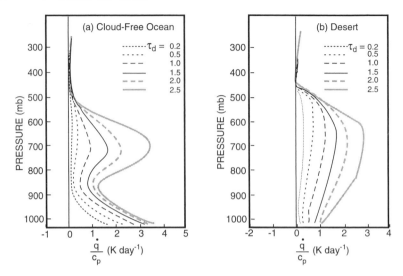

Figure 9.41 Vertical profiles of heating rate for Saharan dust that is distributed between 1000 and 500 mb and has optical depths τ_d over (a) cloud-free ocean and (b) desert. Adapted from Carlson and Benjamin (1982). Copyright by Spectrum Press (A. Deepak Publishing).

amount and type of cloud, with stratiform being favored. Therefore, changes in the concentration, size, or composition of atmospheric aerosol can introduce important changes in cloud cover and radiative energetics.

9.5.2 Involvement in Chemical Processes

Clouds also support important interactions with chemical species. Precipitation is the primary removal mechanism for atmospheric pollutants. By serving as CCN, aerosol particles are incorporated into cloud droplets and precipitated to the earth's surface. This process scavenges air of most nuclei activated and others absorbed through collision with cloud droplets. The efficiency of washout accounts for the short residence time of tropospheric aerosol (~1 day). Washout also operates on gaseous pollutants that are water soluble.

Aerosol particles support heterogeneous chemistry, which requires the presence of multiple phases. Cloud particles comprising PSCs are directly involved in the formation of the Antarctic ozone hole (Chapter 17). Beyond that role, PSCs introduce an important transport mechanism that can also influence ozone depletion. If they attain sufficient size, PSC particles undergo sedimentation—like cirrus ice particles. Water vapor and nitrogen are then removed from the lower stratosphere.

Chemical reactions similar to those occurring inside PSCs take place on natural aerosol, the concentration of which increases sharply following major volcanic eruptions (Fig. 9.1). Enhanced aerosol surface area, which figures centrally in heterogeneous chemistry, is then comparable to that normally observed over the Antarctic in connection with PSCs. This mechanism may have been at least partly responsible for anomalously low ozone abundances that were observed following the eruption of El Chichon (Hoffmon and Solomon, 1989).

Suggested Reading

Atmospheric Science: An Introductory Survey (1977) by Wallace and Hobbs contains a nice overview of cloud formation. Advanced treatments of cloud microphysics are given in *Microphysical Processes in Clouds* (1993) by Young and in *Microphysics of Clouds and Precipitation* (1978) by Pruppacher and Klett, which also includes observations of macrophysical cloud properties.

Production of atmospheric aerosol is discussed in the monograph *Atmospheric Aerosols: Their Formation, Optical Properties, and Effects* (1982), edited by Deepak. *Aerosol and Climate* (1988) edited by Hobbs and McCormick discusses possible impacts of increased aerosol. Stratospheric aerosol is reviewed jointly with polar stratospheric clouds in *Report of the International Ozone Trends Panel: 1988* (WMO, 1988).

International Cloud Atlas (WMO, 1969) presents a complete classification of clouds and a photographic survey.

Cloud Dynamics (1993) by Houze is a comprehensive treatment of the subject that includes numerous illustrations of cloud formation and structure. An advanced treatment of the fluid dynamics surrounding cloud formation and of laboratory simulations is presented in *Environmental Aerodynamics* (1978) by Scorer.

Classical Electrodynamics (1975) by Jackson provides a rigorous development of Rayleigh scattering. Mie scattering is formally discussed in *An Introduction to Atmospheric Radiation* (1980) by Liou.

Radiation and Cloud Processes in the Atmosphere (1990) by Liou is a comprehensive treatment of cloud radiative processes and their roles in global energetics. The radiative properties of water clouds and their parameterization are developed in Stephens (1978a,b).

Hartmann *et al.* (1986) contains a nice overview of how cloud and surface effects figure in the Earth's energy budget.

Climate Change: The IPCC Scientific Assessment (Houghton *et al.*, 1990) discusses historical variations of climate, surveying trends in atmospheric aerosols and trace gases and their possible involvement in climate feedback mechanisms.

Problems

9.1. Contrast the characteristic vertical scale of aerosol in the troposphere with that in the stratosphere, in light of removal processes.

9.2. Why is nitrate an inevitable by-product of combustion?

9.3. In terms of the size distribution $n(a)$, derive an expression for the size spectrum of (a) aerosol surface area $S(a)$ and (b) aerosol mass $M(a)$, if particles have uniform density ρ.

9.4. A population of cloud droplets has the log-normal size spectrum

$$\frac{dn}{d(\log a)} = \frac{n_0}{\sigma\sqrt{2\pi}} \exp\left[-\frac{\ln^2\left(\frac{a}{a_m}\right)}{2\sigma^2}\right],$$

where $n_0 = 300$ cm^{-3}, $a_m = 5$ μm is the median droplet radius, and $\sigma = 4$ μm is the rms deviation of $\ln(a)$. (a) Plot the size spectrum $dn/d(\log a)$, the surface area spectrum $dS/d(\log a)$, and the mass spectrum $dM/d(\log a)$. (b) Contrast the distributions of those properties over droplet radius a.

9.5. Discuss how the introduction of volcanic dust can affect the mean surface temperature of the earth.

9.6. Derive expression (9.11) for the growth rate of a droplet through condensation.

9.7. Show that the rate a droplet's mass increases via coalescence is described by (9.13).

9.8. Provide an expression for the potential barrier that must be overcome for a water droplet to be activated.

9.9. (a) Determine the critical radius at 0.2% supersaturation for a droplet of pure water. (b) How large must a droplet containing 10^{-16} g of sodium chloride become to grow spontaneously?

9.10. For the cloud described in Fig. 5.5, calculate the liquid water content at 650 mb, presuming adiabatic conditions.

9.11. A cumulus cloud forms inside a moist thermal. The base of the thermal is at 1000 mb, where $T_0 = 30°$C and $r_0 = 10$ g kg^{-1}. (a) If, at all elevations, the thermal is 2°C warmer in its core than near its periphery (other factors being equivalent), determine the height difference between the cloud base at its center versus at its periphery. (b) How would mixing of moisture with the environment modify this result?

9.12. Discuss why clouds often have bases at, not one, but several distinct levels.

9.13. Convection is favored over elevated terrain. In terms of θ and hydrostatic stability, discuss why convection tends to form earlier during the day and become more intense over mountains.

9.14. In relation to cloud dissipation, contrast entrainment of environmental air from above versus below an established cumulus.

9.15. The flow about a falling sphere of radius a and velocity w is *laminar* if the dimensionless Reynolds number

$$Re = \frac{\rho a w}{\mu}$$

is less than 5000 (Chapter 13), where $\mu = 1.7 \times 10^{-5}$ kg m^{-1} s^{-1} is the coefficient of viscosity for air. Under these circumstances, the sphere experiences a viscous force

$$D = -6\pi\mu a w$$

known as *Stokes drag*. (a) Calculate the terminal velocity w_t of a spherical cloud droplet of radius a. (b) For what range of a does the preceding result apply if the flow becomes turbulent for $Re > 5000$? (c) If the flow about them remains laminar, how large must droplets grow before they can no longer be supported by an updraft of 1 m s^{-1}?

9.16. Consider a population of cloud droplets having size distribution $n(a)$ under the conditions of Problem 9.15. (a) Derive an expression for the specific drag force exerted on air by the falling droplet population. (b) What is required to maintain an updraft in the presence of those droplets?

9.17. Estimate the time for volcanic dust particles of density 2.6×10^3 kg m^{-3} to reach the earth's surface from 50 mb if the tropopause is at 200 mb and the particles are spherical with a mean radius of (a) 1.0 μm and (b) 0.1 μm.

9.18. (a) Use typical observed values of ρ_l / ρ_l^{ad} to characterize the fractional composition of a nonprecipitating cumulus by surface air. (b) Discuss the thermodynamics of an air parcel inside the cloud and its interaction with environmental air.

9.19. Estimate the entrainment height (Sec. 7.4.2) for a thermal of diameter (a) 100 m and (b) 1 km.

9.20. Describe the evolution of a stack of lenticular clouds if they form from moisture anomalies upstream that are sheared and advected through the mountain wave by the prevailing flow.

9.21. Plot the saturation mixing ratio with respect to water and with respect to ice as functions of temperature for cloud at 400 mb.

9.22. Why does the sky appear brighter when the ground is snow covered?

9.23. Discuss why the milky appearance of the sky, which is often observed from the surface, disappears at an altitude of a kilometer or two.

9.24. Other factors identical, why does a nimbus cloud appear darker than others?

9.25. Use the scattered intensity (9.19) to obtain the total scattered power (9.20) for Rayleigh scattering.

9.26. Transmission of diffuse SW radiation through a scattering layer is described in terms of the phase function $P(\cos \Theta)$, where Θ is the three-dimensional scattering angle. Express $\cos \Theta$ in terms of the zenith angles θ and θ' and the azimuthal angles ϕ and ϕ' of scattered and incident radiation, respectively.

9.27. Calculate the atmospheric albedo due to Rayleigh scattering for overhead sun and (a) $\lambda = 0.7$ μm, (b) $\lambda = 0.5$ μm, and (c) $\lambda = 0.3$ μm. The refractive index of air at 1000 mb and 273 K is approximated by

$$(m - 1)10^6 = 6.4328 \ 10^1 + \frac{2.94981 \times 10^4}{146 - \lambda^{-2}} + \frac{2.554 \times 10^2}{41 - \lambda^{-2}},$$

with λ in micrometers and $m - 1$ proportional to density.

9.28. Scattering of microwave radiation by cloud droplets and precipitation enables cloud properties to be measured by radar. (a) Verify that Rayleigh scattering is a valid description of such behavior. (b) The backscattering coefficient (8.6), which measures the reflected power per unit length of scattering medium, follows from the scattering cross section as

$$\beta_s(\pi) = n\hat{\sigma}_s P(\pi),$$

where n is the particle number density and $P(\pi)$ is the phase function for backward scattering. Derive an expression for $\beta_s(\pi)$ in terms of the radius a of scatterers and the fractional volume $\eta = n \cdot \frac{4}{3}\pi a^3$ occupied by them to show that the reflectivity is proportional to a^6.

9.29. Derive the radiative transfer equation governing azimuthal–mean diffuse radiation (9.36) from the full radiative transfer equation.

9.30. Sunsets were modified for several years following the eruption of Krakatoa. How small must particles have been?

9.31. Show that, for single scattering, the diffuse intensity $I(\mu, \tau)$ is proportional to the phase function $P(\mu; \mu_s)$.

9.32. Derive the system (9.37) governing upwelling and downwelling fluxes of diffuse radiation.

9.33. Show that the blackbody function $B^*[T]$ is a particular solution of (9.46).

9.34. Show that transmission of diffuse SW radiation is governed by the system (9.39).

9.35. For the cloud droplet population in Problem 9.4, compute (a) the scattering-equivalent mean droplet radius, (b) the liquid water path for a cloud 4 km deep, (c) the cloud's optical depth, and (d) the cloud's absorptivity for $\lambda = 0.5$ μm and a solar zenith angle of 30°.

9.36. Calculate the optical depth posed to SW radiation by a 5-km cumulus congestus (an organization of cumulus), which has an average liquid water content of 0.66 $g m^{-3}$ and a scattering-equivalent mean droplet radius of 12 μm.

9.37. Satellite measurements of SW radiation enable the albedo of clouds to be determined, from which the cloud optical depth can be inferred. (a) Plot cloud albedo as a function of optical depth for a solar zenith angle of 30°, presuming a homogeneous scattering layer and a black underlying surface. (b) Discuss the sensitivity of cloud albedo to changes in optical depth for $\tau_c > 50$. What implications does this have for inferring optical depth?

9.38. A satellite measuring backscattered SW radiation and LW radiation emitted at 11 μm observes a tropical cloud with an albedo of 0.9 and a brightness temperature of 220 K. If the solar zenith angle is 30° and a scattering-equivalent radius of 5 μm applies, use the result of Problem 9.37 to estimate (a) the liquid water path of the cloud and (b) the cloud's average liquid water content if it is 1 km thick.

9.39. A homogeneous cloud layer of optical depth τ_c is illuminated by the solar flux F_s at a zenith angle $\theta_s = \cos^{-1} \mu_s$. If the surface is gray with absorptivity a, (a) derive expressions for the upwelling and downwelling diffuse fluxes inside the scattering layer and (b) plot the albedo and transmissivity, as functions of a, for overhead sun and $\tau_c = 20$.

9.40. Consider the radiative–convective equilibrium for the cloudy atmosphere in Fig. 9.33. (a) Explain why the surface temperature in Fig. 9.34 increases linearly with the cloud's height once the cloud is elevated sufficiently above the surface. (b) Nearly identical results are obtained for other cloud optical depths. Why?

9.41. Satellite observations reveal the following scene properties for a region of maritime convection in the tropics: a fractional cloud cover of 0.5, average clear-sky and overcast albedos of 0.2 and 0.8, respectively, and average clear-sky and overcast LW fluxes of 350 and 100 $W m^{-2}$, respectively. Calculate the average SW, LW, and net cloud forcing for the region.

9.42. Use the mean latent heat flux to the atmosphere, 90 $W m^{-2}$ (Fig. 1.27), to estimate the global–mean precipitation rate.

9.43. Estimate the fraction of the water vapor column in the tropics that is processed by convection during one day. Reconcile this fraction with the characteristic timescale and depth of individual convective cells.

9.44. Use the mean annual precipitation in Fig. 9.38 and a characteristic tropopause height of 16 km to estimate the latent heating rate inside the ITCZ.

9.45. Consider a horizontal interface separating two conservative media with real refractive indices n_1 and n_2. Use the condition that the horizontal trace speed of electromagnetic waves (Sec. 14.1.2) must be identical on either side of the interface to derive Snell's law:

$$\frac{\sin \theta_2}{\sin \theta_1} = \frac{n_1}{n_2},$$

where θ_1 and θ_2 denote the zenith angles of incident and refracted radiation and $n_1 = c_0/c_1$ and $n_2 = c_0/c_2$ are the corresponding indices of refraction, which are defined in terms of the speeds c_1 and c_2 at which light propagates in media 1 and 2.

9.46. Scattering of SW radiation is enhanced in polar regions, where solar zenith angles are large. As a simple model, consider conservative scattering of SW radiation from a wind-blown ocean surface, which randomly occupies all angles α from the horizontal between $\pm 90°$. (a) Use Snell's law and conservation of energy crossing the surface to determine that fraction of incident radiation arriving at zenith angle $\theta_s > 0$ which is scattered into the zenith angle θ. (b) For $\theta_s = 30°$, compare the intensity of light which is forward scattered ($\theta_s < 0$) against that which is backward scattered ($\theta_s > 0$).

9.47. Use the directionality of scattering to assess the change in overall SW transmission accompanying the reduction of direct transmission following the eruption of El Chichon in Fig. 9.5.

Chapter 10 | Atmospheric Motion

Under radiative equilibrium, the first law reduces to a balance between radiative transfers of energy, which determines the radiative–equilibrium thermal structure. This simple balance is valid for a resting atmosphere, but breaks down for an atmosphere in motion because heat is then also transferred mechanically by air movement. Discrepancies of the radiative–equilibrium thermal structure (Fig. 8.21) from the observed global–mean temperature (Fig. 1.2) point to the importance of mechanical heat transfer. Radiative–convective equilibrium, which accounts for vertical heat transfer by convection, reconciles the equilibrium thermal structure with that observed, but only in the global–mean.

Radiative and radiative–convective equilibrium both produce a meridional temperature gradient that is too steep compared to observed thermal structure (Fig. 10.1). Each accounts only for vertical transfers of energy. According to the distribution of net radiation (Fig. 1.29), low latitudes experience radiative heating, whereas middle and high latitudes experience radiative cooling. Thermal equilibrium then requires those radiative imbalances to be compensated by a poleward transfer of heat. That horizontal heat transfer is accomplished in large part by the atmospheric circulation. Thus, understanding how observed thermal structure is maintained requires an understanding of atmospheric motion and the horizontal transfers of energy that it supports.

10.1 Descriptions of Atmospheric Motion

Until now, the development has focused on an infinitesimal fluid element that defines an individual air parcel. Thermodynamic and hydrostatic influences acting on that discrete system must now be complemented by dynamical considerations that control the movement of individual bodies of air. Together, these properties constitute the individual or *Lagrangian description* of fluid motion, which represents atmospheric behavior in terms of the collective properties of material elements comprising the atmosphere. Because the basic laws of physics apply to a discrete bounded system, they are developed most simply within the Lagrangian framework. However, the Lagrangian description of atmospheric behavior requires representing not only the thermal,

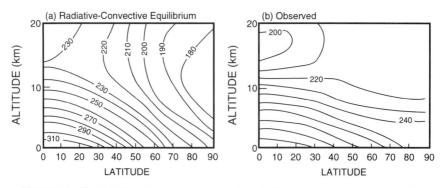

Figure 10.1   Zonal–mean temperature as a function of latitude and height (a) under radiative–convective equilibrium and (b) observed during northern winter. Without horizontal heat transfer, radiative–convective equilibrium establishes a meridional temperature gradient that is much stronger than observed. Sources: Liou (1990) and Fleming *et al.* (1988).

mechanical, and chemical histories of individual air parcels, but also tracking their positions—not to mention their distortion as they move through the circulation.

Like any fluid system, the atmosphere is a continuum. Hence, it is comprised of infinitely many such discrete systems, all of which must be formally represented in the Lagrangian description of atmospheric behavior. For this reason, it is more convenient to describe atmospheric behavior in terms of field variables that represent the distributions of properties at particular instants. The distribution of temperature $T(x, t)$, where $x = (x, y, z)$ denotes three-dimensional position, is a scalar field variable. So is the distribution of mixing ratio $r(x, t)$ and the component of motion in the ith coordinate direction $v_i(x, t)$. Collecting the scalar components of motion in the three coordinate directions gives the three-dimensional motion field $v(x, t) = v_i(x, t)$; $i = 1, 2, 3$, which is a vector field variable that has both magnitude and direction at each location. Like thermodynamic and chemical variables, the motion field $v(x, t)$ is a property of the fluid system, one that describes the circulation. Because air motion transfers heat and chemical constituents, $v(x, t)$ is coupled to other field properties. Collectively, the distributions of such properties constitute the *field* or *Eulerian description* of fluid motion, which is governed by the equations of continuum mechanics.

The Eulerian description simplifies the representation of atmospheric behavior. However, physical laws governing atmospheric behavior, such as the first and second laws of thermodynamics and Newton's laws of motion, apply directly to a fixed collection of matter. For this reason, the equations governing atmospheric behavior are developed most intuitively in the Lagrangian framework. In the Eulerian framework, the field property at a specified location involves different material elements at different times. Despite this complication, the Eulerian description is related to the Lagrangian description by a fundamental kinematic constraint: The field property at a given location and time must equal the property possessed by the material element occupying that position at that instant.

10.2 Kinematics of Fluid Motion

A material element (Fig. 10.2) can be identified by its initial position

$$\boldsymbol{\xi} = \left(x_0, y_0, z_0 \right), \tag{10.1.1}$$

which is referred to as the *material coordinate* of that element. The locus of points

$$\boldsymbol{x}(t) = \left[x(t), y(t), z(t) \right] \tag{10.1.2}$$

traced out by the element defines a *material* or *parcel trajectory*, which is uniquely determined by the material coordinate $\boldsymbol{\xi}$ and the motion field $\boldsymbol{v}(\boldsymbol{x}, t)$. With $\boldsymbol{\xi}$ held fixed, the vector $d\boldsymbol{x} = (dx, dy, dz)$ describes the incremental displacement of the material element during the time interval dt. The element's

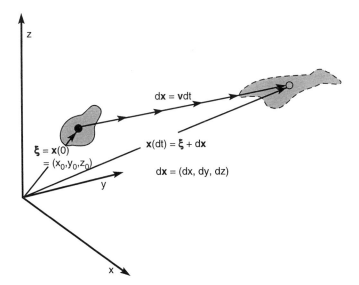

Figure 10.2 Positions $\boldsymbol{x}$ of a material element initially and after an interval dt. The element's initial position $\boldsymbol{x}_0 = (x_0, y_0, z_0)$ defines its *material coordinate* $\boldsymbol{\xi}$. During the interval dt, the element traces out an increment of *parcel trajectory* $d\boldsymbol{x} = \boldsymbol{v}dt$, where $\boldsymbol{v}$ is the element's instantaneous velocity.

velocity is then

$$v = \frac{dx}{dt},$$

(10.2.1)

with components

$$u = \frac{dx}{dt},$$

$$v = \frac{dy}{dt},$$

(10.2.2)

$$w = \frac{dz}{dt}.$$

Because it is evaluated for fixed ξ, the time derivative in (10.2) corresponds to an individual material element and therefore to the Lagrangian time rate of change on which the development has focused until now. By the kinematic constraint relating Eulerian and Lagrangian descriptions, the velocity in (10.2) must then equal the field value $v(x, t)$ at the material element's position $x(t)$ at time t.

At any instant, the velocity field defines a family of *streamlines* (Fig. 10.3) that are everywhere tangential to $v(x, t)$. If the motion field is *steady*, namely, if local velocities do not change with time, parcel trajectories coincide with streamlines. This follows from (10.2) because a material element is then always displaced along the same streamline, which remains fixed. Under unsteady conditions, parcel trajectories do not coincide with streamlines, which then evolve to new positions. Initially tangential to one streamline, the velocity of a material element will displace it to a different streamline, with which its velocity is again temporarily tangential. A *streakline*, which also characterizes the motion, is the locus of points traced out by a dye that is released into the flow at a particular location. As illustrated in Fig. 10.3, a streakline represents the positions at a given time of contiguous material elements that have passed through the location where the tracer is released. Like parcel trajectories, streaklines coincide with streamlines only if the motion field is steady.

In addition to translation, a material element can undergo rotation and deformation as it moves through the circulation. These effects are embodied in the velocity gradient

$$\nabla v = \frac{\partial v_i}{\partial x_j}, \qquad i, j = 1, 2, 3,$$

(10.3)

which is a two-dimensional tensor with the subscripts i and j referring to three Cartesian coordinate directions. The local velocity gradient $\nabla v(x)$ determines the relative motion between material coordinates and hence the distortion experienced by the material element located at x. Two material coordinates

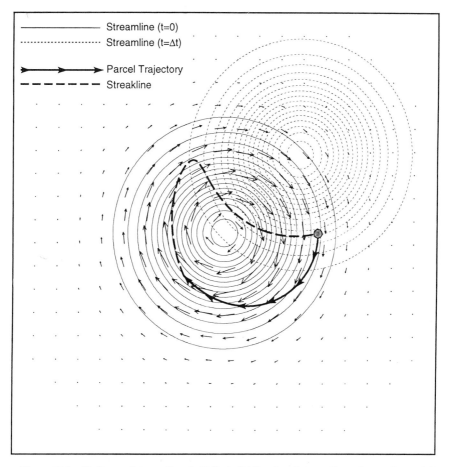

Figure 10.3   Motion and streamlines initially (solid lines) and streamlines after an interval Δt (dashed lines) for a vortex translating uniformly. Superposed is (1) the *parcel trajectory* at time Δt (bold solid line) of a material element that is initially positioned at the shaded circle and (2) the *streakline* at time Δt (bold dashed line) that originates from the same position.

that are separated initially by dx_j experience relative motion

$$
\begin{aligned}
dv_i &= \frac{\partial v_i}{\partial x_j} dx_j, \\
&= \frac{\partial v_i}{\partial x_1} dx_1 + \frac{\partial v_i}{\partial x_2} dx_2 + \frac{\partial v_i}{\partial x_3} dx_3,
\end{aligned} \tag{10.4}
$$

where repeated indices in the same term imply summation and the vector product—the so-called *Einstein notation*. How the material element evolves through relative motion is elucidated by separating the velocity gradient into symmetric and antisymmetric components:

$$\frac{\partial v_i}{\partial x_j} = e_{ij} + \omega_{ij}, \tag{10.5.1}$$

where

$$e_{ij} = \frac{1}{2}\left(\frac{\partial v_i}{\partial x_j} + \frac{\partial v_j}{\partial x_i}\right), \tag{10.5.2}$$

$$\omega_{ij} = \frac{1}{2}\left(\frac{\partial v_i}{\partial x_j} - \frac{\partial v_j}{\partial x_i}\right). \tag{10.5.3}$$

The symmetric component $e(x, t)$ of the velocity gradient defines the *rate of strain* or *deformation tensor* acting on the material element at x. Diagonal components of e, which describe variations of velocity "in the direction of motion," characterize the rate of longitudinal deformation, or "stretching," in the three coordinate directions. For example, e_{11} describes the rate of elongation of a material segment aligned in the direction of coordinate 1 (Fig. 10.4a). Collecting all three components of stretching gives the *trace* of e and the divergence of the velocity field

$$\frac{\partial v_i}{\partial x_i} = \nabla \cdot v, \tag{10.6}$$

which reflects the rate of increase in volume or *dilatation* of the material element (Problem 10.4). In the case of pure shear, for which the field $v(x, t)$ varies only in directions transverse to the motion, diagonal elements of e vanish. The material element may then undergo distortion, but no change of volume.

Off-diagonal components of $e(x, t)$, which describe variations of velocity "orthogonal to the motion," characterize the rate of transverse deformation, or "shear," experienced by the material element at x. As pictured in Fig. 10.4b, $2e_{12}$ represents the rate of decrease of the angle between two material segments aligned initially in the directions of coordinates 1 and 2 (Problem 10.5). In the case of rectilinear motion that varies only longitudinally, off-diagonal components of $e(x, t)$ vanish. The material element at x may then experience stretching, but no shear.

The antisymmetric component of $\nabla v(x, t)$ also describes relative motion, but not deformation. Because it contains only three independent components, the tensor ω defines a vector

$$\hat{\omega} = \begin{bmatrix} \omega_{23} \\ -\omega_{13} \\ \omega_{12} \end{bmatrix}. \tag{10.7}$$

The vector $\hat{\omega}(x, t)$ is related to the curl of the motion field or *vorticity*

$$2\hat{\omega} = \nabla \times v, \tag{10.8}$$

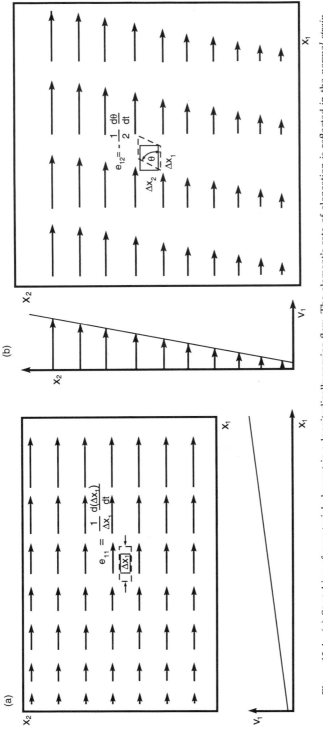

Figure 10.4 (a) Stretching of a material element in a longitudinally varying flow. The element's rate of elongation is reflected in the *normal strain rate* e_{11}. (b) Shearing of a material element in a transversely varying flow. The rate at which the material angle θ decreases is reflected in the *shear strain rate* e_{12}.

which represents twice the local angular velocity. Thus, $\hat{\omega}(x, t)$ represents the angular velocity of the material element at position x. If symmetric components of $\nabla v(x, t)$ vanish, so does the local deformation tensor $e(x, t)$. The material element's motion then reduces to translation plus rigid body rotation.

10.3   The Material Derivative

To transform to the Eulerian description, the Lagrangian derivative appearing in conservation laws must be expressed in terms of field properties. Consider a field variable $\psi = \psi(x, y, z, t)$. The incremental change of property ψ is described by the total differential

$$d\psi = \frac{\partial \psi}{\partial t} dt + \frac{\partial \psi}{\partial x} dx + \frac{\partial \psi}{\partial y} dy + \frac{\partial \psi}{\partial z} dz$$

$$= \frac{\partial \psi}{\partial t} dt + \nabla \psi \cdot dx, \tag{10.9}$$

where dx, dy, dz, and dt are suitably defined increments in space and time. Let $dx = (dx, dy, dz)$ and dt in (10.9) denote increments of space and time with the material coordinate ξ held fixed. Then dx represents the displacement of the material element ξ during the time interval dt and the increment of parcel trajectory shown in Fig. 10.2. With position and time related in this manner, the total differential describes the incremental change of property ψ observed "in a frame moving with the material element." Differentiating with respect to time (with ξ still held fixed) gives the time rate of change of ψ "following a material element" or the time rate of change "following the motion"

$$\frac{d\psi}{dt} = \frac{\partial \psi}{\partial t} + u \frac{\partial \psi}{\partial x} + v \frac{\partial \psi}{\partial y} + w \frac{\partial \psi}{\partial z}$$

$$= \frac{\partial \psi}{\partial t} + v \cdot \nabla \psi, \tag{10.10}$$

which defines the *Lagrangian* or *material derivative* of the field variable $\psi(x, t)$.

According to (10.10), the material derivative includes two contributions. The first, $\partial \psi / \partial t$, which is referred to as the *Eulerian* or *local derivative*, represents the rate of change of the material property ψ introduced by temporal changes in the field variable at the position x where the material element is located. The second contribution, $v \cdot \nabla \psi$, which is referred to as the *advective contribution*, represents the change of the material property ψ due to motion of the material element to positions of different field values. Even if the field is steady, namely, if $\psi(x, t) = \psi(x)$ so that local values do not evolve with time, the property of a material element will change if that element moves across contours of the field property $\psi(x)$. The rate that ψ changes for the material element is then given by its velocity in the direction of the gradient of ψ times that gradient.

10.4 Reynolds' Transport Theorem

Owing to the kinematic relationship between Lagrangian and Eulerian descriptions, the material derivative emerges naturally in the laws governing field properties. Changes of ψ described by (10.10) apply to an infinitesimal material element. Consider now changes for a finite material volume $V(t)$, i.e., one containing a fixed collection of matter. Because $V(t)$ has finite dimension, we must account for variations of velocity v and property ψ across the material volume. Consider the integral property

$$\int_{V(t)} \psi(x, y, z, t) dV'$$

over the finite system. The time rate of change of this property follows by differentiating the volume integral. However, the time derivative d/dt cannot be commuted inside the integral because the limits of integration (namely, the position and form of the material volume) are themselves variable.

The time rate of change of the integral material property has two contributions (Fig. 10.5), analogous to those treated earlier for an infinitesimal material element:

1. Values of $\psi(x, t)$ within the instantaneous material volume change temporally due to unsteadiness of the field.
2. The material volume moves to regions of different field values. Relative to a frame moving with $V(t)$, such motion introduces a flux of property ψ across the material volume's surface $S(t)$.

The first contribution is just the collective time rate of change of $\psi(x, t)$ within the material volume:

$$\int_{V(t)} \frac{\partial \psi}{\partial t} dV'.$$

The second is the net transfer of ψ across the boundary of $V(t)$. If $S(t)$ has the local outward normal n and local velocity v, the local flux of ψ relative to that section of material surface is $-\psi v$. Then the flux of ψ into the material volume is given by $-\psi v \cdot -n = \psi v \cdot n$. Integrating over $S(t)$ gives the net rate that ψ is transferred into the material volume $V(t)$ due to motion across contours of ψ:

$$\int_{S(t)} \psi v \cdot n dS'.$$

Collecting the two contributions and applying Gauss' theorem obtains

$$\frac{d}{dt} \int_{V(t)} \psi dV' = \int_{V(t)} \left\{ \frac{\partial \psi}{\partial t} + \nabla \cdot (v\psi) \right\} dV'$$

$$= \int_{V(t)} \left\{ \frac{d\psi}{dt} + \psi \nabla \cdot v \right\} dV' \qquad (10.11)$$

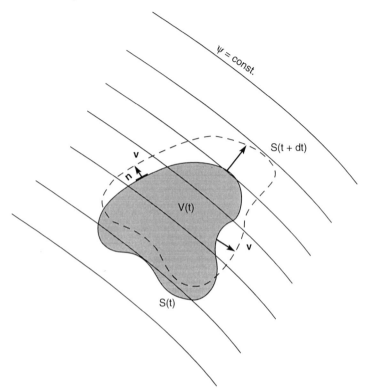

Figure 10.5 Finite material volume $V(t)$, containing a fixed collection of matter, that is displaced across a field property ψ.

for the time rate of change of the integral material property. Known as *Reynolds' transport theorem*, (10.11) relates the time rate of change of some property of a finite body of fluid to the corresponding field variable and to the motion field $v(x, t)$. As such, it constitutes a transformation between the Lagrangian and Eulerian descriptions of fluid motion.[1] To develop the equations governing field variables in the Eulerian description of atmospheric motion, we apply Reynolds' theorem to properties of an arbitrary material volume.

10.5　Conservation of Mass

Let $\psi = \rho(x, t)$. An arbitrary material volume $V(t)$ then has mass

$$\int_{V(t)} \rho(x, t) dV'.$$

[1] Mathematically, the transport theorem is a generalization to a moving body of fluid of Leibniz's rule for differentiating an integral with variable limits. It enables the time derivative of the integral property to be expressed in terms of an integral of time and space derivatives.

Because the system is comprised of a fixed collection of matter, the time rate of change of its mass must vanish

$$\frac{d}{dt} \int_{V(t)} \rho(x, t) dV' = 0. \tag{10.12}$$

Applying Reynolds' transport theorem transforms (10.12) into

$$\int_{V(t)} \left\{ \frac{d\rho}{dt} + \rho \nabla \cdot v \right\} dV' = 0. \tag{10.13}$$

This relation must hold for "arbitrary" material volume $V(t)$. It follows that the quantity in braces must vanish identically. Thus, conservation of mass for individual bodies of air requires

$$\frac{d\rho}{dt} + \rho \nabla \cdot v = 0 \tag{10.14.1}$$

or

$$\frac{\partial \rho}{\partial t} + \nabla \cdot (\rho v) = 0, \tag{10.14.2}$$

which provides a constraint on the field variables $\rho(x, t)$ and $v(x, t)$. Known as the *continuity equation*, (10.14) must hold pointwise throughout the domain and continuously in time.

Consider a specific field property f (namely, one referenced to a unit mass). Then $\psi = \rho f$ represents the absolute concentration of that property. Reynolds' transport theorem implies

$$\frac{d}{dt} \int_{V(t)} \rho f dV' = \int_{V(t)} \left\{ \frac{d}{dt}(\rho f) + \rho f (\nabla \cdot v) \right\} dV'$$

$$= \int_{V(t)} \left\{ \rho \frac{df}{dt} + f \left[\frac{d\rho}{dt} + \rho \nabla \cdot v \right] \right\} dV'. \tag{10.15}$$

By the continuity equation, the term in square brackets vanishes, so (10.15) reduces to the identity

$$\frac{d}{dt} \int_{V(t)} \rho f dV' = \int_{V(t)} \rho \frac{df}{dt} dV'. \tag{10.16}$$

Hence, for an absolute concentration, the time rate of change of an integral material property assumes this simpler form.

10.6 The Momentum Budget

10.6.1 Cauchy's Equations of Motion

In an inertial reference frame, Newton's second law of motion applied to the material volume $V(t)$ can be expressed

$$\frac{d}{dt} \int_{V(t)} \rho v dV' = \int_{V(t)} \rho f dV' + \int_{S(t)} \tau \cdot n dS', \tag{10.17}$$

where ρv is the absolute concentration of momentum, f is the specific body force acting internal to the material volume, and τ is the *stress tensor* acting on its surface (Fig. 10.6). The stress tensor τ, which is a counterpart of the deformation tensor e, represents the vector force per unit area exerted on surfaces normal to the three coordinate directions. Then $\tau \cdot n$ is the vector force per unit area exerted on the section of material surface with unit normal n.

For commonly considered fluids like air, the stress tensor is symmetric

$$\tau_{ji} = \tau_{ij}. \tag{10.18}$$

Like $e(x, t)$, the local stress tensor contains two basic contributions. Diagonal components of $\tau(x, t)$ define *normal stresses* that act orthogonal to surfaces with normals in the three coordinate directions. The component $\tau_{11}(x, t)$ describes the force per unit area acting in the direction of coordinate 1 on an element of surface with normal in the same direction (Fig. 10.6). Off-diagonal components of $\tau(x, t)$ define *shear stresses* that act tangential to those surfaces. The component τ_{21} describes the force per unit area acting in the direction of coordinate 2 on an element of surface with normal in the direction of coordinate 1.

Each of the stresses in τ also represents a flux of momentum. The normal stress τ_{ii} (summation suspended) represents the flux of i momentum in the longitudinal i direction. The shear stress τ_{ji} represents the flux of j momentum in the transverse i direction. Formally, these fluxes are accomplished by molecular diffusion of momentum, which gives fluid viscosity. However, in applications to large-scale atmospheric motion, mixing by small-scale turbulence accomplishes similar fluxes of momentum. Far greater than molecular diffusion, turbulent fluxes are treated in analogous fashion. Both render the behavior of an individual material element diabatic because they transfer heat and momentum across its boundary (e.g., Fig. 2.1).

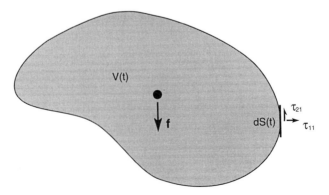

Figure 10.6 Finite material volume $V(t)$, which experiences a specific body force f internally and a stress tensor τ on its surface.

In the atmosphere, the body force and stress tensor are either prescribed or determined by the motion field. Incorporating Reynolds' transport theorem for an absolute concentration (10.16) and Gauss' theorem transforms (10.17) into

$$\int_{V(t)} \rho \frac{dv}{dt} dV' = \int_{V(t)} \left\{ \rho f + \nabla \cdot \tau \right\} dV'$$

or

$$\int_{V(t)} \left\{ \rho \frac{dv}{dt} - \rho f - \nabla \cdot \tau \right\} dV' = 0. \tag{10.19}$$

As before, (10.19) must hold for arbitrary material volume, so the quantity in braces must vanish identically. Thus, Newton's second law for individual bodies of air requires

$$\rho \frac{dv}{dt} = \rho f + \nabla \cdot \tau, \tag{10.20}$$

which provides constraints on the field properties $\rho(x, t)$ and $v(x, t)$. Known as *Cauchy's equations*, (10.20) hold pointwise throughout the domain and continuously in time.[2]

The body force relevant to the atmosphere is gravity

$$f = g, \tag{10.21}$$

which is prescribed. On the other hand, internal stresses are determined autonomously by the motion. A *Newtonian fluid* like air has a stress tensor that is linearly proportional to the local rate of strain $e(x, t)$ and that, in the absence of motion, reduces to the normal stresses exerted by pressure. For our purposes, it suffices to define the stress tensor as

$$\tau = \begin{bmatrix} -p & & 2\mu e_{ij} \\ & -p & \\ 2\mu e_{ji} & & -p \end{bmatrix}, \tag{10.22}$$

where μ is the *coefficient of viscosity*.[3] Equation (10.22) is the form assumed by the stress tensor for an incompressible fluid (Problem 10.16). Shear stresses internal to the fluid then arise from shear in the motion field $\partial v_i / \partial x_j$. When expressed in terms of specific momentum, the governing equations involve the *kinematic viscosity* $\nu = \mu/\rho$, which is also termed the molecular *diffusivity* and has dimensions length/time^2. Turbulent momentum transfer is frequently modeled as diffusion, but with an eddy diffusivity in place of ν (Chapter 13).

[2] In the absence of friction, these are called *Euler's equations*.

[3] Strictly, normal stresses also include a contribution from viscosity, but that effect is small enough to be neglected for most applications (see, e.g., Aris, 1962).

With τ defined by (10.22), Cauchy's equations reduce to

$$\frac{d\boldsymbol{v}}{dt} = \boldsymbol{g} - \frac{1}{\rho}\nabla p - \boldsymbol{D}, \qquad (10.23.1)$$

where

$$
\begin{aligned}
\boldsymbol{D} &= -\frac{1}{\rho}\nabla\cdot\tau \\
&= -\frac{1}{\rho}\nabla\cdot(\mu\nabla\boldsymbol{v}) = -\frac{1}{\rho}\frac{\partial}{\partial x_j}\mu\frac{\partial v_i}{\partial x_j} \qquad (10.23.2)
\end{aligned}
$$

denotes the specific *drag force* exerted on the material element at location $\boldsymbol{x}$. Known as the *momentum equations*, (10.23) are a simplified form of the *Navier Stokes equations*, which embody the full representation of τ in terms of $\boldsymbol{e}$ and constitute the formal description of fluid motion. Conceptually, (10.23) asserts that the momentum of a material element changes according to the resultant force exerted on it by gravity, pressure gradient, and frictional drag.

10.6.2 Momentum Equations in a Rotating Reference Frame

Because they follow from Newton's laws of motion, the momentum equations apply in an inertial reference frame. The reference frame of the earth (in which we observe atmospheric motion), however, is rotating and therefore noninertial. Consequently, the momentum equations must be modified to apply in that reference frame. Scalar quantities like $\rho(\boldsymbol{x}, t)$ appear the same in inertial and noninertial reference frames, as do their Lagrangian derivatives. However, vector quantities differ between those reference frames. Vector variables describing a material element's motion (e.g., $\boldsymbol{x}$, $\boldsymbol{v} = d\boldsymbol{x}/dt$, and $\boldsymbol{a} = d\boldsymbol{v}/dt$) must therefore be corrected to account for acceleration of the earth's reference frame.

Consider a reference frame rotating with angular velocity $\boldsymbol{\Omega}$ (Fig. 10.7). A vector $\boldsymbol{A}$ that is constant in that frame must rotate when viewed in an inertial reference frame. During an interval dt, $\boldsymbol{A}$ will change by a vector increment $d\boldsymbol{A}$, which is perpendicular to the plane of $\boldsymbol{A}$ and $\boldsymbol{\Omega}$ and has magnitude

$$|d\boldsymbol{A}| = A\sin\theta\cdot\Omega\,dt,$$

where θ is the angle between $\boldsymbol{A}$ and $\boldsymbol{\Omega}$. Hence, in an inertial reference frame, the vector $\boldsymbol{A}$ changes at the rate

$$\left|\frac{d\boldsymbol{A}}{dt}\right| = A\Omega\sin\theta$$

and in a direction perpendicular to the plane of $\boldsymbol{A}$ and $\boldsymbol{\Omega}$. It follows that the time rate of change of $\boldsymbol{A}$ apparent in an inertial reference frame is described by

$$\left(\frac{d\boldsymbol{A}}{dt}\right)_i = \boldsymbol{\Omega}\times\boldsymbol{A}. \qquad (10.24)$$

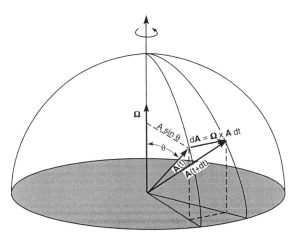

Figure 10.7 A vector A, which is fixed in a rotating reference frame, changes in an inertial reference frame during an interval dt by the increment $|dA| = A \sin \theta \cdot \Omega \, dt$ in a direction orthogonal to the plane of A and Ω, or by the vector increment $dA = \Omega \times A \, dt$.

More generally, a vector A that has the time rate of change dA/dt in a rotating reference frame has the time rate of change

$$\left(\frac{dA}{dt}\right)_i = \frac{dA}{dt} + \Omega \times A \tag{10.25}$$

in an inertial reference frame.

Consider the position x of a material element. By (10.25), the element's velocity $v = dx/dt$ apparent in an inertial reference frame is

$$v_i = v + \Omega \times x. \tag{10.26}$$

Similarly, the acceleration apparent in the inertial frame is given by

$$\left(\frac{dv_i}{dt}\right)_i = \frac{dv_i}{dt} + \Omega \times v_i$$

Incorporating the velocity apparent in the inertial frame (10.26) obtains

$$\left(\frac{dv_i}{dt}\right)_i = \left(\frac{dv}{dt} + \Omega \times v\right) + \Omega \times (v + \Omega \times x)$$

$$= \frac{dv}{dt} + 2\Omega \times v + \Omega \times (\Omega \times x) \tag{10.27}$$

for the acceleration apparent in the inertial frame.

According to (10.27), two corrections to the material acceleration arise to account for rotation of the reference frame. The first, $2\Omega \times v$ is the *Coriolis acceleration*. Following from an artificial force in the rotating frame of the earth, the Coriolis acceleration is perpendicular to an air parcel's motion and to the *planetary vorticity* $2\mathbf{\Omega}$. It is important for motions with timescales comparable to that of the earth's rotation (Chapter 12). The second correction: $\mathbf{\Omega} \times (\mathbf{\Omega} \times \mathbf{x})$ is the *centrifugal acceleration* of an air parcel due to the earth's rotation, which was treated in Chapter 6. When geopotential coordinates are used, the artificial force corresponding to this correction is automatically absorbed into effective gravity.

Incorporating the material acceleration apparent in an inertial reference frame (10.27) transforms the momentum equations into a form valid in the rotating frame of the earth:

$$\frac{d\mathbf{v}}{dt} + 2\mathbf{\Omega} \times \mathbf{v} = -\frac{1}{\rho}\nabla p - g\mathbf{k} - \mathbf{D}, \tag{10.28}$$

where g is understood to denote effective gravity and $\mathbf{k}$ the upward normal in the direction of increasing geopotential (i.e., $g\mathbf{k} = \nabla\Phi$). The correction $2\mathbf{\Omega} \times \mathbf{v}$ enters as the "Coriolis acceleration" when it appears on the left-hand side of the momentum balance. Moved to the right-hand side, it enters as the "Coriolis force" $-2\mathbf{\Omega} \times \mathbf{v}$, a fictitious force that appears to act on a material element in the rotating frame of the earth. Because it acts orthogonal to the displacement, the Coriolis force performs no work on the material element.

10.7 The First Law of Thermodynamics

Applied to the material volume $V(t)$, the first law can be expressed

$$\frac{d}{dt}\int_{V(t)} \rho c_v T dV' = -\int_{S(t)} \mathbf{q} \cdot \mathbf{n} dS' - \int_{V(t)} \rho p \frac{d\alpha}{dt} dV' + \int_{V(t)} \rho \dot{q} dV', \tag{10.29}$$

where $c_v T$ represents the specific internal energy. Forcing it on the right-hand side, $\mathbf{q}$ is the local heat flux so $-\mathbf{q} \cdot \mathbf{n}$ represents the heat flux "into" the material volume, $\alpha = 1/\rho$ is the specific volume so $p(d\alpha/dt)$ represents the specific work rate, and $\dot{q}$ denotes the specific rate of internal heating (e.g., associated with the latent heat release and frictional dissipation of motion).

The local rate of expansion work is related to the dilatation of the material element occupying that position. Because

$$\frac{1}{\rho}\frac{d\rho}{dt} = \alpha\frac{d\left(\frac{1}{\alpha}\right)}{dt}$$

$$= -\frac{1}{\alpha}\frac{d\alpha}{dt}, \tag{10.30}$$

the continuity equation (10.14) can be expressed

$$\frac{1}{\alpha}\frac{d\alpha}{dt} = \nabla \cdot \mathbf{v}. \tag{10.31}$$

Incorporating (10.31) along with Reynolds' transport theorem (10.16) and Gauss' theorem transforms (10.29) into

$$\int_{V(t)} \left\{ \rho c_v \frac{dT}{dt} + \nabla \cdot \boldsymbol{q} + p\nabla \cdot \boldsymbol{v} - \rho \dot{q} \right\} dV' = 0.$$

Again, since $V(t)$ is arbitrary, the quantity in braces must vanish identically. Therefore, the first law applied to individual bodies of air requires

$$\rho c_v \frac{dT}{dt} = -\nabla \cdot \boldsymbol{q} - p\nabla \cdot \boldsymbol{v} + \rho \dot{q}, \tag{10.32}$$

which provides a constraint on the field variables involved. Of the three thermodynamic properties represented, only two are independent because they must also satisfy the gas law. Similarly, the heat flux $\boldsymbol{q}$ and the internal heating rate $\dot{q}$ are determined autonomously by properties already represented, so they introduce no additional unknowns.

It is convenient to separate the heat flux into radiative and diffusive components:

$$\boldsymbol{q} = \boldsymbol{q}_R + \boldsymbol{q}_T$$

$$= \boldsymbol{F} - k\nabla T, \tag{10.33}$$

where $\boldsymbol{F}$ is the net radiative flux (Sec. 8.2) and k denotes the thermal conductivity in Fourier's law of heat conduction. The first law then becomes

$$\rho c_v \frac{dT}{dt} + p\nabla \cdot \boldsymbol{v} = -\nabla \cdot \boldsymbol{F} + \nabla \cdot (k\nabla T) + \rho \dot{q}. \tag{10.34}$$

Known as the *thermodynamic equation*, (10.34) expresses the rate that a material element's internal energy changes in terms of the rate that work is performed on it and the net rate it absorbs heat through convergence of radiative and diffusive energy fluxes.

The thermodynamic equation can be expressed more compactly in terms of potential temperature. For an individual air parcel, the fundamental relation for internal energy

$$du = Tds - pd\alpha \tag{10.35.1}$$

relates the change of entropy to other material properties. Incorporating the identity between entropy and potential temperature (3.25) transforms this into

$$c_p Td \ln \theta = du + pd\alpha. \tag{10.35.2}$$

Differentiating with respect to time, with the material coordinate $\boldsymbol{\xi}$ held fixed, obtains

$$c_p T \frac{d \ln \theta}{dt} = \frac{du}{dt} + p \frac{d\alpha}{dt}, \tag{10.36}$$

where d/dt represents the Lagrangian derivative (10.10). Multiplying by ρ and introducing the continuity equation (10.4) transforms this into

$$\frac{\rho c_p T}{\theta} \frac{d\theta}{dt} = \rho c_v \frac{dT}{dt} + p\nabla \cdot \boldsymbol{v}. \tag{10.37}$$

Then incorporating (10.37) into (10.34) absorbs the compression work into the time rate of change of θ to yield the thermodynamic equation

$$\rho \frac{c_p T}{\theta} \frac{d\theta}{dt} = -\nabla \cdot \boldsymbol{F} + \nabla \cdot (k\nabla T) + \rho\dot{q}. \tag{10.38}$$

Collectively, the continuity, momentum, and thermodynamic equations represent five partial differential equations in five dependent field variables: three components of motion and two independent thermodynamic properties. Advective contributions to the material derivative, like advection of momentum $\boldsymbol{v} \cdot \nabla\boldsymbol{v}$ and of temperature $\boldsymbol{v} \cdot \nabla T$, make the governing equations quadratically nonlinear. Their solution requires initial conditions that specify the preliminary state of the atmosphere and boundary conditions that specify its properties along physical borders. Referred to as the *equations of motion*, these equations govern the behavior of a compressible, stratified atmosphere in a rotating reference frame. As such, they constitute the starting point for dynamical investigations, as well for investigations of chemistry and radiation in the presence of motion.

Suggested Reading

Vectors, Tensors, and the Basic Equations of Fluid Mechanics (1962) by Aris includes an excellent development of the kinematics of fluid motion and of the governing equations from a Lagrangian perspective.

An Introduction to Dynamic Meteorology (1992) by Holton contains alternate derivations of the continuity and thermodynamic equations from the Eulerian perspective. It includes an illuminating derivation of the total energy balance for a material element in terms of contributions from mechanical energy and internal energy.

Problems

10.1. Show that the antisymmetric component of the velocity gradient tensor can be expressed in terms of the angular velocity vector $\hat{\boldsymbol{\omega}}$ in (10.7).

10.2. Show that the vorticity equals twice the local angular velocity (10.8).

10.3. Derive the integral identity

$$\int_{V(t)} \left\{ \rho c_v \frac{dT}{dt} + \nabla \cdot \boldsymbol{q} + p\nabla \cdot \boldsymbol{v} - \rho\dot{q} \right\} dV' = 0$$

governing a material volume $V(t)$.

10.4. Demonstrate that the trace of the deformation tensor represents the fractional rate at which a material element's volume increases (Fig. 10.4a).

10.5. Demonstrate that the shear strain rate e_{12} represents one-half the rate of decrease of the angle between two material segments that are aligned initially in the directions of coordinates 1 and 2 (Fig. 10.4b).

10.6. The equations of motion determine three components of velocity and two thermodynamic properties. (a) Why are only two thermodynamic properties explicitly determined? (b) Describe the geometric form those thermodynamic properties assume for a particular atmospheric state. (c) As in part (b), but for the special case when one thermodynamic property can be expressed in terms of the other, for example, $\alpha = \alpha(p)$.

10.7. Derive the momentum equations (10.23) from Cauchy's equations of motion.

10.8. Prove that the relationship $f(x, t) = $ const describes a material surface if and only if f satisfies $df/dt = 0$.

10.9. Consider the two-dimensional motion

$$u(x, y, t) = -\frac{\partial \psi}{\partial y},$$

$$v(x, y, t) = \frac{\partial \psi}{\partial x},$$

where

$$\psi(x, y, t) = 4\exp\left(-\left\{[x - x_0(t)]^2 + [y - y_0(t)]^2\right\}\right)$$

defines a family of streamlines and

$$x_0(t) = y_0(t) = t.$$

Plot the parcel trajectory and streakline from $t = 0$ to 4 beginning at (a) $(x, y) = (0, 1)$, (b) $(x, y) = (-1, 0)$, and (c) $(x, y) = (0, 0)$.

10.10. For the motion in Problem 10.9, plot at $t = 1, 2, 3, 4$ the material volume that, at $t = 0$, is defined by the radial coordinates:

$$1 - \Delta r < r < 1 + \Delta r,$$
$$-\Delta\phi < \phi < \Delta\phi$$

for (a) $\Delta r = 0.5$ and $\Delta\phi = \pi/4$ and (b) $\Delta r = 0.1$ and $\Delta\phi = 0.1$. (c) Contrast the deformations experienced by the material volumes in parts (a) and (b) and use them to infer the limiting behavior for Δr and $\Delta\phi \to 0$.

10.11. Use the vector identity (D.14) in Appendix D to show that the material acceleration can be expressed

$$\frac{d\boldsymbol{v}}{dt} = \frac{\partial \boldsymbol{v}}{\partial t} + \nabla\frac{|\boldsymbol{v}|^2}{2} + \boldsymbol{\zeta} \times \boldsymbol{v},$$

where $\boldsymbol{\zeta} = \nabla \times \boldsymbol{v}$.

10.12. Demonstrate that the Coriolis force does not enter the budget of specific kinetic energy $|\boldsymbol{v}|^2/2$.

10.13. Show that the vorticity $\zeta = \boldsymbol{k} \cdot \nabla \times \boldsymbol{v}$ is a conserved property for two-dimensional nondivergent motion in an inertial reference frame.

10.14. Simplify the equations of motion for the special case of (a) incompressible motion, wherein the volume of a material element is conserved, and (b) adiabatic motion.

10.15. Describe the circumstances under which a property ψ is conserved, yet varies spatially in a steady flow.

10.16. The stress tensor in a Newtonian fluid is described by

$$\tau_{ij} = -p\delta_{ij} + \left(\mu' - \frac{2}{3}\mu\right)\delta_{ij}\nabla \cdot \boldsymbol{v} + 2\mu e_{ij},$$

where μ' is the so-called *bulk viscosity*. (a) Determine the average of the normal stresses in the three coordinate directions and discuss its relationship to the thermodynamic pressure p. (b) As in part (a), but for $\mu' = 0$, which describes a *Stokesian fluid*. (c) Show that the stress tensor reduces to (10.22) and the drag to (10.23.2) if the motion is incompressible.

10.17. A free surface is one that moves to alleviate any stress and maintain $\tau = 0$. If the rate of strain tensor is dominated by the vertical shears $\partial u/\partial z$ and $\partial v/\partial z$, describe how the horizontal velocity varies adjacent to a free surface.

10.18. A geostationary satellite is positioned over 30° latitude and 0° longitude. From it, a projectile is fired northward at a speed v_0. By assuming the deflection of the projectile's trajectory to be small and ignoring sphericity and the satellite's altitude, estimate the longitude where the projectile crosses 45° latitude if (a) $v_0 = 1000 \text{ m s}^{-1}$, (b) $v_0 = 100 \text{ m s}^{-1}$, and (c) $v_0 = 10 \text{ m s}^{-1}$. (d) For each of the preceding results, evaluate the dimensionless timescale for the traversal scaled by that for rotation of the earth.

10.19. Consider a parcel with local speed v, the natural coordinate s measured along the parcel's trajectory, and a unit vector $\boldsymbol{s}$ that is everywhere tangential to the trajectory. (a) Express the material derivative in terms of v, s, and $\boldsymbol{s}$. (b) Show that the material acceleration is described by

$$\frac{d\boldsymbol{v}}{dt} = \frac{dv}{dt}\boldsymbol{s} + v\frac{d\boldsymbol{s}}{dt}.$$

(c) Interpret the two accelerations appearing on the right.

10.20. In terms of Newton's third law, explain how turbulent mixing can exert drag on a moving air parcel.

Chapter 11 | Atmospheric Equations of Motion

In vector form, the equations of motion are valid in any coordinate system. However, those equations do not lend themselves to application and standard methods of solution. To be useful, the governing equations must be expressed in scalar form, which then depend on the coordinate system. We develop the scalar equations of motion within the general framework of curvilinear coordinates. In addition to accounting for geometric distortions inherent to spherical coordinates, that framework allows a straightforward development of the equations in thermodynamic coordinates, which afford a number of simplifications.

11.1 Curvilinear Coordinates

Consider the *Cartesian coordinates*

$$x = (x_1, x_2, x_3), \tag{11.1.1}$$

which are measured from planar coordinate surfaces: $x_i = $ const. If those planes are perpendicular, (x_1, x_2, x_3) are referred to as *rectangular Cartesian coordinates*. More generally, the *curvilinear coordinates*

$$\hat{x} = (\hat{x}_1, \hat{x}_2, \hat{x}_3) \tag{11.1.2}$$

are measured from coordinate surfaces that need not be planar nor mutually orthogonal (Fig. 11.1). Insofar as x and $\hat{x}$ are both viable coordinate systems, there exists a transformation

$$\hat{x} = \hat{x}(x) \tag{11.2}$$

from one representation to the other, which constitutes a mapping between all coordinates x in the original system and coordinates $\hat{x}$ in the curvilinear coordinate system. The mapping (11.2) is unique and therefore invertible, provided that the *Jacobian* of the transformation

$$J(x, \hat{x}) = \left| \frac{\partial \hat{x}_i}{\partial x_j} \right| \tag{11.3}$$

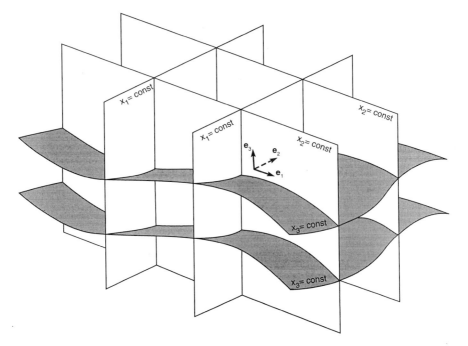

Figure 11.1 Coordinate surfaces $x_i = $ const for curvilinear coordinates. Coordinate vectors e_i point in the directions of increasing x_i.

is nonzero for all x. The Jacobian (which applies irrespective of whether or not the original coordinate system is Cartesian) has the reciprocal property

$$J(x, \hat{x}) \cdot J(\hat{x}, x) = 1 \qquad (11.4)$$

(Problem 11.1).

Unlike Cartesian coordinates, the curvilinear coordinates $\hat{x}_i$ need not represent length. For example, the spherical coordinates longitude λ and latitude ϕ represent angular displacements. The physical length ds_i corresponding to an increment of the curvilinear coordinate $\hat{x}_i$ is accounted for in the *metric scale factor* h_i:

$$ds_i = h_i d\hat{x}_i \qquad (11.5.1)$$

$$h_i^{-1} = |\nabla \hat{x}_i|, \qquad (11.5.2)$$

where the summation convention is not in force and ∇ refers to differentiation with respect to the original coordinates x_i. Similarly, the physical volume corresponding to the incremental element $dV_{\hat{x}} = d\hat{x}_1 d\hat{x}_2 d\hat{x}_3$ in the curvilinear coordinate system is given by

$$dV_x = J(\hat{x}, x)dV_{\hat{x}}$$

$$= \frac{1}{J(x, \hat{x})}dV_{\hat{x}}. \tag{11.6}$$

The curvilinear coordinate system is said to be *orthogonal* if coordinate surfaces are mutually perpendicular. Under those circumstances, coordinate vectors $\hat{e}_i$, which point in the directions of increasing $\hat{x}_i$ (Fig. 11.1), are likewise mutually perpendicular

$$\hat{e}_i \cdot \hat{e}_j = \delta_{ij}.$$

Then the Jacobian reduces to

$$J(\hat{x}, x) = h_1 h_2 h_3 \tag{11.7}$$

(Problem 11.2) and the physical volume corresponding to the element $d\hat{x}_1 d\hat{x}_2 d\hat{x}_3$ in the curvilinear system is just the product of the corresponding physical lengths (11.5).

The Jacobian accounts for distortions of physical length in all three curvilinear coordinates. In the special case when only the third coordinate is transformed

$$\hat{x}(x) = (x_1, x_2, \hat{x}_3) \tag{11.8}$$

(e.g., when surfaces of constant height are replaced by isobaric surfaces), the Jacobian is given by

$$J(x, \hat{x}) = \begin{vmatrix} 1 & 0 & 0 \\ 0 & 1 & 0 \\ \frac{\partial \hat{x}_3}{\partial x_1} & \frac{\partial \hat{x}_3}{\partial x_2} & \frac{\partial \hat{x}_3}{\partial x_3} \end{vmatrix}$$

$$= \frac{\partial \hat{x}_3}{\partial x_3}. \tag{11.9}$$

For this special class of transformations, (11.4) reduces to the simple reciprocal property

$$\frac{\partial \hat{x}_3}{\partial x_3} = \left(\frac{\partial x_3}{\partial \hat{x}_3}\right)^{-1}, \tag{11.10}$$

which is not true in general.

Vector operations in arbitrary curvilinear coordinates can now be expressed in terms of the corresponding metric scale factors:

$$\nabla \psi = \frac{1}{h_1} \frac{\partial \psi}{\partial \hat{x}_1} \hat{e}_1 + \frac{1}{h_2} \frac{\partial \psi}{\partial \hat{x}_2} \hat{e}_2 = \frac{1}{h_3} \frac{\partial \psi}{\partial \hat{x}_3} \hat{e}_3, \tag{11.11.1}$$

$$\nabla \cdot A = \frac{1}{h_1 h_2 h_3} \left[\frac{\partial}{\partial \hat{x}_1} (h_2 h_3 A_1) + \frac{\partial}{\partial \hat{x}_2} (h_1 h_3 A_2) + \frac{\partial}{\partial \hat{x}_3} (h_1 h_2 A_3) \right], \tag{11.11.2}$$

$$\nabla \times A = \begin{vmatrix} h_1 \hat{e}_1 & h_2 \hat{e}_2 & h_3 \hat{e}_3 \\ \frac{\partial}{\partial \hat{x}_1} & \frac{\partial}{\partial \hat{x}_2} & \frac{\partial}{\partial \hat{x}_3} \\ h_1 A_1 & h_2 A_2 & h_3 A_3 \end{vmatrix}, \tag{11.11.3}$$

$$\nabla^2 \psi = \frac{1}{h_1 h_2 h_3} \left[\frac{\partial}{\partial \hat{x}_1} \left(\frac{h_2 h_3}{h_1} \frac{\partial \psi}{\partial \hat{x}_1} \right) \right.$$
$$\left. + \frac{\partial}{\partial \hat{x}_2} \left(\frac{h_1 h_3}{h_2} \frac{\partial \psi}{\partial \hat{x}_2} \right) + \frac{\partial}{\partial \hat{x}_3} \left(\frac{h_1 h_2}{h_3} \frac{\partial \psi}{\partial \hat{x}_3} \right) \right]. \tag{11.11.4}$$

11.2 Spherical Coordinates

Consider the rectangular Cartesian coordinates $x = (x_1, x_2, x_3)$ having origin at the center of the earth and the spherical coordinates

$$\hat{x} = (\lambda, \phi, r),$$

both fixed with respect to the earth (Fig. 11.2). Spherical coordinate surfaces $\lambda = $ const, $\phi = $ const, and $r = $ const are then (1) vertical planes intersecting the Cartesian x_3 axis, (2) conical shells with apex at the origin $x = 0$, and (3) concentric spherical shells, respectively. The corresponding unit vectors

$$e_\lambda = i,$$
$$e_\phi = j, \tag{11.12}$$
$$e_r = k,$$

are everywhere perpendicular to those coordinate surfaces. Because coordinate surfaces are mutually perpendicular, so are the unit vectors, and the spherical coordinates (λ, ϕ, r) constitute an orthogonal coordinate system.

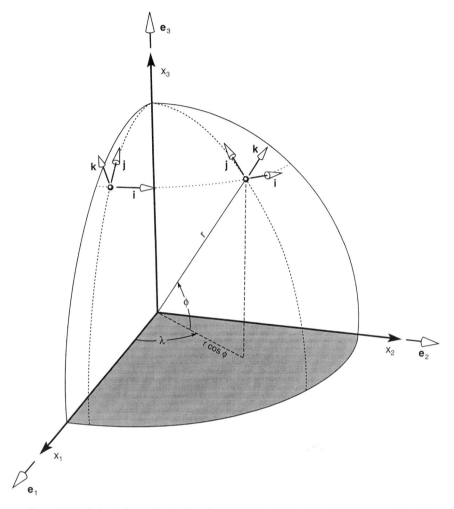

Figure 11.2   Spherical coordinates: longitude λ, latitude ϕ, and radial distance r. Coordinate vectors $e_\lambda = i$, $e_\phi = j$, and $e_r = k$ change with position (e.g., relative to fixed coordinate vectors e_1, e_2, and e_3 of rectangular Cartesian coordinates).

The rectangular Cartesian coordinates can be expressed in terms of the spherical coordinates as

$$x_1 = r\cos\phi\cos\lambda,$$
$$x_2 = r\cos\phi\sin\lambda, \qquad (11.13.1)$$
$$x_3 = r\sin\phi,$$

which may be inverted for the spherical coordinates:

$$\lambda = \tan^{-1}\left(\frac{x_2}{x_1}\right),$$

$$\phi = \tan^{-1}\left(\frac{x_3}{x_1^2 + x_2^2}\right), \tag{11.13.2}$$

$$r = \sqrt{x_1^2 + x_2^2 + x_3^2}.$$

Metric scale factors for the spherical coordinates then follow as

$$h_1 = r \cos \phi,$$

$$h_2 = r, \tag{11.14}$$

$$h_3 = 1$$

(Problem 11.4). Because $\hat{x}$ constitutes an orthogonal coordinate system, the Jacobian of the transformation follows from (11.7) as

$$J(\hat{x}, x) = r^2 \cos \phi, \tag{11.15}$$

which is nonzero except at the poles. At $\phi = \pm 90°$, coordinate surfaces $\phi = $ const converge onto the x_3 axis, so the incremental volume encased by two such surfaces vanishes and the transformation is singular. Vector operations follow from (11.11) as

$$\nabla \psi = \frac{1}{r \cos \phi} \frac{\partial \psi}{\partial \lambda} i + \frac{1}{r} \frac{\partial \psi}{\partial \phi} j + \frac{\partial \psi}{\partial r} k, \tag{11.16.1}$$

$$\nabla \cdot A = \frac{1}{r \cos \phi} \frac{\partial A_\lambda}{\partial \phi} + \frac{1}{r \cos \phi} \frac{\partial}{\partial \phi} (\cos \phi A_\phi) + \frac{1}{r^2} \frac{\partial}{\partial r} (r^2 A_r), \tag{11.16.2}$$

$$\nabla \times A = \frac{1}{r^2 \cos \phi} \begin{vmatrix} r \cos \phi i & rj & k \\ \frac{\partial}{\partial \lambda} & \frac{\partial}{\partial \phi} & \frac{\partial}{\partial r} \\ r \cos \phi A_\lambda & rA_\phi & A_r \end{vmatrix}, \tag{11.16.3}$$

$$\nabla^2 \psi = \frac{1}{r^2 \cos^2 \phi} \frac{\partial^2 \psi}{\partial \lambda^2} + \frac{1}{r^2 \cos \phi} \left[\cos \phi \frac{\partial \psi}{\partial \phi} \right] + \frac{1}{r^2} \frac{\partial}{\partial r} \left(r^2 \frac{\partial \psi}{\partial r} \right). \tag{11.16.4}$$

From (11.14), physical displacements in the directions of increasing longitude, latitude, and radial distance are described by

$$dx = r \cos \phi d\lambda,$$

$$dy = r d\phi, \tag{11.17.1}$$

$$dz = dr,$$

in which height

$$z = r - a, \tag{11.17.2}$$

where a denotes the mean radius of the earth, is used to measure vertical distance.[1] Physical velocities in the spherical coordinate system are then expressed by

$$
\begin{aligned}
u &= \frac{dx}{dt} = r \cos \phi \frac{d\lambda}{dt}, \\
v &= \frac{dy}{dt} = r \frac{d\phi}{dt}, \\
w &= \frac{dz}{dt} = \frac{dr}{dt}.
\end{aligned}
\tag{11.18}
$$

The vector equations of motion can now be cast in terms of spherical coordinates. Derivatives of scalar quantities transform directly with the above expressions. However, Lagrangian derivatives of vector quantities are complicated by the dependence on position of the coordinate vectors i, j, and k. Each rotates in physical space under a displacement of longitude or latitude. For example, an air parcel moving along a latitude circle at a constant speed u has a velocity $v = ui$, which appears constant in the spherical coordinate representation, but which actually rotates in physical space (Fig. 11.2). Consequently, the parcel experiences an acceleration that must be accounted for in the equations of motion.

Consider the velocity

$$v = ui + vj + wk.$$

Because the spherical coordinate vectors i, j, and k are functions of position $\hat{x}$, the material acceleration is actually

$$
\begin{aligned}
\frac{dv}{dt} &= \frac{du}{dt}i + \frac{dv}{dt}j + \frac{dw}{dt}k + u\frac{di}{dt} + v\frac{dj}{dt} + w\frac{dk}{dt} \\
&= \left(\frac{dv}{dt}\right)_C + u\frac{di}{dt} + v\frac{dj}{dt} + w\frac{dk}{dt},
\end{aligned}
\tag{11.19.1}
$$

where the subscript refers to the basic form of the material derivative in Cartesian geometry

$$\left(\frac{d}{dt}\right)_C = \frac{\partial}{\partial t} + v \cdot \nabla. \tag{11.19.2}$$

To evaluate corrections on the right-hand side of (11.19.1), the spherical coordinate vectors are expressed in terms of the fixed rectangular Cartesian

---

[1] This representation is only approximate, because it ignores departures from sphericity of height surfaces (Sec. 6.2).

coordinate vectors e_1, e_2, e_3

$$i = -\sin \lambda e_1 + \cos \lambda e_2,$$
$$j = -\sin \phi \cos \lambda e_1 - \sin \phi \sin \lambda e_2 + \cos \phi e_3, \qquad (11.20)$$
$$k = \cos \phi \cos \lambda e_1 + \cos \phi \sin \lambda e_2 + \sin \phi e_3,$$

the demonstration of which is left as an exercise. From these and reciprocal expressions for e_1, e_2, and e_3 in terms of i, j, and k, Lagrangian derivatives of the spherical coordinate vectors follow as

$$\frac{di}{dt} = u\left(\frac{\tan \phi}{r}j - \frac{1}{r}k\right),$$

$$\frac{dj}{dt} = -u\frac{\tan \phi}{r}i - \frac{v}{r}k, \qquad (11.21)$$

$$\frac{dk}{dt} = \frac{u}{r}i + \frac{v}{r}j$$

(see Problem 11.5). Then material accelerations in the spherical coordinate directions become

$$\frac{du}{dt} = \left(\frac{du}{dt}\right)_C - \frac{uv \tan \phi}{r} + \frac{uw}{r}, \qquad (11.22.1)$$

$$\frac{dv}{dt} = \left(\frac{dv}{dt}\right)_C + \frac{u^2 \tan \phi}{r} + \frac{vw}{r}, \qquad (11.22.2)$$

$$\frac{dw}{dt} = \left(\frac{dw}{dt}\right)_C - \left(\frac{u^2 + v^2}{r}\right). \qquad (11.22.3)$$

Corrections appearing on the right-hand sides of (11.22), which are referred to as *metric terms*, describe accelerations that result from curvature of the coordinate system.

Because the atmosphere occupies a thin shell about the earth, it is customary to simplify the radial dependence by neglecting small fractional changes of r in (11.16) through (11.22). The *shallow atmosphere approximation* makes use of the fact that $z \ll a$ to take $r = a$ and ignore the geometric divergence associated with vertical displacements. The vector operations (11.16) then reduce to

$$\nabla \psi = \frac{1}{a \cos \phi}\frac{\partial \psi}{\partial \lambda}i + \frac{1}{a}\frac{\partial \psi}{\partial \phi}j + \frac{\partial \psi}{\partial z}k, \qquad (11.23.1)$$

$$\nabla \cdot A = \frac{1}{a \cos \phi}\frac{\partial A_\lambda}{\partial \phi} + \frac{1}{a \cos \phi}\frac{\partial}{\partial \phi}(\cos \phi A_\phi) + \frac{\partial A_z}{\partial z}, \qquad (11.23.2)$$

$$\nabla \times A = \frac{1}{a^2 \cos \phi} \begin{vmatrix} a \cos \phi i & a j & k \\ \frac{\partial}{\partial \lambda} & \frac{\partial}{\partial \phi} & \frac{\partial}{\partial z} \\ a \cos \phi A_\lambda & a A_\phi & A_z \end{vmatrix}, \quad (11.23.3)$$

$$\nabla^2 \psi = \frac{1}{a^2 \cos^2 \phi} \frac{\partial^2 \psi}{\partial \lambda^2} + \frac{1}{a^2 \cos \phi} \left[\cos \phi \frac{\partial \psi}{\partial \phi} \right] + \frac{\partial^2 \psi}{\partial z^2}, \quad (11.23.4)$$

in which height has been formally adopted as the vertical coordinate.

With this approximation and material accelerations (11.22), the equations of motion can be cast into component form. Expressing the planetary vorticity in terms of horizontal and vertical components of the earth's rotation

$$2\boldsymbol{\Omega} = 2\Omega(\cos \phi \boldsymbol{j} + \sin \phi \boldsymbol{k}) \quad (11.24)$$

(Fig. 11.3) yields the scalar equations of motion in spherical coordinates:

$$\frac{du}{dt} - 2\Omega(v \sin \phi - w \cos \phi) = -\frac{1}{\rho a \cos \phi} \frac{\partial p}{\partial \lambda} + uv \frac{\tan \phi}{a} - \frac{uw}{a} - D_\lambda, \quad (11.25.1)$$

$$\frac{dv}{dt} + 2\Omega u \sin \phi = -\frac{1}{\rho a} \frac{\partial p}{\partial \phi} - \frac{u^2 \tan \phi}{a} - \frac{uw}{a} - D_\phi, \quad (11.25.2)$$

$$\frac{dw}{dt} - 2\Omega u \cos \phi = -\frac{1}{\rho} \frac{\partial p}{\partial z} - g + \frac{u^2 + v^2}{a} - D_z, \quad (11.25.3)$$

$$\frac{d\rho}{dt} + \rho \nabla \cdot \boldsymbol{v} = 0, \quad (11.25.4)$$

$$\rho c_v \frac{dT}{dt} + p \nabla \cdot \boldsymbol{v} = \dot{q}_{net}, \quad (11.25.5)$$

where

$$\frac{d}{dt} = \frac{\partial}{\partial t} + \frac{u}{a \cos \phi} \frac{\partial}{\partial \lambda} + \frac{v}{a} \frac{\partial}{\partial \phi} + w \frac{\partial}{\partial z} \quad (11.25.6)$$

and

$$\dot{q}_{net} = -\nabla \cdot \boldsymbol{F} + \nabla \cdot (k \nabla T) + \rho \dot{q} \quad (11.25.7)$$

denotes the net heating rate from all diabatic sources.

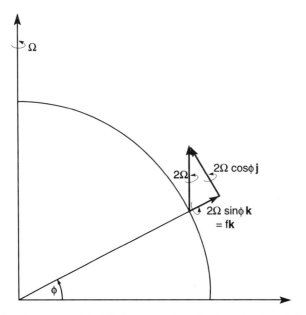

Figure 11.3   Planetary vorticity 2Ω decomposed into horizontal and vertical components.

11.2.1 The Traditional Approximation

Although it simplifies the mathematics, using height as the vertical coordinate with r replaced by the constant value a has an important drawback: The resulting equations do not possess an angular momentum principle, like the one satisfied by the equations in full spherical geometry (Problem 11.6). On conceptual grounds alone, this is an important deficiency because much of atmospheric dynamics concerns how angular momentum is concentrated into strong jets that characterize the general circulation. The failure of (11.25) to properly represent the angular momentum of individual air parcels follows from an inconsistency in the treatment of horizontal and vertical displacements. Geometric variations in the radial direction are neglected, but Coriolis and metric terms proportional to $\cos \phi$, which accompany radial displacements, are retained.

Terms proportional to $2\Omega \cos \phi$ in (11.25) are associated with the horizontal component of planetary vorticity (11.24) and vertical motion. Those terms are much smaller than others and are often neglected on the basis of scaling arguments (Sec. 11.4). Known as the *traditional approximation* (Eckart, 1960), the neglect of terms proportional to $2\Omega \cos \phi$ is formally valid in the limit of strong stratification, wherein

$$\frac{N^2}{\Omega^2} \to \infty. \tag{11.26}$$

Air is then constrained to move quasi-horizontally, which makes the vertical component of planetary vorticity $2\Omega \sin \phi$ dominant in the Coriolis acceleration $2\boldsymbol{\Omega} \times \boldsymbol{v}$. For a lapse rate of $\Gamma = 6.5 \, \mathrm{K\,km^{-1}}$, (7.10) gives N^2 of order 10^{-4} s^{-2} and $\Omega^2 \cong 5 \times 10^{-9} \, \mathrm{s}^{-2}$, so $N^2/\Omega^2 \cong 2 \times 10^4$ is in good agreement with (11.26). Weak or unstable stratification is controlled by convection, which operates on timescales short enough to render both components of the Coriolis acceleration unimportant.

An alternate derivation of the equations of motion (Phillips, 1966) provides a rationale for neglecting the aforementioned terms. The component equations assume a simpler form if derived from the momentum equations in vector-invariant form. In an inertial reference frame, the vector equations of motion can be expressed

$$\left(\frac{\partial \boldsymbol{v}_i}{\partial t}\right)_i + \nabla \left(\frac{\boldsymbol{v}_i \cdot \boldsymbol{v}_i}{2}\right) + (\nabla \times \boldsymbol{v}_i) \times \boldsymbol{v}_i = -\frac{1}{\rho} \nabla p - g\boldsymbol{k} - \boldsymbol{D}, \qquad (11.27.1)$$

where the subscript refers to the velocity and local time rate of change apparent in the inertial frame (Problem 11.7). Evaluating metric scale factors at $r = a$ before derivatives are taken recovers the vector operations (11.23). Then expressing the velocity and time derivative in terms of those apparent in the rotating frame of the earth

$$\boldsymbol{v}_i = \boldsymbol{v} + \Omega a \cos \phi \boldsymbol{i}, \qquad (11.27.2)$$

$$\left(\frac{\partial}{\partial t}\right)_i = \frac{\partial}{\partial t} - \Omega \frac{\partial}{\partial \lambda} \qquad (11.27.3)$$

and incorporating (11.23) produces the simplified momentum equations

$$\frac{du}{dt} - \left(f + \frac{u \tan \phi}{a}\right) v = -\frac{1}{\rho a \cos \phi} \frac{\partial p}{\partial \lambda} - D_\lambda, \qquad (11.28.1)$$

$$\frac{dv}{dt} + \left(f + \frac{u \tan \phi}{a}\right) u = -\frac{1}{\rho a} \frac{\partial p}{\partial \phi} - D_\phi, \qquad (11.28.2)$$

$$\frac{dw}{dt} = -\frac{1}{\rho} \frac{\partial p}{\partial z} - g - D_z, \qquad (11.28.3)$$

where the *Coriolis parameter*

$$f = 2\Omega \sin \phi \qquad (11.28.4)$$

represents the vertical component of planetary vorticity (Fig. 11.3).

Equations (11.28) are free of metric and Coriolis terms proportional to w in the horizontal momentum equations and all such terms in the vertical momentum equation. The above system possesses the conservation principle

$$\frac{d}{dt}[(u + \Omega a \cos \phi)a \cos \phi] = -a \cos \phi \left(\frac{1}{\rho a \cos \phi} \frac{\partial p}{\partial \lambda} + D_\lambda\right), \qquad (11.29)$$

wherein the angular momentum of an air parcel changes according to the torque about the axis of rotation exerted on it by longitudinal forces (Problem 11.8). Consistent with the shallow atmosphere approximation, angular momentum is treated as though the air parcel remains at $r = a$.

11.3 Special Forms of Motion

Certain conditions simplify the governing equations. For *incompressible motion*, the specific volume of an individual material element is conserved, so

$$\frac{d\rho}{dt} = 0.$$

Then the continuity equation (11.25.4) reduces to

$$\nabla \cdot \boldsymbol{v} = 0. \tag{11.30}$$

Because ρ is conserved, a material element that coincides initially with a particular isochoric surface, $\rho = $ const, remains on that surface—despite movement of that surface. Thus isochoric surfaces are material surfaces (namely, they are comprised of a fixed collection of matter). If the motion is steady, ρ surfaces are also stream surfaces, to which the motion is tangential. These conditions hold automatically for an incompressible fluid like water. They also hold for a compressible fluid like air if the motion is steady and orthogonal to the gradient of density. Because large-scale motions are quasi-horizontal, the latter condition is approximately satisfied by atmospheric circulations.

For *adiabatic motion*, individual material elements experience no heat transfer with their surroundings (Sec. 3.6.1). The thermodynamic equation (10.38) then reduces to

$$\frac{d\theta}{dt} = 0, \tag{11.31}$$

which asserts that the potential temperature of a material element is conserved. Thus, an air parcel that coincides initially with an isentropic surface, $\theta = $ const, remains on that surface (refer to Fig. 2.9) and isentropic surfaces are material surfaces.

The cases just discussed are particular examples of a conserved property, which behaves as a material tracer. More generally, a property r that is conserved for individual material elements obeys the continuity equation

$$\frac{dr}{dt} = 0. \tag{11.32}$$

Mixing ratios of long-lived chemical species approximately satisfy (11.32) because the rates of production and destruction for such species, which formally belong on the right-hand side, are much slower than advective changes in the Lagrangian derivative, which appear on the left-hand side. Ozone, which has a photochemical lifetime in the lower stratosphere of several weeks, and water vapor, which is conserved away from clouds and the earth's surface, behave approximately in this manner (see Fig. 1.15). Hence, to a first approximation, particular values of mixing ratio track the movement of individual bodies of air and the surfaces $r =$ const are material surfaces.

11.4 Prevailing Balances

The preceding equations govern motion in a compressible, stratified, and rotating atmosphere, so they apply to a wide range of phenomena. While describing planetary-scale circulations that involve timescales of days and longer, the equations of motion also describe small-scale acoustic waves that have timescales of only fractions of a second. This generality needlessly complicates the description of phenomena that figure in the general circulation. To elucidate essential balances controlling large-scale atmospheric motions, it is useful to examine the relative sizes of various terms.

11.4.1 Motion-Related Stratification

The stratification, which is represented in the distributions of pressure and density, is comprised of two components: (1) a basic component associated with static conditions and (2) a small departure from it that is related directly to motion:

$$
\begin{aligned}
p_{\text{tot}} &= p_0(z) + p(x, t), \\
\rho_{\text{tot}} &= \rho_0(z) + \rho(x, t),
\end{aligned}
\tag{11.33}
$$

where $x = (x, y, z)$ follows from (11.17). The static components p_0 and ρ_0, which are functions of height alone, are symbolized by the global–mean pressure and density. Those components overshadow vertical variations associated with motion, so it is instructive to eliminate them from the governing equations.

Incorporating (11.33) into the horizontal momentum equation transforms the pressure gradient force into

$$
-\frac{1}{(\rho_0 + \rho)} \nabla_h (p_0 + p) \cong -\frac{1}{\rho_0} \left(1 - \frac{\rho}{\rho_0} \right) \nabla_h p
$$

$$
\cong -\frac{1}{\rho_0} \nabla_h p,
\tag{11.34}
$$

where ∇_h denotes the horizontal gradient and $\rho/\rho_0 \ll 1$ and $p/p_0 \ll 1$ have been used in combination with the binomial expansion to ignore higher order terms. Because $p_0(z)$ depends on height alone, it drops out of (11.34) to leave the horizontal momentum equations in their original form (11.28).

In the vertical momentum equation, the basic components of stratification introduce vertical gradients that cannot be ignored. The right-hand side of (11.28.3) contains the net buoyancy force acting on an air parcel (7.2). Incorporating (11.33) transforms the buoyancy force into

$$f_b = -\frac{1}{(\rho_0 + \rho)}\frac{\partial}{\partial z}(p_0 + p) - g \cong -\frac{1}{\rho_0}\left(1 - \frac{\rho}{\rho_0}\right)\left(\frac{\partial p_0}{\partial z} + \frac{\partial p}{\partial z}\right) - g$$

$$\cong -\frac{1}{\rho_0}\left(\frac{\partial p}{\partial z} + \rho g\right), \tag{11.35}$$

wherein the basic stratification automatically satisfies hydrostatic equilibrium. Then the vertical momentum equation becomes

$$\frac{dw}{dt} = -\frac{1}{\rho_0}\frac{\partial p}{\partial z} - \frac{\rho}{\rho_0}g - D_z. \tag{11.36}$$

11.4.2 Scale Analysis

With dependent variables related directly to motion, terms in the momentum equations can now be evaluated for relative importance. Away from the earth's surface and regions of organized convection, the frictional drag D is small enough to be ignored. Large-scale atmospheric motions are then characterized by the scales

$$U = 10 \text{ m s}^{-1} \qquad L = 10^3 \text{ km} \qquad P = 10 \text{ mb} = 10^3 \text{ Pa}$$
$$W = 10^{-2} \text{ m s}^{-1} \qquad H = 10 \text{ km} \qquad f_0 = 10^{-4} \text{ s}^{-1},$$

where U and W refer to horizontal and vertical motion, respectively, L and H are horizontal and vertical length scales that characterize the motion field, and $f_0 = 2\Omega \sin \phi_0$ is the planetary vorticity at a representative latitude ϕ_0. The pressure P characterizes the departure from pure static conditions, namely, the pressure variation associated directly with motion. If the Lagrangian derivative is dominated by advective changes, L/U represents a timescale for advection that can be used to characterize d/dt.

Scaling velocities, lengths, and the Lagrangian derivative by the scales given above (e.g., $u \to Uu$, $x \to Lx$, $d/dt \to (U/L)d/dt$, with u, x, and d/dt then nondimensional) transforms the horizontal momentum equations (11.28.1) and (11.28.2) into the dimensionless forms

$$\text{Ro}\frac{du}{dt} - \sin \phi v - \text{Ro}\left(\frac{L}{a}\right)\tan \phi uv = -\frac{P}{f_0\rho_{00}UL}\cdot\frac{1}{\rho_0}\frac{\partial p}{\partial x}, \tag{11.37.1}$$

$$\text{Ro}\frac{dv}{dt} + \sin \phi u + \text{Ro}\left(\frac{L}{a}\right)\tan \phi u^2 = -\frac{P}{f_0\rho_{00}UL}\cdot\frac{1}{\rho_0}\frac{\partial p}{\partial y}, \tag{11.37.2}$$

where $\rho_{00} = \rho_0(0)$ and all variables are nondimensional, of order unity, and are related to their dimensional counterparts through multiplication by the above scale factors. The *Rossby number*

$$Ro = \frac{U}{f_0 L} \qquad (11.38)$$

appearing in the scaled momentum equations is a dimensionless parameter that represents the ratio of the rotational timescale f_0^{-1} to the advective timescale L/U.

With the preceding scales, dimensionless factors in (11.37) are characterized by the following orders of magnitude:

$$Ro = 10^{-1} \qquad Ro\left(\frac{L}{a}\right) = 10^{-2} \qquad \frac{P}{f_0 \rho_{00} UL} = 1.$$

Hence, the horizontal momentum equations are dominated by a balance between the Coriolis acceleration associated with the vertical component of planetary vorticity and the horizontal pressure gradient force, both of which are of order unity. Relative to these, the material acceleration, which is of order Ro, is small because the timescale for advection is long compared to the rotational timescale. Therefore, horizontal motion of an individual air parcel is dominated by rotation. Accelerations associated with metric terms and with vertical advection of momentum in d/dt are even smaller, whereas those proportional to w that were eliminated in the derivation of (11.28) are much smaller.

Similar treatment transforms the vertical momentum equation (11.36) into the dimensionless form

$$Ro\left(\frac{W}{U}\right)\frac{dw}{dt} = -\left(\frac{L}{H}\right)\frac{P}{f_0 \rho_{00} UL} \cdot \frac{1}{\rho_0}\frac{\partial p}{\partial z} - \left(\frac{P}{\rho_{00}}\right)\frac{g}{f_0 U} \cdot \frac{\rho}{\rho_0}, \qquad (11.39)$$

where $\rho_{00} = \rho_0(0)$. The preceding scales imply the orders of magnitude

$$Ro\left(\frac{W}{U}\right) = 10^{-3} \qquad \left(\frac{L}{H}\right)\frac{P}{f_0 \rho_{00} UL} = 10^2 \qquad \left(\frac{P}{\rho_{00}}\right)\frac{g}{f_0 U} = 10^2.$$

This scale analysis demonstrates that the vertical momentum equation for large-scale motions is dominated by a balance between the vertical pressure gradient and gravitational forces. Hence, the motion-related component of stratification also satisfies hydrostatic equilibrium, other vertical forces being much smaller. Vertical forces involved in hydrostatic equilibrium are also two orders of magnitude greater than accelerations in the horizontal momentum equations, which illustrates the comparative influences of gravity and rotation (11.26).

The foregoing analysis allows the equations of motion to be simplified. Inclusive of the basic stratification, the dimensional equations governing large-scale atmospheric motion in spherical coordinates then become

$$\frac{du}{dt} - \left(f + \frac{u \tan \phi}{a}\right) v = -\frac{1}{\rho a \cos \phi} \frac{\partial p}{\partial \lambda} - D_\lambda, \tag{11.40.1}$$

$$\frac{dv}{dt} + \left(f + \frac{u \tan \phi}{a}\right) u = -\frac{1}{\rho a} \frac{\partial p}{\partial \phi} - D_\phi, \tag{11.40.2}$$

$$\frac{1}{\rho} \frac{\partial p}{\partial z} = -g, \tag{11.40.3}$$

$$\frac{d\rho}{dt} + \rho \nabla \cdot \boldsymbol{v} = 0, \tag{11.40.4}$$

$$\rho c_v \frac{dT}{dt} + p \nabla \cdot \boldsymbol{v} = \dot{q}_{net}, \tag{11.40.5}$$

which are termed the *primitive equations* because they constitute the starting point for investigations of large-scale atmospheric dynamics. Metric terms proportional to $\tan \phi$ and horizontal drag have been retained in (11.40), but, while often included in numerical integrations, they are small enough to be omitted for many analyses. In chemical applications, the primitive equations are augmented by continuity equations of the form

$$\frac{dr_i}{dt} = \dot{P}_i - \dot{D}_i \tag{11.40.6}$$

for the mixing ratio of the ith species, where $\dot{P}_i$ and $\dot{D}_i$ denote the local rates of production and destruction of that species.

The governing equations must be closed with boundary conditions. Spherical geometry is periodic, so continuity suffices for horizontal boundary conditions. Vertical boundary conditions constrain the upward motion. At the ground, the motion must be tangential to that lower boundary. If the surface has elevation $z_s(\lambda, \phi, t)$,[2] this is equivalent to requiring an air parcel initially in contact with it to track along the same elevation:

$$w = \frac{dz_s}{dt}, \tag{11.41}$$

where d/dt is given by (11.25.6).[3] In the presence of viscosity, the motion must also satisfy the *no-slip condition*, which requires the velocity at the ground to vanish. Upper boundary conditions are more difficult to apply because the

[2] The lower boundary condition can be applied on a surface internal to the fluid, in which case the elevation is time dependent.

[3] Even though a fluid element at the ground must remain in contact with that boundary, neighboring elements can achieve large vertical displacements (e.g., in convection; refer to Fig. 5.4) through deformation of finite bodies of fluid.

domain is unbounded, but similar constraints are usually imposed at a finite height.

11.5 Thermodynamic Coordinates

11.5.1 Isobaric Coordinates

Because they involve hydrostatic equilibrium, the primitive equations simplify when formulated with pressure as the vertical coordinate. Hence, we consider a transformation from the standard spherical coordinates $x = (x, y, z)$ given by (11.17) to the modified coordinates $x_p = (x, y, p)$. Then isobaric surfaces $p = \text{const}$ replace constant height surfaces $z = \text{const}$ as coordinate surfaces, along which horizontal derivatives must be evaluated. Using p as a vertical coordinate is possible because hydrostatic equilibrium (11.40.3) ensures that pressure varies with height monotonically and therefore is a single-valued function of z.

In casting the equations into isobaric coordinates, the variables p and z are interchanged. Pressure then becomes an independent variable and z becomes a dependent variable. For a specified pressure, $z(p)$ represents the height of an isobaric surface, which is contoured in Fig. 1.9. Isobaric surfaces are not perpendicular to the other coordinate surfaces, so this representation does not constitute an orthogonal coordinate system.

Consider a scalar quantity

$$\psi(x, y, z, t) = \hat{\psi}[x, y, p(x, y, z, t), t]. \tag{11.42}$$

Hereafter, the caret will be omitted with the pressure dependence understood. By the chain rule, the vertical derivative can be expressed

$$\left(\frac{\partial \psi}{\partial z}\right)_{xyt} = \left(\frac{\partial \psi}{\partial p}\right)_{xyt} \left(\frac{\partial p}{\partial z}\right)_{xyt}, \tag{11.43}$$

where subscripts denote variables that are held fixed. Likewise, differentiation with respect to x becomes

$$\left(\frac{\partial \psi}{\partial x}\right)_{yzt} = \left(\frac{\partial \psi}{\partial x}\right)_{ypt} + \left(\frac{\partial \psi}{\partial p}\right)_{xyt} \left(\frac{\partial p}{\partial x}\right)_{yzt}$$

and similarly for differentiation with respect to y. Thus, the horizontal gradient evaluated on surfaces of constant height ∇_z translates into

$$\nabla_z \psi = \nabla_p \psi + \left(\frac{\partial \psi}{\partial p}\right)_{xyt} \nabla_z p, \tag{11.44.1}$$

where

$$\nabla_p = \left(\frac{\partial}{\partial x}\right)_{ypt} \mathbf{i} + \left(\frac{\partial}{\partial y}\right)_{xpt} \mathbf{j} \tag{11.44.2}$$

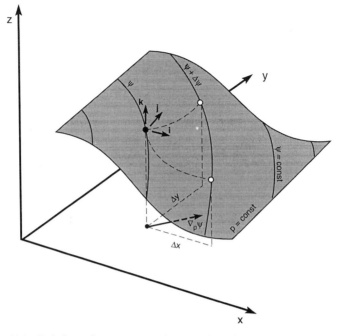

Figure 11.4 Variation of property ψ along an isobaric surface. Horizontal derivatives $(\partial\psi/\partial x)_{ypt} = \lim_{\Delta x \to 0}(\Delta\psi/\Delta x)$ and $(\partial\psi/\partial y)_{xpt} = \lim_{\Delta y \to 0}(\Delta\psi/\Delta y)$ are evaluated with changes of ψ along the isobaric surface $p = $ const. (For clarity, Δx and Δy have been chosen to place the incremented values of ψ on the same contour.) Those derivatives imply the horizontal gradient evaluated along the isobaric surface $\nabla_p\psi = (\partial\psi/\partial x)_{ypt}\boldsymbol{i} + (\partial\psi/\partial y)_{xpt}\boldsymbol{j}$, which lies in the horizontal plane but reflects how ψ varies along the isobaric surface.

denotes the horizontal gradient evaluated on an isobaric surface (Fig. 11.4). The local time derivative becomes

$$\left(\frac{\partial\psi}{\partial t}\right)_{xyz} = \left(\frac{\partial\psi}{\partial t}\right)_{xyp} + \left(\frac{\partial\psi}{\partial p}\right)_{xyt}\left(\frac{\partial p}{\partial t}\right)_{xyz}. \qquad (11.45)$$

Incorporating these expressions transforms the Lagrangian derivative into

$$\frac{d\psi}{dt} = \left(\frac{\partial\psi}{\partial t}\right)_{xyp} + \boldsymbol{v}_h \cdot \nabla_p\psi + \left(\frac{\partial\psi}{\partial p}\right)_{xyt}\left[\left(\frac{\partial p}{\partial t}\right)_{xyz} + \boldsymbol{v}_h \cdot \nabla_z p + w\left(\frac{\partial p}{\partial z}\right)_{xyt}\right]$$

$$= \left(\frac{\partial\psi}{\partial t}\right)_{xyp} + \boldsymbol{v}_h \cdot \nabla_p\psi + \frac{dp}{dt}\left(\frac{\partial\psi}{\partial p}\right)_{xyt}, \qquad (11.46)$$

where

$$\boldsymbol{v}_h = u\boldsymbol{i} + v\boldsymbol{j} \qquad (11.47)$$

denotes the horizontal velocity. By (11.44.2), the second term in (11.46) represents horizontal advective changes evaluated on an isobaric surface—not to be confused with advective changes due to motion along that surface. The third term has the form of vertical advection if

$$\omega = \frac{dp}{dt} \qquad (11.48)$$

is identified as the vertical velocity in pressure coordinates. With p as the vertical coordinate, ω denotes the Lagrangian derivative of vertical position and is positive in the direction of increasing p, that is, *downward*. Although ω itself does not have dimensions of velocity, the product $\omega(\partial\psi/\partial p)$ has dimensions of vertical advection of the property ψ. Then the material derivative can be expressed

$$
\begin{aligned}
\frac{d\psi}{dt} &= \frac{\partial\psi}{\partial t} + (\boldsymbol{v} \cdot \nabla\psi)_p \\
&= \frac{\partial\psi}{\partial t} + \boldsymbol{v}_h \cdot \nabla_p\psi + \omega\frac{\partial\psi}{\partial p},
\end{aligned} \qquad (11.49)
$$

in which p is held fixed unless otherwise noted and $(\boldsymbol{v} \cdot \nabla\psi)_p$ denotes three-dimensional advection of ψ in pressure coordinates.

By taking $\psi = z(x, y, p, t)$ in (11.43) and (11.44), which for fixed p represents the height of an isobaric surface as a function of horizontal position and time, we obtain

$$\frac{\partial z}{\partial z} = 1$$

$$\nabla_z z = \left(\frac{\partial z}{\partial x}\right)_{yzt} \boldsymbol{i} + \left(\frac{\partial z}{\partial y}\right)_{xzt} \boldsymbol{j} = 0.$$

Then (11.43) reduces to the simple reciprocal property

$$
\begin{aligned}
\left(\frac{\partial z}{\partial p}\right)^{-1} &= \frac{\partial p}{\partial z} \\
&= -\rho g,
\end{aligned} \qquad (11.50.1)
$$

which followed earlier from the Jacobian (11.10) for transformations involving the vertical coordinate alone. Similarly, (11.44) reduces to

$$
\begin{aligned}
\nabla_z p &= -\left(\frac{\partial z}{\partial p}\right)^{-1} \nabla_p z \\
&= \rho g \nabla_p z.
\end{aligned} \qquad (11.50.2)
$$

Incorporating (11.50.2) into the horizontal momentum equation transforms the pressure gradient force into

$$
\begin{aligned}
-\frac{1}{\rho}\nabla_z p &= -g\nabla_p z \\
&= -\nabla_p \Phi,
\end{aligned} \qquad (11.51)
$$

where the geopotential Φ is given by (6.7). Then with (11.49), the horizontal momentum equations become

$$\frac{\partial v_h}{\partial t} + (v \cdot \nabla v_h)_p + \left(f + u \frac{\tan \phi}{a} \right) k \times v_h = -\nabla_p \Phi - D_h, \qquad (11.52)$$

where D_h denotes the horizontal component of drag. Simple substitution transforms the hydrostatic equation into

$$\frac{\partial \Phi}{\partial p} = -\alpha. \qquad (11.53)$$

The continuity equation follows from its expression in terms of specific volume, which holds irrespective of coordinate system. In isobaric coordinates, (10.31) becomes

$$\frac{d\alpha_p}{dt} = \alpha_p (\nabla \cdot v)_p, \qquad (11.54.1)$$

where α_p refers to the incremental material volume dV_p and

$$(\nabla \cdot v)_p = \nabla_p \cdot v_h + \frac{\partial \omega}{\partial p}. \qquad (11.54.2)$$

Then (11.6) implies

$$\alpha_p = J(x, x_p) \alpha. \qquad (11.55)$$

According to (11.9),

$$J(x, x_p) = \left| \frac{\partial p}{\partial z} \right|$$
$$= \rho g, \qquad (11.56)$$

so

$$\alpha_p = \rho \alpha \cdot g$$
$$= g. \qquad (11.57)$$

It follows that

$$\frac{d\alpha_p}{dt} = 0, \qquad (11.58)$$

which asserts that the volume of a material element is conserved in isobaric coordinates. Then (11.54) implies

$$\nabla_p \cdot v_h + \frac{\partial \omega}{\partial p} = 0. \qquad (11.59)$$

Requiring the three-dimensional divergence to vanish, the continuity equation in isobaric coordinates has the same form as that for incompressible motion (11.30).

The thermodynamic equation can be developed from the first law for an individual air parcel (2.22), which implies

$$c_p \frac{dT}{dt} - \alpha \frac{dp}{dt} = \dot{q}_{net}. \tag{11.60}$$

With (11.49), this becomes

$$\frac{\partial T}{\partial t} + (v_h \cdot \nabla T)_p - \frac{\alpha \omega}{c_p} = \frac{\dot{q}_{net}}{c_p}, \tag{11.61.1}$$

in which expansion work is proportional to the motion across isobaric surfaces. As before, the thermodynamic equation assumes a more compact form when expressed in terms of potential temperature:

$$\frac{\partial \theta}{\partial t} + (v_h \cdot \nabla \theta)_p = \frac{\theta}{c_p T} \dot{q}_{net}, \tag{11.61.2}$$

the derivation of which is left as an exercise.

Collectively, the equations of motion in isobaric coordinates are then given by

$$\frac{dv_h}{dt} + \left(f + u \frac{\tan \phi}{a} \right) k \times v_h = -\nabla_p \Phi - D_h, \tag{11.62.1}$$

$$\frac{\partial \Phi}{\partial p} - -\alpha, \tag{11.62.2}$$

$$(\nabla \cdot v)_p = 0, \tag{11.62.3}$$

$$\frac{d\theta}{dt} = \frac{\theta}{c_p T} \dot{q}_{net}, \tag{11.62.4}$$

with

$$\frac{d}{dt} = \frac{\partial}{\partial t} + v_h \cdot \nabla_p + \omega \frac{\partial}{\partial p}. \tag{11.62.5}$$

The lower boundary condition requires air to maintain the surface elevation z_s, so

$$\frac{d\Phi}{dt} = g \frac{dz_s}{dt} \tag{11.63}$$

at $p = p_s(x, y, t)$.

Simplifications introduced by transforming the equations into this coordinate system include the following:

1. The pressure gradient force has been linearized through elimination of ρ.
2. The continuity equation has reduced to a statement of three-dimensional nondivergence. In addition to being linear, it is now "diagnostic" (i.e., it involves no time derivatives).

These simplifications are not acquired without a price. Coordinate surfaces $p = \text{const}$ are now time dependent, so their positions evolve with the circulation. Further, isobaric surfaces need not coincide with the ground. The lower boundary condition (11.63) must then be prescribed on different values of the vertical coordinate $p = p_s(x, y, t)$, which complicates its application.

Complications in the lower boundary condition can be averted by introducing the modified pressure coordinate

$$\sigma = \frac{p}{p_s}, \tag{11.64}$$

which preserves a constant value at the ground. Although they pass the complication into the governing equations, σ *coordinates* are advantageous from a computational standpoint and are used in general circulation models (GCMs) (see, e.g., Haltiner and Williams, 1980).

11.5.2   Log–Pressure Coordinates

Combining some of the virtues of height and pressure is the modified vertical coordinate

$$z^* = -H \ln\left(\frac{p}{p_{00}}\right), \tag{11.65.1}$$

with

$$H = \frac{RT_0}{g} \tag{11.65.2}$$

treated as constant. At this level of approximation, the basic pressure and density associated with static conditions are described by

$$\begin{aligned} p_0(z^*) &= p_0(0)e^{-\frac{z^*}{H}}, \\ \rho_0(z^*) &= \rho_0(0)e^{-\frac{z^*}{H}}. \end{aligned} \tag{11.66}$$

Log–pressure height, which is formally constant on isobaric surfaces, is based on the stratification under static conditions (e.g., global–mean properties), so it only approximates geopotential height. However, discrepancies between z^* and z are typically small and the two are identical under isothermal conditions.

In terms of z^*, the hydrostatic equation becomes

$$\frac{\partial \Phi}{\partial z^*} = \frac{RT}{H}. \tag{11.67}$$

The vertical velocity in log–pressure coordinates is given by

$$\begin{aligned} w^* &= \frac{dz^*}{dt} \\ &= -\frac{H}{p}\frac{dp}{dt} = -\frac{H}{p}\omega. \end{aligned} \tag{11.68}$$

Then the Lagrangian derivative translates into

$$\frac{d}{dt} = \frac{\partial}{\partial t} + \boldsymbol{v}_h \cdot \nabla_{z^*} + w^* \frac{\partial}{\partial z^*}, \tag{11.69}$$

where ∇_{z^*} reflects the horizontal gradient evaluated on an isobaric surface. Likewise, vertical divergence in pressure coordinates becomes

$$\frac{\partial \omega}{\partial p} = \frac{\partial w^*}{\partial z^*} - \frac{w^*}{H}$$

$$= \frac{1}{\rho_0} \frac{\partial}{\partial z^*} (\rho_0 w^*), \tag{11.70}$$

which transforms the continuity equation into

$$\nabla_{z^*} \cdot \boldsymbol{v}_h + \frac{1}{\rho_0} \frac{\partial}{\partial z^*} (\rho_0 w^*) = 0. \tag{11.71}$$

The thermodynamic equation is transformed in similar fashion, which is left as an exercise.

With the foregoing expressions, the equations of motion in log–pressure coordinates become

$$\frac{d\boldsymbol{v}_h}{dt} + \left(f + u \frac{\tan \phi}{a} \right) \boldsymbol{k} \times \boldsymbol{v}_h = -\nabla_{z^*} \Phi - \boldsymbol{D}_h, \tag{11.72.1}$$

$$\frac{\partial \Phi}{\partial z^*} = \frac{RT}{H}, \tag{11.72.2}$$

$$\nabla_{z^*} \cdot \boldsymbol{v}_h + \frac{1}{\rho_0} \frac{\partial}{\partial z^*} (\rho_0 w^*) = 0, \tag{11.72.3}$$

$$\left(\frac{\partial}{\partial t} + \boldsymbol{v}_h \cdot \nabla_{z^*} \right) \left(\frac{\partial \Phi}{\partial z^*} \right) + N^{*2} w^* = \frac{\kappa}{H} \dot{q}_{\text{net}}, \tag{11.72.4}$$

where

$$\frac{d}{dt} = \frac{\partial}{\partial t} + \boldsymbol{v}_h \cdot \nabla_{z^*} + w^* \frac{\partial}{\partial z^*}, \tag{11.72.5}$$

and

$$N^{*2} = \frac{R}{H} \left(\frac{\partial T}{\partial z^*} + \frac{\kappa}{H} T \right) \tag{11.72.6}$$

represents the static stability in log–pressure coordinates. In the troposphere, N^{*2} varies with height only weakly, so it can be treated as constant to first order. The lower boundary condition, in which geometric vertical velocity is specified (11.63), becomes

$$\frac{\partial \Phi}{\partial t} + \boldsymbol{v}_h \cdot \nabla_{z^*} \Phi + \frac{RT}{H} w^* = g \frac{dz_s}{dt}. \tag{11.73}$$

At the level of approximation inherent to log–pressure coordinates, this can be evaluated at the constant elevation $z^* = z_s^*$, but with variations of z_s^* accounted for in the right-hand side of (11.73).

The equations in log–pressure coordinates have several advantages. Variables are analogous to those in physical coordinates, so they are easily interpreted. Yet, the pressure gradient force, hydrostatic equation, and continuity equation retain nearly the same simplified forms as in isobaric coordinates. In addition, mathematical complications surrounding comparatively small variations of temperature are ignored.

11.5.3   Isentropic Coordinates

The nearly adiabatic nature of air motion simplifies the governing equations when θ is treated as the vertical coordinate. Isentropic surfaces are then coordinate surfaces, to which air motion is nearly tangential (Fig. 2.9). Hence, we consider a transformation from the standard spherical coordinates $x = (x, y, z)$ to the modified coordinates $x_\theta = (x, y, \theta)$. For it to serve as a vertical coordinate, potential temperature must vary monotonically with altitude. Hydrostatic stability requires $\partial \theta / \partial z > 0$, so we are ensured of a single-valued relationship between potential temperature and height as long as the stratification remains stable. As is true for isobaric coordinates, using isentropic surfaces as coordinate surfaces leads to a coordinate system that is nonorthogonal.

Consider the scalar variable

$$\psi = \hat{\psi}[x, y, \theta(x, y, z, t), t]. \tag{11.74}$$

Proceeding as in the development of (11.43) and (11.44) transforms vertical and horizontal derivatives into

$$\frac{\partial \psi}{\partial z} = \frac{\partial \psi}{\partial \theta} \frac{\partial \theta}{\partial z}, \tag{11.75.1}$$

$$\nabla_z \psi = \nabla_\theta \psi + \frac{\partial \psi}{\partial \theta} \nabla_z \theta, \tag{11.75.2}$$

where θ is held fixed unless otherwise noted and

$$\nabla_\theta = \left(\frac{\partial}{\partial x} \right)_{y \theta t} i + \left(\frac{\partial}{\partial y} \right)_{x \theta t} j \tag{11.75.3}$$

represents the horizontal gradient evaluated on an isentropic surface. Then the Lagrangian derivative becomes

$$\frac{d\psi}{dt} = \frac{\partial \psi}{\partial t} + (v \cdot \nabla \psi)_\theta$$

$$= \frac{\partial \psi}{\partial t} + v_h \cdot \nabla_\theta \psi + \frac{d\theta}{dt} \frac{\partial \psi}{\partial \theta}, \tag{11.76}$$

in which we identify

$$\omega_\theta = \frac{d\theta}{dt} \tag{11.77}$$

as the vertical velocity in potential temperature coordinates. Positive upward, ω_θ represents the Lagrangian derivative of vertical position in this coordinate system.

Taking $\psi = z(x, y, \theta, t)$ in (11.75), which for fixed θ represents the height of an isentropic surface, leads to the identities

$$\frac{\partial\theta}{\partial z} = \left(\frac{\partial z}{\partial\theta}\right)^{-1}, \tag{11.78.1}$$

$$\nabla_\theta z = -\frac{\partial z}{\partial\theta}\nabla_z\theta. \tag{11.78.2}$$

Then substituting (11.78.2) transforms (11.75.2) into

$$\nabla_z\psi = \nabla_\theta\psi - \frac{\partial\psi}{\partial z}\nabla_\theta z. \tag{11.79}$$

Taking $\psi = p$ and incorporating (11.79) transforms the pressure gradient force in the horizontal momentum equations into

$$-\frac{1}{\rho}\nabla_z p = -\frac{1}{\rho}\nabla_\theta p + \frac{1}{\rho}\frac{\partial p}{\partial z}\nabla_\theta z$$

$$= -\frac{1}{\rho}\nabla_\theta p - g\nabla_\theta z. \tag{11.80}$$

Poisson's relationship for potential temperature (2.31) implies the identity

$$\ln\theta = \ln T - \kappa(\ln p - \ln p_0).$$

By applying the horizontal gradient evaluated on an isentropic surface, we obtain

$$c_p\nabla_\theta T = \frac{RT}{p}\nabla_\theta p$$

$$= \frac{1}{\rho}\nabla_\theta p. \tag{11.81}$$

Then the pressure gradient force (11.80) reduces to

$$-\frac{1}{\rho}\nabla_z p = -c_p\nabla_\theta T - g\nabla_\theta z$$

$$= -\nabla_\theta\Psi, \tag{11.82}$$

where the *Montgomery streamfunction*

$$\Psi = c_p T + gz$$

$$= c_p T + \Phi \tag{11.83}$$

plays a role in isentropic coordinates analogous to the one played by geopotential in isobaric coordinates. With (11.82) and (11.76), the horizontal momentum equations become

$$\frac{\partial \boldsymbol{v}_h}{\partial t} + (\boldsymbol{v} \cdot \nabla \boldsymbol{v}_h)_\theta + \left(f + u\frac{\tan \phi}{a}\right)\boldsymbol{k} \times \boldsymbol{v}_h = -\nabla_\theta \Psi - \boldsymbol{D}_h. \qquad (11.84)$$

To transform the hydrostatic equation, we consider (11.75.1) with $\psi = p$, which gives

$$-\rho g\frac{\partial z}{\partial \theta} = \frac{\partial p}{\partial \theta}. \qquad (11.85)$$

Poisson's relationship implies

$$\frac{1}{\theta} = \frac{1}{T}\frac{\partial T}{\partial \theta} - \frac{\kappa}{p}\frac{\partial p}{\partial \theta}.$$

Then substituting into (11.85) obtains

$$c_p\frac{\partial T}{\partial \theta} + g\frac{\partial z}{\partial \theta} = \frac{c_p T}{\theta}$$

or

$$\frac{\partial \Psi}{\partial \theta} = c_p\left(\frac{p}{p_0}\right)^\kappa \qquad (11.86)$$

for the hydrostatic equation in isentropic coordinates.

The continuity equation follows from its expression in terms of specific volume. In isentropic coordinates, this becomes

$$\frac{d\alpha_\theta}{dt} = \alpha_\theta(\nabla \cdot \boldsymbol{v})_\theta, \qquad (11.87.1)$$

where α_θ refers to the incremental material volume dV_θ and

$$(\nabla \cdot \boldsymbol{v})_\theta = \nabla_\theta \cdot \boldsymbol{v}_h + \frac{\partial \omega_\theta}{\partial \theta}. \qquad (11.87.2)$$

According to (11.6),

$$\alpha_\theta = J(\boldsymbol{x}, \boldsymbol{x}_\theta)\alpha, \qquad (11.88)$$

where

$$J(\boldsymbol{x}, \boldsymbol{x}_\theta) = \left|\frac{\partial \theta}{\partial z}\right|. \qquad (11.89)$$

By incorporating (11.85) with (11.78.1), we obtain

$$\alpha_\theta = \rho\alpha \cdot g\left(\frac{\partial p}{\partial \theta}\right)^{-1}$$

$$= g\left(\frac{\partial p}{\partial \theta}\right)^{-1}. \qquad (11.90)$$

Then (11.87) yields

$$\frac{d}{dt}\left(\frac{\partial p}{\partial \theta}\right)^{-1} = \left(\frac{\partial p}{\partial \theta}\right)^{-1} (\nabla \cdot \boldsymbol{v})_\theta,$$

or

$$\frac{d}{dt}\left(\frac{\partial p}{\partial \theta}\right) + \left(\frac{\partial p}{\partial \theta}\right)(\nabla \cdot \boldsymbol{v})_\theta = 0 \qquad (11.91)$$

for the continuity equation in isentropic coordinates. Note that (11.91) is identical to the continuity equation in physical coordinates (11.25) if $(\partial p/\partial \theta)$ is identified with density. The thermodynamic equation has the same form as earlier.

Collectively, the equations of motion in isentropic coordinates are then given by

$$\frac{d\boldsymbol{v}_h}{dt} + \left(f + u\frac{\tan \phi}{a}\right)\boldsymbol{k} \times \boldsymbol{v}_h = -\nabla_\theta \Psi - \boldsymbol{D}_h, \qquad (11.92.1)$$

$$\frac{\partial \Psi}{\partial \theta} = c_p \left(\frac{p}{p_0}\right)^\kappa, \qquad (11.92.2)$$

$$\frac{d}{dt}\left(\frac{\partial p}{\partial \theta}\right) + \left(\frac{\partial p}{\partial \theta}\right)(\nabla \cdot \boldsymbol{v})_\theta = 0, \qquad (11.92.3)$$

$$\omega_\theta = \frac{\theta}{c_p T}\dot{q}_{\text{net}}, \qquad (11.92.4)$$

with

$$\frac{d}{dt} = \frac{\partial}{\partial t} + \boldsymbol{v}_h \cdot \nabla_\theta + \omega_\theta \frac{\partial}{\partial \theta}. \qquad (11.92.5)$$

The lower boundary condition becomes (Problem 11.22)

$$\frac{d\Psi}{dt} - \frac{d}{dt}\left(\theta \frac{\partial \Psi}{\partial \theta}\right) = g\frac{dz_s}{dt}. \qquad (11.93)$$

These equations simplify the description of vertical motion, which is related directly to the rate at which heat is absorbed by an air parcel (11.92.4). Under adiabatic conditions, there is no motion across coordinate surfaces, so ω_θ and vertical advection vanish. Under diabatic conditions, the system is still advantageous because it relates vertical motion (which is difficult to measure due to its smallness) to quantities that are more reliably determined. Another advantage of isentropic coordinates is enhanced resolution in regions of strong temperature gradient, such as those typical of frontal zones. Figure 11.5 shows a frontal surface represented in terms of isobaric and isentropic coordinates. In the isobaric representation (Fig. 11.5a), motion and potential temperature vary sharply across the frontal zone, which then requires a fine computational mesh to be properly represented. These sharp gradients de-

VERTICAL COORDINATE REPRESENTATIONS

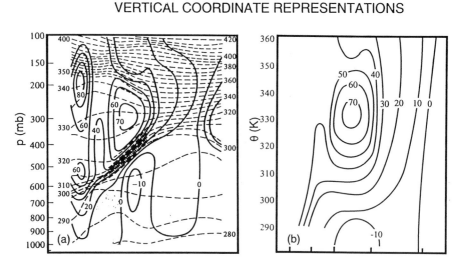

Figure 11.5 Cross section through a frontal zone represented in (a) isobaric coordinates and (b) isentropic coordinates. Sources: Shapiro and Hastings (1973) and R. Bleck (U. Miami), personal communication.

velop through nearly adiabatic flow deformation (Sec. 10.2) that concentrates θ surfaces into narrow regions during the amplification of extratropical cyclones (Chapter 16). Isentropic coordinates maintain a fixed separation of θ surfaces, so they magnify the region of sharp behavior (Fig. 11.5b), which then varies smoothly and poses much less of a computational challenge. While offering these advantages, isentropic coordinates make the continuity equation more complex. They also suffer from variations of θ along the ground, which complicate the lower boundary condition.

Suggested Reading

Vectors, Tensors, and the Basic Equations of Fluid Mechanics (1962) by Aris provides a complete treatment of curvilinear coordinates. It includes a general derivation of the continuity equation, valid for an arbitrary coordinate system, in terms of the Jacobian of the transformation between Lagrangian and Eulerian coordinates.

A treatment of generalized vertical coordinates is given in *The Ceaseless Wind* (1986) by Dutton and in Bleck (1978a,b), which also discusses their computational advantages in numerical applications.

Numerical Prediction and Dynamic Meteorology (1980) by Haltiner and Williams describes σ coordinates and their application in large-scale numerical integrations.

Problems

11.1. Demonstrate that the Jacobian satisfies the reciprocal property (11.4).

11.2. Show that the Jacobian reduces to the product of the metric scale factors (11.7) if the curvilinear coordinate system is orthogonal.

11.3. Express spherical coordinates in terms of the rectangular Cartesian coordinates (11.13).

11.4. Show that metric scale factors for spherical coordinates are given by (11.14).

11.5. (a) Express the spherical coordinate vectors in terms of the Cartesian coordinate vectors (11.20). (b) Show that the spherical coordinate vectors have spatial derivatives

$$\frac{\partial i}{\partial \lambda} = \sin \phi j - \cos \phi k \quad \frac{\partial i}{\partial \phi} = 0 \quad \frac{\partial i}{\partial z} = 0$$

$$\frac{\partial j}{\partial \lambda} = -\sin \phi i \quad \frac{\partial j}{\partial \phi} = -k \quad \frac{\partial j}{\partial z} = 0$$

$$\frac{\partial k}{\partial \lambda} = \cos \phi i \quad \frac{\partial k}{\partial \phi} = j \quad \frac{\partial k}{\partial z} = 0.$$

(*Hint:* Obtain reciprocal identities for e_1, e_2, and e_3 in terms of i, j, and k.) (c) Evaluate the Lagrangian derivatives of the spherical coordinate vectors to obtain (11.21).

11.6. Derive the angular momentum principle implied by the vector equations of motion in full spherical geometry.

11.7. Show that the equations of motion in an inertial reference frame are expressed by (11.27). (*Hint:* See Problem 10.11.)

11.8. Show that the approximate system (11.28) has the angular momentum principle (11.29).

11.9. (a) Derive the approximate expression (11.35) for the specific buoyancy force. (b) Discuss the validity of this approximation.

11.10. Perform a dimensional analysis to demonstrate that, for scales representative of the midlatitude stratospheric circulation (e.g., the polar-night jet disturbed by planetary waves; refer to Fig. 1.10), metric and Coriolis terms proportional to $\cos \phi$ appearing in the equations of motion are negligible.

11.11. The equations of motion are expressed conveniently in terms of the *Exner function* $\pi = c_p(p/p_0)^\kappa$ as the vertical coordinate, where $p_0 = 1000$ mb. Transform the equations from isobaric coordinates to Exner coordinates.

11.12. In isobaric coordinates, derive the thermodynamic equation in terms of potential temperature (11.61.2).

11.13. Show that isochoric surfaces are material if the motion is steady and v is everywhere orthogonal to $\nabla \rho$.

11.14. (a) Discuss the circumstances under which isobaric and isentropic coordinates provide a valid description of atmospheric motion. (b) Discuss the circumstances under which applying those coordinate representations is straightforward.

11.15. Derive the thermodynamic equation in log–pressure coordinates (11.72.4).

11.16. The relationship

$$\frac{\partial y}{\partial x} = \left(\frac{\partial x}{\partial y}\right)^{-1}$$

is not true in general. Show that, for the special circumstances underlying thermodynamic coordinates, this identity is in fact true, thereby validating (11.50.1) and (11.78.1).

11.17. Develop an analogue of the hypsometric equation in isentropic coordinates.

11.18. (a) For a characteristic velocity of 10 m s^{-1}, at what horizontal scale does the earth's rotation become important? (b) How will the earth's rotation be manifested in the streamlines of steady flow?

11.19. The *geostrophic wind* v_g is the horizontal velocity that follows from a balance between the Coriolis and pressure gradient forces. Provide expressions for u_g and v_g in (a) isobaric coordinates and (b) isentropic coordinates. (c) How does v_g behave as the equator is approached?

11.20. Use the result in Problem 11.19 along with the hydrostatic equation to derive an expression in isobaric coordinates for the vertical variation of geostrophic wind.

11.21. Frictional drag beneath a cyclone produces horizontal convergence which varies inside the planetary boundary layer as

$$-\nabla \cdot v_h = \zeta e^{-\frac{z^*}{h}},$$

where $\zeta = (\nabla \times v_g) \cdot k$ is the (constant) vorticity of the geostrophic wind above the boundary layer and z^* is log–pressure height. (a) Determine the vertical motion as a function of height if $h/H \ll 1$. (b) Describe the divergent component of motion that must exist above the boundary layer ($z^* \gg h$) if vertical motion eventually vanishes above the tropopause.

11.22. Derive the lower boundary condition in isentropic coordinates (11.93).

11.23. In the zonal mean, isentropic surfaces slope meridionally more steeply than isobaric surfaces. Use the zonal–mean surface temperature in Fig. 10.1b and a uniform lapse rate of 6.5 K km^{-1} to estimate the heights of (a) the 300 K isentropic surface over the equator and pole and (b) the 700-mb isobaric surface over the equator and pole.

Chapter 12 | Large-Scale Motion

   Scale analysis (Sec. 11.4) indicates that large-scale motion is dominated by a balance between the Coriolis acceleration and the pressure gradient force. Material acceleration, which is of order Rossby number [denoted $O(Ro)$], is an order of magnitude smaller, whereas metric terms are typically much smaller. Then the horizontal momentum equations can be expressed in terms of the horizontal velocity $\boldsymbol{v}_h$

$$\frac{d\boldsymbol{v}_h}{dt} + f\boldsymbol{k} \times \boldsymbol{v}_h = -\nabla_p \Phi - \boldsymbol{D}. \tag{12.1}$$

We use the prevalence of certain terms in (12.1) to illustrate the essential balances controlling large-scale atmospheric motion. In the spirit of an asymptotic series representation, dependent variables in (12.1) can be expanded in a power series in the small parameter Ro (Problem 12.1). The momentum equations can then be balanced with the expansion in Ro truncated to a finite number of terms. In principle, this procedure can be carried out recursively, obtaining the momentum balance to successively higher order in Ro based on the balance to lower order. However, $Ro \ll 1$ makes terms omitted much smaller than those retained in the approximate momentum balance, so retaining only a few terms in the expansion often provides sufficient accuracy.

12.1 Geostrophic Equilibrium

To zero order in Ro, the momentum equations reduce to

$$f\boldsymbol{k} \times \boldsymbol{v}_g = -\nabla_p \Phi, \tag{12.2.1}$$

where frictional drag $\boldsymbol{D}$ is presumed to be $O(Ro)$ or smaller. Reflecting a balance between the Coriolis acceleration and the horizontal pressure gradient force, (12.2.1) defines *geostrophic equilibrium* (*geo* referring to "earth" and *strophic* to "turning"). The horizontal velocity satisfying (12.2.1) is the *geostrophic velocity*:

$$\boldsymbol{v}_g = \frac{1}{f}\boldsymbol{k} \times \nabla_p \Phi \tag{12.2.2}$$

(Problem 12.2). Similar expressions follow in isentropic coordinates with the Montgomery streamfunction (Problem 12.3).

According to (12.2.2), the geostrophic velocity is tangential to contours of height $z = \Phi/g$, with low height on the left (right) in the Northern (Southern) Hemisphere. This is equivalent to motion along isobars on a constant height surface (Sec. 6.3). Apparent in the observed circulation (Fig. 1.9), such motion is perpendicular to fluid motion normally observed in an inertial reference frame (e.g., from high toward low pressure). This peculiarity of large-scale atmospheric motion is known as the *geostrophic paradox*, which was not resolved until the the earth's rotation was incorporated into its explanation.

To illustrate how geostrophic equilibrium is established, consider a hypothetical initial value problem describing an atmosphere that is initially at rest and in which contours of isobaric height are nearly straight, uniformly spaced, and, for the sake of illustration, fixed (Fig. 12.1). At $t = 0$, an air parcel has $v_h = 0$, so the Coriolis force acting on it vanishes. In response to the pressure gradient force, the parcel accelerates parallel to $-\nabla_p \Phi$ toward low height. As soon as motion develops, the Coriolis force $-f\mathbf{k} \times v_h$ acts perpendicular to the horizontal velocity. In the Northern Hemisphere, where $f = 2\Omega \sin \phi$ is positive, the Coriolis force deflects the parcel's motion toward the right. Were it in the Southern Hemisphere, the parcel would be deflected to the left. In

GEOSTROPHIC EQUILIBRIUM

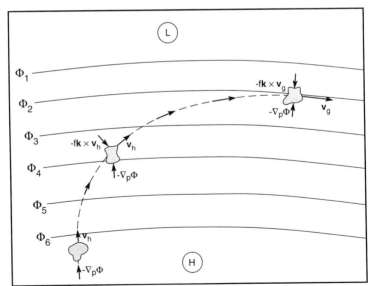

Figure 12.1 Evolution of a hypothetical air parcel that is initially motionless in the Northern Hemisphere and in stratification characterized by contours of isobaric height $z = \Phi/g$ that are nearly straight, uniformly spaced, and, for the sake of illustration, fixed.

either event, the Coriolis force performs no work on the parcel because it acts orthogonal to the instantaneous motion. While the parcel's trajectory veers to the right, the pressure gradient force remains unchanged because $\nabla_p \Phi$ is invariant of position under the foregoing conditions. However, the Coriolis force changes continually to remain proportional and orthogonal to v_h. As the parcel accelerates under the pressure gradient force, its trajectory veers increasingly to the right, until eventually v_h has become tangential to contours of Φ. The Coriolis force then acts in direct opposition to the pressure gradient force, both of which are perpendicular to the parcel's velocity. If v_h satisfies (12.2), the motion is then in geostrophic equilibrium and the parcel experiences no further acceleration.[1]

Once this mechanical equilibrium is established, the parcel's motion is steady—except for gradual adjustments allowed for in contours of Φ. According to (12.2.2), geostrophic wind speed v_g is proportional to $\nabla_p \Phi$ and hence inversely proportional to the spacing of height contours (Fig. 12.2). Geostrophic wind speed increases into a region where height contours converge and decreases into a region where they diverge, v_g remaining tangential to Φ contours. In the absence of vertical motion, this behavior automatically satisfies conservation of mass (Problem 12.12), which makes height contours streamlines of geostrophic motion.

Geostrophic equilibrium implies circular motion about a closed center of height. The balance (12.2) is valid for curved motion so long as an anomaly's scale is large enough and its velocity slow enough to render the material ac-

---

[1] If the Coriolis and pressure gradient forces do not balance exactly, the motion will undergo an oscillation about contours of Φ, which upon being damped out leaves the parcel in geostrophic equilibrium.

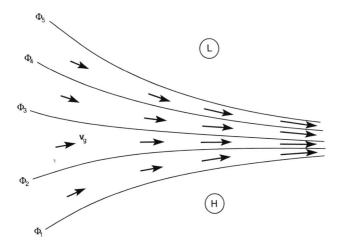

Figure 12.2 Variation of geostrophic velocity with changes of isobaric height.

celeration negligible (e.g., $Ro \ll 1$) relative to the pressure gradient and Coriolis forces (Sec. 11.4). Because low height lies to the left (right) in the Northern (Southern) Hemisphere, geostrophic flow about a low is counterclockwise (clockwise). In each hemisphere, motion about a closed low has the same sense as the planetary vorticity $f\mathbf{k}$, which is termed *cyclonic*. About a high, geostrophic motion is clockwise (counterclockwise) in the Northern (Southern) Hemisphere. Opposite to the planetary vorticity in each hemisphere, motion about a high is termed *anticyclonic*.

Such behavior is apparent in the 500-mb circulation (Fig. 1.9a), which is punctuated by anomalies of low height with counterclockwise flow about them. In fact, the circumpolar flow itself can be regarded as a cyclonic vortex. By the hypsometric relationship (6.12), poleward-decreasing temperature in the troposphere produces low isobaric height over the pole. Geostrophic equilibrium then establishes a cyclonic circulation about the pole, as manifested in the subtropical jet (Fig. 1.9b). The time–mean circulation at 10 mb (Fig. 1.10b) is also cyclonic, but intensified. Ozone heating at low latitudes produces poleward-decreasing temperature across the winter stratosphere, which establishes the polar-night vortex (see Fig. 1.8). Outside the vortex, the *Aleutian high* is accompanied by anticyclonic motion. Even though the instantaneous circulation can be highly disturbed from zonal symmetry (Figs. 1.9a and 1.10a), motion remains nearly tangential to contours of height.

12.1.1 Motion on an f Plane

Consider motion in which meridional displacements are sufficiently narrow to ignore variations of f in its description. Expanding f in a Taylor series about the reference latitude ϕ_0 and truncating to zeroth order gives the constant Coriolis parameter

$$f_0 = 2\Omega \sin \phi_0. \tag{12.3.1}$$

In the same framework, it is convenient to neglect spherical curvature in favor of simple Cartesian geometry. A Cartesian plane tangent to the earth at latitude ϕ_0 (Fig. 12.3) describes horizontal position in terms of the distances

$$x = a \cos \phi_0 \cdot \lambda,$$
$$y = a(\phi - \phi_0), \tag{12.3.2}$$

but with the earth's sphericity ignored. Together, these simplifications comprise the *f-plane* description of atmospheric motion.

On an f plane, the geostrophic velocity defines a nondivergent vector field. Tangential paths of $\mathbf{v}_g$ are given by height contours, which then serve as

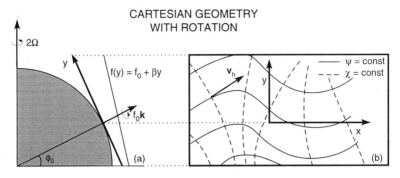

Figure 12.3 Approximation of the rotating spherical earth by Cartesian geometry. On a plane tangent to the earth at latitude ϕ_0, distance is measured by the coordinates $x = a\cos\phi_0 \cdot \lambda$ and $y = a(\phi - \phi_0)$, which increase to the east and north, respectively. (a) Approximating the Coriolis parameter by its zeroth-order variation, $f(y) = f_0 = 2\Omega\sin\phi_0$, yields the *f-plane* description of atmospheric motion. Approximating it up to first order: $f(y) = f_0 + \beta y$, where $\beta = (df/dy)_{y=0}$, yields the *β-plane* description of atmospheric motion. (b) The horizontal motion field at any instant $v_h(x, y, t)$ can be expressed in terms of a *streamfunction* $\psi(x, y, t)$ and a *velocity potential* $\chi(x, y, t)$: $v_h = k \times \nabla\psi + \nabla\chi$, which represent the rotational and divergent components of v_h, respectively.

streamlines. These and other implications of nondivergence follow from the general representation of a vector field.

THE HELMHOLTZ THEOREM

Any vector field v can be represented in terms of a divergent or *irrotational* component and a rotational or *solenoidal* component:

$$v = \nabla\chi + \nabla \times \psi, \tag{12.4}$$

where χ is a *scalar potential* (analogous to the potential function in Chapter 2) and ψ is a *vector potential*. The divergent component

$$v_d = \nabla\chi \tag{12.5.1}$$

possesses zero vorticity

$$\nabla \times v_d = 0. \tag{12.5.2}$$

The solenoidal component

$$v_s = \nabla \times \psi \tag{12.6.1}$$

possesses zero divergence

$$\nabla \cdot v_s = 0. \tag{12.6.2}$$

For the two-dimensional field of horizontal motion, (12.4) reduces to

$$v_h = \nabla \chi + k \times \nabla \psi, \tag{12.7.1}$$

where χ is the *velocity potential* and ψ is the *streamfunction*. The horizontal motion field then has divergence

$$\nabla \cdot v_h = \nabla^2 \chi \tag{12.7.2}$$

and vorticity

$$\nabla \times v_h = \nabla^2 \psi. \tag{12.7.3}$$

The divergent component of motion v_d is orthogonal to contours of χ, whereas the solenoidal component v_s is tangential to contours of ψ.

According to (12.2.2), geostrophic motion on an f plane has the form of a solenoidal vector field, one characterized by the *geostrophic streamfunction*

$$\psi = \frac{1}{f_0} \Phi. \tag{12.8}$$

Because the divergence of v_g vanishes,[2] the continuity equation implies little or no vertical motion. If pressure variations along the surface can be ignored, integrating (11.54.2) upward to some isobaric surface obtains

$$\omega(x, y, p, t) = -\int_p^{p_s} \nabla \cdot v_h \, dp, \tag{12.9}$$

which implies zero vertical motion under pure geostrophic equilibrium (12.6.2). Air parcels then simply exchange horizontal positions with no vertical rearrangement. The solenoidal character of large-scale atmospheric motion follows from the earth's rotation, which maintains the circulation close to geostrophic equilibrium. Vertical motion does occur in the large-scale circulation, but it is higher order in Rossby number and therefore small, following from the ageostrophic component of horizontal motion v_d.

Owing to its rotational character, geostrophic motion possesses a high degree of deformation. Horizontal shear associated with vorticity also introduces deformation (10.5), which distorts fluid bodies into complex forms. Figure 12.4 shows the evolution of a material volume as it is advected though a two-dimensional cyclone. Shear strains deform the body into an elongated shape, wherein the transverse separation of boundaries collapses to small scales and accompanying gradients steepen. On those scales, turbulence and nonconservative processes (Chapter 13) act efficiently to homogenize sharp contrasts that have developed from shear strain. This consequence of rotational fluid motion is the basis for mixing. Analogous to cream being stirred into a cup of coffee, shear strains exaggerate fluid gradients until they are smoothed

---

[2] Hereafter, the term *divergence* will be understood to refer to the divergence of the horizontal velocity component.

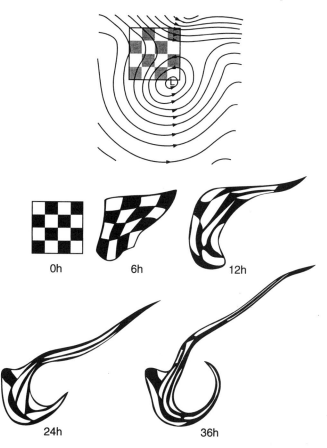

Figure 12.4 Deformation of a material volume at successive times inside a two-dimensional cyclone obtained by integrating the barotropic nondivergent vorticity equation. Adapted from Welander (1955). Copyright (1955) Munksgaard International Publishers Ltd., Copenhagen, Denmark.

out by turbulent and eventually molecular diffusion. In the atmosphere, diffusion is accomplished by horizontal eddy motions and by vertical motions in convection (refer to Figs. 1.15 and 1.23).

12.2 Vertical Shear of the Geostrophic Wind

Geostrophic balance determines the horizontal structure of motion. Together with hydrostatic balance, it also determines the vertical structure. A fundamental principle of rotating fluid mechanics known as the *Taylor–Proudman theorem* asserts that the motion of a homogeneous incompressible fluid cannot vary along the axis of rotation. Motion then occurs in so-called *Taylor–*

Proudman columns.[3] The atmosphere is not homogeneous nor incompressible. Nevertheless, it possesses an analogue of Taylor–Proudman behavior, which provides an essential relationship between horizontal motion and stratification.

12.2.1   Classes of Stratification

The circulation is closely related to the stratification, which is represented in the distributions of thermodynamic properties. For dry air, only two such properties are independent (Sec. 2.1.4). Therefore, any two families of thermodynamic surfaces uniquely describe the stratification (Fig. 12.5). Mathematically, this is expressed by $\theta = \theta(p, T)$, where p and T are functions of space and time.

Should isentropic surfaces coincide with isobaric surfaces (Fig. 12.5a), $\theta = \theta(p)$ and the stratification is said to be *barotropic*. Under barotropic stratification, the circulation possesses only one thermodynamic degree of freedom, which is reflected in the single independent family of thermodynamic surfaces. Because other thermodynamic surfaces coincide with that family, specifying p uniquely determines θ, which, in turn, determines the thermodynamic state and all other thermodynamic properties.

More generally, two families of thermodynamic surfaces do not coincide (Fig. 12.5b), so $\theta = \theta(p, T)$. The stratification is then said to be *baroclinic*. Under baroclinic stratification, the circulation possesses two thermodynamic degrees of freedom. Consequently, along any thermodynamic surface, other thermodynamic properties vary. The geostrophic velocity then changes with elevation and, as demonstrated below, it does so in direct proportion to the variation of temperature along isobaric surfaces.

12.2.2   Thermal Wind Balance

Under barotropic stratification, temperature variations along isobaric surfaces vanish. Integrating the hypsometric relationship (6.12) upward from the ground then gives Φ as a contribution from the lower boundary (which may vary horizontally) plus a function of pressure alone (Problem 12.20). The distribution of height contours then does not change from one isobaric surface to another. It follows that the geostrophic velocity (12.1.2) is independent of elevation. Invariant in the direction of $f\boldsymbol{k}$, geostrophic motion under barotropic stratification (wherein density is uniform along isobaric surfaces) is analogous to Taylor–Proudman flow for a homogeneous incompressible fluid.

Under baroclinic stratification, T varies along isobaric surfaces. By differentiating (12.1.2) with respect to p, we obtain

$$\frac{\partial \boldsymbol{v}_g}{\partial p} = \frac{1}{f}\boldsymbol{k} \times \nabla_p \frac{\partial \Phi}{\partial p}. \tag{12.10}$$

---

[3] See Greenspan (1968) for a laboratory demonstration.

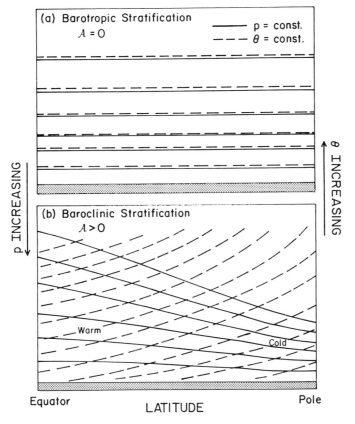

Figure 12.5   Thermal structure corresponding to (a) barotropic stratification, wherein isentropic surfaces coincide with isobaric surfaces and available potential energy $\mathscr{A}$ is zero (Sec. 15.1.3), and (b) baroclinic stratification, wherein isentropic surfaces do not coincide with isobaric surfaces and $\mathscr{A}$ is positive. The rotation of isobaric and isentropic surfaces from their positions under barotropic stratification is symbolic of atmospheric heating at low latitude and cooling at middle and high latitudes (refer to Fig. 1.29c).

Incorporating the hydrostatic equation (11.62.2) then yields for the vertical gradient of geostrophic velocity

$$-\frac{\partial \boldsymbol{v}_g}{\partial \ln p} = \frac{R}{f} \boldsymbol{k} \times \nabla_p T. \tag{12.11}$$

Known as *thermal wind balance*, (12.11) asserts that vertical shear of the geostrophic velocity is directly proportional to the horizontal temperature gradient along isobaric surfaces. The term *thermal wind* applies to the change of velocity $\Delta \boldsymbol{v}_g$ across an incremental layer of thickness $-H \Delta \ln p$, which is proportional to the horizontal gradient of temperature in that layer (Problem 12.28). Thermal wind balance, which follows from hydrostatic and geostrophic equilibrium, couples the circulation to the stratification and makes vertical shear a measure of the departure from barotropic stratification.

From (12.11), an equatorward temperature gradient is accompanied by positive shear of the zonal flow. Motion then becomes increasingly westerly with elevation. Examples are found in the troposphere and winter stratosphere (Fig. 1.7). Radiative heating at low latitude and cooling at middle and high latitudes establish an equatorward temperature gradient in the troposphere in both hemispheres. Thermal wind balance then implies westerlies that intensify upward to form the subtropical jets (Fig. 1.8). In the winter stratosphere, ozone heating at low latitude and longwave (LW) cooling to space at high latitude (e.g., in polar night) establish a temperature gradient of the same sense over a deep layer. Westerlies then intensify upward in the polar-night jet to a maximum near the stratopause. Opposite behavior occurs in the summer stratosphere, where solar insolation maximizes at high latitudes (Fig. 1.28) to produce a poleward temperature gradient. The zonal flow then becomes easterly above the tropopause and intensifies upward to a maximum near the stratopause.

These features can also be inferred directly from geostrophic equilibrium and the hypsometric relationship, which is a statement of hydrostatic balance. In the presence of a horizontal temperature gradient, vertical spacing of isobaric surfaces is compressed in cold air and expanded in warm air (Fig. 6.2). The meridional gradient of isobaric height in the troposphere then steepens with elevation, which, through geostrophic equilibrium, produces zonal motion that intensifies upward. Strong zonal motion favored by the distribution of radiative heating (Fig. 1.29) and geostrophic equilibrium transfers neither heat nor chemical constituents meridionally. Consequently, the earth's rotation tends to stratify properties meridionally, just as gravity tends to stratify them vertically.

Thermal wind balance also applies to zonally asymmetric motion. The polar front, which separates warm tropical air from cold polar air (Fig. 6.3), coincides with strong westerly shear and the instantaneous position of the jet stream (Fig. 1.9a). Inside a cold-core low, as typifies extratropical cyclones (Problem 6.7), cyclonic shear reinforces circumferential motion, which then intensifies upward to a maximum near the tropopause. The reverse occurs inside a warm-core low, as typifies hurricanes (Problem 6.9). Anticyclonic shear then opposes circumferential motion to produce maximum winds near the surface (Problem 12.34).

Geostrophic equilibrium governs motion that is sufficiently steady, slow, and weakly curved for material acceleration to be negligible relative to the Coriolis and pressure gradient forces (i.e., for $Ro << 1$). It also requires frictional drag to be negligible. These conditions are well satisfied by large-scale extratropical motion away from the surface, which implies that v_h is nearly nondivergent. However, the conditions for geostrophic equilibrium break down as the equator is approached or as the horizontal scale of advection becomes small, which render terms $O(Ro)$ nonnegligible. By introducing an ageostrophic component to v_h, these effects introduce divergence and hence vertical motion (12.9).

12.3 Frictional Geostrophic Motion

Inside the planetary boundary layer, frictional drag is large enough to invalidate geostrophic equilibrium. Turbulent eddy motions within a kilometer of the surface, which are produced by strong vertical shear (see Fig. 13.3), mix momentum between bodies of air. The accompanying momentum flux exerts shear stresses on individual air parcels that render D of the same order as the Coriolis and pressure gradient forces (Sec. 10.6).

If drag is represented as *Rayleigh friction* with linear drag coefficient K,

$$D = K v_h, \tag{12.12}$$

then mechanical equilibrium is expressed by

$$f k \times v_h = -\nabla_p \Phi - K v_h, \tag{12.13}$$

which can be solved for the components of horizontal motion (Problem 12.13). Friction modifies both the magnitude and the direction of v_h. The ensuing mechanical equilibrium can be motivated by gradually introducing drag into the geostrophic balance in Fig. 12.1. Introducing D reduces the velocity of the air parcel, which reduces the Coriolis force $-f k \times v_h$ acting on it, which, in turn, allows the pressure gradient force to drive the parcel's motion across isobars toward low pressure (Fig. 12.6). Eventually, the motion achieves an angle δ from isobars such that the resultant of the three forces acting on the parcel vanishes. The air parcel is then in *frictional geostrophic equilibrium*.

Unlike simple geostrophic motion, frictional geostrophic motion is not solenoidal. The divergent component of v_h then introduces vertical motion via continuity (12.9). Frictional convergence into a center of low surface pressure (Fig. 12.7a) is compensated by rising motion overhead. By organizing moisture near the surface and reinforcing upward displacements, such motion favors cloud formation in a cyclone. It also reduces vertical stability (Sec. 7.4.4), which likewise provides conditions favorable to convection. Conversely, frictional divergence out of a center of high surface pressure (Fig. 12.7b) is compensated by subsidence. By opposing upward displacements, sinking motion inhibits cloud formation and favors the formation of a subsidence inversion that traps pollutants near the surface.

The foregoing behavior is a consequence of vertical motion, which does not develop under simple geostrophic equilibrium. Divergence accompanying frictional geostrophic motion is inversely proportional to f (Problem 12.14). Therefore, vertical motion is favored in the tropics, where small f allows large departures from geostrophic equilibrium. The latter support thermally direct circulations in which air ascends over low pressure and descends over high pressure (Fig. 1.30).

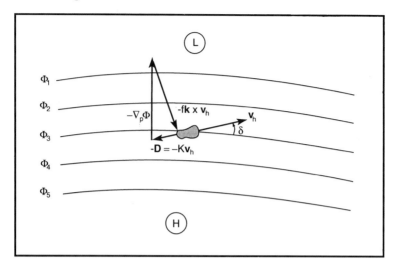

Figure 12.6   Frictional geostrophic balance in a flow characterized by nearly straight height contours and containing Rayleigh friction.

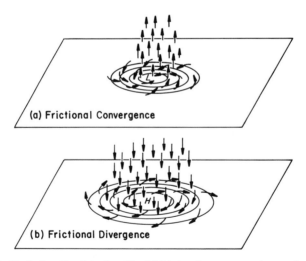

Figure 12.7   Vertical motion introduced by (a) frictional convergence into surface low pressure and (b) frictional divergence out of surface high pressure.

12.4   Curvilinear Motion

The other factor driving the circulation out of geostrophic equilibrium is material acceleration. If its trajectory is sufficiently curved, an air parcel will experience a centripetal acceleration (which is embodied in the advective contribution $v_h \cdot \nabla_p v_h$) that renders the Rossby number nonnegligible. Unsteadiness (which can introduce the same effect through the local time derivative $\partial \mathbf{v}_h/\partial t$) is typically less important.

Consider steady motion in which parcels move along curved trajectories characterized by the local radius of curvature R. Parcel trajectories then coincide with streamlines. It is convenient to introduce the *trajectory coordinates* s and n, which increase along and orthogonal to the path of an individual parcel (Fig. 12.8). Then the coordinate vectors $\mathbf{s}$ and $\mathbf{n}$ are tangential and orthogonal to the local velocity $\mathbf{v}_h$, with $\mathbf{n}$ pointing to the left of $\mathbf{s}$. R is defined to be positive if the center of curvature lies in the positive $\mathbf{n}$ direction. Curvature is then cyclonic if R is of the same sign as f ($fR > 0$) and anticyclonic if it is of opposite sign ($fR < 0$).

In this coordinate system, the horizontal velocity is described by

$$\mathbf{v}_h = v\mathbf{s}, \tag{12.14.1}$$

where the speed is simply

$$v = \frac{ds}{dt}. \tag{12.14.2}$$

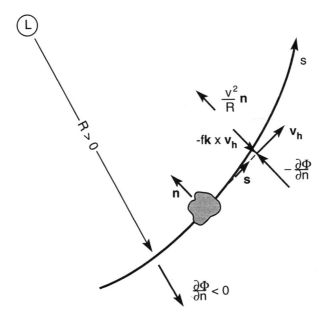

Figure 12.8 Trajectory coordinates s and n, which increase along and orthogonal to the path of an individual parcel. The coordinate vectors $\mathbf{s}$ and $\mathbf{n}$ are then tangential and orthogonal to the local velocity $\mathbf{v}_h$, with $\mathbf{n}$ pointing to the left of $\mathbf{s}$ to form a right-handed coordinate system: $\mathbf{s} \times \mathbf{n} = \mathbf{k}$. The trajectory's curvature is represented in the radius of curvature R, which is defined to be positive (negative) if the center of curvature lies in the positive (negative) n direction. The local velocity is then described by $\mathbf{v}_h = v\mathbf{s}$, with speed $v = ds/dt$. Centripetal acceleration $(v^2/R)\mathbf{n}$ follows from an imbalance between the pressure gradient force $-\partial\Phi/\partial n$ and the Coriolis force $-f\mathbf{k} \times \mathbf{v}_h$.

Then the material acceleration of an individual parcel follows as

$$\frac{d\boldsymbol{v}_h}{dt} = \frac{dv}{dt}\boldsymbol{s} + v\frac{d\boldsymbol{s}}{dt}. \tag{12.15}$$

The first term on the right-hand side represents longitudinal acceleration along the parcel's trajectory. The second represents centripetal acceleration, which acts transverse to the parcel's motion and follows from curvature of its trajectory.

By the chain rule,

$$\frac{d\boldsymbol{s}}{dt} = \frac{d\boldsymbol{s}}{ds}\frac{ds}{dt}$$

$$= v\frac{d\boldsymbol{s}}{ds}. \tag{12.16}$$

Analysis similar to that in Sec. 10.6.2 shows that the unit vector $\boldsymbol{s}$ changes at the rate

$$\frac{d\boldsymbol{s}}{dt} = \frac{v}{R}\boldsymbol{n} \tag{12.17}$$

(Problem 12.21). Substituting (12.16) then results in

$$\frac{d\boldsymbol{v}_h}{dt} = \frac{dv}{dt}\boldsymbol{s} + \frac{v^2}{R}\boldsymbol{n} \tag{12.18}$$

for the material acceleration in terms of components longitudinal and transverse to the motion. As $R \to \infty$, the centripetal acceleration approaches zero and the parcel's motion becomes rectilinear.

Incorporating (12.18) transforms the horizontal momentum equation into

$$\frac{dv}{dt}\boldsymbol{s} + \frac{v^2}{R}\boldsymbol{n} = -fv\boldsymbol{n} - \frac{\partial \Phi}{\partial s}\boldsymbol{s} - \frac{\partial \Phi}{\partial n}\boldsymbol{n}, \tag{12.19}$$

which implies the component equations in the $\boldsymbol{s}$ and $\boldsymbol{n}$ directions:

$$\frac{dv}{dt} = -\frac{\partial \Phi}{\partial s}, \tag{12.20.1}$$

$$\frac{v^2}{R} + fv = -\frac{\partial \Phi}{\partial n}. \tag{12.20.2}$$

Consider now motion that is tangential to contours of height, which then serve as streamlines. Then $\partial \Phi / \partial s$ vanishes and (12.20.1) implies that $v =$ const along an individual streamline. Accordingly, we focus on the momentum budget in the normal direction. The centripetal acceleration v^2/R reflects the material acceleration in (11.37) and is $O(Ro)$, where the Rossby number

$$Ro = \frac{U}{fR}$$

is based on the length scale R. According to (12.20.2), centripetal acceleration follows as an imbalance between the Coriolis and pressure gradient forces.

12.4.1 Inertial Motion

Consider a balance between the centripetal and Coriolis accelerations alone

$$\frac{v^2}{R} + fv = 0, \tag{12.21.1}$$

which results in the absence of a pressure gradient. The radius of curvature follows as

$$R = -\frac{v}{f}. \tag{12.21.2}$$

On an f plane, $R = $ const, so streamlines form circular orbits that are anticyclonic. The period of revolution is given by

$$\tau = \frac{\pi}{\Omega \sin \phi}, \tag{12.22}$$

which is half a *sidereal day*: the time for a Foucalt pendulum to sweep through $180°$ of azimuth. The motion (12.21) describes an *inertial oscillation*, in which parcels revolve opposite to the vertical component of planetary vorticity. Inertial oscillations have been observed in the oceans, but they do not play a major role in the atmospheric circulation.

12.4.2 Cyclostrophic Motion

Consider motion in the limit of either R or f approaching zero. Then $Ro \rightarrow \infty$, so the material acceleration dominates the Coriolis acceleration. Under these conditions, (12.20.2) reduces to the balance

$$\frac{v^2}{R} = -\frac{\partial \Phi}{\partial n}, \tag{12.23.1}$$

which defines *cyclostrophic equilibrium*. Motion with the *cyclostrophic wind speed*

$$v_c = \sqrt{-R\frac{\partial \Phi}{\partial n}} \tag{12.23.2}$$

can be either cyclonic or anticyclonic, but always about low pressure (e.g., $\partial \Phi / \partial n$ of opposite sign to R).

Cyclostrophic equilibrium applies to (1) strong flow curvature, wherein the advective acceleration is large, or (2) slow planetary rotation, wherein the Coriolis force is weak. The former is typical of small-scale vortices such as tornadoes, whereas the latter is important in tropical cyclones. Slow rotation is also a feature of the Venusian atmosphere, which is governed by cyclostrophic equilibrium instead of geostrophic equilibrium.

12.4.3   Gradient Motion

To first order in Ro, all three terms in (12.20.2) are retained, which defines *gradient equilibrium*. Gradient balance describes horizontal motion when no two of the terms dominate. This often applies to tropical motions, for which fR is small.

Equation (12.20.2) is quadratic in v and can be solved for the *gradient wind speed*

$$v_{gr} = -\frac{fR}{2} \pm \sqrt{\frac{f^2 R^2}{4} - R\frac{\partial \Phi}{\partial n}}. \tag{12.24}$$

Not all roots of (12.24) are physically meaningful, real wind speed requiring

$$\frac{\partial \Phi}{\partial n} \begin{cases} < \frac{f^2 R}{4} & (R > 0) \\ > \frac{f^2 R}{4} & (R < 0) \end{cases}. \tag{12.25}$$

We restrict subsequent analysis to regular cyclonic or anticyclonic motion in the Northern Hemisphere, for which $\partial \Phi / \partial n < 0$ (Fig. 12.8). The second inequality in (12.25) limits the pressure gradient that can be sustained by anticyclonic motion. Because $\partial \Phi / \partial n < 0$ and $R < 0$, $\partial \Phi / \partial n$ is sandwiched between 0 and $f^2 R / 4$. Near the center of an anticyclone, where $R \to 0$, the height distribution must therefore become flat and the accompanying motion weak. No corresponding limitation applies to a cyclone because, with $R > 0$, $\partial \Phi / \partial n < 0$ automatically satisfies the first inequality in (12.25). Hence, cyclones are free to intensify, but amplification of anticyclones is limited. This distinction, which explains why winds are comparatively light inside an anticyclone, follows from the fact that the pressure gradient force acts in the same direction as the centripetal acceleration in a cyclone but opposite to it in an anticyclone (Problem 12.32).

Letting $Ro \to 0$ recovers geostrophic equilibrium from (12.20.2), whereas $Ro \to \infty$ recovers cyclostrophic equilibrium. Intermediate to these extremes, the geostrophic wind speed only approximates the gradient wind. Substituting the geostrophic wind speed

$$v_g = -\frac{1}{f}\frac{\partial \Phi}{\partial n}$$

for the pressure gradient transforms (12.20.2) into

$$\frac{v_g}{v_{gr}} = 1 + \frac{v_{gr}}{fR}. \tag{12.26}$$

The geostrophic wind speed overestimates the gradient wind speed in cyclonic motion ($fR > 0$) and underestimates it in anticyclonic motion ($fR < 0$), the discrepancy being proportional to Ro. For large-scale extratropical motions, the discrepancy is of order 10 to 20%. However, discrepancies as large as 50 to 100% can occur inside intense cyclones and in tropical systems characterized by small fR.

12.5   Weakly Divergent Motion

Geostrophic balance and the idealized curvilinear balance considered above are useful for describing the structure of large-scale motion, but they provide no information on how the circulation evolves (Problem 12.22). Such equations are termed *diagnostic*. Unsteadiness and divergence, although comparatively small, allow the circulation to change from one state to another. Equations describing such changes are termed *prognostic*. A simple framework for describing how the circulation evolves can be developed in the absence of vertical motion.

12.5.1   Barotropic Nondivergent Motion

Consider motion that is horizontally nondivergent and under barotropic stratification. Continuity (12.9) then implies that vertical motion vanishes. By thermal wind balance (12.11), v_h is independent of height, so the motion is also two-dimensional. The governing equations in physical coordinates then reduce to

$$\frac{d v_h}{dt} + f k \times v_h = -\alpha \nabla p, \tag{12.27.1}$$

$$\nabla \cdot v_h = 0, \tag{12.27.2}$$

where friction is ignored and ∇ is understood to refer to the horizontal gradient. With vector identity (D.14) from Appendix D, (12.27.1) can be written

$$\frac{\partial v_h}{\partial t} + \frac{1}{2} \nabla (v_h \cdot v_h) + (\zeta + f) k \times v_h = -\alpha \nabla p, \tag{12.28.1}$$

where

$$\zeta = (\nabla \times v_h) \cdot k$$
$$= \frac{\partial v}{\partial x} - \frac{\partial u}{\partial y} \tag{12.28.2}$$

is the vertical component of *relative vorticity*. Consistent with prior convention, relative vorticity is cyclonic if it has the same sign as the planetary vorticity

f and anticyclonic otherwise. Applying the curl to (12.28.1) and making use of other vector identities in Appendix D yields the vorticity budget for an air parcel

$$\frac{\partial \zeta}{\partial t} + \boldsymbol{v}_h \cdot \nabla \zeta + \boldsymbol{v}_h \cdot \nabla f = (\nabla p \times \nabla \alpha) \cdot \boldsymbol{k}$$

or

$$\frac{d\zeta}{dt} + v\frac{\partial f}{\partial y} = 0 \tag{12.29}$$

because $\alpha = \alpha(p)$ under barotropic stratification. Since

$$v\frac{\partial f}{\partial y} = \frac{df}{dt},$$

the vorticity budget can be expressed

$$\frac{d(\zeta + f)}{dt} = 0. \tag{12.30}$$

Equation (12.30) asserts that the *absolute vorticity* $(\zeta + f)$ (e.g., that apparent to an observer in an inertial reference frame) is conserved for an individual air parcel. Thus, under barotropic nondivergent conditions, absolute vorticity behaves as a tracer of horizontal air motion and is rearranged by the circulation. Air moving southward must compensate for decreasing f by spinning up cyclonically (see the outbreak of cold polar air in the cyclone west of Africa in Figs. 1.23 and 1.24). Air moving northward must evolve in the opposite manner. On large scales, the circulation tends to remain close to barotropic stratification, which makes absolute vorticity approximately conserved.

Because it is nondivergent, this motion can be represented in terms of a streamfunction (12.6), which automatically satisfies the continuity equation (12.27.2). Substituting the vorticity (12.7.3) then transforms (12.30) into the *barotropic nondivergent vorticity equation*:

$$\left(\frac{\partial}{\partial t} - \frac{\partial \psi}{\partial y}\frac{\partial}{\partial x} + \frac{\partial \psi}{\partial x}\frac{\partial}{\partial y}\right)\{\nabla^2 \psi + f\} = 0, \tag{12.31}$$

where ψ is defined by (12.8). Equation (12.31) provides a closed prognostic system for the single unknown ψ that describes how the motion field evolves. Advection terms in the material derivative make (12.31) quadratically nonlinear in the dependent variable ψ. With suitable initial conditions and boundary conditions, the barotropic nondivergent vorticity equation can be integrated for the motion at some later time.

12.5.2 Vorticity Budget under Baroclinic Stratification

Under more general circumstances, absolute vorticity is not conserved. However, the governing equations still possess a conservation principle, albeit more complex. Baroclinic stratification renders the motion fully 3-dimensional. Even though vorticity is then a 3-dimensional vector quantity, the quasi-horizontal nature of large-scale motion makes its vertical component dominant (Problem 12.41).

Exclusive of friction, the horizontal momentum equations in physical coordinates are

$$\frac{\partial u}{\partial t} + \boldsymbol{v} \cdot \nabla u - fv = -\alpha \frac{\partial p}{\partial x}, \tag{12.32.1}$$

$$\frac{\partial v}{\partial t} + \boldsymbol{v} \cdot \nabla v + fu = -\alpha \frac{\partial p}{\partial y}, \tag{12.32.2}$$

where $\boldsymbol{v}$ and ∇ denote the full 3-dimensional velocity and gradient. Cross differentiating (12.32) with respect to x and y and subtracting results in

$$\left(\frac{\partial}{\partial t} + \boldsymbol{v} \cdot \nabla \right) \left(\frac{\partial v}{\partial x} - \frac{\partial u}{\partial y} \right) + \frac{\partial \boldsymbol{v}}{\partial x} \cdot \nabla v - \frac{\partial \boldsymbol{v}}{\partial y} \cdot \nabla u$$

$$+ f \left(\frac{\partial u}{\partial x} + \frac{\partial v}{\partial y} \right) + v \frac{\partial f}{\partial y} = \frac{\partial p}{\partial x} \frac{\partial \alpha}{\partial y} - \frac{\partial p}{\partial y} \frac{\partial \alpha}{\partial x},$$

or

$$\frac{d(\zeta + f)}{dt} + f \nabla_z \cdot \boldsymbol{v}_h + \left(\frac{\partial \boldsymbol{v}}{\partial x} \cdot \nabla v - \frac{\partial \boldsymbol{v}}{\partial y} \cdot \nabla u \right) = (\nabla_z p \times \nabla_z \alpha) \cdot \boldsymbol{k}. \tag{12.33}$$

The third term in (12.33) can be written

$$\frac{\partial \boldsymbol{v}}{\partial x} \cdot \nabla v - \frac{\partial \boldsymbol{v}}{\partial y} \cdot \nabla u = \frac{\partial u}{\partial x} \left(\frac{\partial v}{\partial x} - \frac{\partial u}{\partial y} \right) + \frac{\partial v}{\partial y} \left(\frac{\partial v}{\partial x} - \frac{\partial u}{\partial y} \right) + \frac{\partial w}{\partial x} \frac{\partial v}{\partial z} - \frac{\partial w}{\partial y} \frac{\partial u}{\partial z}$$

$$= \zeta \nabla_z \cdot \boldsymbol{v}_h + \frac{\partial w}{\partial x} \frac{\partial v}{\partial z} - \frac{\partial w}{\partial y} \frac{\partial u}{\partial z}. \tag{12.34}$$

Similarly, the last term in (12.34) can be expressed

$$\frac{\partial w}{\partial x} \frac{\partial v}{\partial z} - \frac{\partial w}{\partial y} \frac{\partial u}{\partial z} = -\left(\xi \frac{\partial w}{\partial x} + \eta \frac{\partial w}{\partial y} \right), \tag{12.35.1}$$

where

$$\xi = \frac{\partial w}{\partial y} - \frac{\partial v}{\partial z} \tag{12.35.2}$$

$$\eta = \frac{\partial u}{\partial z} - \frac{\partial w}{\partial x} \tag{12.35.3}$$

are the horizontal components of vector vorticity $\boldsymbol{\zeta} = (\xi, \eta, \zeta)$. Then the budget for the vertical component of vorticity becomes

$$\frac{d(\zeta + f)}{dt} = -(\zeta + f)\nabla_z \cdot \boldsymbol{v}_h + \left(\xi\frac{\partial w}{\partial x} + \eta\frac{\partial w}{\partial y}\right) + (\nabla_z p \times \nabla_z \alpha) \cdot \boldsymbol{k}, \quad (12.36)$$

wherein the Lagrangian derivative includes full three-dimensional advection.

According to (12.36), the absolute vorticity of a material element can change through three mechanisms (Fig. 12.9):

1. *Vertical stretching*: By continuity, horizontal convergence $-\nabla_z \cdot \boldsymbol{v}_h$ compensates vertical stretching of a material element (Fig. 12.9a). Its reduced moment of inertia then leads to that fluid body spinning up (in the same sense as its absolute vorticity) to conserve angular momen-

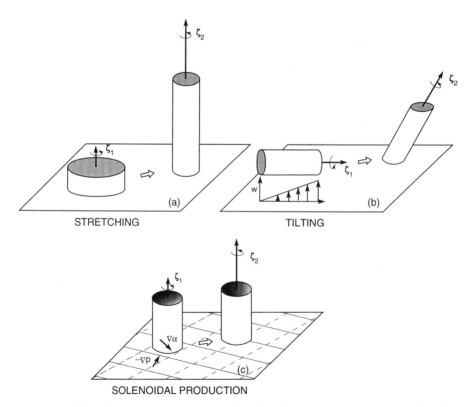

Figure 12.9 Forcing of the absolute vorticity $(f + \zeta)$ of a material element by (a) vertical stretching, which is compensated by horizontal convergence, (b) tilting, which exchanges horizontal and vertical vorticity, and (c) solenoidal production, which results from variation of density (shaded) across the pressure gradient force. The latter exerts a torque on the material element that is located inside a *solenoid* defined by two intersecting isobars (solid lines) and isochores (dashed lines).

tum. Horizontal divergence leads to the material element spinning down to conserve angular momentum.

2. *Tilting*: Horizontal shear of the vertical motion $\nabla_z w$ deflects a material element and its vector angular momentum (Fig. 12.9b). This process transfers vorticity from horizontal components into the vertical component. Strong vertical shear of the horizontal motion (e.g., as accompanies a sharp temperature gradient) magnifies the components ξ and η, which then make tilting potentially important as a source of vertical vorticity.

3. *Solenoidal production*: Under baroclinic stratification, isobars on a surface of constant height do not coincide with isochores. Two sets of intersecting isobars and isochores then define a *solenoid*, across which pressure varies in one direction and density varies in another (Fig. 12.9c). Variation of density across the pressure gradient force then introduces a torque, which changes the angular momentum of the material element occupying a solenoid. Under barotropic stratification, density is uniform across the pressure gradient, so the torque and solenoidal production of vorticity $\nabla_z p \times \nabla_z \alpha$ vanish.

Each of these forcing mechanisms reflects a rearrangement of angular momentum—not dissipation, which has been excluded by considering adiabatic and inviscid conditions. Scale analysis indicates that vertical stretching dominates the forcing of absolute vorticity in large-scale motion (Problem 12.44). Because other mechanisms on the right-hand side of (12.36) follow from baroclinic stratification, this is another reflection of the nearly barotropic nature of the large-scale circulation.

Under adiabatic conditions, tilting and solenoidal production disappear formally when the vorticity budget is expressed in isentropic coordinates. Making use of vector identities in Appendix D allows the horizontal momentum equation (11.92.1) to be expressed

$$\frac{\partial \boldsymbol{v}_h}{\partial t} + \frac{1}{2}\nabla_\theta (\boldsymbol{v}_h \cdot \boldsymbol{v}_h) + (\zeta_\theta + f)\boldsymbol{k} \times \boldsymbol{v}_h + \omega_\theta \frac{\partial \boldsymbol{v}_h}{\partial \theta} = -\nabla_\theta \Psi, \qquad (12.37)$$

where $\zeta_\theta = (\nabla_\theta \times \boldsymbol{v}_h) \cdot \boldsymbol{k}$ is the relative vorticity evaluated on an isentropic surface. Under adiabatic conditions, the term representing vertical advection disappears. Applying $\boldsymbol{k} \cdot \nabla_\theta \times$ to (12.37) then obtains the vorticity budget

$$\frac{\partial (\zeta_\theta + f)}{\partial t} + \boldsymbol{v}_h \cdot \nabla_\theta (\zeta_\theta + f) = -(\zeta_\theta + f)\nabla_\theta \cdot \boldsymbol{v}_h$$

or

$$\frac{d(\zeta_\theta + f)}{dt} = -(\zeta_\theta + f)\nabla_\theta \cdot \boldsymbol{v}_h, \qquad (12.38)$$

in which the Lagrangian derivative involves only horizontal advection. Because vertical motion vanishes identically, so does the tilting apparent in physical coordinates (12.36). Likewise, solenoidal production vanishes because the vertical vorticity ζ_θ is evaluated on an isentropic surface (Problem 12.48). The continuity equation in isentropic coordinates (11.92.3) reduces to

$$\frac{d}{dt}\left(\frac{\partial p}{\partial \theta}\right) + \left(\frac{\partial p}{\partial \theta}\right) \nabla_\theta \cdot \boldsymbol{v}_h = 0 \tag{12.39}$$

under adiabatic conditions. Combining this with (12.38) yields

$$\frac{1}{(\zeta + f)}\frac{d(\zeta + f)}{dt} = \frac{1}{\left(\frac{\partial p}{\partial \theta}\right)}\frac{d}{dt}\left(\frac{\partial p}{\partial \theta}\right),$$

or

$$\frac{dQ}{dt} = 0, \tag{12.40.1}$$

where

$$Q = \frac{\zeta_\theta + f}{-\frac{1}{g}\left(\frac{\partial p}{\partial \theta}\right)} \tag{12.40.2}$$

defines the *potential vorticity*, also termed the *Ertel potential vorticity* (Ertel, 1942) and the *isentropic potential vorticity*.

A generalization of the conservation principle for barotropic nondivergent motion, (12.40) asserts that the potential vorticity of an air parcel is conserved under inviscid adiabatic conditions. The parcel's absolute vorticity $(\zeta_\theta + f)$ can change, but in direct proportion to the vertical spacing $-(1/g)\partial p/\partial \theta$ of isentropic surfaces (11.85), which are material surfaces under adiabatic conditions. Despite its name, the conserved property Q does not have dimensions of vorticity, the denominator in (12.40.2) introducing other dimensions.

Potential vorticity is a dynamical tracer of horizontal motion. Under the same conditions that Q is conserved, vertical motion is determined by variations in the elevation of an isentropic surface, on which Q is evaluated. Unlike chemical tracers, potential vorticity is not advected "passively" by the circulation. Rather, the vorticity at a point induces motion about it, which makes Q and the motion field completely dependent on one another. According to the so-called *invertibility principle* (Hoskins et al., 1985), the distribution of Q at some instant uniquely determines the circulation. The conserved property Q can therefore be regarded as being self-advected.

In practice, friction and diabatic effects are slow enough away from the surface and outside convection for Q to be conserved over many advection times. Synoptic disturbances are marked by anomalies of potential vorticity (see Fig. 12.10). Hence, forecasting such disturbances amounts to predicting the distribution of Q, which is controlled primarily by advection. Over timescales much longer than a day, dissipative processes lead to production and destruction of Q (e.g., inside an individual air parcel). Nonconservative effects associated with turbulence, radiative transfer, and condensation eventually render the behavior of an air parcel diabatic, which introduces motion across isentropic surfaces (Sec. 3.6.1). Irreversibility associated with production and destruction

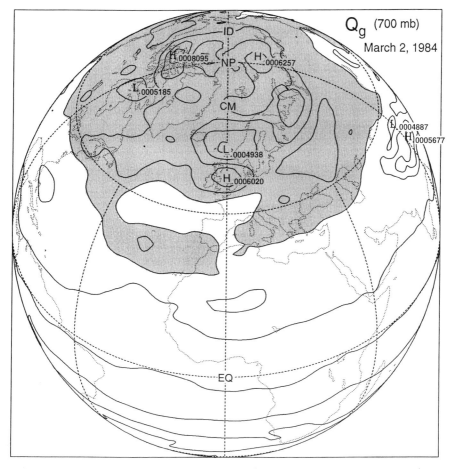

Figure 12.10 Distribution of quasi-geostrophic potential vorticity at 700 mb on March 2, 1984. A cyclone off the coast of Africa (Figs. 1.9a and 1.24) rearranges air to produce a dipole pattern, with high Q_g (shaded) folded south of low Q_g (compare Fig. 16.6). A similar pattern appears over Europe, where Q_g has been folded by another cyclone.

of Q then leads to a vertical drift of air, one ultimately manifested in the mean meridional circulation.

Unlike barotropic nondivergent motion (12.31), behavior described by (12.40) cannot be represented solely in terms of a streamfunction. Therefore, the vorticity budget alone does not provide sufficient information to describe the motion. Divergence, although comparatively small, interacts with vorticity in (12.38). A closed prognostic system is formed in large-scale numerical integrations by augmenting the vorticity budget with the budget of divergence, which is obtained in a similar fashion (see, e.g., Haltiner and Williams, 1980).

12.5.3   Quasi-Geostrophic Motion

An approximate prognostic system governing large-scale motion, which is useful for many applications, can be derived via an expansion in Ro—as was used to introduce the diagnostic description of geostrophic motion. For the resulting system to remain prognostic, the expansion must then retain terms higher order than those in (12.2) and, in particular, it must retain horizontal divergence that forces changes of absolute vorticity (12.38).

We consider motion with meridional displacements that are $O(Ro)$ and hence sufficiently narrow to ignore the earth's curvature. Analogous to the f-plane development, spherical geometry is approximated in a Cartesian system tangent to the earth at latitude ϕ_0 (Fig. 12.3). For consistency with other terms in the equations, the variation with latitude of the Coriolis force must now be retained to order Ro. Expanding f in a Taylor series in y (which reflects the meridional displacement of an air parcel) and truncating to first order yields the *β-plane approximation*

$$f = f_0 + \beta y, \tag{12.41.1}$$

$$\beta = \left.\frac{df}{dy}\right|_{y=0}, \tag{12.41.2}$$

where f_0 and y are given by (12.3). The motion is also presumed to satisfy the so-called *Boussinesq approximation*, in which variations of density are ignored except where accompanying gravity in the buoyancy force. The Boussinesq approximation neglects compression of an air parcel introduced by changes of pressure, retaining only density variations that are introduced thermally. Vertical displacements of air must then be shallow (e.g., compared to H).

Following the development in Sec. 11.4, all variables are nondimensionalized by characteristic scale factors. Dependent variables are then expanded in a power series in Rossby number. The dimensionless horizontal velocity becomes

$$v_h = v_g + Ro v_a + \dots, \tag{12.42}$$

where v_g satisfies geostrophic balance with $f = f_0$ and the *ageostrophic velocity* where v_a defines an $O(Ro)$ correction to it. Both are $O(1)$, as are all dependent variables in a formal scale analysis (see, e.g., Charney, 1973), in which magnitudes of various terms are contained in dimensionless parameters like Ro. To streamline the development, we depart from the formal procedure by absorbing dimensionless parameters other than Ro into the dependent variables, the magnitudes of which can then be inferred from the ensuing equations. This allows the dimensionless equations to be expressed in forms nearly identical to their dimensional counterparts.

In log–pressure coordinates, the horizontal momentum equation exclusive of friction becomes

$$Ro \left[\frac{\partial}{\partial t} + (v_g + Ro v_a + \ldots) \cdot \nabla_z \right]$$
$$\cdot (v_g + Ro v_a + \ldots) + Row \frac{\partial}{\partial z}(v_g + Ro v_a + \ldots) \quad (12.43)$$
$$+ (f_0 + Ro\beta y)k \times (v_g + Ro v_a + \ldots)$$
$$= -\nabla_z(\Phi_g + Ro\Phi_a + \ldots),$$

where z denotes log–pressure height (11.65) and factors such as β are understood to be dimensionless and $O(1)$. Because v_g satisfies (12.2), the Ro^0 balance drops out of (12.43). Under the Boussinesq approximation, w is smaller than $O(Ro)$ [denoted $o(Ro)$], so vertical advection of momentum also drops out to $O(Ro)$. Then the momentum balance at $O(Ro)$ is given by

$$\frac{d_g v_g}{dt} + f_0 k \times v_a + \beta y k \times v_g = -\nabla_z \Phi_a, \quad (12.44.1)$$

where

$$\frac{d_g}{dt} = \frac{\partial}{\partial t} + v_g \cdot \nabla_z \quad (12.44.2)$$

is the time rate of change moving horizontally with the geostrophic velocity. According to (12.44), first-order departures from simple geostrophic equilibrium provide a time rate of change or *tendency* to v_g, which is absent at $O(Ro^0)$.

The continuity equation (11.72.3) becomes

$$\frac{1}{\rho_0} \frac{\partial}{\partial z}(\rho_0 w) = -\nabla_z \cdot (v_g + Ro v_a + \ldots)$$
$$= -Ro\nabla_z \cdot v_a, \quad (12.45)$$

so w follows directly from v_a. The ageostrophic velocity introduces a *secondary circulation*, which, unlike the $O(Ro^0)$ geostrophic velocity, involves vertical motion. Under adiabatic conditions, the thermodynamic equation (11.72.4) becomes

$$\left[\frac{\partial}{\partial t} + (v_g + Ro v_a + \ldots) \cdot \nabla_z \right] \frac{\partial}{\partial z}(\Phi_g + Ro\Phi_a + \ldots) + N^2 w = 0$$

or

$$\frac{d_g}{dt}\left(\frac{\partial \Phi_g}{\partial z} \right) + N^2 w = 0, \quad (12.46)$$

where N^2 is evaluated with the basic stratification T_0. The Boussinesq approximation requires N^2 to be large enough to render vertical displacements small, so the second term in (12.46) must be retained to leading order.

Equations (12.44), (12.45), and (12.46) describe *quasi-geostrophic motion* on a β plane. Accurate to $O(Ro)$, they have dimensional counterparts that are nearly identical. By expressing the momentum budget as in (12.28) and applying $k \cdot \nabla_z \times$, we obtain the vorticity equation

$$\frac{d_g \zeta_g}{dt} + v\beta + f_0 \nabla_z \cdot v_a = 0$$

or

$$\frac{d_g(\zeta_g + f)}{dt} = -f_0 \nabla_z \cdot v_a, \qquad (12.47.1)$$

where

$$\zeta_g = \nabla_z^2 \psi$$

$$= \frac{1}{f_0} \nabla_z^2 \Phi_g \qquad (12.47.2)$$

is the geostrophic vorticity. This is similar to the full vorticity equation in isentropic coordinates (12.38), but the conditions of quasi-geostrophy make the relative vorticity ζ_g negligible compared to the planetary vorticity f_0 in the divergence term. With the continuity equation, (12.47) can be written[4]

$$\frac{d_g(\zeta_g + f)}{dt} = f_0 \frac{1}{\rho_0} \frac{\partial}{\partial z}(\rho_0 w).$$

Then applying the operator $(1/\rho_0)\partial/\partial z(\rho_0)$ to the thermodynamic equation and eliminating w yields the *quasi-geostrophic potential vorticity equation*:

$$\frac{d_g Q_g}{dt} = 0, \qquad (12.48.1)$$

where

$$Q_g = \left[\nabla_z^2 \psi + f_0 + \beta y + \frac{1}{\rho_0} \frac{\partial}{\partial z}\left(\frac{f_0^2}{N^2} \rho_0 \frac{\partial \psi}{\partial z} \right) \right] \qquad (12.48.2)$$

is the *quasi-geostrophic potential vorticity*. Since d_g/dt involves only v_g, (12.48) can be expressed in terms of the single dependent variable ψ

$$\left(\frac{\partial}{\partial t} - \frac{\partial \psi}{\partial y}\frac{\partial}{\partial x} + \frac{\partial \psi}{\partial x}\frac{\partial}{\partial y} \right)\left[\nabla_z^2 \psi + f_0 + \beta y + \frac{1}{\rho_0}\frac{\partial}{\partial z}\left(\frac{f_0^2}{N^2}\rho_0\frac{\partial \psi}{\partial z} \right) \right] = 0,$$

$$(12.49)$$

which is a generalization of the barotropic nondivergent vorticity equation (12.31) to weakly divergent motion.

Equation (12.49) constitutes a closed prognostic system for the streamfunction ψ. Like (12.31), it is quadratically nonlinear and can be integrated for the motion field's evolution with suitable initial and boundary conditions. The conserved property Q_g is an analogue of isentropic potential vorticity (12.40.2),

---

[4] Under fully Boussinesq conditions, the stretching term simplifies to $f_0 \partial w/\partial z$—the same as under incompressible conditions. With the density weighting retained, the approximation is referred to as *quasi-Boussinesq* and accounts for vertical structure associated with the amplification of atmospheric waves (Chapter 14).

but is invariant following v_g rather than the actual horizontal motion (see Problem 12.53).

Figure 12.10 shows the distribution of Q_g at 700 mb two days prior to the analysis in Fig. 1.9. In the eastern Atlantic, air of high potential vorticity (shaded) is being drawn southeastward in a cyclone that matures into the one evident in Figs. 1.23 and 1.24. That cold air is being exchanged with warm air of low potential vorticity, which is being drawn in the opposite sense. This rearrangement of air introduces a dipole pattern with high Q_g folded south of low Q_g. Another appears over Europe, where the distribution of Q_g has been folded by a cyclone over the North Sea. The conserved property Q_g must be derived indirectly from temperature observations (12.48.2), so it tends to be noisy. But the same basic signature emerges clearly in satellite imagery of water vapor (compare Fig. 16.6).

The distribution of Q_g carries the same essential information as its formal counterpart Q. However, the two conserved properties deviate at low latitude, where v_h departs from v_g. The quasi-geostrophic equations describe motion in which the rotational component is dominant and only weakly coupled to the divergent component. As the equator is approached, f becomes small enough to invalidate the underpinnings of quasi-geostrophy and render the divergent component of motion comparable to the rotational component. For this reason, tropical circulations must be described with the primitive equations or with forms specialized to the tropics that include divergence and vorticity to leading order.

Suggested Reading

An Introduction to Dynamic Meteorology (1992) by Holton gives a complete treatment of gradient wind solutions and provides an excellent introduction to the concepts of vorticity and circulation. An advanced treatment of those concepts is presented in *Geophysical Fluid Dynamics* (1979) by Pedlosky.

An Introduction to Fluid Dynamics (1977) by Batchelor contains a formal development of vorticity and its relationship to angular momentum. The interdependence of vorticity and stratification is discussed in *The Ceaseless Wind* (1986) by Dutton.

Hoskins *et al.* (1985) provide an overview of isentropic potential vorticity and different applications. Comparisons with quasi-geostrophic potential vorticity are given in *Atmosphere–Ocean Dynamics* (1982) by Gill and in *Middle Atmosphere Dynamics* (1987) by Andrews *et al.*

Introduction to Circulating Atmospheres (1993) by James discusses vertical motion in large-scale flows and contains a concise summary of quasi-geostrophic relationships.

Problems

12.1. With the scaling in Sec. 11.4, perform an asymptotic series expansion of dependent variables in the horizontal momentum equations to obtain (12.1) accurate to $O(Ro^0)$ and then to $O(Ro^1)$, where drag may be regarded $O(Ro^0)$ inside the boundary layer and smaller elsewhere.

12.2. Recover the geostrophic velocity (12.2.2) from the vector equations of motion.

12.3. Obtain expressions for the geostrophic velocity in (a) height coordinates and (b) isentropic coordinates.

12.4. Write down explicit expressions for the geostrophic velocity components in terms of geopotential for (a) Cartesian coordinates, (b) spherical coordinates, and (c) vertical cylindrical coordinates.

12.5. Discuss why the Coriolis force is not apparent for laboratory-scale flow in the reference frame of the earth.

12.6. For geostrophic motion in spherical geometry and isobaric coordinates, (a) determine an expression for the vertical vorticity, (b) determine an expression for the horizontal divergence, and (c) interpret the significance of the divergence in part (b) in light of observed circulations, wherein divergence is strongly correlated to lows and highs.

12.7. The bullet train moves over straight and level track at $40°$ latitude. If the train's mass is 10^5 kg, how fast must it move to experience a transverse force equal to 0.1% of its weight?

12.8. Interpret the velocity potential in relation to the general concept of a potential function introduced in Sec. 2.1.

12.9. Demonstrate that the Helmholtz theorem reduces to (12.7.1) for a two-dimensional vector field.

12.10. Consider a steady vortex with circular streamlines and azimuthal velocity v. (a) Determine the form of the velocity profile $v(r)$ that possesses no vorticity for $r > 0$. (b) Calculate the *circulation* $\Gamma(r) = \oint v(r) \cdot ds = \int_0^{2\pi} v(r)r\,d\phi$ about the vortex. (c) Use Stokes' theorem to interpret the distribution of vorticity.

12.11. (a) Discuss the time during which a fluid body may be regarded as a compact closed system (e.g., a finite-dimensional air parcel) in relation to its size and deformation (see Problem 10.10). (b) After sufficient time, what additional factors intervene to invalidate the description of part (a)?

12.12. Show that geostrophic motion on an f plane automatically satisfies conservation of mass.

12.13. (a) Determine the components of frictional geostrophic motion tangential and orthogonal to contours of height, in terms of the corresponding geostrophic velocity. (b) Construct an expression for the angle δ by which v_h deviates from height contours. Estimate δ for a characteristic wind speed of 5 m s^{-1} at a latitude of (c) 45° and (d) 15°.

12.14. (a) Derive an expression for the divergence of frictional geostrophic motion on an f plane to show that, in the presence of Rayleigh friction, $\nabla \cdot v_h$ is proportional to the $\nabla^2 \Phi$. (b) From this expression, infer how vertical motion over a surface low depends on its latitude.

12.15. Consider a surface low that is narrow enough for motion about it to be treated on an f plane. For Rayleigh friction with a timescale of 1 day, determine the latitude at which horizontal divergence inside the boundary layer becomes comparable to vertical vorticity.

12.16. The surface motion beneath a cyclone is described by the stream function and velocity potential

$$\psi(x, y) = \Psi \left\{ 1 - \exp\left[-\frac{(x^2 + y^2)}{L^2} \right] \right\}$$

$$\chi(x, y) = X \exp\left[-\frac{(x^2 + y^2)}{L^2} \right],$$

where Ψ and X are constants. Determine (a) the horizontal velocity, (b) the horizontal divergence, and (c) the vertical vorticity. (d) Sketch $v_h(x, y)$ and discuss it in relation to $\psi(x, y)$ and $\chi(x, y)$. (e) Indicate the horizontal trajectory of an air parcel and discuss it in relation to three-dimensional air motion.

12.17. (a) Express the vertical velocity ω in isobaric coordinates in terms of the geometric vertical velocity w under the conditions of large-scale, steady motion above the boundary layer. (b) Estimate the geometric vertical velocity that corresponds to $\omega = 100$ mb day^{-1} at 500 mb.

12.18. The 500-mb surface has height

$$z_{500} = \begin{cases} \overline{\overline{z}}_{500} - \overline{z}_{500} \left(\dfrac{y}{a} \right) - z'_{500} \left[1 + \cos\left(\pi \dfrac{x^2 + y^2}{L^2} \right) \right] & (x^2 + y^2)^{\frac{1}{2}} \leq L \\ \overline{\overline{z}}_{500} - \overline{z}_{500} \left(\dfrac{y}{a} \right) & (x^2 + y^2)^{\frac{1}{2}} > L \end{cases},$$

where the constants $\overline{\overline{z}}_{500}$, $\overline{z}_{500}$, and z'_{500} reflect global–mean, zonal–mean, and perturbation contributions to the height field, respectively, a is the radius of the earth, and y is measured on midlatitude f plane centered at latitude ϕ_0. (a) Provide an expression for the 500-mb geostrophic flow v_{500}. (b) Sketch $z_{500}(x, y)$ and $v_{500}(x, y)$, presuming $z'_{500} \ll \overline{z}_{500}$. Discuss the contributions to v_{500}, noting how the flow is influenced by the thermal perturbation. (c) Provide an expression for the absolute vorticity at 500 mb and discuss its contributions.

12.19. (a) Determine the geostrophic motion accompanying the upper-level depression in Problem 6.7. (b) Plot the vertical profile of horizontal motion at $(\lambda, \phi) = (0, \phi_0 - L)$.

12.20. Show that, under barotropic stratification, $\nabla_p \Phi$ is invariant with elevation.

12.21. Show that the Lagrangian derivative of the coordinate vector s tangential to a parcel's trajectory is given by (12.17). (*Hint:* Express the vectorial change of s during an interval dt in terms of its angular velocity ω, which, in turn, can be expressed in terms of v and R; see Fig. 10.7.)

12.22. Determine the circumstances under which the special class of curvilinear motion considered in Sec. 12.4 is solenoidal.

12.23. Show that the relative error introduced by treating steady motion geostrophically is $O(Ro)$.

12.24. A midlatitude cyclone centered at $45°N$ produces a radial height gradient at 500 mb of 0.16 m km^{-1} at a characteristic radial distance of $10°$. If the motion is sufficiently steady, calculate the corresponding (a) geostrophic wind and (b) gradient wind.

12.25. As in Problem 12.24, but for a tropical cyclone centered at $10°N$, which produces a radial height gradient at 850 mb of 1.88 m km^{-1} at a characteristic radial distance of $1°$.

12.26. Estimate the relative error introduced by treating motion geostrophically for (a) an *omega block*, in which the jet stream is deflected poleward at $40°$, where it has a velocity of 30 m s^{-1} and assumes a radius of curvature of 2000 km, (b) a tropical depression at $10°$ characterized by a radius of 200 km and a velocity of 30 m s^{-1}, and (c) the undisturbed polar-night vortex at $60°$, wherein the motion is zonal and has a velocity of 80 m s^{-1}.

12.27. The *Kelvin wave* plays an important role in tropical circulations, having the form of a vertical overturning in the equatorial plane. Horizontal motion in the Kelvin wave is zonal and described by

$$v_h(x, y) = \exp\left[-\left(\frac{y}{Y}\right)^2\right]\exp[i(kx - \sigma t)]i,$$

where y is measured on an equatorial β plane ($\phi_0 = 0$) and Y corresponds to $10°$ of latitude. (a) Sketch the velocity field over one wavelength in the x direction. (b) Calculate the vorticity and divergence of the wave, sketching each on the equatorial β plane. (c) Derive a system of equations that determines the streamfunction and velocity potential of the wave. Outline the solution of this system.

12.28. Meteorologists refer to the vector change of horizontal wind across a layer as the *thermal wind*. Show that the thermal wind is proportional to the horizontal gradient of layer–mean temperature.

12.29. Derive an expression for thermal wind balance in height coordinates.

12.30. A vertical sounding at 45°N reveals isothermal conditions with $T = 0°C$ below 700 mb, but vertical shear characterized by the following wind profile:

$$\mathbf{v} = \begin{cases} 10 \text{ m s}^{-1} & \text{S} & 950 \text{ mb} \\ 14.14 \text{ m s}^{-1} & \text{SW} & 850 \text{ mb} \\ 10 \text{ m s}^{-1} & \text{S} & 750 \text{ mb} \end{cases},$$

where the wind blows from the direction indicated. (a) Calculate the mean horizontal temperature gradient in the layers between 950 and 850 mb and between 850 and 750 mb. (b) Estimate the rate at which the preceding layers would warm/cool locally through temperature advection (e.g., were temperature approximately conserved). (c) Use the result of part (b) to estimate the lapse rate created between those layers after 1 day, noting the physical implications.

12.31. In terms of horizontal forces, discuss why cyclostrophic motion can be either cyclonic or anticyclonic, but always about low pressure.

12.32. Use the horizontal force balance to explain why, physically, the strength of curvilinear motion is limited for anticyclonic curvature but not for cyclonic curvature.

12.33. Demonstrate that

$$v_{gr} \begin{cases} < v_g & \text{cyclonic motion} \\ > v_g & \text{anticyclonic motion.} \end{cases}$$

12.34. A circular hurricane has azimuthal velocity

$$v(r, p) = v_0(p) \left(\frac{r_0}{r}\right)^2 \qquad r > r_0,$$

where r is measured from the center of the system, $v_0(1000 \text{ mb}) = 50$ m s^{-1}, and the hurricane's characteristic dimension, $r_0 = 100$ km, is small enough for it to be treated on an f plane with $f_0 = 5.0 \times 10^{-5}$ s^{-1}. (a) As a function of p, determine the range of r where the motion can be treated as geostrophic, providing justification for this approximation. (b) In the foregoing range of r, determine the geopotential field $\Phi(r, p)$, which approaches an undisturbed value Φ_∞ as $r \to \infty$. (c) Hurricanes are warm-core depressions (see Problem 6.9), so the radial temperature gradient is negative. It follows that the azimuthal velocity decreases with height. If the system has an average horizontal temperature gradient of 1 K km^{-1}, in the foregoing range of r, estimate the level $p(r)$ where v becomes negligible.

12.35. Consider the core of a tornado that is isothermal and rotates with uniform angular velocity Ω. Derive an expression for the radial profile of pressure inside the tornado's core in terms of the pressure p_0 at its center.

12.36. Isobaric height in the Venusian atmosphere varies locally as

$$\Phi(x, y) = \overline{\Phi}\left(1 + \epsilon \frac{x^2 + y^2}{L^2}\right),$$

where the length scale L is of arbitrary dimension. (a) If the motion is steady and frictionless, provide an expression for the horizontal velocity. (b) Can a similar result be obtained if the flow is transient? Why?

12.37. Satellite measurements reveal the following thermal structure:

$$T(x, y, p) = \overline{\overline{T}} - \overline{T}\left(\frac{y}{a}\right) - \Delta T(p)\exp\left(-\frac{x^2 + y^2}{L^2}\right)$$

where y is measured from $45°$, a is the radius of the earth, the length scale L is of order 1000 km or longer, $\overline{\overline{T}}$ and $\overline{T}$ are constants that refer to the undisturbed zonally symmetric state, and

$$\Delta T(p) = A + B\ln\left(\frac{p_{00}}{p}\right)$$

describes a cold-core cyclonic disturbance with A and B constants and $p_{00} = 1000$ mb. (a) Determine the height of the 500-mb surface as a function of position. Sketch the distribution of $z_{500}(x, y)$, presuming ΔT is not large enough to reverse the equatorward temperature gradient. (b) Determine the horizontal motion if meridional excursions are small enough for it to be treated on an f plane. Justify the validity of this expression based on a characteristic velocity of 10 m s^{-1}. Sketch the motion. (c) For undisturbed conditions ($\Delta T = 0$), describe the vertical profile of horizontal wind. If motion vanishes at the surface, determine the pressure at which the horizontal wind speed reaches 20 m s^{-1}.

12.38. A stratospheric sudden warming is accompanied by a reversal of the westerly circumpolar flow corresponding to the polar-night vortex (Fig. 1.10). Developing within a couple of days, this dramatic change in motion is attended by sharply warmer temperatures over the polar cap that follow from a rearrangement of air by planetary waves (see Fig. 14.26). If the initial motion at middle and high latitudes is approximated by rigid body rotation with angular velocity 0.25Ω, if the reversed circumpolar flow at high latitude is also characterized by rigid body rotation, but with zonal easterlies of 20 m s^{-1} at $75°$N, and if the motion leading to this state is approximately nondivergent, estimate where air prevailing over the polar cap must have originated from.

12.39. Consider steady, barotropic, nondivergent motion on a midlatitude β plane. The motion is zonal and uniform with $v = \overline{u}i$ upstream of an obstacle at $x = 0$, which deflects streamlines meridionally from their undisturbed latitudes. If the disturbed motion downstream deviates from the upstream motion by a small meridional perturbation v',

(a) construct a differential equation governing the trajectory $y(x)$ of an air parcel initially at $y = 0$, (b) obtain the general solution downstream of the obstacle, (c) determine the trajectory if the parcel is deflected at $(0, 0)$ with a slope of $\tan \alpha$, and (d) interpret the motion downstream of the obstacle in terms of a restoring force.

12.40. Derive the vertical vorticity budget for barotropic nondivergent motion:

$$\frac{\partial \zeta}{\partial t} + \mathbf{v}_h \cdot \nabla \zeta + \mathbf{v}_h \cdot \nabla f = (\nabla p \times \nabla \alpha) \cdot \mathbf{k}.$$

12.41. Perform a scale analysis to show that the vertical component of vorticity dominates horizontal components for large-scale, three-dimensional motion under generally baroclinic conditions.

12.42. Derive the vorticity budget (12.33).

12.43. Show that a state of static equilibrium ($\mathbf{v} \equiv 0$) can persist only under barotropic stratification.

12.44. Perform a scale analysis to show that vertical stretching dominates other mechanisms forcing absolute vorticity in (12.36).

12.45. Rising motion is an essential feature of organized convection in the tropics, where f is small enough to permit comparatively large horizontal divergence yet still dominates relative vorticity. In terms of vorticity forcing, explain why motion in convective centers is anticyclonic near the tropopause (see Fig. 15.8).

12.46. Explain why cyclogenesis would be favored in a cold continental air mass that has moved over a warmer maritime region, such as occurs in the north Pacific and Atlantic storm tracks (refer to Fig. 9.38).

12.47. The zonal circulation is disrupted by a planetary wave that advects air northward from 10°N to 50°N. If winds at 10°N are easterly with speed of 5 m s^{-1} and relative vorticity of 1.2×10^{-5} s^{-1}, estimate the column-averaged vorticity at 50°N if (a) air motion is barotropic nondivergent and (b) air motion is confined between an isentropic surface near 1000 mb and another near the tropopause, which slopes downward towards the pole from 100 mb in the tropics to 300 mb at 50°N.

12.48. (a) Explain why solenoidal production of vorticity vanishes in isentropic coordinates (12.38). (*Hint:* How many thermodynamic degrees of freedom are present in general? Along an isentropic surface?) (b) Apply this argument to the budget of vorticity in isobaric coordinates.

12.49. Use (11.27) to derive the equivalent expression for potential vorticity

$$Q = \frac{(\nabla \times \mathbf{v}_i) \cdot \nabla \theta}{\rho},$$

which is likewise conserved under inviscid adiabatic conditions.

12.50. Consider a material element which is bounded above and below by (nearly horizontal) isentropic surfaces that are separated incrementally by $d\theta$. The element has cross-sectional area dA on those surfaces. (a) Express the element's incremental mass in terms of $\partial p/\partial \theta$. (b) Use potential vorticity together with Stokes' theorem to derive an expression governing the circulation $\Gamma = \oint \boldsymbol{v}_h \cdot d\boldsymbol{s}$ about the element.

12.51. Consider an idealized model of the jet stream and its interaction with a transverse mountain range like the Rockies: A two-dimensional ridge oriented north–south on a midlatitude β plane presents a topographic barrier to an upstream zonal flow. The motion is inviscid and adiabatic. An isentropic surface coincides with the topography (which displaces it to smaller pressure) and another remains level near the tropopause. If the upstream flow is uniform, use conservation of potential vorticity to (a) sketch a vertical section of the flow and topography, noting the sign of relative vorticity for an air column as it is advected past different stations, and (b) determine the horizontal trajectory of an air column over and downstream of the barrier.

12.52. Derive the conservation principle (12.48) for quasi-geostrophic motion.

12.53. Vertical stretching, which forces changes in absolute vorticity, appears in the numerator of Q_g, rather than the position it assumes in the isentropic potential vorticity Q. Argue this form from the shallowness of vertical displacements, which makes relative changes of an air parcel's depth small. See Andrews *et al.* (1987) for a formal comparison of Q_g and Q.

12.54. Satellite observations provide thermal structure from the surface upward and the mixing ratio of a long-lived chemical species above the tropopause—both at successive times. (a) Construct an algorithm to deduce the three-dimensional trajectories of individual air parcels on the timescale of a week, noting the level of approximation and sources of error. (b) Where does this approximation break down? How could this limitation be averted with additional chemical tracers and what property would then be required of those tracers? (c) How would the application of this algorithm be complicated in the troposphere? (*Hint:* See Fig. 13.2.)

12.55. In practice, satellite observations are available *asynoptically*, that is, different positions are observed at different instants. (a) How might this feature of tracer observations be accommodated to implement the algorithm in Problem 12.54? (b) Discuss the practical limitations of inferring global air motion in light of the timescale for advection versus that for the earth to be sampled completely.

Chapter 13 | The Planetary Boundary Layer

Previous development treats motion as a smoothly varying field property. However, atmospheric motion is always partially turbulent. Laboratory experiments provide a criterion for the onset of turbulence in terms of the dimensionless *Reynolds number*

$$Re = \frac{LU}{\nu}, \tag{13.1}$$

where L and U are scales characterizing a flow and $\nu = \mu/\rho$ is the kinematic viscosity (Sec. 10.6). For Re greater than a critical value of about 5000, smooth laminar motion undergoes a transition to turbulent motion, which is inherently unsteady, three-dimensional, and involves a spectrum of space and time scales. Even for a length scale as short as 1 m, the critical Reynolds number with $\nu = 10^{-5}$ m^2 s^{-1} is exceeded for velocities of $U > 0.05$ m s^{-1}. Consequently, atmospheric motion inevitably contains some turbulence.[1]

Until now, turbulent transfers have been presumed small enough to treat an individual air parcel as a closed system. This approximation is valid away from the surface and outside convection because the timescale for turbulent exchange in the *free atmosphere* is much longer than the 1-day timescale characterizing advective changes of an air parcel. Budgets of momentum, internal energy, and constituents for an individual parcel can then be treated exclusive of turbulent exchanges with the parcel's environment—at least over timescales comparable to advection. This feature of the circulation is reflected in the spectrum of kinetic energy (Fig. 13.1). In the free atmosphere, kinetic energy is concentrated at periods longer than a day, where it is associated with large-scale disturbances and seasonal variations. However, in a neighborhood of the surface, as much as half of the kinetic energy lies at periods of order minutes. Inside the *planetary boundary layer*, turbulent eddies are generated mechanically from strong shear as the flow adjusts sharply to satisfy the no-slip condition at the ground (see Fig. 13.3). Turbulence is also generated thermally through buoyancy when the stratification is destabilized.

---

[1]Stratification of mass leads to ν increasing exponentially with height. Therefore, a level is eventually reached where ν is sufficiently large to drive Re below the critical value and suppress turbulence (Problem 13.3). This transition occurs at the turbopause, as pictured in Fig. 1.4.

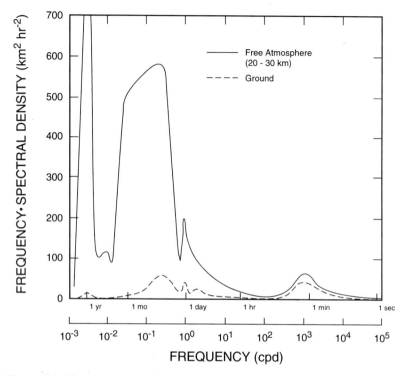

Figure 13.1   Power spectrum of atmospheric kinetic energy. Adapted from Vinnichenko (1970). Copyright (1970) Munksgaard International Publishers Ltd., Copenhagen, Denmark.

Mixing between air parcels cannot be ignored inside the boundary layer because the timescale for turbulent exchange there is comparable to that for advection by the large-scale flow. Turbulent exchanges of mass, heat, and momentum then make an air parcel an open system and render its behavior diabatic. Turbulent mixing transfers heat and moisture between the surface and the atmosphere. Because it destroys gradients inherent to large-scale motion, turbulent mixing also acts to dissipate the circulation.

13.1   Description of Turbulence

Turbulent fluctuations need not be hydrostatic, so the equations of motion are expressed most conveniently in physical coordinates. The limited vertical scale of turbulent eddies makes compressibility nonessential. However, variations of density associated with temperature fluctuations play a key role in stratified turbulence because, through buoyancy, they couple the motion to gravity. These features are embodied in the Boussinesq approximation (Sec. 12.5). Equivalent to incompressibility, the Boussinesq approximation allows density to be treated as constant, except where accompanied by gravity in the buoyancy force.

Consider the motion-related stratification (Sec. 11.4), which is regarded as a small departure from static conditions. In terms of the static density $\rho_0(z)$, the specific buoyancy force experienced by an air parcel (11.35) is

$$f_b = -\frac{1}{\rho_0}\frac{\partial p}{\partial z} - \frac{\rho}{\rho_0}g. \tag{13.2}$$

The gas law and Poisson's relation for potential temperature, along with the neglect of compressibility, then imply

$$\frac{\rho}{\rho_0} = -\frac{T}{T_0} + \frac{p}{p_0}$$

$$\cong -\frac{\theta}{\theta_0}. \tag{13.3}$$

Hence, the vertical momentum budget becomes

$$\frac{dw}{dt} - \frac{1}{\rho_0}\frac{\partial p}{\partial z} + g\frac{\theta}{\theta_0} - D_z. \tag{13.4}$$

Similarly, the thermodynamic equation becomes

$$\frac{d\theta}{dt} + w\frac{d\theta_0}{dz} = \frac{\theta_0}{c_p T_0}\dot{q}_{net}, \tag{13.5}$$

where $\theta_0(z)$ describes the basic stratification and θ the small departure from it. Under the Boussinesq approximation, the continuity equation reduces to a statement of three-dimensional nondivergence. Then the equations governing turbulent motion become

$$\frac{du}{dt} - fv = -\frac{1}{\rho_0}\frac{\partial p}{\partial x} - D_x, \tag{13.6.1}$$

$$\frac{dv}{dt} + fu = -\frac{1}{\rho_0}\frac{\partial p}{\partial y} - D_y, \tag{13.6.2}$$

$$\frac{dw}{dt} = -\frac{1}{\rho_0}\frac{\partial p}{\partial z} + g\frac{\theta}{\theta_0} - D_z, \tag{13.6.3}$$

$$\frac{\partial u}{\partial x} + \frac{\partial v}{\partial y} + \frac{\partial w}{\partial z} = 0, \tag{13.6.4}$$

$$\frac{d\theta}{dt} + w\frac{d\theta_0}{dz} = \frac{\theta_0}{c_p T_0}\dot{q}_{net}, \tag{13.6.5}$$

where

$$\frac{d}{dt} = \frac{\partial}{\partial t} + \boldsymbol{v} \cdot \nabla$$

$$= \frac{\partial}{\partial t} + u\frac{\partial}{\partial x} + v\frac{\partial}{\partial y} + w\frac{\partial}{\partial z} \qquad (13.6.6)$$

includes three-dimensional advection. The frictional drag $\boldsymbol{D} = -(1/\rho)\nabla \cdot \boldsymbol{\tau}$, which follows from the deformation tensor $\boldsymbol{e}$ (10.23), reduces under the Boussinesq approximation to

$$\boldsymbol{D} = -\nu\nabla^2\boldsymbol{v}. \qquad (13.6.7)$$

Compared to the timescale of turbulent eddies, radiative heating in $\dot{q}_{net}$ is slow enough to be ignored. However, thermal diffusion (e.g., conduction) operates efficiently on small scales that are created by turbulent deformations (refer to Fig. 12.4). Diffusion of momentum and heat are both involved in dissipation of turbulence, which takes place on the smallest scales of motion. Sharp gradients produced when anomalies are strained down to small dimensions are acted on efficiently by molecular diffusion.[2] When those gradients are destroyed, so too is the energy transferred to small scales. The foregoing process reflects a cascade of energy from large scales of organized motion to small scales, where it is dissipated by molecular diffusion; see Tennekes and Lumley (1972) for a formal development of turbulent energetics.

13.1.1   Reynolds Decomposition

Of relevance to the large-scale circulation is the time-averaged motion and how it is influenced by turbulent fluctuations. To describe such interactions, each field variable is separated into a slowly varying mean component, which is denoted by an overbar, and a fluctuating component, which is denoted by a prime, for example,

$$\boldsymbol{v} = \bar{\boldsymbol{v}} + \boldsymbol{v}', \qquad (13.7.1)$$

with

$$\overline{\boldsymbol{v}'} = 0. \qquad (13.7.2)$$

For this decomposition to be meaningful, timescales of the two components must be widely separated. Only then can averaging be performed over an interval long enough for a stable mean to be recovered from fluctuating properties yet short enough for mean properties to be regarded as steady.

---

[2] The scale dependence of diffusion can be inferred from the viscous drag (13.6.7), which increases quadratically with the inverse scale of motion anomalies.

The continuity equation (13.6.4) allows the material acceleration to be expressed in terms of the divergence of a momentum flux

$$\frac{du}{dt} = \frac{\partial u}{\partial t} + \nabla \cdot (vu), \tag{13.8.1}$$

$$\frac{dv}{dt} = \frac{\partial v}{\partial t} + \nabla \cdot (vv), \tag{13.8.2}$$

$$\frac{dw}{dt} = \frac{\partial w}{\partial t} + \nabla \cdot (vw), \tag{13.8.3}$$

where terms on the far right describe the three-dimensional momentum carried by the velocity v. Likewise, the material rate of change of temperature in (13.6.7) can be expressed in terms of the divergence of a heat flux

$$\frac{d\theta}{dt} = \frac{\partial \theta}{\partial t} + \nabla \cdot (v\theta). \tag{13.9}$$

Expanding as in (13.5) and averaging then obtains the equations governing the time–mean motion

$$\frac{d\overline{u}}{dt} - f\overline{v} = -\frac{1}{\rho_0}\frac{\partial \overline{p}}{\partial x} - \left(\frac{\partial \overline{u'u'}}{\partial x} + \frac{\partial \overline{u'v'}}{\partial y} + \frac{\partial \overline{u'w'}}{\partial z}\right), \tag{13.10.1}$$

$$\frac{d\overline{v}}{dt} + f\overline{u} = -\frac{1}{\rho_0}\frac{\partial \overline{p}}{\partial y} - \left(\frac{\partial \overline{u'v'}}{\partial x} + \frac{\partial \overline{v'v'}}{\partial y} + \frac{\partial \overline{v'w'}}{\partial z}\right), \tag{13.10.2}$$

$$\frac{d\overline{w}}{dt} = -\frac{1}{\rho_0}\frac{\partial \overline{p}}{\partial z} + g\frac{\overline{\theta}}{\theta_0} - \left(\frac{\partial \overline{u'w'}}{\partial x} + \frac{\partial \overline{v'w'}}{\partial y} + \frac{\partial \overline{w'w'}}{\partial z}\right), \tag{13.10.3}$$

$$\frac{\partial \overline{u}}{\partial x} + \frac{\partial \overline{v}}{\partial y} + \frac{\partial \overline{w}}{\partial z} = 0, \tag{13.10.4}$$

$$\frac{d\overline{\theta}}{dt} + \overline{w}\frac{d\theta_0}{dz} = -\left(\frac{\partial \overline{u'\theta'}}{\partial x} + \frac{\partial \overline{v'\theta'}}{\partial y} + \frac{\partial \overline{w'\theta'}}{\partial z}\right), \tag{13.10.5}$$

where

$$\frac{d}{dt} = \frac{\partial}{\partial t} + \overline{v}\cdot\nabla \tag{13.10.6}$$

is the time rate of change following the mean motion, and diffusive transfers of momentum and heat are slow enough, on scales of the mean flow, to be ignored.

Terms in (13.10.1) through (13.10.3) involving time-averaged products of fluctuating velocities represent the convergence of an eddy momentum flux (e.g., of eddy momentum carried by the eddy velocity). Mean x momentum

(13.10.1) is forced by the convergence of eddy x momentum flux and so forth. Then, according to the discussion in Sec. 10.6, fluxes of x momentum carried by the three components of fluctuating velocity

$$\tau_{xx} = -\rho_0 \overline{u'u'} \tag{13.11.1}$$

$$\tau_{xy} = -\rho_0 \overline{u'v'} \tag{13.11.2}$$

$$\tau_{xz} = -\rho_0 \overline{u'w'} \tag{13.11.3}$$

constitute stresses in the x direction exerted on the mean flow by turbulent motions. Referred to as *Reynolds stresses*, they are responsible for turbulent drag on the mean motion of an air parcel. Collectively, the Reynolds stresses define a turbulent stress tensor $\tau = -\rho_0 \overline{v'v'}$, which is a counterpart of the stress tensor in (10.22) and the divergence of which forces $\overline{v}$ according to (13.10.1) through (13.10.3). A similar interpretation applies to the mean potential temperature (13.10.5), which is forced by the convergence of eddy heat flux

$$\frac{q}{c_p} = \rho_0 \overline{v'\theta'}. \tag{13.12}$$

13.1.2   Turbulent Diffusion

Reynolds decomposition provides a framework for describing how turbulent motions interact with the mean flow. However, it offers little clue as to how the turbulent fluxes of momentum and heat forcing the mean circulation can be determined. The description for mean motion (13.10) can be closed only by resorting to empirical or *ad hoc* relationships between the eddy fluxes and mean fields.

The closure scheme adopted most widely is inspired by molecular diffusion, which smooths out gradients. Although more complex, turbulence has a similar effect on mean properties. Figure 13.2 shows the evolution of a turbulent spot that has been introduced into a uniform flow at large Re. Streaklines of smoke captured by the spot are dispersed across the turbulent region, which spreads laterally with distance downstream. Therefore, mean concentrations inside the turbulent region decrease steadily and have progressively weaker gradients. Turbulent dispersion is responsible for mean concentrations of pollutants diminishing following the breakdown of the nocturnal inversion (Sec. 7.6.3) and for the expansion of buoyant thermals (Sec. 9.3).

Turbulent dispersion acts on all conserved properties and is, at least qualitatively, analogous to molecular diffusion. The eddy momentum flux can then be expressed in terms of the gradient of mean momentum (13.6.7)

$$\tau = \rho_0 K_M \cdot \nabla \overline{v}, \tag{13.13.1}$$

Figure 13.2 Turbulent spot introduced into a laminar flow (from right to left) at $Re =$ 4.0×10^5. After Van Dyke (1982).

where the *eddy diffusivity* of momentum K_M is an analogue of the kinematic viscosity ν for molecular diffusion.[3] As far as the boundary layer is concerned, it suffices to consider mean motion that is horizontally homogeneous, in which case horizontal diffusion of momentum vanishes and K_M pertains to the vertical flux alone. The vertical eddy flux of momentum

$$\overline{w'\boldsymbol{v}'} = -K_M \frac{\partial \overline{\boldsymbol{v}}}{\partial z} \qquad (13.13.2)$$

is then proportional to the vertical gradient of mean momentum. Likewise, the vertical heat flux becomes

$$\overline{w'\theta'} = -K_H \frac{\partial \overline{\theta}}{\partial z}, \qquad (13.13.3)$$

where K_H is the eddy diffusivity of heat.

Expressions (13.13) comprise the so-called *flux–gradient relationship*. Turbulent transfer in (13.10) then assumes the form of diffusion of mean properties (e.g., $K_M \nabla^2 \overline{\boldsymbol{v}}$). Eddy diffusivity, although it can be defined as above, depends on the unknown fluctuating field properties. So, in practice, K_M and K_H must be evaluated empirically in terms of the eddy fluxes $\overline{w'u'}$ and $\overline{w'\theta'}$. Measured eddy diffusivities range between 1 and 10^2 m^2 s^{-1} inside the boundary layer. These values of K_M are orders of magnitude greater than its counterpart for molecular diffusion ($\nu = 10^{-5}$ m^2 s^{-1}), which makes turbulent diffusion the chief source of drag and mechanical heating exerted on the mean flow.

[3] For anisotropic turbulence (e.g., wherein statistical properties vary with direction) K_M is a tensor, the components of which refer to turbulent dispersion in the horizontal and vertical directions; see Andrews *et al.* (1987) for applications to two-dimensional models.

13.2   Structure of the Boundary Layer

13.2.1   The Ekman Layer

In terms of eddy diffusivity, the horizontal momentum equations inside a boundary layer in which mean properties are steady and horizontally homogeneous become

$$-f\overline{v} = -\frac{1}{\rho_0}\frac{\partial \overline{p}}{\partial x} + K\frac{\partial^2 \overline{u}}{\partial z^2}, \tag{13.14.1},$$

$$f\overline{u} = -\frac{1}{\rho_0}\frac{\partial \overline{p}}{\partial y} + K\frac{\partial^2 \overline{v}}{\partial z^2}, \tag{13.14.2}$$

where $K = K_M$ is regarded as constant and the Boussinesq approximation has been used to ignore vertical advection of mean momentum. Eliminating the pressure gradient in favor of the geostrophic velocity transforms (13.14) into

$$K\frac{\partial^2 \overline{u}}{\partial z^2} + f(\overline{v} - v_g) = 0, \tag{13.15.1}$$

$$K\frac{\partial^2 \overline{v}}{\partial z^2} - f(\overline{u} - u_g) = 0. \tag{13.15.2}$$

Multiplying (13.15.1) by $i = \sqrt{-1}$ and adding to (13.15.2) results in the single equation

$$K\frac{\partial^2 \chi}{\partial z^2} - if\chi = -if\chi_g \tag{13.16.1}$$

in terms of the consolidated variables

$$\chi = \overline{u} + i\overline{v}, \tag{13.16.2}$$

$$\chi_g = u_g + iv_g. \tag{13.16.3}$$

If v_g does not vary through the boundary layer and is oriented parallel to the x axis, (13.16) has the general solution

$$\chi(z) = A\exp[(1+i)\gamma z] + B\exp[-(1+i)\gamma z] + u_g \tag{13.17.1}$$

where

$$\gamma = \sqrt{\frac{f}{2K}}. \tag{13.17.2}$$

The no-slip condition requires the horizontal velocity to vanish at the surface, whereas the motion must asymptotically approach the geostrophic velocity above the boundary layer:

$$\overline{v} = 0 \qquad z = 0,$$
$$\overline{v} \sim u_g i \qquad z \rightarrow \infty. \tag{13.17.3}$$

Incorporating these boundary conditions and collecting real and imaginary parts then yields the component velocities

$$\overline{u}(z) = u_g \left(1 - e^{-\gamma z} \cos \gamma z\right), \tag{13.18.1}$$

$$\overline{v}(z) = v_g e^{-\gamma z} \sin \gamma z, \tag{13.18.2}$$

which define an *Ekman layer*.

The mean velocity in (13.18) describes an *Ekman spiral*, which is plotted in Fig. 13.3a as a function of height. Moving downward through the boundary layer witnesses a rotation of the mean velocity, to the left in the Northern Hemisphere and across isobars toward low pressure (Sec. 12.3), until a limiting deflection $\delta = 45°$ is reached at the ground. The deflection from isobars is an indication of turbulent drag exerted on the mean flow, which increases with mean vertical shear (Fig. 13.3b) and drives the motion out of geostrophic equilibrium. The effective depth of the Ekman layer is given by π/γ, where $\overline{v}$ reverses sign and above which $\overline{v}$ remains close to v_g (Fig. 13.3c). Values of $K = 10$ m^2 s^{-1} and $f = 10^{-4}$ s^{-1} give $\pi/\gamma \cong 1$ km. In practice, a pure Ekman spiral is seldom observed.[4] Nonetheless, it captures the salient structure of the planetary boundary layer, a deflection from surface isobars $\delta_{\mathrm{obs}} \cong 25°$ being typical.

13.2.2   The Surface Layer

Within the lowest 15% of the boundary layer, the assumption that $K \cong$ const breaks down. Instead, K, which is determined by the characteristic velocity and length scales of turbulent eddies, varies linearly with height. In this region, the no-slip condition forces eddies to have a characteristic length scale proportional to the distance z from the boundary and a characteristic velocity scale

$$u_* = \left(-\overline{u'w'}\right)^{\frac{1}{2}}, \tag{13.19}$$

which is known as the *friction velocity* because it reflects the turbulent shear stress near the ground. The flux–gradient relationship (13.13) then requires

$$u_*^2 = kzu_* \frac{\partial \overline{u}}{\partial z}, \tag{13.20}$$

---

[4]The high Re characteristic of the atmosphere makes simple Ekman flow dynamically unstable.

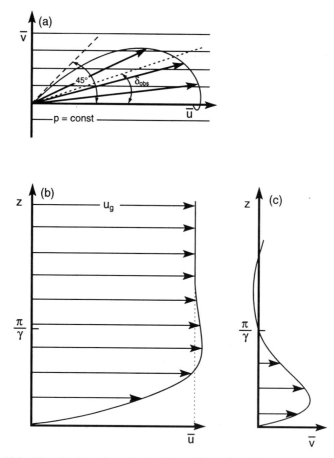

Figure 13.3 Mean horizontal motion inside an Ekman layer in which isobars are oriented parallel to the x axis, with p decreasing northward. (a) Planform view of the Ekman spiral, with $\overline{\boldsymbol{v}} = (\overline{u}, \overline{v})$ shown at different levels. Vertical profiles of (b) $\overline{u}$ and (c) $\overline{v}$ through the boundary layer.

where $\overline{\boldsymbol{v}} = \overline{u}\boldsymbol{i}$ near the surface has been presumed parallel to the x axis and $k = 0.4$ is the so-called *von Karman constant*. Integrating (13.20) gives the logarithmic law

$$\left(\frac{\overline{u}}{u_*}\right) = \frac{1}{k}\ln\left(\frac{z}{z_0}\right) \tag{13.21}$$

governing the surface layer, where z_0 is a *roughness length* that characterizes the surface. The depth of the surface layer varies in relation to its roughness through z_0, which has values smaller than 5 cm for level vegetated terrain.

The structure (13.21) is an exact counterpart of the *viscous sublayer* in turbulent flow over a flat plate (see, e.g., Schlicting, 1968), where fluxes of momentum and heat are independent of height. For neutral stability, the logarithmic behavior predicted by (13.21) is well-obeyed over a wide range of level terrain (Fig. 13.4). The same is true for stable stratification, within some distance of the surface.

13.3 Influence of Stratification

Under more general circumstances, static stability sharply modifies the character of the boundary layer by influencing the production and destruction of turbulence. Vertical shear of the mean flow leads to mechanical production of turbulent kinetic energy. Under neutral stability, that energy cascades equally into all three components of motion. Turbulent kinetic energy is also produced through buoyancy. Contrary to mechanical production, buoyancy can represent either a source or a sink of turbulent kinetic energy. Under unstable stratification, turbulent vertical velocities w' are reinforced by buoyancy, whereas they are damped out under stable stratification. In either event, buoyancy acts selectively on the vertical component of motion. Even though some of that energy cascades into the horizontal components, buoyancy makes the turbulent motion field anisotropic. In the presence of strong positive stability, the vertical component may contain very little of the turbulent kinetic energy.

The degree of anisotropy is reflected in the dimensionless *flux Richardson number*

$$R_f = \frac{\frac{g}{\theta}\overline{w'\theta'}}{\overline{w'u'}\frac{\partial \bar{u}}{\partial z}}, \tag{13.22}$$

which represents the ratio of thermal to mechanical production of turbulent kinetic energy. Unlike other dimensionless parameters, R_f is a function of position. Inside the surface layer, downward flux of positive x momentum, which exerts drag on the ground, makes the denominator of (13.22) negative. Under unstable stratification, the vertical heat flux must be positive to transfer heat from lower to upper levels and drive the stratification toward neutral stability (Sec. 7.5). Then $R_f < 0$. Large negative values of R_f, such as those that occur in the presence of weak shear, imply turbulence that is driven chiefly by buoyancy, which is termed *free convection*. If the heat flux vanishes, $R_f = 0$ and turbulence is driven solely by shear. Under stable stratification, the vertical heat flux must be negative (Problem 13.12), so $R_f > 0$. Buoyancy then opposes vertical eddy motions and damps out turbulent kinetic energy. For turbulence to be maintained, production by shear must exceed damping by buoyancy.

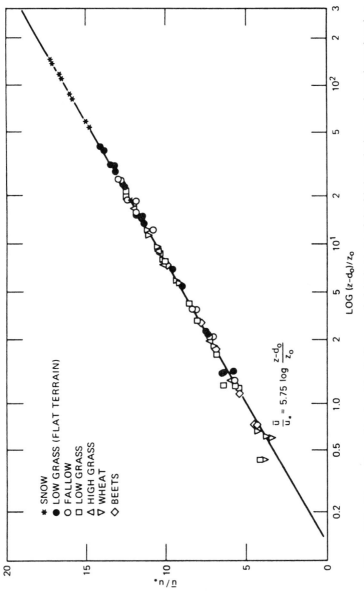

Figure 13.4 Observations of horizontal velocity versus height in terms of the friction velocity u_* and roughness length z_0 for different terrain characterized by the height d_0. After Paeschke (1937).

Analogous, but more readily evaluated, is the *gradient Richardson number*

$$Ri = \frac{\frac{g}{\theta}\frac{\partial\bar{\theta}}{\partial z}}{\left(\frac{\partial\bar{u}}{\partial z}\right)^2}$$

$$= \frac{N^2}{\left(\frac{\partial\bar{u}}{\partial z}\right)^2}. \tag{13.23.1}$$

With (13.13), the gradient Richardson number is related to the flux Richardson number as

$$R_f = \frac{K_H}{K_M}Ri. \tag{13.23.2}$$

If $K_M = K_H$, the two are equivalent. The gradient Richardson number reflects the dynamical stability of the mean flow, so it provides a criterion for the onset of turbulence. For $N^2 > 0$, Ri is positive and buoyancy stabilizes the motion by opposing vertical displacements. Disturbances having vertical motion are then damped out if the destabilizing influence of shear in the denominator is sufficiently small. Stability analysis (see, e.g., Thorpe, 1969) reveals sheared flow to be dynamically unstable if

$$Ri < \frac{1}{4}. \tag{13.24}$$

Supported by laboratory experiments (Thorpe, 1971), this criterion corresponds to shear production of turbulent kinetic energy exceeding buoyancy damping by a factor of 4. Small disturbances then amplify into fully developed turbulence.[5] Similar behavior occurs for $Ri < 0$, in which case turbulence is also driven buoyantly. For $Ri > \frac{1}{4}$, small disturbances are damped out by buoyancy, so turbulence is not favored.

Stable stratification inhibits turbulence inside the boundary layer. But increasing shear near the surface eventually makes mechanical production of turbulent kinetic energy large enough to overcome damping by buoyancy. Consequently, there inevitably exists a turbulent layer close to the ground in which $\bar{u}$ varies logarithmically and fluxes of momentum and heat are independent of height. The depth of this layer is controlled by surface roughness and by stability. Strong stability limits the layer to a few tens of meters above the ground, whereas weak stability allows a deeper layer with smaller $\partial\bar{u}/\partial z$.

Strong surface heating, such as that occuring under cloud-free conditions or over warm sea surface temperature (SST), favors buoyantly driven turbulence

---

[5] Outside the boundary layer, the criterion for the onset of turbulence can be satisfied locally, for example, in association with fractus rotor clouds (Fig. 9.22) and severe turbulence found in the mountain wave (see Fig. 14.24).

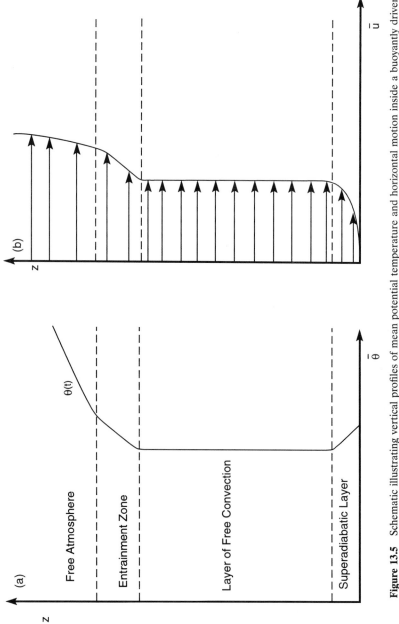

Figure 13.5 Schematic illustrating vertical profiles of mean potential temperature and horizontal motion inside a buoyantly driven boundary layer. Air parcels emerging from the superadiabatic layer near the surface ascend through the layer of free convection, which extends to the stable layer aloft and in which $\bar{\theta}$ and $\bar{u}$ are homogenized. Continued heat supply at the surface enables the convective layer to advance upward by eroding the stable layer from below.

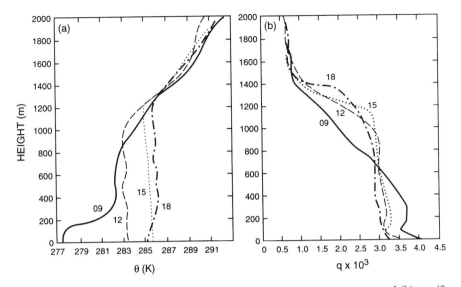

Figure 13.6 Observed profiles during the day of (a) potential temperature and (b) specific humidity. Adapted from Sun and Chang (1986).

by destabilizing the stratification. These conditions lead to the formation of a superadiabatic layer immediately above the surface (Fig. 13.5a). Buoyant parcels from the superadiabatic layer enter a deeper layer of free convection that is characterized by nearly uniform stability ($\bar{\theta}$ = const) and strong overturning. Vertical mixing of momentum makes the profile of mean motion likewise uniform in this region (Fig. 13.5b).[6] The height of convective plumes is eventually limited by stable air overhead, which interacts with convection through entrainment and subsequent mixing (Sec. 7.4.1). Continued heat supply at the surface enables the convective layer to advance upward by eroding the stable layer from below.

The foregoing process is responsible for the diurnal cycle of the boundary layer (Sec. 7.6.3) and is revealed by vertical soundings (Fig. 13.6). In early morning, increasing θ below 300 m (Fig. 13.6a) marks the nocturnal inversion, which has been neutralized in a shallow layer near the surface. Specific humidity (Fig. 13.6b) decreases sharply with height. By mid-day, surface convection has broken through the inversion and thoroughly mixed potential temperature and moisture across the lowest kilometer. Mean motion (not shown) is likewise independent of height inside this layer. Later in the day, the convective layer advances a few hundred meters higher, until, by early evening, the collapse of surface heating begins to reestablish the nocturnal inversion.

---

[6] Since the flux–gradient relationship breaks down in this well-mixed layer, turbulent drag can be represented as Rayleigh friction (Sec. 12.3), the coefficient of which is then determined by $\bar{v}$; see Holton (1992).

13.4   Ekman Pumping

Turbulent drag inside the boundary layer drives the large-scale motion out of geostrophic equilibrium and introduces a divergent component (Sec. 12.3). To satisfy continuity, that divergence must be compensated by vertical motion, which imposes a secondary circulation on the $O(Ro^0)$ geostrophic flow in the free atmosphere. Frictional convergence (divergence) inside the boundary layer leads to rising (sinking) motion over surface low (high) pressure (Fig. 12.7).

Vertical motion atop the boundary layer is related to the circulation in the free atmosphere through the boundary condition (13.17.3). By continuity, it must also be related to the convergence of horizontal mass flux inside the Ekman layer, which follows from the cross-isobaric component $\bar{v}$. Integrating vertically over the boundary layer obtains the column-integrated mass flux across isobars

$$M = \rho_0(0) \int_0^\infty \bar{v}\,dz$$

$$= \rho_0(0) \int_0^\infty u_g e^{-\gamma z} \sin \gamma z\,dz, \qquad (13.25.1)$$

which reduces to

$$M = \frac{\rho_0(0)}{2\gamma} u_g. \qquad (13.25.2)$$

Since $\bar{v}$ is directed toward low pressure, the column-integrated mass flux across isobars can be expressed

$$M = \frac{\rho_0(0)}{2\gamma} k \times v_g. \qquad (13.26)$$

By integrating the continuity equation (13.10.4) in a similar fashion, we obtain

$$\nabla_z \cdot M + \rho_0(0)w_0 = 0, \qquad (13.27)$$

where w_0 denotes the vertical velocity or *Ekman pumping* atop the boundary layer. Incorporating (13.26) then yields

$$w_0 = \frac{1}{2\gamma} \zeta_g(0), \qquad (13.28.1)$$

where

$$\zeta_g(0) = \nabla_z^2 \psi(0) \qquad (13.28.2)$$

is the geostrophic vorticity at the base of the free atmosphere. According to (13.28), Ekman pumping is directly proportional to the vorticity above the boundary layer. For values used previously to evaluate γ and for $\zeta_g = 10^{-5}\,\text{s}^{-1}$ typical of quasi-geostrophic motion, (13.28) gives w_0 of order 5 mm s^{-1}.

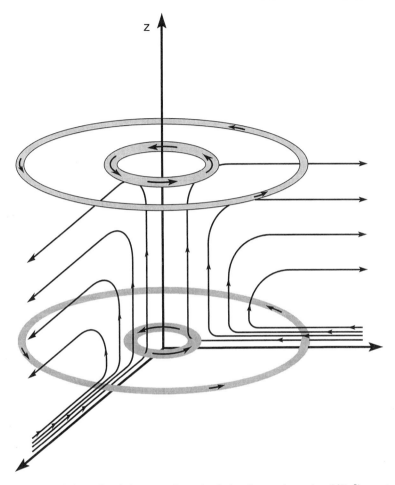

Figure 13.7 Schematic of the secondary circulation imposed on the $O(Ro^0)$ geostrophic motion above the boundary layer. Frictional convergence in the boundary layer beneath a cyclone is compensated by ascending motion and divergence in the free atmosphere. A material ring diverging from the center of the cyclone must then spin down to conserve angular momentum, which decelerates organized motion above the boundary layer.

Frictional convergence and divergence inside the boundary layer act to equalize pressure variations and thus destroy gradients supporting large-scale motion. They also have an indirect but more important influence. Upward motion out of the boundary layer must be compensated by horizontal divergence in the free atmosphere (Fig. 13.7). From the vorticity budget (12.36), that divergence acts to reduce the absolute vorticity of air in the free atmosphere. A material ring spins down as it expands radially to conserve angular momentum, which weakens organized motion about a cyclone.

Under barotropic stratification, the horizontal velocity in the free atmosphere is independent of height. Integrating the continuity equation (13.10.4) from the top of the boundary layer to a height h where w has been driven to zero then yields

$$w_0 = h\nabla_z \cdot v$$

$$= h\nabla_z \cdot v_a, \tag{13.29}$$

where v_a is the ageostrophic velocity associated with the secondary circulation. The vorticity equation for quasi-geostrophic motion (12.47) restricted to an f plane reduces to

$$\frac{d_g \zeta_g}{dt} = -f_0 \nabla_z \cdot v_a. \tag{13.30}$$

With (13.29) and (13.28), this becomes

$$\frac{d_g \zeta_g}{dt} = -\frac{f_0}{2\gamma h} \zeta_g$$

or

$$\frac{1}{\zeta_g}\frac{d_g \zeta_g}{dt} = -\left(\frac{f_0 K}{2h^2}\right)^{\frac{1}{2}}. \tag{13.31}$$

A vorticity anomaly like that accompanying a cyclone then decays exponentially as

$$\zeta_g = \exp\left(-\frac{t}{\tau_E}\right), \tag{13.32.1}$$

where

$$\tau_E = \left(\frac{f_0 K}{2h^2}\right)^{-\frac{1}{2}} \tag{13.32.2}$$

is an Ekman e-folding time. Hence, the secondary circulation imposed by Ekman pumping causes the primary circulation to spin down. This process involves no actual dissipation, but rather follows from a redistribution of angular momentum when absolute vorticity is diluted over a wider region. By Stokes' theorem, the circulation along an expanding material contour $s(x, y)$

$$\Gamma = \oint v_g \cdot ds$$

$$= \int_A \zeta_g dA \tag{13.33}$$

remains unchanged. (Should a disturbance's vertical structure make $v_g = 0$ at the surface, Ekman pumping and divergence in the free atmosphere vanish, eliminating spin down.) In practice, the depth h over which spin down occurs is shaped by static stability in the free atmosphere. For values representative of the troposphere, (13.32) gives an e-folding time of order days, which is an order of magnitude shorter than the timescale for direct frictional dissipation in the free atmosphere.

Suggested Reading

A First Course in Turbulence (1972) by Tennekes and Lumley presents an introduction to the subject at the graduate level and includes a formal treatment of turbulent energetics. *Statistical Fluid Mechanics* (1973) by Monin and Yaglom is an advanced treatment that includes a wide range of observations and a Lagrangian description of turbulent dispersion.

An Introduction to Boundary Layer Meteorology (1988) by Stull is a nice development at the graduate level. An extensive collection of recent observations is included in *The Structure of the Atmospheric Boundary Layer* (1989) by Sorbjan.

Hydrodynamic Instability (1985) by Drazin and Reid contains a formal treatment of shear instability at the graduate level.

Problems

13.1. Calculate the Reynolds number for the following flows: (a) Stokes flow about a spherical cloud droplet (Problem 9.15) of radius 10 μm and fall speed of 3 mm s^{-1}, (b) cumulus convection of characteristic dimension 1 km and velocity 1 m s^{-1}, and (c) a midlatitude cyclone of characteristic dimension 10^3 km and velocity 5 m s^{-1}.

13.2. Express the Reynolds number as a ratio of timescales and then use them to interpret the criterion for the onset of turbulence.

13.3. Estimate the height of the turbopause based on characteristic length and velocity scales of 1 km and 10 m s^{-1}, respectively.

13.4. In terms of temperature, water vapor mixing ratio, and ozone mixing ratio, contrast the state of an air parcel before and after encountering (a) the boundary layer and (b) cumulus convection.

13.5. Frictional dissipation is an essential feature of turbulence because it cascades energy from large scales of organized motion to small scales, on which molecular diffusion operates efficiently. The specific energy dissipation rate for homogeneous isotropic turbulence is described by

$$\epsilon \cong \frac{U^3}{L},$$

where U and L are the characteristic velocity and length scales of the large-scale motion that drives the cascade. Hence, the rate of energy dissipation, which occurs almost entirely at small scales, is dictated by the motion at large scales, which is nearly inviscid. Use this result to estimate the rate at which energy is dissipated inside a cumulonimbus cloud of characteristic dimension 10 km and velocity 10 m s^{-1}. Compare this value with the 1000 MW generated by a large urban power plant.

13.6. For the cumulonimbus cell in Problem 13.5, estimate the precipitation rate necessary to sustain the motion.

13.7. Derive the approximate vertical momentum balance (13.4).

13.8. Establish the flux form of the momentum equations (13.8).

13.9. Contrast the signs of the momentum flux in (13.11) and the heat flux in (13.12) in light of their appearance in the primitive equations.

13.10. Verify that (13.18) is the solution of the boundary-layer equations, subject to appropriate boundary conditions.

13.11. Recover (13.20) governing mean motion inside the surface layer.

13.12. Demonstrate that the turbulent vertical heat flux must be positive (negative) under unstable (stable) stratification.

13.13. Show that divergence associated with the secondary circulation above the boundary layer does not alter the circulation $\Gamma = \oint \boldsymbol{v} \cdot d\boldsymbol{s}$ along a material contour.

13.14. Derive expression (13.27) for vertically integrated conservation of mass.

13.15. A midlatitude cyclone is associated with the disturbance height in the lower troposphere

$$z' = -Z \exp\left(-\frac{x^2 + y^2}{2L^2}\right),$$

where $Z = 200$ m, $L = 1000$ km, y is measured from $45°$, and the boundary layer has an eddy diffusivity of 10 m^{2}s^{-1}. Calculate the maximum vertical velocity atop the boundary layer.

13.16. Observations reveal a nearly logarithmic variation with height of mean wind below 100 m above prairie grass land, which has a roughness length of 4 cm. If the mean wind is 5 m s^{-1} at the top of this layer, calculate the stress exerted on the ground.

13.17. Inside a *well-mixed boundary layer* of depth h, convection makes mean horizontal motion and potential temperature invariant with height (Fig. 13.5), rendering the flux–gradient relationship inapplicable. The vertical momentum flux then decreases linearly with height from a surface maximum described by

$$\overline{w'v'}\big|_s = -c_d|\overline{v}|\overline{v},$$

where c_d is a dimensionless *drag coefficient*. (a) Construct a counterpart of the Ekman equations (13.15) governing mean motion inside the well-mixed layer to show that the effect of turbulence is then equivalent to Rayleigh friction (Sec. 12.3). (b) Determine $\overline{v}$ if $v_g = u_g i$ above the well-mixed boundary layer. (c) Calculate the timescale of Rayleigh friction for $h = 1$ km, $c_d = 5.0 \times 10^{-3}$, and a mean wind speed of 5 m s^{-1}.

13.18. Estimate how strong vertical shear must become before turbulence forms under stability representative of (a) mean conditions in the troposphere: $N^2 = 1.0 \times 10^{-4} \text{s}^{-2}$ and (b) mean conditions in the stratosphere: $N^2 = 6.0 \times 10^{-4} \text{s}^{-2}$. (c) How would these values be affected if, locally, isentropic surfaces are steeply deflected from the horizontal?

13.19. Calculate the characteristic spin-down time due to Ekman pumping for a cyclone at $45°$, vertical motion extending to the tropopause at 10 km, and an eddy diffusivity inside the boundary layer of $10 \text{ m}^2\text{s}^{-2}$. Compare this spin-down time with the timescale of radiative damping.

13.20. The contrail formed from the exhaust of a commercial jet flying at 200 mb contains liquid water. At a distance x from the jet, the contrail has a temperature $T(x)$ and a mixing ratio for total water $r_t(x)$. The contrail dissolves some distance x_d downstream of the jet through entrainment with ambient air, of temperature T_a and mixing ratio r_a. Immediately aft of the jet, the exhaust plume has temperature $T(0) = T_0$ and total water mixing ratio $r_t(0) = r_0$. (a) Formulate a differential equation governing $r_t(x)$, in terms of the entrainment length $L^{-1} = d(\ln m)/dx$, which characterizes the rate at which the contrail expands through mixing with ambient air (Sec. 7.4.2). (b) Derive expressions for $r_t(x)$ and $T(x)$. (c) Derive an expression for the saturation mixing ratio of the contrail $r_w(x)$. (d) Determine the distance x_d at which the contrail dissolves for $L = 100$ m, $r_0 = 4 \text{ g kg}^{-1}$, $r_a = 0.05 \text{ g kg}^{-1}$, $T_0 = -20°$ C, and $T_a = -50°$ C. (e) As in part (d), but under the warm moist conditions: $T_a = -30°$ C, $r_a = 1.25 \text{ g kg}^{-1}$, $T_0 = 0°$ C, symbolic of conditions inside the warm sector of an advancing cyclone.

Chapter 14 | Atmospheric Waves

The governing equations support several forms of wave motion. Atmospheric waves are excited when air is disturbed from equilibrium (e.g., mechanically when air is displaced over elevated terrain or thermally when air is heated inside convection). By transferring momentum, wave motions communicate the influence of one region to another. This mechanism enables the troposphere to drive the stratosphere out of radiative equilibrium (Fig. 8.27) and tropical convection to influence the extratropical circulation.

14.1 Description of Wave Propagation

Wave motions are possible in the presence of a positive restoring force, which, by opposing disturbances from equilibrium, supports local oscillations in field properties. Buoyancy provides such a restoring force under stable stratification (Sec. 7.3). The compressibility of air provides another. Variations with latitude of the Coriolis force provide yet another restoring force, which will be seen to support large-scale atmospheric disturbances.

14.1.1 Surface Water Waves

The description of wave motion is illustrated with an example under nonrotating conditions that will serve as a model of buoyancy oscillations in the atmosphere. Consider disturbances to a layer of incompressible fluid of uniform density ρ and depth H (Fig. 14.1). The layer is bounded below by a rigid surface and above by a *free surface*, namely, one that adjusts position to relieve any stress. Motions inside this layer are governed by

$$\rho \frac{d\mathbf{v}}{dt} = -\nabla p - g\mathbf{k} \tag{14.1.1}$$

$$\nabla \cdot \mathbf{v} = 0, \tag{14.1.2}$$

with $\mathbf{v}$ and ∇ referring to 3-dimensional vector quantities.

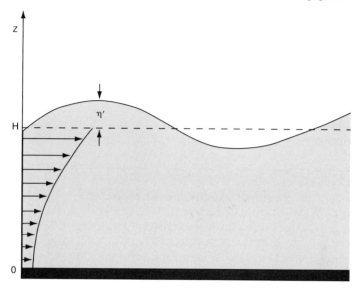

Figure 14.1 A layer of incompressible fluid of uniform density ρ and depth H. The layer is bounded below by a rigid surface and above by a free surface that has displacement η' from its mean elevation. The profile of horizontal motion for surface water waves is indicated.

We are interested in how the fluid layer responds to disturbances that are imposed through a deflection of its free surface. It is convenient to decompose the behavior into a zonal–mean basic state, which reflects undisturbed conditions and is denoted by an overbar, and perturbations from it, which are denoted by a prime, for example,

$$v = \bar{v} + v' \tag{14.2.1}$$

with

$$\overline{v'} = 0, \tag{14.2.2}$$

analogous to the treatment of turbulence (Chapter 13). For simplicity, the undisturbed zonal motion $\bar{v} = \bar{u}i$ is taken to be independent of x, analogous to motion in Fig. 1.8.

Incorporating (14.2) into equations (14.1) produces terms involving mean properties, perturbation properties, and products thereof. In the spirit of the asymptotic analysis for small Rossby number (Chapter 12), the resulting system can be solved recursively to successively higher order in wave amplitude ϵ, which symbolizes small departures from equilibrium [e.g., u', v', $p' = O(\epsilon)$]. To zero order in ϵ (namely, in the absence of waves), the governing equations reduce to a statement of hydrostatic equilibrium

$$\frac{\partial \bar{p}}{\partial z} = -\rho g \tag{14.3}$$

that must be satisfied by the mean state. To first order in ϵ and with the basic state balance (14.3) eliminated, the governing equations reduce to

$$\frac{D\boldsymbol{v}'}{Dt} = -\nabla p' \tag{14.4.1}$$

$$\nabla \cdot \boldsymbol{v}' = 0, \tag{14.4.2}$$

where

$$\frac{D}{Dt} = \frac{\partial}{\partial t} + \bar{u}\frac{\partial}{\partial x} \tag{14.4.3}$$

defines the material derivative following the mean motion. Linear in perturbation quantities, equations (14.4) govern the wave field. However, because they neglect products of perturbation quantities resulting from the advective terms $\boldsymbol{v}' \cdot \nabla \boldsymbol{v}'$, the first-order equations ignore influences the waves exert on the mean state. Such feedback is reflected in the equations accurate to second order in ϵ, which, upon eliminating the zero- and first-order balances and averaging zonally, yields

$$\frac{\partial \bar{u}}{\partial t} = -\overline{(\boldsymbol{v}' \cdot \nabla \boldsymbol{v}')}$$

$$= -\frac{\partial \overline{(v'u')}}{\partial y} - \frac{\partial \overline{(w'u')}}{\partial z}. \tag{14.5}$$

Quadratic terms appearing on the right-hand side of (14.5) represent the convergence of an eddy momentum flux that forces the zonal-mean motion—just like the turbulent eddy fluxes forcing mean motion in (13.10). Corresponding changes in $\bar{u}$ then reflect second-order corrections to the zeroth-order steady motion.

In the limit of small wave amplitude, to which the development is now restricted, second-order terms in (14.5) are small, so the basic state may be regarded as steady. Motion in the fluid layer is then described by the equations accurate to first order in wave amplitude, which are said to have been *linearized*. If the basic state is in uniform motion, so that $\bar{u}$ is independent of position, applying $\nabla\cdot$ to (14.4.1) and incorporating (14.4.2) leads to the consolidated equation for the perturbation pressure

$$\nabla^2 p' = 0. \tag{14.6}$$

With fixed boundaries, Laplace's equation cannot describe wave propagation because it is diagnostic. However, boundary conditions governing the free surface introduce time dependence to make the complete system prognostic.

The free surface, which is an interface between fluids. must satisfy two conditions. The *dynamic condition* requires the stress and hence total pressure to vanish. With hydrostatic equilibrium for the mean state (14.3), the total pressure on the free surface is

$$p = -\rho g \eta' + p' \qquad z = H + \eta', \tag{14.7.1}$$

where η' is the deflection of the free surface from its undisturbed elevation. Then the perturbation pressure at the top of the layer is given by

$$p' = \rho g \eta' \qquad z = H, \qquad (14.7.2)$$

which may be evaluated at the mean elevation of the free surface to the same degree of approximation inherent in (14.6). The *kinematic condition* requires the free surface to be a material surface, so its vertical displacement must satisfy

$$\frac{d\eta'}{dt} = w \qquad z = H + \eta'. \qquad (14.8.1)$$

In the limit of small amplitude, this reduces to

$$\frac{D\eta'}{Dt} = w' \qquad z = H. \qquad (14.8.2)$$

Combining the dynamic condition (14.7.2) and the kinematic condition (14.8.2) obtains

$$\frac{1}{\rho g} \frac{Dp'}{Dt} = w' \qquad z = H, \qquad (14.9.1)$$

which, when incorporated into the vertical momentum equation, gives the upper boundary condition

$$\frac{1}{g} \frac{D^2 p'}{Dt^2} = -\frac{\partial p'}{\partial z} \qquad z = H. \qquad (14.9.2)$$

Another boundary condition applies at the base of the layer, where vertical motion must vanish identically. Then the vertical momentum equation requires

$$\frac{\partial p'}{\partial z} = 0 \qquad z = 0. \qquad (14.10)$$

14.1.2 Fourier Synthesis

Consider a small disturbance to the free surface. The system is positively stratified because the density of the liquid is much greater than that of the overlying air, which is ignored. Buoyancy then provides a positive restoring force, which drives fluid back toward its undisturbed position and sets up oscillations. As a result, a compact initial disturbance radiates away in the form of waves.

Wave activity radiating away from the initial disturbance may assume a complex form (e.g., because the structure of the disturbance imposes that form initially or because individual components of the wave field propagate differently). Equations governing the wave field are linear in perturbation quantities and have coefficients that are independent of position and time.

These features allow the wave field to be constructed from a superposition of plane waves, which is expressed in the Fourier integral

$$p'(x, t) = \frac{1}{(2\pi)^3} \int \int \int_{-\infty}^{\infty} P_{kl}^{\sigma}(z)e^{i(kx+ly-\sigma t)} \, dk \, dl \, d\sigma$$

$$= \frac{1}{(2\pi)^3} \int \int \int_{-\infty}^{\infty} P_k^{\sigma}(z)e^{i(k\cdot x-\sigma t)} \, dk \, d\sigma, \qquad (14.11.1)$$

wherein the z dependence must satisfy boundary conditions. Equation (14.11.1) describes a spectrum of monochromatic plane waves, individual components of which have angular frequency σ, wavenumber vector $k = (k, l)$, and complex amplitudes $P_k^{\sigma}(z)$ that contain the magnitude and phase of oscillations. An individual component of the wave spectrum (Fig. 14.2) has lines of constant phase: $kx + ly - \sigma t = $ const, which propagate

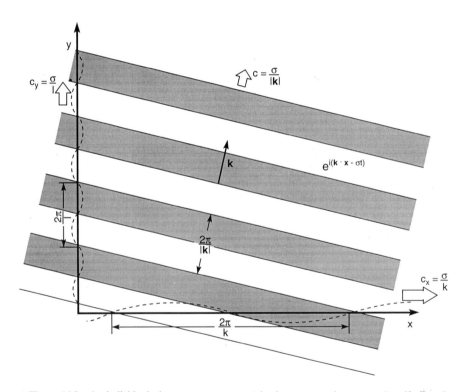

Figure 14.2   An individual plane wave component having wavenumber vector $k = (k, l)$ and frequency σ. Lines of constant phase: $kx + ly - \sigma t = $ const, propagate parallel to k ($\sigma > 0$) or antiparallel to k ($\sigma < 0$), with *phase speed* $c = \sigma/|k|$. Phase propagation in the x and y directions is described by the *trace speeds* $c_x = \sigma/k$ and $c_y = \sigma/l$, which equal or exceed c and are related to it through $1/c_x^2 + 1/c_y^2 = 1/c^2$, so they do not represent its components. As k becomes orthogonal to a direction, the trace speed in that direction becomes infinite.

in the direction of k with the *phase speed*

$$c = \frac{\sigma}{|k|}, \tag{14.11.2}$$

where $|k|^2 = k^2 + l^2$. Phase propagation in the x and y directions is described by the *trace speeds*

$$c_x = \frac{\sigma}{k},$$
$$c_y = \frac{\sigma}{l}. \tag{14.11.3}$$

Note that $c_x^2 + c_y^2 \neq c^2$. According to Fig. 14.2, the trace speed in a given direction equals or exceeds the phase speed and approaches infinity as k becomes orthogonal to that direction.

By the Fourier integral theorem (see, e.g., Rektorys, 1969), the complex amplitudes in (14.11.1) follow as

$$P_{kl}^{\sigma}(z) = \int \int \int_{-\infty}^{\infty} p'(x, y, z, t) e^{-i(kx+ly-\sigma t)} \, dx dy dt, \tag{14.12.1}$$

which defines the *Fourier transform* of the variable p' at level z. Equation (14.12.1) describes a spectrum of complex wave amplitude over the space–time scales k, l, and σ. Real p' requires the amplitude spectrum to satisfy the Hermitian property

$$P_{-k-l}^{-\sigma} = P_{kl}^{\sigma *}. \tag{14.12.2}$$

Because (14.11.1) and (14.12.1) constitute a one-to-one mapping, $p'(x, t)$ in physical space and its counterpart in Fourier space, $P_{kl}^{\sigma}(z)$, are equivalent representations of the wave field. The *power spectrum*, $|P_{kl}^{\sigma}|^2$, then describes the distribution over wavenumber and frequency of variance $|p'|^2$, which reflects the energy in the wave field. In general, a disturbance excites a continuum of space and timescales, so wave variance follows as an integral over wavenumber and frequency:

$$\int \int \int_{-\infty}^{\infty} |p'(x, t)|^2 dx dt = \frac{1}{(2\pi)^3} \int \int \int_{-\infty}^{\infty} |P_k^{\sigma}|^2 dk d\sigma, \tag{14.13}$$

which is known as *Parseval's theorem*.

Differentiating p' in physical space is equivalent to multiplying its transform P_{kl}^{σ} by the corresponding space and time scales:

$$\frac{\partial p'}{\partial x} \leftrightarrow ik P_{kl}^{\sigma},$$

$$\frac{\partial p'}{\partial y} \leftrightarrow il P_{kl}^{\sigma}, \tag{14.14}$$

$$\frac{\partial p'}{\partial t} \leftrightarrow -i\sigma P_{kl}^{\sigma},$$

the demonstration of which is left as an exercise. Then applying (14.12.1) transforms the material derivative into

$$\frac{Dp'}{Dt} \rightarrow -i\omega P_{kl}^\sigma,$$ (14.15.1)

where

$$\omega = \sigma - k\overline{u}$$ (14.15.2)

is the *intrinsic frequency*—that relative to the medium, for the component with frequency σ and zonal wavenumber k. The term $-k\overline{u}$ represents a Doppler shift of the intrinsic frequency by background motion. A component propagating opposite to $\overline{u}$ is Doppler-shifted to higher intrinsic frequency: $\omega > \sigma$, whereas one propagating in the same direction as $\overline{u}$ is Doppler-shifted to lower intrinsic frequency: $\omega < \sigma$.

Applying (14.12.1) transforms the governing equations in physical space: (14.6), (14.9.2), and (14.10), into their counterparts in Fourier space:

$$\frac{\partial^2 P_{kl}^\sigma}{\partial z^2} - |k|^2 P_{kl}^\sigma = 0,$$ (14.16.1)

$$\frac{\partial P_{kl}^\sigma}{\partial z} = 0 \qquad z = 0,$$ (14.16.2)

$$\frac{\partial P_{kl}^\sigma}{\partial z} - \frac{\omega^2}{g} P_{kl}^\sigma = 0 \qquad z = H.$$ (14.16.3)

Space–time dependence involving derivatives with respect to x, y, and t has been reduced to algebraic dependence on the scales k, l, and σ for individual modes, which can be resynthesized via (14.11.1). The general solution of (14.16.1) satisfying the lower boundary condition (14.16.2) is of the form

$$P_{kl}^\sigma(z) = A \cosh(|k|z)$$ (14.17)

(Fig. 14.1). Substituting into the upper boundary condition (14.16.3) then yields the algebraic identity

$$\omega^2 = g|k| \tanh(|k|H).$$ (14.18)

Referred to as the *dispersion relation*, (14.18) must be satisfied by each component (k,l,σ) if the wave spectrum is to obey the governing equations. The dispersion relation expresses one of the space–time scales (e.g., frequency) in terms of the others. Therefore, the spatial scales of an individual wave component determine its frequency.

14.1.3 Limiting Behavior

According to (14.18), different horizontal scales propagate relative to the medium with different intrinsic phase speeds

$$\hat{c} = \frac{\omega}{|k|}$$

$$= \left[gH \frac{\tanh(|k|H)}{|k|H} \right]^{\frac{1}{2}},$$ (14.19)

which are plotted as functions of $|k|H$ in Fig. 14.3. Phase speed decreases monotonically with wavenumber, so the longest waves travel fastest.

In the limit of long wavelength,

$$\hat{c} \sim \sqrt{gH} \qquad |k|H \to 0. \tag{14.20}$$

The dispersion relation (14.20) describes *shallow water waves*, which have wavelengths $\lambda = 2\pi/|k|$ long compared to the depth H of the fluid. Phase speed is then independent of wavelength, so different components of a shallow water disturbance propagate in unison. An ocean 5 km deep has $\hat{c} > 200$ m s^{-1} (comparable to the speed of sound), which enables large-scale components of a tsunami to traverse an ocean in only hours.

In the longwave limit, vertical structure (14.17) is independent of z (Fig. 14.4a). Then, by continuity (14.4.2) and the lower boundary condition, shallow water waves possess no vertical motion. Under these circumstances,

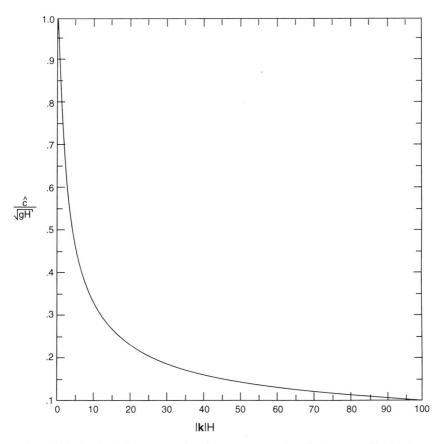

Figure 14.3 Intrinsic phase speed of surface water waves (equals phase speed in the absence of mean motion), as a function of horizontal wavenumber.

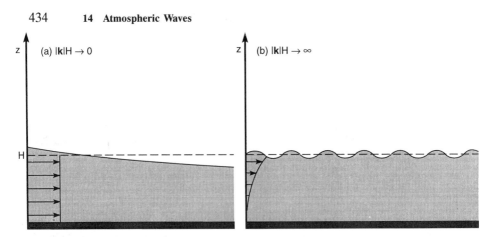

Figure 14.4 Surface water waves (a) in the longwave limit, where they assume the form of *shallow water waves*, with horizontal wavelengths long compared to the fluid depth and horizontal motion that is invariant with elevation, and (b) in the shortwave limit, where they assume the form of *deep water waves*, with horizontal wavelengths short compared to the fluid depth and motion that decays exponentially away from the free surface as an edge wave.

the vertical momentum equation reduces to a statement of hydrostatic equilibrium. Hence, the longwave limit is equivalent to invoking hydrostatic balance, which follows from the shallowness of vertical displacements relative to horizontal displacements.

In the limit of short wavelength,

$$\hat{c} \sim \sqrt{\frac{g}{|k|}} \qquad |k|H \to \infty. \tag{14.21}$$

The dispersion relation (14.21) describes *deep water waves*, which have wavelengths that are short compared to the depth of the fluid. In this limit, vertical structure (14.17) decreases exponentially away from the free surface in the form of an *edge wave* (Fig. 14.4b). Deep water waves do not feel the lower boundary because their energy is negligible there. Unlike shallow water waves, they have phase speeds that vary with wavelength, so longer components of a deep water disturbance leave behind shorter ones.

14.1.4   Wave Dispersion

Because a spectrum of waves is involved, interference among components can modify wave activity as it radiates away from the imposed disturbance. To illustrate such behavior, consider a *group* of plane waves defined by an incremental band of wavenumber centered at k, with l fixed (Fig. 14.5a). Wave activity in this band can be approximated by the rectangle shown, in which case variance is shared equally by two discrete components with wavenumbers $k \pm dk$. Those components have frequencies $\sigma \pm d\sigma$ which are related to

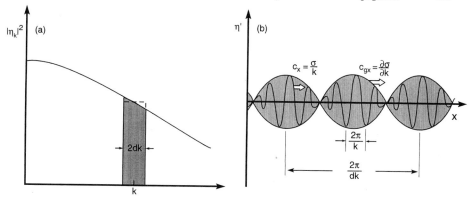

Figure 14.5 Power spectrum of surface water waves as a function of zonal wavenumber. A *wave group* is defined by variance in the incremental band centered at k (shaded), which may be approximated by the rectangle shown. Wave activity is then carried by two discrete components at $k \pm dk$. (b) Wave activity in the band produces a disturbance of wavenumber k that is modulated by an envelope of wavenumber dk. Oscillations propagate with the trace speed $c_x = \sigma/k$ but their envelope propagates with the *group speed* $c_{gx} = \partial\sigma/\partial k$.

wavenumber through the dispersion relation (14.18). For unit amplitude, wave activity in the band is described by

$$\eta'_k(x, t) = \frac{1}{2}\{\cos[(k - dk)x - (\sigma - d\sigma)t] + \cos[(k + dk)x - (\sigma + d\sigma)t]\}$$
$$= \cos(dk \cdot x - d\sigma \cdot t) \cdot \cos(kx - \omega t), \qquad (14.22.1)$$

which follows from (14.11.1) with the amplitude spectrum approximated as above and with the Hermitian property (14.12.2). Wave activity in the band can then be expressed

$$\eta'_k(x, t) = \cos\left[dk\left(x - \frac{\partial\sigma}{\partial k}t\right)\right] \cdot \cos(kx - \sigma t). \qquad (14.22.2)$$

As shown in Fig. 14.5b, (14.22) describes a disturbance with wavenumber k and trace speed $c_x = \sigma/k$ that is modulated by an envelope of wavenumber dk, which propagates in the x direction with the *group speed*

$$c_{gx} = \frac{\partial\sigma}{\partial k}$$
$$= \frac{\partial\omega}{\partial k} + \overline{u}. \qquad (14.23.1)$$

Since it is concentrated inside the envelope, wave activity in this band propagates, not with the phase speed, but with the group speed. A group defined by a continuous Gaussian window (instead of the two discrete components used to approximate the rectangular window) produces a Gaussian envelope, outside of which oscillations vanish (Problem 14.8). Clearly, wave activity must then move with the envelope—irrespective of the movement of individual crests and troughs inside it. A similar analysis applied to wavenumber l with k fixed yields the group speed in the y direction:

$$c_{gy} = \frac{\partial \sigma}{\partial l}$$

$$= \frac{\partial \omega}{\partial l}. \qquad (14.23.2)$$

Then wave activity is propagated in the x–y plane with the *group velocity*

$$c_g = \left(\frac{\partial \sigma}{\partial k}, \frac{\partial \sigma}{\partial l} \right)$$

$$= \left(\frac{\partial \omega}{\partial k} + \bar{u}, \frac{\partial \omega}{\partial l} \right). \qquad (14.23.3)$$

A function of k and l, c_g describes the propagation of wave activity in individual bands of the spectrum (14.11.1). According to (14.23.3), background motion advects wave activity with the basic flow. The intrinsic group velocity

$$\hat{c}_g = c_g - \bar{u}i \qquad (14.23.4)$$

gives the same information relative to the medium and is recognized as just the gradient of ω with respect to $k = (k, l)$.

Let us return now to the limiting forms of surface water waves described previously and in the absence of background motion: $\bar{u} = 0$. In the longwave limit, phase speed (14.20) is independent of k. Since individual wave components all propagate with the same speed, a disturbance comprised of shallow water waves retains its initial shape. Such waves are said to be *nondispersive* because a disturbance that is initially compact remains so. The group speed of shallow water waves follows from (14.23) as

$$c_g = \sqrt{gH}$$

$$= c. \qquad (14.24)$$

Because all of its components propagate with identical phase speed, a group of shallow water waves centered at wavenumber k propagates in the same direction and with the same speed.

An example is presented in Figs. 14.6 and 14.7. The free surface is initialized with a compact disturbance (Fig. 14.6a)

$$\eta'(x, y, 0) = \frac{1}{\sqrt{2\pi}L} e^{-\frac{x^2+y^2}{2L^2}}, \qquad (14.25.1)$$

with characteristic horizontal scale L. Because it is not simple harmonic, that disturbance excites a continuum of waves. Transforming (14.25.1) via (14.12.1) with $t = 0$ gives its power spectrum

$$|\eta_{kl}|^2 = e^{-L^2(k^2+l^2)} \qquad (14.25.2)$$

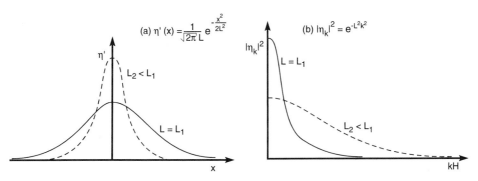

Figure 14.6   (a) A Gaussian disturbance imposed initially to the free surface. The length L characterizes the width of the disturbance. (b) The power spectrum of the disturbance, as a function of zonal wavenumber k. As L increases, variance becomes concentrated at small k in a *red* spectrum, which excites waves in the longwave limit. As L decreases, variance becomes distributed widely over k, approaching a *white* spectrum with most of the variance at large k, which excites waves in the shortwave limit.

(Fig. 14.6b), the demonstration of which is left as an exercise. The excited wave field is then described by (14.11.1), with individual components subject to the dispersion relation (14.18).

If the scale of the imposed disturbance is sufficiently long (e.g., $L >> H$), the power spectrum of η' (Fig. 14.6b) is very *red*: Variance is concentrated at small $|k|H$, so the excited wave field is comprised of shallow water waves. Far from the impulse, where waves can be treated as planar, the disturbance (Fig. 14.7a) then translates outward jointly with its envelope at a uniform speed $c = c_g$ and retains its initial form. Therefore, wave activity (shaded) remains confined to the same range of x as initially.

In the shortwave limit, the phase speed (14.21) depends on $|k|$. Individual components (k, l) then propagate with different phase speeds, so a disturbance comprised of deep water waves cannot retain its initial shape. The difference in phase speed between components with small and large $|k|$ leads to a group unraveling with time. Such waves are said to be *dispersive* because a disturbance that is initially compact is eventually smeared over a wider domain. The group speed for deep water waves follows from (14.21) as

$$c_g = \frac{1}{2}\sqrt{\frac{g}{|k|}}$$
$$= \frac{c}{2}. \tag{14.26}$$

Thus, a group of deep water waves propagates at exactly half their median phase speed, both c and c_g being functions of $|k|$.

The behavior predicted by (14.26) is readily observed with the toss of a pebble into a pond and is recovered from (14.25) if the horizontal scale of the disturbance is sufficiently short (e.g., $L << H$). The power spectrum of

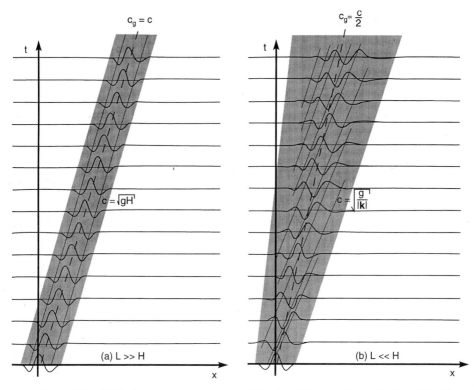

Figure 14.7 (a) Surface water waves produced by the initial disturbance in Fig. 14.6, with $L \gg H$. Shallow water waves radiate *nondispersively* away from the initial disturbance: Wave components with different k propagate at identical phase speed $c = \sqrt{gH}$ (solid lines). The envelope of wave activity then propagates at the same speed $c_g = c$ (dashed line). Under these circumstances, the shape of the initial waveform is preserved, so wave activity (shaded) remains confined to the same range of x as initially. (b) As in part (a) but for $L \ll H$ and a different x scale. Deep water waves radiate *dispersively* away from the initial disturbance: Wave components with different k propagate at different phase speeds $c = \sqrt{g/|k|}$ (solid line), so the initial waveform unravels into a series of oscillations. Occupying a progressively wider range of x, the envelope of wave activity (shaded) propagates at exactly half the median phase speed of individual components $c_g = c/2$ (dashed line). Individual crests and troughs therefore overrun the envelope, disappearing at its leading edge, to be replaced by new ones at its trailing edge.

η' (Fig. 14.6b) is then nearly *white*: Variance decreases slowly with $|k|$, so most of the excited wave spectrum lies at large $|k|H$ and is comprised of deep water waves. Under these conditions, the initial disturbance unravels into a series of oscillations (Fig. 14.7b), with wave activity occupying an increasingly wider range of x. Contrary to shallow water waves (Fig. 14.7a), the envelope of wave activity translates outward slower than individual crests and troughs. Waves inside it then overrun the envelope and disappear at its leading edge, to be replaced by new ones that appear at its trailing edge.

14.2 Acoustic Waves

The simplest wave motions supported by the governing equations are sound waves. Compressibility provides the restoring force for simple acoustic waves, which have timescales short enough to ignore rotation, heat transfer, friction, and buoyancy. Further, because they are longitudinal disturbances, in which fluid displacements are parallel to the propagation vector $\boldsymbol{k}$, we may consider motion in the x direction. Under these circumstances, air motion is governed by

$$\frac{du}{dt} = -\frac{1}{\rho}\frac{\partial p}{\partial x} \qquad (14.27.1)$$

$$\frac{d\rho}{dt} + \rho\frac{\partial u}{\partial x} = 0 \qquad (14.27.2)$$

$$p\rho^{-\gamma} \quad \text{conserved,} \qquad (14.27.3)$$

where Poisson's relation (2.30) serves as a statement of the first law under adiabatic conditions. Applying the logarithm to (14.27.3) followed by the material derivative obtains

$$\frac{1}{\gamma}\frac{d\ln p}{dt} = \frac{1}{\rho}\frac{d\rho}{dt},$$

which, on substitution into (14.27.2), yields

$$\frac{1}{\gamma}\frac{d\ln p}{dt} + \frac{\partial u}{\partial x} = 0. \qquad (14.28)$$

For a homogeneous basic state (e.g., one that is isothermal and in uniform motion), the first-order perturbation equations are then

$$\frac{Du'}{Dt} = -\frac{1}{\overline{\rho}}\frac{\partial p'}{\partial x}$$

$$\frac{Dp'}{Dt} + \gamma\overline{p}\frac{\partial u'}{\partial x} = 0. \qquad (14.29)$$

These may be consolidated into a single second-order equation

$$\frac{D^2 p'}{Dt^2} - \gamma R\overline{T}\frac{\partial^2 p'}{\partial x^2} = 0. \qquad (14.30)$$

Equation (14.30) is a one-dimensional wave equation for the perturbation pressure p', from which other field properties follow. Its coefficients are independent of x and t, so we may consider solutions of the form $\exp[i(kx - \sigma t)]$, which is an implicit application of (14.12.1). Substituting then transforms (14.30) into the dispersion relation

$$\hat{c}^2 = \gamma R\overline{T}, \qquad (14.31)$$

which defines the speed of sound. Because the intrinsic phase speed is independent of k, sound waves are nondispersive and their group velocity equals their phase velocity.

14.3   Buoyancy Waves

More important to the atmosphere are disturbances supported by the positive restoring force of buoyancy under stable stratification. Known as *gravity waves*, these disturbances typically have timescales short enough to ignore rotation, heat transfer, and friction. Vertical motions induced by gravity waves need not be hydrostatic, so the governing equations are expressed most conveniently in physical coordinates. Likewise, because they involve motion transverse to the propagation vector k, gravity waves require two dimensions to be described. In the x–z plane, air motion is then described by

$$\frac{du}{dt} = -\frac{1}{\rho}\frac{\partial p}{\partial x}, \tag{14.32.1}$$

$$\frac{dw}{dt} = -\frac{1}{\rho}\frac{\partial p}{\partial z} - g, \tag{14.32.2}$$

$$\frac{1}{\rho}\frac{d\rho}{dt} + \frac{\partial u}{\partial x} + \frac{\partial w}{\partial z} = 0, \tag{14.32.3}$$

$$\frac{d\ln p}{dt} - \gamma\frac{d\ln\rho}{dt} = 0. \tag{14.32.4}$$

Since they account for compressibility, equations (14.32) also describe acoustic waves, which, at low frequency, are modified by buoyancy.

For a basic state that is isothermal, in uniform motion, and in hydrostatic equilibrium, the perturbation equations become

$$\frac{Du'}{Dt} = -gH\frac{\partial \hat{p}'}{\partial x}, \tag{14.33.1}$$

$$\frac{Dw'}{Dt} = -gH\frac{\partial \hat{p}'}{\partial z} + g\hat{p}' - g\hat{\rho}', \tag{14.33.2}$$

$$\frac{D\hat{\rho}'}{Dt} + \frac{\partial u'}{\partial x} + \frac{\partial w'}{\partial z} - \frac{w'}{H} = 0, \tag{14.33.3}$$

$$\frac{D}{Dt}(\hat{p}' - \gamma\hat{\rho}') + \gamma\frac{N^2}{g}w' = 0, \tag{14.33.4}$$

where

$$\hat{p}' = \frac{p'}{\overline{p}} \qquad \hat{\rho}' = \frac{\rho'}{\overline{\rho}} \tag{14.33.5}$$

are scaled by the basic state stratification,

$$H = \frac{R\overline{T}}{g},$$ (14.33.6)

and

$$\frac{N^2}{g} = \frac{\partial \ln \overline{\theta}}{\partial z}$$

$$= \frac{\kappa}{H}$$ (14.33.7)

gives a buoyancy period of about 5 min.

Equations (14.33) constitute a closed system for the four unknowns: u', w', $\hat{p}'$, and $\hat{\rho}'$. Because it has constant coefficients, solutions can be expressed in terms of plane waves (14.11.1). However, owing to the stratification of mass, a group of waves propagating vertically must adjust in amplitude to conserve energy. For quadratic quantities like $\overline{\rho} u'^2$ (e.g., the kinetic energy density) to remain constant following the group, u', w', $\hat{p}'$, and $\hat{\rho}'$ must amplify vertically like $\overline{\rho}^{-\frac{1}{2}} = \exp[z/2H]$. Then considering solutions of the form $\exp[(z/2H) + i(kx + mz - \sigma t)]$, where m denotes vertical wavenumber, transforms the differential system (14.33) into an algebraic one. Consolidating that system yields the dispersion relation

$$m^2 = k^2 \left(\frac{N^2}{\omega^2} - 1 \right) + \frac{\omega^2 - \omega_a^2}{c_s^2},$$ (14.34.1)

where

$$c_s^2 = \gamma R\overline{T}$$ (14.34.2)

is the speed of sound and

$$\omega_a = \frac{c_s}{2H}$$ (14.34.3)

is the *acoustic cutoff frequency* for vertical propagation.

Equation (14.34) describes *acoustic–gravity waves*, which involve buoyancy as well as compression. Their vertical propagation is controlled by the sign of m^2. For $m^2 > 0$, individual components of the wave spectrum oscillate in z. Vertically propagating, such waves are referred to as *internal* because their oscillatory structure occurs in the interior of a bounded domain.[1] Internal waves amplify upward like $\exp[z/2H]$ (Fig. 14.8a) to offset the stratification of mass and make energy densities invariant with height. For $m^2 = -\hat{m}^2 < 0$, individual wave components vary exponentially in the vertical (Fig. 14.8b). Oscillations then occur in-phase throughout the column. For the column energy to remain bounded, the energy density of these waves must decrease exponentially with height. Such waves are referred to as *evanescent* or *external* because

[1] This terminology is a misnomer because internal modes are actually trapped vertically between boundaries.

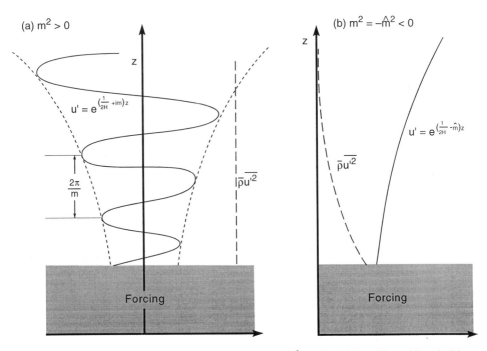

Figure 14.8 Vertical structure of (a) *internal waves* ($m^2 > 0$), which oscillate with z (solid line) and amplify upward like $e^{z/2H}$ to make energy density (dashed line) invariant with height, and (b) *external waves* ($m^2 < 0$), whose energy density (dashed line) decays exponentially with height but whose amplitude (solid line) can either amplify or decay depending on m.

they form along the exterior of a boundary. The edge wave structure of surface water waves in the deep water limit (Fig. 14.4b) is an example. Even though their energy decreases, external waves in the range $1/4H^2 < m^2 < 0$ amplify upward like $\exp[(1/2H - \hat{m})z]$ due to stratification of mass.

The dispersion relation assumes two limiting forms associated with vertical propagation, as seen when $m^2(k, \omega)$ is plotted (Fig. 14.9). For $\omega^2 \gg N^2$, (14.34) reduces to

$$k^2 + m^2 + \frac{1}{4H^2} \cong \frac{\omega^2}{c_s^2} \qquad \frac{\omega^2}{N^2} \to \infty, \qquad (14.35.1)$$

which describes high-frequency acoustic waves modified by stratification. This limiting behavior can also be recovered by letting $N \to 0$, which eliminates buoyancy. Acoustic waves propagate vertically for ω greater than those on the upper curve for $m^2 = 0$ in Fig. 14.9. Their intrinsic phase speeds approach the speed of sound for $k^2 + m^2 \to \infty$. In the longwave limit $k \to 0$, the minimum ω for vertical propagation equals the acoustic cutoff ω_a. In the shortwave limit $k \to \infty$, the minimum ω for vertical propagation tends to infinity and

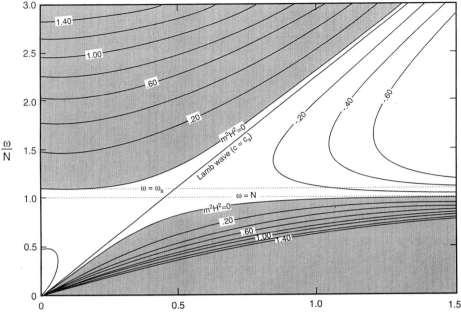

Figure 14.9   Vertical wavenumber squared contoured as a function of horizontal wavenumber and intrinsic frequency. Vertical propagation (shaded) occurs for acoustic waves at ω greater than the acoustic cutoff ω_a and for gravity waves at ω less than the buoyancy frequency N, with $\omega_a \cong 1.1N$ under isothermal conditions. The Lamb wave, which connects the limiting behavior of acoustic waves for $m^2 = 0$ and $k \to \infty$ to the limiting behavior of gravity waves for $m^2 = 0$ and $k \to 0$, propagates horizontally at the speed of sound.

the corresponding waves have horizontal phase speeds approaching the speed of sound.

For $\omega^2 << \omega_a^2$, (14.34) reduces to

$$k^2 + m^2 + \frac{1}{4H^2} \cong \frac{N^2}{\omega^2} k^2 \qquad \frac{\omega^2}{\omega_a^2} \to 0, \qquad (14.35.2)$$

which describes low-frequency gravity waves modified by stratification of mass. This limiting behavior can also be recovered by letting $c_s \to \infty$, which renders air motion incompressible (Problem 14.14). Gravity waves propagate vertically for ω less than those on the lower curve for $m^2 = 0$ in Fig. 14.9. Unlike most vibrating systems, internal gravity waves have a high-frequency cutoff for vertical propagation, N. In the shortwave limit $k \to \infty$, the maximum ω for vertical propagation approaches N. Since the horizontal phase speed then vanishes, the corresponding behavior describes column oscillations at the buoyancy frequency (7.11). In the longwave limit $k \to 0$, the maximum ω for vertical propagation approaches zero and the corresponding waves have horizontal phase speeds approaching the speed of sound.

The regions of vertical propagation in Fig. 14.9 (shaded) are well separated. Between them, $m^2 < 0$ and waves are external. Acoustic–gravity waves then propagate horizontally, with energy that decreases exponentially above the forcing and no net influence in the vertical (Problem 14.12). Connecting the limiting behavior of internal acoustic waves for $k \to \infty$ and internal gravity waves for $k \to 0$ is the external *Lamb wave* (Lamb, 1910), which propagates horizontally at the speed of sound for all k.

Dispersion characteristics of the acoustic and gravity branches of (14.34) are illustrated in Fig. 14.10, which plots ω as a function of $\mathbf{k} = (k, m)$. The

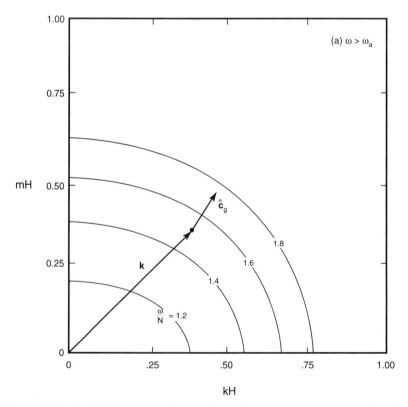

Figure 14.10 Intrinsic frequency of acoustic–gravity waves contoured as a function of wavenumber. Group velocity relative to the medium is directed orthogonal to contours of intrinsic frequency toward increasing ω. (a) Acoustic waves ($\omega > \omega_a$) have group velocity relative to the medium $\hat{c}_g$ that is nearly parallel to phase propagation $\mathbf{k}$. (*continues*)

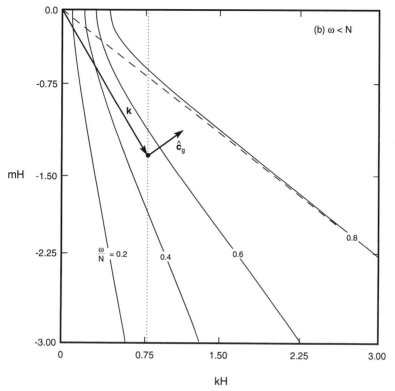

Figure 14.10 (*Continued*) (b) Gravity waves ($\omega < N$) have $\hat{c}_g$ that is nearly orthogonal to k, except close to $m = 0$, where both become horizontal.

intrinsic group velocity (14.23) is directed orthogonal to contours of intrinsic frequency toward increasing ω. On the other hand, phase propagates with the velocity $\hat{c}(k/|k|)$ parallel ($\omega > 0$) or antiparallel ($\omega < 0$) to the wavenumber vector. For acoustic waves (Fig. 14.10a), contours of ω are elliptical, modified only slightly from their forms under unstratified conditions. The group velocity is then close to the phase velocity for all k, so those waves are only weakly dispersive.

In the case of gravity waves (Fig. 14.10b), contours of ω are hyperbolic and the situation is quite different. For ω decreasing with k fixed (e.g., moving downward along the dotted line), contours of ω become straight lines that pass through the origin. The intrinsic group velocity is then orthogonal to the phase velocity, so the waves are highly dispersive. A wave group defined by the vector k in Fig. 14.10b has phase that propagates downward and to the right, but its envelope propagates upward and to the right. A packet of gravity waves (Fig. 14.11) propagates parallel to lines of constant phase (e.g.,

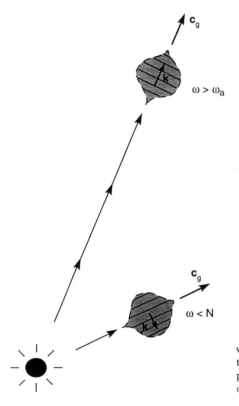

Figure 14.11 A packet of simple gravity waves ($\omega < N$) propagates along crests and troughs—just perpendicular to a packet of simple acoustic waves ($\omega > \omega_a$), which propagates orthogonal to crests and troughs.

along crests and troughs), rather than perpendicular to them, as is observed of more familiar acoustic waves.[2] Consequently, downward propagation of phase translates into upward propagation of wave activity. This peculiarity follows from the dispersive nature of gravity waves and applies to all waves whose vertical restoring force is buoyancy. For ω increasing with k fixed (e.g., moving upward along the dotted line in Fig. 14.10b), contours of ω become parallel to the m axis. Intrinsic group velocity is then in the same direction as the phase velocity—both horizontal.

At very low frequency, rotation cannot be ignored. The Coriolis force then modifies the dispersion relation by making parcel trajectories three-dimensional. Air motion (u, w) associated with buoyancy oscillations experiences a Coriolis force fu, which drives parcel trajectories out of the x-z plane and introduces v. Disturbances in this range of frequency are referred to as *inertio–gravity waves* (see Gill, 1982).

[2] See Lighthill (1978) for a laboratory demonstration.

14.3.1 Shortwave Limit

In the limit $k \to \infty$ with ω fixed, the dispersion relation (14.34) reduces to

$$\frac{\omega^2}{N^2} = \frac{k^2}{k^2 + m^2}$$
$$= \cos^2 \alpha, \tag{14.36}$$

where α is the angle k makes with the horizontal. Contours of ω are then straight lines, with $\hat{c}_g$ orthogonal to k, as for the limiting behavior in Fig. 14.10b. Simple gravity waves propagate at progressively shallower angles as ω is increased (Fig. 14.12), until, at $\omega = N$, k is horizontal with phase lines oriented vertically. External waves with $\omega > N$ are vertically trapped because c_{gz} then vanishes.

The limiting behavior (14.36) applies to mountain waves (Fig. 9.22), which have horizontal wavelengths of about 10 km. Generated by forced ascent over elevated terrain, these waves are fixed with respect to the earth, so they have frequencies $\sigma \cong 0$. Their intrinsic frequencies, however, do not vanish, but rather are determined by the Doppler shift $-k\overline{u}$. Westerly flow makes contours of phase propagate westward relative to the medium. Since upward propagation of energy requires downward phase propagation, phase contours

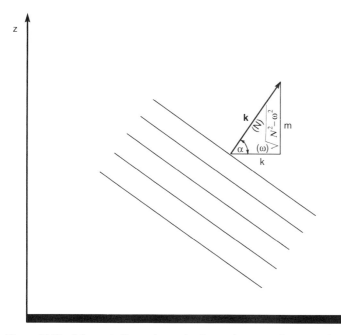

Figure 14.12 Schematic illustrating a simple gravity wave of wavenumber k that is inclined from the horizontal by an angle α.

must tilt westward with height above their forcing. Such behavior is often observed in the leading edge of lenticular clouds (Fig. 9.22a).

Even though their phase velocity relative to the ground vanishes, orographically forced gravity waves have nonvanishing group velocity. As illustrated in Fig. 14.13 for steady motion incident on a two-dimensional ridge, wave activity is transferred upward and downstream. The dispersive nature of gravity waves leads to wave activity occupying a progressively wider sector of the first quadrant with increasing distance from the source. Vertical motion accompanying the oscillations can extend into the stratosphere (see Fig. 14.24), which has enabled sailplanes to reach altitudes of 50,000 ft.

14.3.2   Propagation of Gravity Waves in a Nonhomogeneous Medium

When the basic state is not isothermal and in uniform motion, wave propagation varies with position. Consider the propagation of gravity waves in a basic state that varies with height: $N^2 = N^2(z)$ and $\overline{u} = \overline{u}(z)$, and under the Boussinesq approximation (Sec. 12.5.3). The wave field is then governed by

$$\frac{Du'}{Dt} + w'\overline{u}_z = -gH\frac{\partial \hat{p}'}{\partial x}, \tag{14.37.1}$$

$$\frac{Dw'}{Dt} = -gH\frac{\partial \hat{p}'}{\partial z} - g\hat{\rho}', \tag{14.37.2}$$

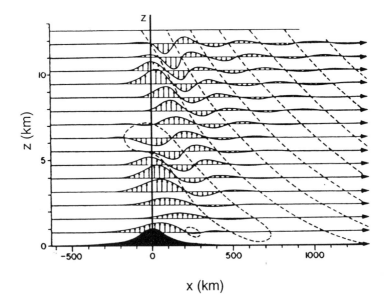

Figure 14.13 Streamlines and vertical motion accompanying a stationary gravity wave pattern that is excited by uniform flow over a two-dimensional ridge 100 km wide. After Queney (1948).

$$\frac{\partial u'}{\partial x} + \frac{\partial w'}{\partial z} = 0, \tag{14.37.3}$$

$$-\frac{D\hat{\rho}'}{Dt} + \frac{N^2}{g}w' = 0, \tag{14.37.4}$$

where (14.37.4) follows from (14.33.4) with the neglect of compressibility. Applying the linearized material derivative to (14.37.2) gives

$$\frac{D^2 w'}{Dt^2} + gH\frac{D}{Dt}\left(\frac{\partial \hat{p}'}{\partial z}\right) = -g\frac{D\hat{\rho}'}{Dt}.$$

With (14.37.4), this becomes

$$\left[\frac{D^2}{Dt^2} + N^2\right]w' + gH\frac{D}{Dt}\left(\frac{\partial \hat{p}'}{\partial z}\right) = 0. \tag{14.38}$$

Considering solutions of the form $\exp[i(kx - \sigma t)]$ (coefficients are still independent of x and t) and combining (14.37.1) with (14.37.3) obtains

$$gH\frac{\partial P_k^\sigma}{\partial z} = \frac{i}{k}\left[(c_x - \overline{u})\frac{\partial^2 W_k^\sigma}{\partial z^2} + \overline{u}_{zz}W_k^\sigma\right], \tag{14.39}$$

where $P_k^\sigma(z)$ denotes the transform (14.12.1) of $\hat{p}'$. Then incorporating (14.39) into (14.38) yields

$$\frac{\partial^2 W_k^\sigma}{\partial z^2} + m^2(z)W_k^\sigma = 0, \tag{14.40.1}$$

in which

$$m^2(z) = \frac{N^2}{(c_x - \overline{u})^2} - k^2 + \frac{\overline{u}_{zz}}{c_x - \overline{u}} \tag{14.40.2}$$

serves as a reduced one-dimensional *index of refraction*.

Known as the *Taylor–Goldstein equation*, (14.40) describes the vertical propagation of wave activity in terms of local refractive properties of the medium. For $m^2 > 0$, $W_k^\sigma(z)$ is oscillatory in z. Wave activity then propagates vertically with a local wavelength $2\pi/m(z)$. Contrary to propagation under the homogeneous conditions described by (14.33), variations of m^2 lead to reflection of wave activity, which couples propagation in the upward and downward directions. Oscillations also do not amplify vertically due to the neglect of compressibility. For $m^2 < 0$, the vertical structure is evanescent. Wave activity incident on such a region must therefore be reflected.

Static stability, which provides the restoring force for vertical oscillations, is the essential positive contribution to m^2 in (14.40.2). The greater N^2, the shorter the vertical wavelength and the more vertical propagation is favored. Since k^2 reduces m^2, short horizontal wavelengths (large k) are less able to propagate vertically than long horizontal wavelengths (small k). Vertical propagation is also controlled by the mean flow, chiefly through the intrinsic trace speed $c_x - \overline{u}$. For small flow curvature ($\overline{u}_{zz} \cong 0$), increasing $c_x - \overline{u}$

inhibits vertical propagation by reducing m^2. The vertical wavelength is then elongated, which makes wave activity increasingly vulnerable to reflection. If m^2 is driven to zero for a particular k, the vertical wavelength becomes infinite (see Fig. 14.14b). Continued propagation is then forbidden and wave activity is fully reflected. The height where this occurs is called the *turning level*, above which wave structure is external ($m^2 < 0$). On the other hand, decreasing $c_x - \bar{u}$ favors vertical propagation by increasing m^2. The vertical wavelength is then shortened, which reduces reflection by making variations in refractive properties comparatively gradual. Should $c_x - \bar{u}$ decrease to zero, (14.40) becomes singular. The refractive index is then unbounded and the vertical wavelength collapses to zero (see Fig. 14.15b). The height where this occurs is called the *critical level*. Contrary to a turning level, which leads to reflection, the condition $m^2 \to \infty$ leads to absorption of wave activity.

14.3.3   The WKB Approximation

Analytical solutions to (14.40) can be obtained only for very idealized profiles of $N^2(z)$ and $\bar{u}(z)$. But an approximate class of solutions that is valid when the basic state varies slowly compared to the phase of oscillations affords great insight into how wave activity propagates—even under more general circumstances. The approximate solutions

$$W^{\pm}(z) = \frac{W_0^{\pm}}{\sqrt{m(z)}} \exp\left[\pm i \int m(z)\,dz\right] \qquad (14.41)$$

satisfy the related equation

$$\frac{\partial^2 W^{\pm}}{\partial z^2} + m^2(z)W^{\pm} = -R(z)m^2(z)W^{\pm}, \qquad (14.42.1)$$

where

$$R(z) = \frac{1}{2m^3}\frac{\partial^2 m}{\partial z^2} - \frac{3}{4m^4}\left(\frac{\partial m}{\partial z}\right)^2 \qquad (14.42.2)$$

symbolizes the interdependence of W^+ and W^- and therefore wave reflection.

In the limit of $R \to 0$ everywhere, (14.42) reduces to the general form of (14.40). Reflection then vanishes and the upward- and downward-propagating waves can be treated independently. In the absence of dissipation, wave activity then propagates conservatively along *rays* that are smoothly refracted through the medium. The solution (14.41) defines the so-called *slowly varying* or *WKB approximation*.[3] Its validity (e.g., $R \ll 1$) rests on refractive properties changing on a scale that is long compared to the wavelength. In practice, (14.41) often provides a qualitatively correct description of the wave field even

---

[3] The letters stand for Wentzel (1926), Kramer (1926), and Brillouin (1926), who rediscovered the procedure in different problems, after being introduced originally by Liouville (1837) and Green (1838).

when it cannot be justified formally. Should R become $O(1)$ somewhere, reflection introduces the other of the components in (14.41). Clearly, $m^2 = 0$
violates the aforementioned condition because the local wavelength is then
infinite, which is attended by strong reflection. Although the WKB approximation breaks down there, the wave field can be matched analytically across
the turning level (see, e.g., Morse and Feschbach, 1953).

14.3.4 Method of Geometric Optics

Under conditions of the WKB approximation, wave activity can be traced from
its source along rays that are defined by the local group velocity c_g. If

$$\bar{u}_{zz} << \frac{N^2}{|c_x - \bar{u}|},$$ (14.43)

(14.40.2) has the same form as the dispersion relation for simple gravity waves
in the short wavelength limit (14.36). Components of c_g then follow from
(14.23) as

$$c_{gx} = \frac{m^2\omega}{k(k^2 + m^2)} + \bar{u},$$

$$c_{gz} = -\frac{m\omega}{(k^2 + m^2)},$$ (14.44)

with $m = m(z)$ and $\bar{u} = \bar{u}(z)$. Tangent to c_g, a ray has local slope in the x-z
plane

$$\frac{dz}{dx} = \frac{c_{gz}}{c_{gx}}.$$ (14.45)

Specifying its initial coordinates then determines the ray corresponding to an
individual component (k, m) of the wave spectrum.

Turning Level

Orographically forced gravity waves involve a spectrum about zero frequency. Figure 14.14 shows rays of two components with $\sigma = 0$ under the
conditions of (14.43) for $N^2 =$const and $\bar{u}(z)$ representative of the troposphere and wintertime stratosphere (Fig. 14.14a). Each ray emanates from a
point source located at the origin of the x-z plane. For component 1, which
has short horizontal wavelength, westerly shear drives ω^2 to N^2 and m^2 to zero
near the tropopause (Fig. 14.14b). The ray for that component (Fig. 14.14c)
is reflected at its turning level z_{t1}, above which wave amplitude decays exponentially.

Since $m = 0$ at the turning level,

$$\begin{aligned} c_{gx} &= \bar{u} \\ &\qquad\qquad z = z_{t1}. \\ c_{gz} &= 0 \end{aligned}$$ (14.46)

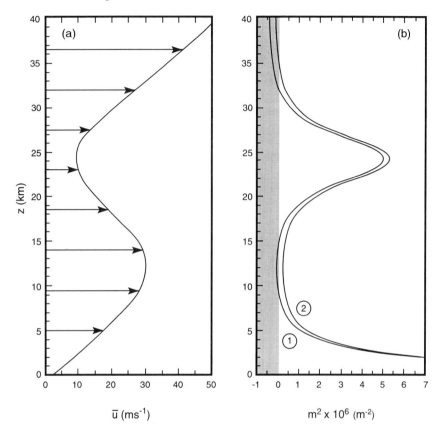

Figure 14.14   (a) Zonal motion representative of the extratropical troposphere and stratosphere during winter. (b) Vertical refractive index squared following from mean motion in part (a) and uniform static stability. (*continues*)

Thus, relative to the medium, wave activity is motionless at $z = z_{t1}$. Consider the time for wave activity to reach the turning level from some neighboring level below. The time for wave activity to propagate from a level z_0 to a level z is given by

$$\Delta t = \int_{z_0}^{z} \frac{dz'}{c_{gz}}. \qquad (14.47)$$

Near the turning level,

$$c_{gz} \sim -\frac{\omega m}{k^2} \qquad z \to z_{t1}. \qquad (14.48.1)$$

Since $\omega = -k\bar{u} < 0$ for $z < z_{t1}$ and $\bar{u}_z > 0$,

$$\omega \sim -N - k\bar{u}_z(z - z_{t1}) \qquad z \to z_{t1}. \qquad (14.48.2)$$

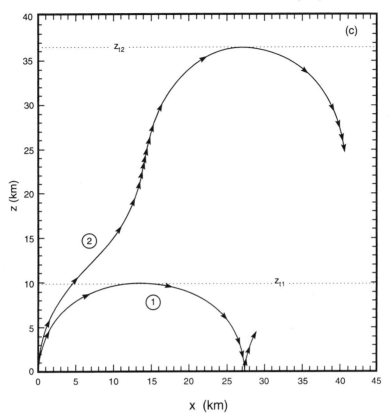

Figure 14.14 (*Continued*) (c) Rays for two stationary wave components with a source at $(x, z) = (0, 0)$. Arrowheads mark uniform increments of time moving along the ray at the group velocity c_g. Component 1, which has short horizontal wavelength, encounters a *turning level* in the troposphere, where it is reflected downward. Component 2, which has a longer horizontal wavelength, propagates into the stratosphere before encountering a turning level.

Substituting into the dispersion relation (14.36) obtains

$$m \sim k \left[2\frac{k}{N}\overline{u}_z(z_{t1} - z) \right]^{\frac{1}{2}} \qquad z \to z_{t1}. \qquad (14.48.3)$$

Then the vertical group speed behaves as

$$c_{gz} \sim \left[2\frac{N}{k}\overline{u}_z(z_{t1} - z) \right]^{\frac{1}{2}} \qquad z \to z_{t1}. \qquad (14.49)$$

Incorporating (14.49) into (14.47) gives for the approach time

$$\Delta t = \left[\frac{2k}{N\overline{u}_z}(z_{t1} - z) \right]^{\frac{1}{2}} \Bigg|_z^{z_0}$$

$$< \infty \qquad z \to z_{t1}. \qquad (14.50)$$

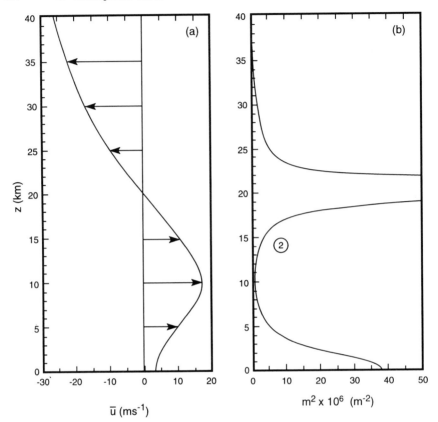

Figure 14.15   As in Fig. 14.14, except for conditions representative of summer and for component 2. The wave is Doppler-shifted to zero intrinsic phase speed: $c_x - \overline{u} = 0$, at the critical level, where $m^2 \to \infty$, group propagation stalls, and the ray terminates. (Frequency of arrowheads near $z = z_{c2}$ reduced for clarity. (*continues*)

Thus, the time for wave activity to reach the turning level is finite, even though c_{gz} vanishes there. It follows that wave activity can rebound from the turning level having suffered only finite dissipation. This is the premise for interpreting behavior near z_{t1} as reflection.

Wave activity trapped between its turning level and the surface is ducted horizontally in the troposphere. This can produce oscillations far downstream of the source, as is revealed by extensive wave patterns in satellite cloud imagery (Fig. 1.11). In contrast, wave activity propagating freely in the vertical (Fig. 14.13) produces only a few oscillations downstream, so cloud bands are confined near the source (Fig. 9.22).

Component 2 of the wave spectrum, which has longer horizontal wavelength, does not encounter a turning level until it has propagated well into the stratospheric jet (Fig. 14.14b). Although m^2 is reduced in the core of the tropospheric jet, it remains positive, which allows component 2 to propagate to

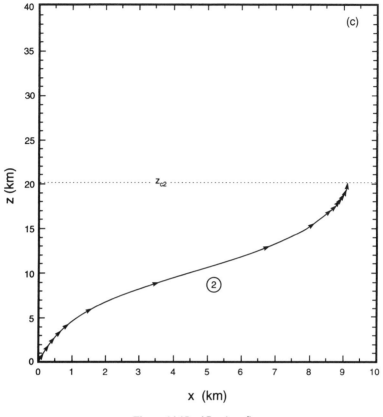

Figure 14.15 (*Continued*)

higher levels. The region of small m^2 in the upper troposphere coincides with small vertical group speed. Propagation is then refracted into a shallower angle (Fig. 14.14c), only to be refracted more steeply again when it encounters weaker zonal flow in the lower stratosphere. Eventually, $\bar{u}$ in the stratospheric jet becomes large enough to drive m^2 to zero, which occurs at the turning level z_{t2} and below which component 2 is vertically trapped.

CRITICAL LEVEL

Consider now conditions representative of summer. Figure 14.15 shows the same features for component 2, but for easterly flow in the stratosphere. Having $c_x = 0$, component 2 is Doppler-shifted to zero at the zero wind line, which lies not far above the tropopause (Fig. 14.15a). At the critical level z_{c2}, $m^2 \to \infty$ and the vertical wavelength collapses to zero (Fig. 14.15b).

Since $\omega = 0$ at the critical level,

$$
\begin{aligned}
c_{gx} &= \bar{u} = 0 \\
c_{gz} &= 0
\end{aligned}
\qquad z = z_{c2},
\qquad (14.51)
$$

so wave activity is again motionless relative to the medium. Consider the time for it to reach the critical level from some neighboring level z_0 below. For z near the critical level,

$$
c_{gz} \sim -\frac{\omega}{m} \qquad z \to z_{c2}. \qquad (14.52)
$$

Since $\omega = -k\bar{u} < 0$ for $z < z_{c2}$ and $\bar{u}_z < 0$,

$$
\omega \sim -k\bar{u}_z(z - z_{c2}) \qquad z \to z_{c2}. \qquad (14.53.1)
$$

Substituting into the dispersion relation gives

$$
m \sim \frac{N}{\bar{u}_z(z - z_{c2})} \qquad z \to z_{c2}. \qquad (14.53.2)
$$

Then the vertical group speed behaves as

$$
c_{gz} \sim \frac{k}{N}\bar{u}_z^2(z - z_{c2})^2 \qquad z \to z_{c2}. \qquad (14.54)
$$

By incorporating (14.54) into (14.47), we obtain for the approach time

$$
\Delta t = \frac{N}{k\bar{u}_z^2}(z - z_{c2})^{-1}\Big|_z^{z_0}
$$

$$
\to \infty \qquad z \to z_{c2}. \qquad (14.55)
$$

Unlike behavior at a turning level, the time for wave activity to reach the critical level is infinite. Thus, the critical level is never actually encountered. Instead, wave activity remains frozen relative to the medium for infinite duration, allowing dissipation to absorb it. The ray for component 2 (Fig. 14.15c) is then refracted into the critical level, where it stalls and terminates.

14.4   The Lamb Wave

Consider oscillations in an unbounded atmosphere and in the absence of forcing. These conditions define an eigenvalue problem for the normal modes of that system, namely, wave modes that can persist without forcing. Should forcing exist at the corresponding eigenfrequencies, normal modes are greatly amplified relative to other wave components.

For oscillations that are hydrostatic, the perturbation equations (14.33) reduce to

$$
\frac{Du'}{Dt} = -gH\frac{\partial \hat{p}'}{\partial x}, \qquad (14.56.1)
$$

$$\left(H\frac{\partial}{\partial z} - 1\right)\hat{p}' + \hat{\rho}' = 0, \tag{14.56.2}$$

$$\frac{D\hat{\rho}'}{Dt} + \frac{\partial u'}{\partial x} + \left(\frac{\partial}{\partial z} - \frac{1}{H}\right)w' = 0, \tag{14.56.3}$$

$$\frac{D}{Dt}(\hat{p}' - \gamma\hat{\rho}') + \gamma\frac{\kappa}{H}w' = 0. \tag{14.56.4}$$

Considering solutions of the form $\exp[i(kx - \sigma t)]$ transforms (14.56) into

$$-i\omega U_k^\sigma = -ikgHP_k^\sigma, \tag{14.57.1}$$

$$\left(H\frac{\partial}{\partial z} - 1\right)P_k^\sigma + \rho_k^\sigma = 0, \tag{14.57.2}$$

$$-i\omega\rho_k^\sigma + ikU_k^\sigma + \left(\frac{\partial}{\partial z} - \frac{1}{H}\right)W_k^\sigma = 0, \tag{14.57.3}$$

$$-i\omega(P_k^\sigma - \gamma\rho_k^\sigma) + \gamma\frac{\kappa}{H}W_k^\sigma = 0. \tag{14.57.4}$$

Using (14.57.2) to eliminate density in (14.57.4) gives

$$W_k^\sigma = \frac{i\omega H}{\kappa}\left(H\frac{\partial}{\partial z} - \kappa\right)P_k^\sigma \tag{14.58}$$

for the vertical velocity. Then, incorporating (14.58), (14.57.1), and (14.57.3) yields a second-order differential equation for the perturbation amplitude $P_k^\sigma(z)$

$$\frac{\partial^2 P_k^\sigma}{\partial z^2} - \frac{1}{H}\frac{\partial P_k^\sigma}{\partial z} + \frac{\kappa g}{H}\frac{k^2}{\omega^2}P_k^\sigma = 0. \tag{14.59}$$

For physically meaningful behavior, solutions to (14.59) must satisfy boundary conditions. At the surface, we require w' to vanish. The upper boundary condition depends on whether the solution is vertically propagating or evanescent. If the solution is evanescent, it must have bounded column energy. If it is vertically propagating, at large distances from the source, the solution must transfer wave activity away from regions of possible excitation (e.g., from finite z). These conditions define the so-called *finite-energy* and *radiation conditions*, respectively, which apply at $z \to \infty$. Collectively, $P_k^\sigma(z)$ must then satisfy

$$\frac{\partial^2 P_k^\sigma}{\partial z^2} - \frac{1}{H}\frac{\partial P_k^\sigma}{\partial z} + \frac{\kappa g}{H}\frac{k^2}{\omega^2}P_k^\sigma = 0, \tag{14.60.1}$$

$$H\left(\frac{\partial}{\partial z} - \kappa\right)P_k^\sigma = 0 \qquad z = 0, \tag{14.60.2}$$

$$\text{finite–energy/radiation} \qquad z \to \infty. \tag{14.60.3}$$

Equations (14.60) define a homogeneous boundary value problem for the vertical structure $P_k^\sigma(z)$, namely, one with no forcing in the interior or on the boundaries of the domain. Nontrivial solutions can exist only at particular frequencies.

Seeking solutions of the form $\exp[(1/2H + im)z]$ transforms (14.60.1) into the dispersion relation

$$m^2 + \frac{1}{4H^2} = \frac{N^2}{\omega^2}k^2. \tag{14.61}$$

Equation (14.61) may be recognized as (14.35.2) in the limit $k \to 0$, wherein the horizontal wavelength is long enough for vertical displacements to be hydrostatic. If $m^2 > 0$, solutions are vertically propagating. By (14.44), upward energy propagation (14.60.3) requires $m\omega < 0$, which corresponds to downward phase propagation. For $\omega > 0$, this requires $m < 0$. It is readily verified that these solutions do not satisfy the homogeneous lower boundary condition (14.60.2). Nor do their counterparts for $\omega < 0$. If $m^2 = -\hat{m}^2 < 0$, the solutions of (14.60.1) are external. The requirement of bounded column energy then selects solutions of the form $\exp[(1/2H - \hat{m})z]$. One of these satisfies the lower boundary condition, namely, $\hat{m} = (\kappa - \frac{1}{2})/H$, which has the vertical structure

$$P_k^\sigma(z) = e^{\kappa(\frac{z}{H})}. \tag{14.62.1}$$

For this particular vertical structure, (14.61) reduces to

$$\hat{c}^2 = \gamma gH$$
$$= c_s^2. \tag{14.62.2}$$

The vertical structure (14.62), which defines the *Lamb mode* and is pictured in Fig. 14.8b, makes w' vanish—not just at the surface, but everywhere (14.58). Buoyancy oscillations then vanish identically. The restoring force for Lamb waves is provided entirely by compressibility so they propagate at the speed of sound (Fig. 14.9). Lamb waves are the normal modes of an unbounded, compressible, stratified atmosphere. Even though their energy decreases upward like an edge wave, Lamb waves amplify vertically, which makes them potentially important in the upper atmosphere.

Because they are normal modes of the atmosphere, Lamb waves are excited preferentially by forcing that is indiscriminate over frequency, for example, an impulsive disturbance (Problems 14.33). The response spectrum to such forcing is unbounded at those wavenumbers and frequencies satisfying (14.62.2), so it is dominated by Lamb waves (Problem 14.32). Historical records include several impulses to the atmosphere that were felt around the earth. Most notable was the eruption of Krakatoa in 1883 (Chapter 8). Disturbances in surface pressure cycled around the globe several times before dissipating. Taylor (1929) used barometric records to infer the vertical structure of atmospheric normal modes. By comparing arrival times at several stations, he showed that the compression wave emanating from Krakatoa propagated at the speed of sound. Then, with the aid of a system like (14.60), Taylor deduced the Lamb structure for atmospheric normal modes. A similar analysis was performed by Whipple (1930) for the impact of the great Siberian meteor in 1908. The Lamb vertical structure applies to atmospheric normal modes at all frequencies, even very low-frequency waves supported by the earth's rotation (Dikii, 1968).

14.5   Rossby Waves

At frequencies below 2Ω, another class of wave motion exists, one possible only in a rotating medium. Named for C. G. Rossby who established their connection to weather phenomena (Rossby *et al.*, 1939), these waves are also referred to as *rotational waves* and, on the gravest dimensions, *planetary waves*. The restoring force for Rossby waves is provided by the variation with latitude of the Coriolis force, which links them directly to the earth's rotation.

14.5.1   Barotropic Nondivergent Rossby Waves

The simplest model of Rossby waves relies on the solenoidal character of large-scale motion (Sec. 12.1). Under nondivergent conditions, atmospheric motion is governed by conservation of absolute vorticity (12.30), which on a beta plane is expressed by

$$\frac{d\zeta}{dt} + \beta v = 0.$$

Linearizing about a basic state that is barotropically stratified and in uniform motion, we obtain the perturbation vorticity equation

$$\frac{D\zeta'}{Dt} + \beta v' = 0. \tag{14.63}$$

The motion is nondivergent, so it can be represented in terms of a streamfunction

$$v'_h = k \times \nabla \psi'. \tag{14.64}$$

Then the perturbation vorticity equation becomes

$$\frac{D}{Dt} \nabla^2 \psi' + \beta \frac{\partial \psi'}{\partial x} = 0, \tag{14.65}$$

which is just the linearized form of the barotropic nondivergent vorticity equation (12.31).

Known as the *Rossby wave equation*, (14.65) reflects a balance between changes in the relative vorticity of an air parcel and changes in its planetary vorticity due to meridional displacement. An eastward-moving parcel deflected equatorward (Fig. 14.16) experiences reduced planetary vorticity f, so it spins up cyclonically to conserve absolute vorticity. Northward motion is then induced ahead of the parcel, which deflects its trajectory poleward—back toward its undisturbed latitude ϕ_0. Upon overshooting ϕ_0, the parcel spins up anticyclonically. Southward motion ahead of the parcel then deflects its trajectory equatorward—again back toward its undisturbed latitude. The variation with latitude of f thus exerts a torque on displaced air (analogous to the positive restoring force of buoyancy under stable stratification), which enables air to cycle back and forth about its undisturbed latitude.

The coefficients of (14.65) are constant, so we consider solutions of the form $\exp[i(kx + ly - \sigma t)]$. Substitution then results in the dispersion relation for Rossby waves

$$c_x - \overline{u} = -\frac{\beta}{k^2 + l^2}. \tag{14.66}$$

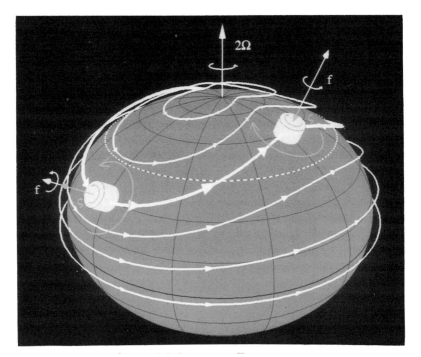

Figure 14.16 Schematic illustrating the reaction of an air parcel to meridional displacement. Displaced equatorward, an eastward-moving parcel spins up cyclonically to conserve absolute vorticity. Northward motion induced ahead of it then deflects the parcel's trajectory poleward back toward its undisturbed latitude. The reverse process occurs when the parcel overshoots and is displaced poleward of its undisturbed latitude.

Relative to the basic state, Rossby waves propagate only westward.[4] Their intrinsic trace speed, $c_x - \bar{u} = \omega/k$, is proportional to the local gradient of planetary vorticity and inversely proportional to the horizontal wavenumber squared. Hence, Rossby waves are dispersive, with the gravest dimensions propagating westward fastest. Small scales with slow phase speeds will be swept eastward by westerly zonal flow, as is typical of synoptic disturbances (Fig. 1.9). Stationary Rossby waves forced orographically propagate only if the basic flow is westerly.

The intrinsic frequency $|\omega|$ of Rossby waves is proportional to β, so it has a high-frequency cutoff of 2Ω—analogous to the cutoff N for internal gravity waves.

The solenoidal character of Rossby waves distinguishes them from gravity waves, which are nearly irrotational. Because Rossby waves are almost non-divergent, the Helmholtz theorem (12.4) implies that horizontal motion can be characterized by the vorticity field, with vertical motion ignored to a first approximation. Conversely, gravity waves are determined chiefly by vertical motion and hence by the divergence field. Although the essential properties of Rossby waves follow from the rotational component of motion, divergence enters by forcing absolute vorticity.

14.5.2 Rossby Wave Propagation in Three Dimensions

To examine three-dimensional wave propagation, divergence must be accounted for. Within the framework of quasi-geostrophy, air motion is governed by conservation of quasi-geostrophic potential vorticity. Linearizing (12.49) about an isothermal basic state in uniform motion leads to the perturbation potential vorticity equation for wave motion on a beta plane

$$\frac{D}{Dt}\left[\nabla^2 \psi' + \left(\frac{f_0^2}{N^2}\right)\frac{1}{\bar{\rho}}\frac{\partial}{\partial z}\left(\bar{\rho}\frac{\partial \psi'}{\partial z}\right)\right] + \beta\frac{\partial \psi'}{\partial x} = 0, \qquad (14.67)$$

where z refers to log–pressure height and $\psi' = (1/f_0)\Phi'$ to the geostrophic streamfunction.

Since coefficients are again constant, we consider solutions of the form $\exp[(z/2H) + i(kx + ly + mz - \sigma t)]$. Substitution into (14.67) recovers the dispersion relation for *quasi-geostrophic Rossby waves*

$$c_x - \bar{u} = -\frac{\beta}{k^2 + l^2 + \left(\frac{f_0^2}{N^2}\right)\left(m^2 + \frac{1}{4H^2}\right)}. \qquad (14.68)$$

[4]The direction of Rossby wave propagation can be deduced from the vorticity pattern in Fig. 14.16, relative to the frame of the material contour shown. Southward motion behind the cyclonic anomaly displaces that segment of the material contour equatorward, shifting the wave trough westward. Northward motion behind the anticyclonic anomaly has the same effect on the wave crest.

If the denominator is identified as the effective total wavenumber squared, (14.68) has a form analogous to (14.66), but modified by the stratification of mass. For many applications, $m^2 \gg 1/4H^2$. Vertical stretching (12.36) is then embodied in the term $(f_0^2/N^2)m^2$. As the vertical wavelength $2\pi/m \to \infty$, divergence forcing of vorticity vanishes and (14.68) reduces to the dispersion relation for barotropic nondivergent Rossby waves.

Because their intrinsic phase speeds depend on wavenumber, Rossby waves are dispersive. Expressing (14.68) as

$$\omega = -\frac{\beta k}{k^2 + l^2 + \left(\frac{f_0^2}{N^2}\right)\left(m^2 + \frac{1}{4H^2}\right)} \tag{14.69}$$

leads to the components of group velocity:

$$c_{gx} - \bar{u} = \beta \frac{k^2 - l^2 - \left(\frac{f_0^2}{N^2}\right)\left(m^2 + \frac{1}{4H^2}\right)}{\left[k^2 + l^2 + \left(\frac{f_0^2}{N^2}\right)\left(m^2 + \frac{1}{4H^2}\right)\right]^2} = \frac{\omega}{k}\left(1 + \frac{2\omega k}{\beta}\right), \tag{14.70.1}$$

$$c_{gy} = \frac{2\beta k l}{\left[k^2 + l^2 + \left(\frac{f_0^2}{N^2}\right)\left(m^2 + \frac{1}{4H^2}\right)\right]^2} = \frac{2l\omega^2}{\beta k}, \tag{14.70.2}$$

$$c_{gz} = \frac{2\beta m k \left(\frac{f_0^2}{N^2}\right)}{\left[k^2 + l^2 + \left(\frac{f_0^2}{N^2}\right)\left(m^2 + \frac{1}{4H^2}\right)\right]^2} = \left(\frac{f_0^2}{N^2}\right)\frac{2m\omega^2}{\beta k}. \tag{14.70.3}$$

Wave activity propagates in the positive y and z directions if $kl > 0$ and $km > 0$, respectively. Since $\omega < 0$, the form adopted for ψ' then implies phase propagation in the opposite directions. Thus, downward and equatorward phase propagation corresponds to energy propagation that is upward and poleward. For stationary waves in the presence of westerly flow, these conditions translate into phase surfaces that tilt westward in the direction of energy propagation, for example, in the upward and equatorward directions for the circumstances just described (Fig. 14.17).

Contrary to c_{gy} and c_{gz}, the sign of $c_{gx} - \bar{u}$ depends on the magnitude of k. For $k < \frac{\beta}{2|\omega|}$ (e.g., long zonal wavelength), group propagation relative to the medium is westward. But $k > \frac{\beta}{2|\omega|}$ (e.g., short zonal wavelength) gives intrinsic group propagation that is eastward. Separating westward and eastward group propagation is the locus of wavenumbers that makes $c_{gx} - \bar{u}$ vanish:

$$k^2 = l^2 + \left(\frac{f_0^2}{N^2}\right)\left(m^2 + \frac{1}{4H^2}\right). \tag{14.71}$$

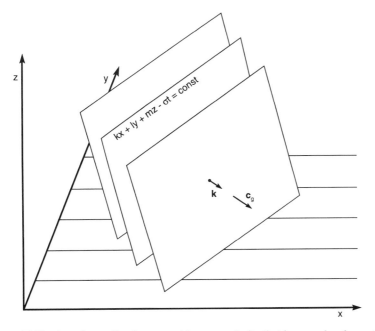

Figure 14.17 A stationary Rossby wave with group velocity that is upward and equatorward tilts westward in the same directions. Phase propagation relative to the medium is antiparallel to k.

These and other properties of Rossby wave propagation can be inferred from the dispersion characteristics $\omega(k, l, m) = $ const. Expressing (14.69) as

$$\left(k - \frac{\beta}{2|\omega|}\right)^2 + l^2 + \left(\frac{f_0^2}{N^2}\right)m^2 = \left(\frac{\beta}{2\omega}\right)^2 - \left(\frac{1}{2H}\right)^2 \qquad (14.72)$$

shows that surfaces of constant ω (Fig. 14.18) are nested ellipsoids centered on the points $(k, l, m) = \left(\frac{\beta}{2}|\omega|, 0, 0\right)$. The intrinsic group velocity (14.23) is directed orthogonal to those surfaces toward increasing ω. Thus, $\hat{c}_g$ is upward for $m > 0$ and equatorward for $l < 0$, corresponding to phase propagation in the opposite directions. For k smaller than values defined by (14.71), which are indicated in Fig. 14.18 by the dashed curve, $\hat{c}_g$ is westward, whereas larger k propagate eastward.

The dispersion relation can be rearranged for the vertical wavenumber

$$m^2 = \left(\frac{N^2}{f_0^2}\right)\left[\frac{\beta}{\overline{u} - c_x} - |k_h|^2\right] - \frac{1}{4H^2}. \qquad (14.73)$$

The gradient of planetary vorticity β, which provides the restoring force for horizontal displacements, is recognized as the essential positive contribution to m^2—analogous to the role played by N^2 for gravity waves (14.40). According to (14.73), Rossby waves propagate vertically only for a restricted range of

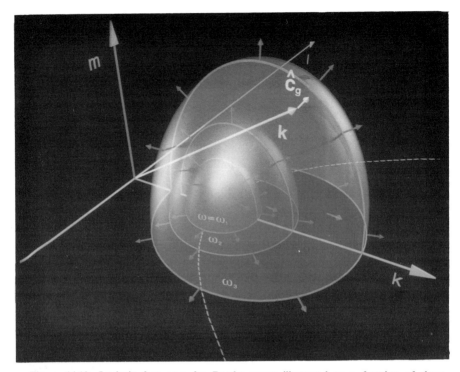

Figure 14.18 Intrinsic frequency for Rossby waves illustrated as a function of three-dimensional wavenumber $k = (k, l, m)$, with $\omega_3 > \omega_2 > \omega_1$. Group velocity relative to the medium is directed orthogonal to surfaces of intrinsic frequency toward increasing ω. For given l, zonal wavenumbers k greater than a critical value (dashed line) have eastward $\hat{c}_g$, whereas smaller k have westward $\hat{c}_g$.

zonal trace speed, namely, for westward phase propagation relative to the mean flow. As for gravity waves, short horizontal scales ($|k_h|$ large) are less able to propagate upward than large horizontal scales ($|k_h|$ small).

For stationary waves, (14.73) admits $m^2 > 0$ only for mean flow in the range

$$0 < \overline{u} < \frac{\beta}{|k_h|^2 + \frac{f_0^2}{4N^2 H^2}}. \tag{14.74}$$

If the mean flow is easterly or if it is westerly and exceeds a critical speed given by the right-hand side of (14.74), Rossby waves are external. Their influence is then exponentially small far above the forcing. Even for zonal wind satisfying (14.74), only the gravest horizontal dimensions ($|k_h|$ small) propagate vertically. The rest have energy that is trapped near the surface.

These features of Rossby wave propagation were advanced by Charney and Drazin (1961) to explain observations from the International Geophysical Year (IGY), which was the first comprehensive campaign to observe the atmosphere. Taken during 1958, observations from the IGY revealed that the synoptic eddies which dominate the tropospheric circulation are conspicuously absent in the stratosphere (refer to Figs 1.9 and 1.10). All eddy scales are absent from the time–mean circulation in the summer hemisphere, which is nearly circumpolar. According to the preceding analysis, only planetary-scale components of the wave spectrum can propagate into the strong westerlies of the winter stratosphere (Fig. 1.8). Disturbances of smaller horizontal dimension encounter turning levels not far above the tropopause (Fig. 14.14), below which they are trapped. During summer, stratospheric easterlies prevent all horizontal scales from propagating vertically because Rossby waves then encounter critical levels (Fig. 14.15), where $c_x = \bar{u}$ and wave activity is absorbed.

Charney and Drazin argued that these features of the stratospheric flow prevent kinetic energy, which is concentrated in the troposphere in synoptic-scale eddies, from propagating to high levels. Were vertical propagation possible for energy-bearing eddies in the troposphere, the $\exp(z/2H)$ amplification associated with stratification would produce enormous temperatures at high altitude to yield a corona similar to the sun.

Easterly flow in the tropical troposphere has a similar effect by blocking wave propagation between hemispheres. Therefore, large orographic features that produce amplified waves in the Northern Hemisphere exert only a minimal influence on the Southern Hemisphere. These interhemispheric differences combine to produce a broken storm track in the Northern Hemisphere (e.g., where amplified wave motions reinforce and interfere with with the zonal–mean jet) and a comparatively uniform one in the Southern Hemisphere.

14.5.3   Planetary Wave Propagation in Sheared Mean Flow

Westerly mean flow during winter and equinox enables planetary-scale Rossby waves to radiate into the middle atmosphere. Due to stratification of mass, those waves amplify vertically. Where planetary wave activity propagates and is absorbed can be understood from a ray-tracing analysis.

Consider zonal–mean flow that varies with latitude and height $\bar{u} = \bar{u}(y, z)$. Linearizing the quasi-geostrophic potential vorticity equation (12.49) and noting that $-\partial\bar{\psi}/\partial y = \bar{u}$ leads to the perturbation potential vorticity equation

$$\frac{D}{Dt}\left[\nabla^2\psi' + \frac{1}{\bar{\rho}}\frac{\partial}{\partial z}\left(\frac{f_0^2}{N^2}\bar{\rho}\frac{\partial\psi'}{\partial z}\right)\right] + \beta_e\frac{\partial\psi'}{\partial x} = 0, \qquad (14.75.1)$$

where

$$\beta_e = \beta - \bar{u}_{yy} - \frac{1}{\bar{\rho}}\frac{\partial}{\partial z}\left(\frac{f_0^2}{N^2}\bar{\rho}\frac{\partial\bar{u}}{\partial z}\right)$$

$$= \frac{\partial\overline{Q}}{\partial y} \qquad (14.75.2)$$

represents the meridional gradient of zonal–mean vorticity and is referred to as *beta effective*. According to (14.75), positive curvature of the basic flow reduces the effective restoring force to meridional displacements of air, whereas negative curvature reinforces it—analogous to gravity wave propagation in sheared flow (14.40).

Considering solutions of the form $\exp[(z/2H) + i(kx - \sigma t)]$ transforms (14.75) into a Helmholtz equation

$$\frac{\partial^2 \Psi_k^\sigma}{\partial y^2} + \frac{\partial}{\partial z}\left(\frac{f_0^2}{N^2}\frac{\partial \Psi_k^\sigma}{\partial z}\right) + \nu^2(y, z)\Psi_k^\sigma = 0, \tag{14.76.1}$$

where

$$\nu^2(y, z) = \frac{\beta_e}{\bar{u} - c_x} - k^2 - \frac{f_0^2}{4N^2 H^2} \tag{14.76.2}$$

represents a reduced two-dimensional index of refraction that describes propagation in the y-z plane. Solutions oscillate in regions where $\nu^2(y, z) > 0$, whereas they are evanescent in regions where $\nu^2(y, z) \leq 0$. Hence, propagation is excluded from regions where β_e is sufficiently small or $\bar{u}$ is sufficiently strong. Should $\bar{u}(y, z) \to c_x$, ν^2 becomes infinite and (14.76) is singular. The critical region defined by this condition differs from that for gravity waves because $\nu^2 \to \pm\infty$ on either side (compare Fig. 14.15b). So wave structure oscillates rapidly on one side of the critical line but is sharply evanescent on the other.

Analytical solutions to (14.76) exist only under very idealized circumstances. But, within the framework of the WKB approximation, wave propagation can be diagnosed in terms of monochromatic behavior along rays. The dispersion relation then assumes the same form as (14.69), but with β_e in place of β. Under these circumstances, propagation in the y-z plane is controlled by

$$l^2 + \left(\frac{f_0^2}{N^2}\right)m^2 = \frac{\beta_e}{\bar{u} - c_x} - k^2 - \frac{f_0^2}{4N^2 H^2}, \tag{14.77}$$

which is analogous to (14.40.2) for gravity waves. Figure 14.19 shows rays introduced into zonal–mean flow representative of the winter stratosphere. Planetary wave activity radiating upward and northward into strengthening westerlies of the polar-night jet encounters a turning line, where $l^2 + m^2$ vanishes, and is reflected. Refracted equatorward toward easterlies, wave activity along that ray then encounters a critical line, where ω and the group velocity vanish, so the ray stalls out. Analysis similar to that in Sec. 14.3.4 for gravity waves shows that, ahead of the critical line, planetary wave activity is frozen in the medium and hence absorbed. Planetary wave activity radiating initially upward and equatorward encounters a critical line and is absorbed even sooner. Dickinson (1968) used the foregoing analysis to explain observed am-

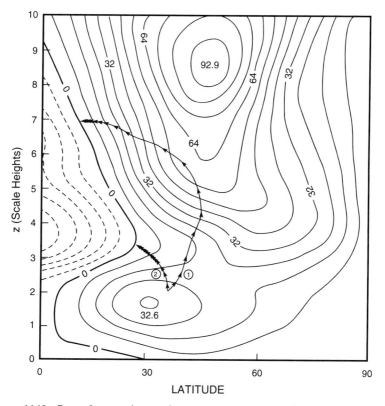

Figure 14.19   Rays of two stationary planetary wave components introduced into the lower stratosphere in zonal-mean winds (contoured in $m\,s^{-1}$) representative of northern winter on a midlatitude beta plane. Arrowheads mark uniform increments of time moving along a ray at the group velocity $\mathbf{c}_g$. The ray for component 1, which initially propagates upward and poleward, encounters a turning line in strong westerlies of the polar-night jet, where wave activity is refracted equatorward. As the ray approaches easterlies, it next encounters a critical line ($\bar{u} = 0$), where $\mathbf{c}_g$ vanishes. Propagation then stalls, leaving wave activity to be absorbed, so the ray terminates. The ray for component 2 is initially directed upward and equatorward, so wave activity encounters the critical line and is absorbed even sooner. Thus, wave activity introduced between tropical easterlies and strong polar westerlies can propagate vertically only a limited distance before being absorbed. (Frequency of arrowheads near the critical line reduced for clarity.)

plitudes of planetary waves, which, according to Charney and Drazin's (1961) one-dimensional analysis, would amplify vertically with little attenuation.

Horizontal wave propagation can be diagnosed in similar fashion. Figure 14.20a shows the anomalous planetary wave pattern observed in the troposphere for disturbed winters during El Niño. A series of positive and negative anomalies marks a planetary wavetrain that radiates poleward from anomalous heating in the equatorial Pacific, which is associated with disturbed convection during El Niño (Chapter 15). Decreasing β along the wavetrain introduces

(a)

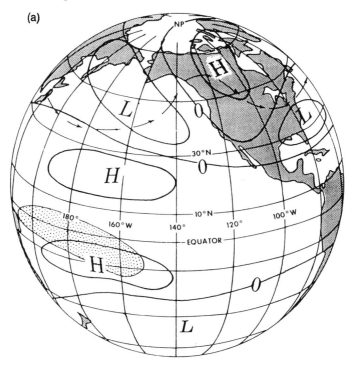

Figure 14.20 (a) Anomalous planetary wave field for northern winters during El Niño, when anomalous convection (stippled) is positioned in the tropical central Pacific. The anomalous ridge over western Canada and trough over the eastern United States characterize the so-called *Pacific North America (PNA) pattern* that upsets the normal track of the jet stream (wavy trajectory) and cyclone activity during El Niño winters. After Horel and Wallace (1981); see Wallace and Gutzler (1981) for a detailed plot. (*continues*)

a turning latitude, where wave activity is reflected equatorward. Calculations with convective heating in the tropical central Pacific (Fig. 14.20b) produce similar structure, in particular, the prominent ridge over western Canada and trough over the eastern United States that characterize the so-called Pacific North America (PNA) pattern during El Niño winters. By interfering with the extratropical planetary wave field (e.g., as manifested in undulations of the jet stream in Fig. 14.20a), those anomalies upset the normal pattern of cyclone development, which controls weather in the troposphere.

14.6   Wave Absorption

In the absence of dissipation, wave activity is conserved following a wavepacket (e.g., moving along a ray with the velocity c_g).[5] When dissipation is present,

[5] Amplitude, however, varies along a ray (14.41); see Lighthill (1978).

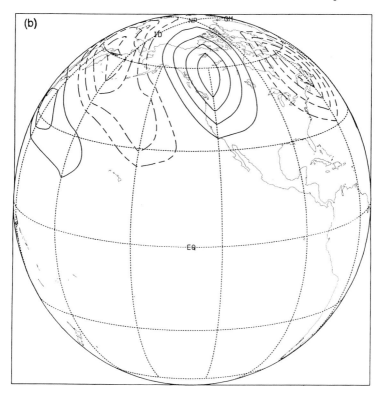

Figure 14.20 (*Continued*) (b) Anomalous 300-mb height field calculated from the linearized primitive equations with convection positioned in the equatorial central Pacific and with zonal-mean winds in Fig. 14.19. See Garcia and Salby (1987) for details of the calculation and Hoskins and Karoly (1981) for a formal ray–tracing analysis on the sphere. Similar structure emerges in calculations that include longitudinal variations of the basic state, albeit modified in detail by zonal variations in background wind (see Jin and Hoskins, 1995).

this is no longer true. Wave activity is then removed from individual wavepackets and deposited in the mean state, which forces $\bar{u}$ through the convergence of momentum flux (14.5). To illustrate how absorption alters the wave field, consider the propagation of planetary waves in the presence of radiative damping.

Planetary waves have vertical wavelengths much longer than a scale height, so radiative transfer reduces to longwave (LW) cooling to space (Sec. 8.6). For small departures from the equilibrium temperature, radiative transfer can then be represented as Newtonian cooling (8.79). The linearized form of the thermodynamic equation (12.46) then becomes

$$\frac{D}{Dt}\left(\frac{\partial \Phi'}{\partial z}\right) + N^2 w' = -\alpha\left(\frac{\partial \Phi'}{\partial z}\right), \qquad (14.78.1)$$

where α is the damping rate of temperature perturbations (units of inverse time), or

$$\left(\frac{D}{Dt} + \alpha\right)\frac{\partial \psi'}{\partial z} + \frac{N^2}{f_0}w' = 0. \tag{14.78.2}$$

Using (14.78) to eliminate w' in the linearized form of the vorticity equation (12.47),

$$\left[\frac{d(\zeta + f)}{dt}\right]' = f_0\frac{1}{\bar\rho}\frac{\partial}{\partial z}(\bar\rho w'), \tag{14.79}$$

where the left-hand side accounts for shear in the mean flow $\bar u = \bar u(y, z)$, yields the perturbation vorticity equation

$$\frac{D}{Dt}\left[\nabla^2\psi' + \frac{1}{\bar\rho}\frac{\partial}{\partial z}\left(\frac{f_0^2}{N^2}\bar\rho\frac{\partial \psi'}{\partial z}\right)\right] + \alpha\frac{1}{\bar\rho}\frac{\partial}{\partial z}\left(\frac{f_0^2}{N^2}\bar\rho\frac{\partial \psi'}{\partial z}\right) + \beta_e\frac{\partial \psi'}{\partial x} = 0, \tag{14.80}$$

which is a generalization of (14.75).

Within the framework of the WKB approximation, considering solutions of the form $\exp[(z/2H) + i(kx + ly + mz - \sigma t)]$ transforms (14.80) into the dispersion relation

$$\left[1 + i\left(\frac{\alpha}{\omega}\right)\right]m^2 = m_0^2, \tag{14.81.1}$$

where

$$m_0^2 = \left(\frac{N^2}{f_0^2}\right)\left[\frac{\beta_e}{\bar u - c_x} - |k_h|^2\right] - \frac{1}{4H^2} \tag{14.81.2}$$

satisfies the dispersion relation in the absence of damping (14.77). Solving for the vertical wavenumber gives

$$m = m_r + im_i = \frac{e^{i\left(\frac{\phi}{2}\right)}}{\left[1 + \left(\frac{\alpha^2}{\omega^2}\right)\right]^{\frac{1}{2}}}m_0, \tag{14.82.1}$$

(Fig. 14.21a), where (since $\omega < 0$)

$$\phi = \tan^{-1}\left(\frac{\alpha}{|\omega|}\right). \tag{14.82.2}$$

An individual wave component has vertical structure

$$e^{(\frac{1}{2H} + im)z} = e^{(\frac{1}{2H} - m_i)z} \cdot e^{im_r z}.$$

Because $m_0 > 0$ for downward phase propagation, (14.82) implies $m_i > 0$. Hence, thermal damping introduces an imaginary component to m that offsets the vertical amplification of wave activity (Fig. 14.21b). Representing the ratio of timescales for wave oscillation and damping, m_i increases with increasing $(\alpha/\omega)m_0$. Wave amplitude then grows with height more slowly than under adiabatic conditions, which corresponds to the kinetic energy density

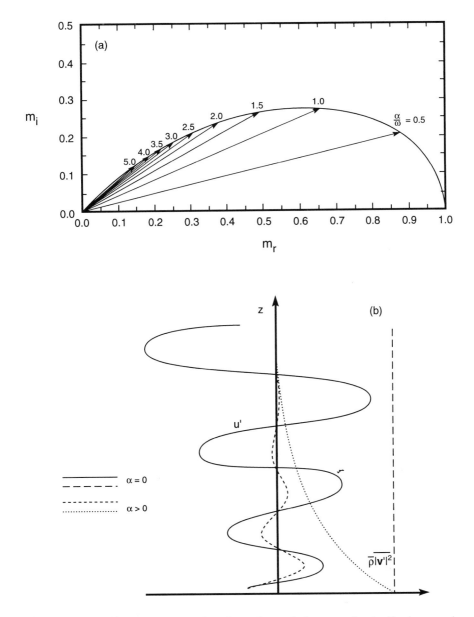

Figure 14.21   (a) Polar representation of complex vertical wavenumber for Rossby waves in the presence of Newtonian cooling with coefficient α. (b) Vertical structure and kinetic energy density of an individual wave component for $\alpha = 0$ and $\alpha > 0$.

(e.g., $\overline{\rho|v_h'|^2}$) decreasing upward as progressively more wave activity is absorbed. For $(\alpha/\omega)m_0$ sufficiently large, amplitude itself decreases with height, so components with small intrinsic frequency or short vertical wavelength decay vertically above their forcing.

In practice, thermal dissipation is introduced whenever the circulation is driven out of radiative equilibrium. Wave motions do this naturally by advecting air from one radiative environment to another. In the stratosphere, the polar-night vortex is displaced out of radiative equilibrium by planetary waves, which drive air across latitude circles (refer to Fig. 1.10). Longwave cooling to space then acts on individual air parcels in proportion to their departure from local radiative equilibrium, which destroys anomalous temperature and the accompanying motion. Air displaced into polar darkness, where the radiative-equilibrium temperature is very low, finds itself anomalously warm, so it cools (Fig. 8.27). Conversely, the vortex in the summer stratosphere remains relatively undisturbed and therefore close to radiative equilibrium.

Because it is proportional to the vertical derivative of ψ', Newtonian cooling leads to stronger damping of short vertical wavelengths than long vertical wavelengths. For vertical wavelengths comparable to H, the cool-to-space approximation breaks down. As shown in Fig. 8.29, LW exchange between neighboring layers then makes α itself strongly scale dependent, being much faster for short vertical wavelengths than long vertical wavelengths.

Absorption can change along a ray because α varies spatially. According to Fig. 8.29, the timescale for thermal damping decreases from several weeks in the troposphere to only a couple of days in the upper stratosphere. Of greater significance, $(\alpha/\omega)m_0 \rightarrow \infty$ if the intrinsic frequency is Doppler-shifted to zero. Near a critical line, oscillations become stationary relative to the medium, allowing wave activity to be fully absorbed.

14.7 Nonlinear Considerations

Under inviscid adiabatic conditions, the equations governing wave propagation become singular at a critical line. Dissipation removes the singularity by absorbing wave activity as the wavelength collapses and c_g approaches zero. Strong gradients near the critical line then magnify thermal damping and diffusion, which, by smoothing out those gradients, act to destroy organized wave motion. These conclusions follow solely from considerations of the first-order equations, in which only simple forms of dissipation are represented. More generally, singular behavior near a critical line invalidates treating the perturbation equations in isolation. In the limit of small dissipation, wave stress exerted on the mean flow (14.5) becomes concentrated at the critical line and unbounded (Dickinson, 1970). Even with plausible dissipation, the convergence of momentum flux implies large accelerations locally, so $\overline{u}$ cannot be regarded as steady. Further, nonlinear interaction between components of the wave spectrum can then also become large.

Recall that the governing equations can be linearized as long as second-order terms such as $\boldsymbol{v}' \cdot \nabla \boldsymbol{v}'$ are much smaller than those first order in wave amplitude. Since, for an individual wave component, the material derivative is

$$
\begin{aligned}
\frac{d}{dt} &= \frac{D}{Dt} + \boldsymbol{v}' \cdot \nabla \\
&= -i\omega + i\boldsymbol{v}' \cdot \boldsymbol{k} \\
&= -ik\left[(c_x - \overline{u}) - \boldsymbol{v}' \cdot \frac{\boldsymbol{k}}{k} \right],
\end{aligned} \tag{14.83}
$$

ignoring second-order behavior is tantamount to requiring the intrinsic trace speed $|c_x - \overline{u}|$ to be large compared to the eddy velocity $|\boldsymbol{v}'|$ induced by the wave. Near a critical line, this requirement breaks down because $|c_x - \overline{u}| \to 0$. Instead of undergoing a small oscillation during the limited time for a wavelength to propagate by, an individual parcel then experiences finite displacements, which can lead to a permanent rearrangement of air.[6]

Attention was first drawn to this issue by Kelvin (1880), who noted the peculiar behavior in Fig. 14.22a for the critical level of a gravity wave under incompressible conditions and neutral stratification. For $N^2 = 0$, (14.40) implies a total streamfunction relative to the medium that is symmetric about $z = z_c$ and contains closed streamlines within some neighborhood of the critical level.[7] Known as *Kelvin's cat's-eye pattern*, this distribution of ψ describes a series of vortices that overturn and eventually wind up air inside the critical region (Fig. 14.22b). Gradients supporting organized wave motion are then sheared down to small scales, where diffusion can destroy them efficiently, culminating in the absorption of wave activity arriving at the critical level. In the limit of weak dissipation, another scenario is possible. The vorticity u_z is then a conserved property (Problem 14.25). Therefore, mixing will homogenize the distribution of vorticity and drive $\overline{u}_{zz}$ to 0, in the same fashion that convective mixing homogenizes the distribution of potential temperature and drives $\partial\theta/\partial z$ to zero (Sec. 7.6). Once this condition is achieved, $m^2 = 0$ in a neighborhood of z_c (14.40.2), so wave activity is subsequently reflected from the critical level.

Rossby waves produce analogous behavior through horizontal advection. Figure 14.23 shows planetary wave activity incident on a critical line at low latitude. Under weakly dissipative conditions (Fig. 14.23a), nonlinear advection near the critical line overturns material lines, which are delineated by contours of potential vorticity Q. Air inside the critical region is rolled up in a series of vortices that destroy organized wave motion on large scales by shearing it down to small scales, where diffusion is effective. Efficient horizontal

---

[6] Dissipation keeps displacements bounded by introducing an imaginary component to $c_x - \overline{u}$ (Problem 14.53).

[7] Rossby waves behave similarly, but with β_e in place of $-\overline{u}_{zz}$.

(a)

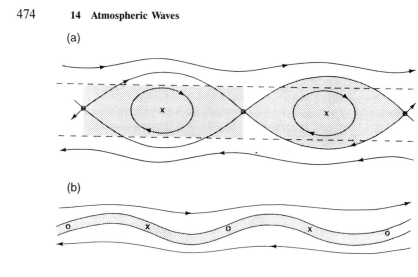

(b)

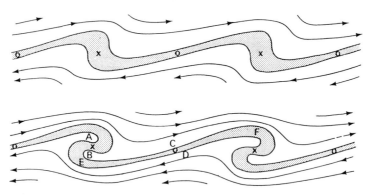

Figure 14.22   (a) Streamfunction at the critical level of a simple gravity wave under neutral stratification, which assumes the form of *Kelvin's cat's-eye pattern.* (b) Evolution of a material contour initialized at the critical level. Adapted from Scorer (1978).

mixing homogenizes Q near the critical line, driving the refractive index there (14.76) to zero. Wave activity is then reflected from the critical line, which can be inferred from the weak phase tilt with latitude (dashed line). Under strongly dissipative conditions (Fig. 14.23b), diffusion prevents the formation of sharp gradients that promote vortex roll-up. Wave activity then continues to propagate equatorward and be absorbed at the critical line, which is evident from the strong westward tilt of phase with decreasing latitude.

Nonlinearity is also introduced by large wave amplitude because it likewise makes $|v'| = O(|c_x - \bar{u}|)$. The conditions presented in Fig. 14.24 were observed during an intense wind storm, when the mountain wave over the Colorado Rockies was greatly amplified. Air moves along isentropic surfaces through large vertical displacements (e.g., descending some 5 km in the lee of

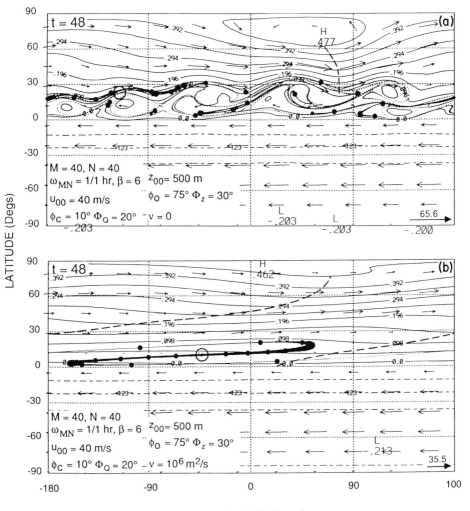

Figure 14.23   Nonlinear evolution of a planetary wave critical layer under barotropic conditions and with zonal wavenumber 1 incident on it. Illustrated in terms of potential vorticity (contours) and the position and thickness of a Lagrangian ensemble (bold/broken curve and dots) that was initialized at the critical latitude. (a) In the presence of weak dissipation, potential vorticity is rolled into a series of vortices that mix air inside the critical layer, driving $\beta_e = \partial \overline{Q}/\partial y$ to zero. Wave activity is then reflected from the critical region, as manifested by the absence of phase tilt (dashed line). (b) In the presence of strong dissipation, diffusion prevents the formation of sharp gradients that support vortex roll-up, so $\partial \overline{Q}/\partial y > 0$ is maintained. The incident planetary wave then continues to propagate equatorward and be absorbed, as is manifested by its systematic meridional tilt. After Salby *et al.* (1990).

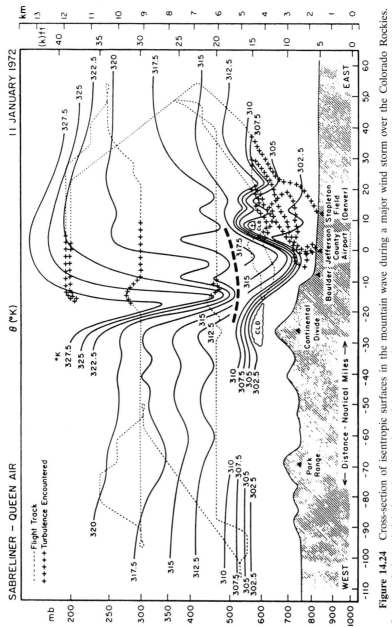

Figure 14.24 Cross-section of isentropic surfaces in the mountain wave during a major wind storm over the Colorado Rockies. Severe turbulence (+) is found at elevations of the mountain cap cloud and below, as well as near the tropopause, but it is visible only in the isolated rotor cloud near 600 mb. After Lilly (1978).

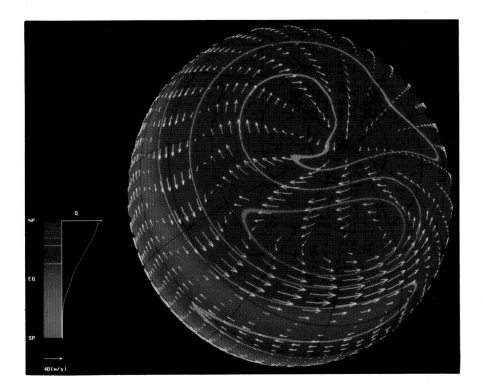

the continental divide). The steep slope of isentropic surfaces reduces $\partial\theta/\partial z$ and static stability locally. In addition, the large amplitude exaggerates vertical shear. Both effects tend to destabilize the organized motion through mechanisms discussed in Sec. 13.3. Richardson numbers less than $\frac{1}{4}$ then produce intense turbulence that damps the wave by cascading energy to small scales, where it is dissipated efficiently. Advected downstream, that turbulence can be violent and is visible only in the isolated rotor cloud at 600 mb (Sec. 9.3). During the event shown, a Boeing 707 on approach to Denver airport sustained loads of 3 g's. Similar conditions were encountered during the 1950s in the Sierra Wave Project, when the jet stream was observed by instrumented sailplanes. The following excerpt from a chronicle of the project (Lincoln, 1972) describes an episode familiar to aviators:

> *Editor's Note: On the day described, Larry Edgar was flying a Pratt-Read, one of the strongest sailplanes ever built, and he had the ultimate experience of flying turbulence in the rotor. This story has probably saved the lives of a number of aviators who have learned the indicated lesson and treat rotor clouds with respect.*
>
> *"Lloyd Licher and I examined the wreckage and compared notes with others. The nose pulled off just at the seats, by what seemed to be a tension failure of the steel tubing. Considering Larry's weight and the instrumentation, this should have required just in excess of 16 g's.... The wreckage showed the left wing to have broken at altitude. It broke downward near the root. The tail boom was broken cleanly from the fuselage pod, at altitude, and appears to have come off upward. The various control cables going from nose to tail were pulled apart completely—in a bunch. The force needed to do this should be far over 10,000 pounds."*

JOACHIM P. KUETTNER

(The pilot survived.)

Large amplitude results naturally from vertical propagation. Upon reaching sufficient height, a gravity wave will have achieved a perturbation potential temperature θ' large enough to drive the vertical gradient of total potential temperature to zero locally

$$\frac{\partial\theta}{\partial z} = \frac{\partial\bar\theta}{\partial z} + \frac{\partial\theta'}{\partial z} = 0. \tag{14.84}$$

Isentropic surfaces are then vertical, which corresponds to zero static stability. The gravity wave may then *break* by overturning the stratification, analogous to a surface water wave incident on a shoreline. Air of low θ is folded over air of higher θ, as shown in Fig. 14.25 for a gravity wave that has reached the mesosphere, so stable stratification is transformed into unstable stratification. Violating the *Ri* criterion for turbulence, the ensuing stratification culminates in vigorous mixing, with wave momentum being deposited at the site of absorption. Momentum deposited by gravity waves is thought to be responsible for driving the zonal–mean circulation of the mesosphere out of radiative

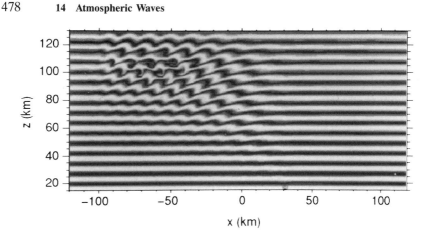

Figure 14.25 Potential temperature distribution in the presence of gravity waves, which are excited in the troposphere by a moving bump analogous to a squall line. Positioned near $x = 30$ km, the disturbance is 10 km wide and migrates eastward. Gravity waves amplify with height sufficiently to break in the mesosphere, where they overturn θ surfaces. Note that wave activity propagates eastward relative to the medium, so phase lines tilt eastward with height. Courtesy of R. Garcia (NCAR).

equilibrium (Fig. 1.7), where temperature increases from the summer pole (in perpetual daylight) to the winter pole (in perpetual darkness).

Planetary waves under disturbed conditions introduce similar behavior by rearranging air horizontally. Figure 14.26 shows the rearrangement of air by planetary waves under amplified conditions typical of the wintertime stratosphere. Anticyclonic motion has displaced the polar-night vortex out of zonal symmetry and advects midlatitude air into the polar cap. To conserve potential vorticity, that air spins up anticyclonically, which tends to establish a reversed circulation over the pole (e.g., easterly circumpolar flow). This behavior may be regarded as an expansion of the critical region in Fig. 14.23a, in which $|v'| = O(|c_x - \bar{u}|)$. Both are characterized by closed streamlines that produce a complex rearrangement of air.

Behavior like that in Fig. 14.26 occurs sporadically in the stratosphere during northern winter (refer to Fig. 17.13). Planetary waves forced by large orography in the Northern Hemisphere often amplify sufficiently to form closed streamlines at middle and high latitudes. Termed *sudden stratospheric warmings*, those episodes are marked by a reversal of the zonal–mean flow and, from

Figure 14.26 Distribution of potential vorticity Q and horizontal motion in a two-dimensional calculation representative of the stratospheric circulation at 10 mb under disturbed conditions. Color and velocity scales shown at left. High-Q polar air (blue), which marks the polar-night vortex, has been displaced well off the pole by an amplified planetary wave and has undergone pronounced distortion. Low-Q air from equatorward (red) that has been advected into the polar cap spins up anticyclonically to form a reversed circulation, with easterly circumpolar flow at high latitudes. These features are characteristic of a *stratospheric sudden warming* (refer to Fig. 17.13).

thermal wind balance, abrupt warming at high latitude—as much as 50 K in just a few days. Air over the winter pole, which is in perpetual darkness, then becomes warmer than air in the tropics, which experiences ozone heating. According to Fig. 1.10a, which shows the 10-mb circulation during a stratospheric warming, this dramatic change in zonal–mean properties actually follows from a complex rearrangement of air.

In the Southern Hemisphere, orographic features at the earth's surface are comparatively small, so planetary waves are weaker. For this reason, sudden warmings and the disturbed conditions accompanying them are rare in the Southern Hemisphere. Less disturbed, the Antarctic polar-night vortex remains closer to radiative equilibrium and therefore much colder than its counterpart over the Arctic.

Suggested Reading

Atmosphere–Ocean Dynamics (1982) by Gill includes an excellent introduction to gravity waves at the graduate level and discusses the role of rotation in adjustments to equilibrium.

The monograph *Atmospheric Waves* (1975) by Gossard and Hooke contains a thorough treatment of acoustic gravity waves, inclusive of rotation.

Waves in Fluids (1978) by Lighthill is an advanced treatment of wave propagation that formally develops wave dispersion and ray tracing analysis.

An Introduction to Dynamic Meteorology (1992) by Holton contains a nice introduction to Rossby waves, drawing examples from orographically forced disturbances. *Middle Atmosphere Dynamics* (1987) by Andrews *et al.* includes an advanced treatment of planetary waves that formally develops the concept of wave activity and discusses its implications for transport.

Problems

14.1. Derive the zonal–mean momentum budget (14.5).

14.2. Show that surface water waves must satisfy (14.9.2) at the free surface.

14.3. Describe the circumstances under which a wave field can be formally represented as a superposition of plane waves.

14.4. Show that transforming the derivative of a function is equivalent to multiplying its transform (14.14).

14.5. Verify that (14.17) is the solution of the homogeneous boundary value problem (14.16).

14.6. Demonstrate that vertical motion vanishes for surface water waves in the long wave limit, $k \rightarrow 0$, to show that shallow water waves represent an analogue of large-scale hydrostatic disturbances in the atmosphere.

14.7. Formally derive the power spectrum (14.25.2) for the Gaussian initial conditions (14.25.1).

14.8. Construct a wavepacket of surface water waves $\eta'(x, t)$ from a Gaussian spectrum of wavenumbers centered at k_0 and of spectral width dk:

$$\eta_k = \frac{1}{\sqrt{2\pi dk}} e^{-\frac{(k-k_0)^2}{2dk^2}}.$$

(a) Use this expression to produce a counterpart of (14.22). (b) Sketch the wavepacket for given k_0 and dk. (c) Recover expressions (14.23) for the group velocity in the limit $dk \to 0$.

14.9. On approach to JFK Airport in New York, winds are gusting nearly along the shoreline. Yet, whitecaps are observed to approach the shore virtually head on. Construct a simple model based on shallow water waves propagating in the x-y plane, with depth decreasing linearly to zero at the shoreline, which coincides with $y = 0$. For a wavepacket characterized initially by $k = (k_0, l_0)$, (a) determine $k(y)$, (b) sketch phase lines corresponding to successive positions of the wavepacket for $l_0 > 0$, and (c) plot rays that are initially oriented 60° and 30° from the shoreline.

14.10. Obtain the dispersion relation (14.34) for acoustic–gravity waves. (*Hint:* What condition must be satisfied for a homogeneous system of linear equations to have nontrivial solution?)

14.11. Show that the dispersion relation for acoustic–gravity waves reduces to that for sound waves in the limit of high frequency and short wavelength.

14.12. Show that, at a given elevation, external gravity waves perform no net work on the overlying column during a complete cycle and therefore transmit no energy vertically. You may restrict the analysis to simple gravity waves in the shortwave limit.

14.13. Lenticular clouds (Chapter 9) form preferentially during winter. Explain their seasonality in relation to conditions favoring gravity waves.

14.14. Recover the limiting dispersion relation for gravity waves (14.35.2) by invoking incompressibility.

14.15. Derive the group velocity (14.44) for simple gravity waves.

14.16. Demonstrate that phase lines of a stationary gravity wave forced at the surface in mean westerlies (e.g., undulations of isentropic surfaces) must slope westward with height.

14.17. Show that the flux of zonal momentum transmitted vertically by simple gravity waves, $\rho_0 \overline{u'w'}$, is positive (negative) if their group velocity is upward and they propagate eastward (westward).

14.18. Show that wave activity for simple gravity waves propagates vertically one vertical wavelength for each horizontal wavelength that it propagates horizontally.

14.19. For gravity waves with horizontal and vertical wavenumbers k and m, respectively, describe the limiting behavior of phase and group propagation for (a) $m/k \to 0$, (b) $m/k \to -\infty$.

14.20. The Manti-La Sal range in southern Utah comprises an isolated ridge that has a characteristic width of 10 km and is oriented N–S. If winds blow from the SW, the range's N–S extent is much longer than 10 km, and static stability is characterized by $N^2 = 2.0 \times 10^{-4}$ s^{-2}, (a) at what polar angle are lenticular clouds likely to be found? (b) What will be their orientation relative to the range?

14.21. Verify that (14.41) satisfies the Taylor–Goldstein equation within the framework of the WKB approximation.

14.22. A gravity wave of zonal wavenumber $k = 0.5$ km^{-1} propagates vertically through mean westerlies of uniform positive shear

$$\bar{u} = \bar{u}_0 + \Lambda z,$$

with $\bar{u}_0 = 10$ m s^{-1}, $\Lambda = 0.1$ m s^{-1} km^{-1}, and constant static stability $N^2 = 2.0 \times 10^{-4}$ s^{-2}. In the framework of the Boussinesq and WKB approximations, sketch the profiles of intrinsic frequency and vertical wavenumber if the gravity wave (a) is stationary, (b) propagates westward at 10 m s^{-1}, and (c) propagates eastward at 10 m s^{-1}.

14.23. Describe the physical circumstances under which reflection may be ignored and the WKB approximation is a valid description of wave propagation.

14.24. A stationary gravity wave of the form

$$w'(x, z) = W(z)e^{ikx}$$

is excited by flow over elevated terrain. The horizontal and vertical scales are short enough for wave propagation to be controlled by local conditions. If the upstream flow $\bar{u}$ is independent of height but the stratification varies as

$$N^2(z) = N_0^2 \left(1 - \frac{z^2}{H^2}\right),$$

(a) characterize the propagation according to levels where wave activity is vertically propagating and external, (b) sketch the vertical wave structure, and (c) determine the range of $\bar{u}$ for which wave activity is trapped at the surface.

14.25. (a) Show that, in the limit of no rotation and weak dissipation, vorticity is a conserved property for incompressible gravity waves and is therefore rearranged by the motion. (b) In light of this feature, describe the limiting behavior ($t \to \infty$) of $\bar{\theta}$ and $\bar{u}$ anticipated near a gravity wave critical level, where the motion assumes the form of a cat's-eye pattern (Fig. 14.22).

14.26. Show that, under the Boussinesq approximation, the linearized continuity equation has the counterpart

$$\frac{D\hat{\rho}'}{Dt} + w\frac{\partial \ln \overline{\rho}}{\partial z} = 0,$$

where $\hat{\rho}' = \rho'/\overline{\rho}$.

14.27. A gravity wave of wavenumber $\mathbf{k} = (k, m)$ propagates into a region where potential temperature varies with height as

$$\frac{\theta}{\theta_0} = -\frac{(z - z_0)^2}{H^2},$$

with $H = \text{const}$. (a) Describe propagation in the x-z plane, sketching the corresponding ray. (b) Sketch the vertical structure of the wave. (c) Describe the behavior under fully nonlinear conditions.

14.28. Flow over oscillating terrain excites a gravity wave of horizontal wavenumber $k = 0.5 \text{ km}^{-1}$ that propagates horizontally and vertically away from its source region at $(x, z) = (0, 0)$. If the wave can be treated within the Boussinesq approximation, the background flow varies vertically as

$$\overline{u} = \overline{u}_0 + \Lambda z,$$

where $\overline{u}_0 = 10 \text{ m s}^{-1}$ and $\Lambda = 1.0 \text{ m s}^{-1} \text{ km}^{-1}$, and $N^2 = 2.0 \times 10^{-4} \text{ s}^{-2} = \text{const}$, (a) plot the ray for a steady wave, (b) plot rays for an unsteady spectrum of wave activity (e.g., generated by low-level flow that fluctuates with time) of the form $\exp(-\sigma^2/2\Sigma^2)$, with $\Sigma = (2 \text{ hr})^{-1}$, for frequencies $\sigma = 0$, $\pm\Sigma$, and $\pm 2\Sigma$, and (c) contrast the behavior for $\Sigma \to \infty$ against that for steady forcing: $\Sigma \to 0$.

14.29. Within the framework of the Boussinesq approximation, consider *inertio-gravity waves*, which are buoyancy waves with horizontal scales long enough to be influenced by rotation. (a) Write down a set of equations governing such motion on an f-plane. (b) Show that these waves satisfy the dispersion relationship

$$\frac{m^2}{k^2} = \frac{N^2 - \omega^2}{\omega^2 - f^2},$$

which is a generalization of (14.36) and illustrates that inertio-gravity waves propagate vertically only for $|\omega| > |f|$. Atmospheric tides, which are generated by diurnal variations of heating, are inertio-gravity waves. Estimate the range of latitude for vertical propagation of (c) the diurnal tide, $\sigma = 1.0 \text{ cpd}$, and (d) the semidiurnal tide, $\sigma = 2.0 \text{ cpd}$.

14.30. Barometric registrations following the eruption of Krakatoa revealed an oscillatory disturbance in surface pressure having an amplitude of 1 mb. If the period of the oscillation was 1 hr, estimate the amplitude of (a) the pressure perturbation at 80 km and (b) the wind perturbation at 80 km.

14.31. Show that the Lamb mode is the only nontrivial solution to the homogeneous boundary value problem (14.60).

14.32. Consider hydrostatic perturbations to a motionless isothermal atmosphere that are introduced by unsteady vertical motion at the surface. If the surface forcing is indiscriminate, with the power spectrum of *white noise*,

$$|W_k^\sigma|^2 = \text{const},$$

(a) determine the wave response in terms of the power spectrum of pressure at some level and (b) for a particular wavenumber, sketch the power spectra of forcing and response as functions of frequency.

14.33. Show that the power spectrum of white noise is recovered from (a) the impulse

$$w(t) = \frac{1}{\sqrt{2\pi}\tau} e^{-\frac{t^2}{2\tau^2}}$$

in the limit $\tau \to 0$ and (b) a superposition of such impulses at random times t_j. (c) Interpret this limiting behavior in light of a characteristic timescale for unsteadiness.

14.34. Interpret the Lamb waves satisfying (14.61) in light of shallow water waves.

14.35. Derive the vertical structure equation (14.59).

14.36. Show that the Lamb mode also satisfies the equations governing nonhydrostatic motion.

14.37. Derive the thermodynamic equation (14.33.4).

14.38. Discuss how propagation in Figs. 14.14 and 14.15 would be modified by realistic variations of N^2.

14.39. Demonstrate that Rossby wave propagation is restricted to intrinsic frequencies smaller than 2Ω.

14.40. See Problem 12.51.

14.41. Consider bartropic nondivergent motion on a beta plane positioned at latitude ϕ_0, which is periodic in x, over length X, and unbounded in y. Steady zonal flow $\bar{u}$ over an isolated mountain excites Rossby waves that radiate away from the topographic feature. The effect of the mountain may be treated as a vorticity source

$$Q(x, y) = \frac{Q_0}{2\pi L^2} \exp\left(-\frac{x^2 + y^2}{2L^2}\right).$$

(a) Provide a rationale for treating surface forcing in this manner. (b) Write down an equation governing the motion. (c) Obtain an approximate solution for the wave field, subject to the condition of radiation away from the source region, by (1) expressing the streamfunction in terms of the Fourier transform and inverse:

$$\psi(x, y) = \frac{1}{2\pi X} \sum_{s=-\infty}^{\infty} \int_{-\infty}^{\infty} dl \Psi_{sl} e^{i(k_s x + ly)}$$

$$\Psi_{sl} = \int_{-X/2}^{X/2} dx \int_{-\infty}^{\infty} dy\, \psi(x, y) e^{-i(k_s x + ly)},$$

$$k_s = s \frac{2\pi}{X} \qquad s = 0, \pm 1, \pm 2, \ldots,$$

which apply to semiperiodic geometry, and (2) treating propagation independently in the positive and negative y half-planes. (d) Plot the wave field for $\phi_0 = 45°$, $\bar{u} = 10$ m s^{-1}, $X = 2\pi a \cos \phi_0$, and $L/X = 1/4\pi$.

14.42. In terms of the vorticity budget (12.36), discuss how volume heating excites planetary waves.

14.43. Derive the group velocity for quasi-geostrophic Rossby waves (14.70).

14.44. Show that stationary planetary wave activity generated by westerly flow over elevated terrain will be found downstream of its topographic forcing.

14.45. Derive the Helmholtz equation governing propagation of quasi-geostrophic planetary waves through sheared zonal flow (14.76).

14.46. Consider quasi-geostrophic motion in spherical geometry. (a) Provide an equation governing the propagation of planetary waves in latitude and height. For a stationary wave that is invariant with height, $\partial \psi'/\partial z = 0$, and propagates horizontally through uniform westerlies of 20 m s^{-1}, estimate the polar turning latitude for (b) zonal wavenumber 1 and (c) zonal wavenumber 4.

14.47. In light of Problem 14.46, describe the wave propagation that could be established with a source at midlatitudes if potential vorticity is homogenized at low latitudes by horizontal mixing (Sec. 14.7).

14.48. Steady flow over oscillating terrain at 45° latitude excites a Rossby wave of zonal wavenumber k that propagates zonally and vertically. If the undisturbed motion,

$$\bar{u}(z) = \bar{u}_0 + \Lambda z,$$

with $\bar{u}_0 = 10$ m s^{-1} and $\Lambda = 0.05$ m s^{-1} km^{-1}, varies slowly enough for wave propagation to be treated locally, (a) plot the refractive index squared as a function of height, (b) sketch the vertical wave structure for $k = 1$–3, (c) plot rays emanating from the source at $(x, y) = (0, 0)$ for wavenumbers $k = 1$–3, and (d) describe vertical propagation of a broad spectrum of stationary wave activity generated near the surface.

14.49. Under the conditions of Problem 14.48, but for $\Lambda = -0.05$ m s^{-1} km^{-1}, $k = 2$, and in the presence of linear dissipation (i.e., Rayleigh friction and Newtonian cooling) with a timescale of 10 days, (a) plot the refractive index squared as a function of height, (b) plot the vertical structure of the wave, (c) plot the ray emanating from the wave source. (d) What condition(s) must be satisfied for the behavior to remain linear?

14.50. Show that propagation of quasi-geostrophic planetary waves in the presence of radiative dissipation is governed by (14.80).

14.51. Consider a gravity wave with potential temperature amplitude $\Theta(z)$ propagating in a region of mean potential temperature $\overline{\theta}(z)$ and static stability $N^2(z)$. (a) Determine the vertical wavenumber m at which the wave will break. (b) Express the breaking wavenumber in terms of the wave's horizontal wavenumber and phase speed to derive an expression governing the breaking height. (c) If the wave's forcing is maintained below, describe the propagation anticipated in the limit $t \to \infty$.

14.52. Under barotropic nondivergent conditions, evaluate the time for a Rossby wavepacket to reach its critical line at $y = 0$ from $y > 0$, where the mean flow varies as

$$\overline{u} = c(1 + \Lambda y).$$

14.53. Within the framework of the WKB approximation, consider a barotropic nondivergent Rossby wave of wavenumber k and phase speed c, which propagates on a beta plane toward its critical line, where the mean flow behaves as

$$\overline{u} = c(1 + \Lambda y)$$

with $\Lambda = 0.01 \text{ km}^{-1}$, and in the presence of small linear dissipation $\epsilon = 0.01kc$. (a) Plot the streamfunction over one wavelength. (b) Plot successive positions of the material contour that coincides initially with $y = 0$. (c) Reversal of the potential vorticity gradient, $\partial Q / \partial y < 0$, provides a condition for dynamical instability (Chapter 16). From advection of the material contour in (b), determine if, when, and where this condition is met.

14.54. Associate the breaking of gravity waves with the breaking of Rossby waves in terms of analogous properties and behavior.

Chapter 15 | The General Circulation

Thermal equilibrium requires net radiative forcing to vanish for the earth–atmosphere system as a whole. Although it applies globally, this requirement need not hold locally. Net radiation (Fig. 1.29c) implies that low latitudes experience radiative heating, whereas middle and high latitudes experience radiative cooling. To preserve thermal equilibrium, that radiative heating and cooling must be compensated by a mechanical transfer of heat from tropical to extratropical regions. The simplest mechanism to transfer heat meridionally is a steady zonally symmetric Hadley cell. This thermally direct overturning (Sec. 1.5), can be driven by heating in the tropics and cooling in the extratropics, which make air rise and sink across isentropic surfaces. Although steady zonally symmetric motion can balance radiative heating and cooling, the observed circulation is more complex (Fig. 1.9).

How meridional heat transfer is actually accomplished in the atmosphere is influenced profoundly by the earth's rotation. Net radiation is almost uniform in longitude, so it tends to establish a steady zonally symmetric thermal structure with temperature and isobaric height decreasing poleward. Since contours of isobaric height are then parallel to latitude circles, geostrophic equilibrium requires steady motion to be nearly zonal—virtually orthogonal to the simple meridional overturning hypothesized above. That zonally symmetric circulation is also virtually orthogonal to the temperature gradient (a reflection of the geostrophic paradox; Sec. 12.1), so it cannot transfer heat poleward efficiently. The meridional temperature gradient must then steepen until heat transferred meridionally by the zonally symmetric Hadley cell balances radiative heating in the tropics and cooling in the extratropics.

This produces a meridional gradient of zonal–mean temperature that is much steeper than observed (e.g., such as that resulting under radiative–convective equilibrium in Fig. 10.1). Thermal wind balance (12.11) then implies a strong zonal jet aloft. The speed of this jet can be inferred from conservation of specific angular momentum $(u + \Omega a \cos \phi)a \cos \phi$ (11.29) for individual air parcels drifting poleward in the hypothetical Hadley cell. An air parcel that is initially motionless at the equator then assumes a zonal velocity at latitude ϕ of

$$u = \Omega a \frac{\sin^2 \phi}{\cos \phi},$$

which predicts wind in excess of 300 m s^{-1} at $45°$.

Discrepancies from observed zonal–mean behavior follow from the absence of zonally asymmetric eddies, which completely alter the extratropical circulation. Those asymmetric motions transfer heat poleward far more efficiently than can a zonally symmetric Hadley cell in the presence of rotation. Intrinsically unsteady, extratropical disturbances are fueled by reservoirs of atmospheric energy and exchanges among them, which maintain the general circulation.

15.1 Forms of Atmospheric Energy

From the time it is absorbed until it is eventually rejected to space, energy undergoes a complex series of transformations between three basic reservoirs:

1. Thermal energy, which is represented in the temperature and moisture content of air.
2. Gravitational potential energy, which is represented in the horizontal distribution of atmospheric mass.
3. Kinetic energy, which is represented in air motion.

Transformations from thermal and potential energy to kinetic energy are responsible for setting the atmosphere into motion and maintaining the circulation against frictional dissipation.

15.1.1 Moist Static Energy

The thermal energy of moist air has two contributions: Sensible heat content reflects the molecular energy of dry air, which is the dominant component of moist air. Latent heat content reflects the energy that can be imparted to the dry air when the vapor component condenses. Both are represented in the specific enthalpy (Sec. 5.3)

$$h = c_p T + l_v r + h_0. \tag{15.1}$$

Here, $c_p T$ measures the sensible heat content of a moist parcel, whereas $l_v r$ measures its latent heat content.

The first law (2.15) implies that the specific enthalpy of an individual air parcel changes according to

$$\frac{dh}{dt} - \alpha \frac{dp}{dt} = \dot{q}_{net}, \tag{15.2.1}$$

where $\dot{q}_{net}$ is the net rate heat is absorbed by the parcel. Incorporating hydrostatic equilibrium allows (15.2.1) to be expressed

$$\frac{dh}{dt} + g\frac{dz}{dt} = \dot{q}_{net}$$

or with (15.1)

$$\frac{d}{dt}(c_p T + l_v r + \Phi) = \dot{q}_{net} \tag{15.2.2}$$

$$= \dot{q}_{rad} + \dot{q}_{mech},$$

where $\dot{q}_{rad}$ and $\dot{q}_{mech}$ denote radiative and mechanical components of heat transfer, respectively. The quantity $(c_p T + l_v r + \Phi)$ defines the *moist static energy* of the air parcel, which includes thermal as well as potential energy. The parcel's kinetic energy is much smaller, so $(c_p T + l_v r + \Phi)$ very nearly equals its total energy. A parcel's moist static energy changes only through heat transfer with its environment. It is unaffected by vaporization and condensation, which involve only internal redistributions of sensible and latent heat contents.[1]

Mechanical heat transfer is dominated by turbulent mixing, which leads to exchanges of sensible heat (temperature) and latent heat (moisture)

$$\dot{q}_{mech} = \dot{q}_{sen} + \dot{q}_{lat}. \tag{15.3}$$

These components of turbulent heat transfer can be represented in terms of mean gradients of temperature and moisture and corresponding eddy diffusivities (13.13).

Equation (15.2.2) describes an air parcel's moist static energy in terms of advection and the distributions of sources and sinks. A parcel moving poleward will have acquired moist static energy at low latitudes through absorption of radiative, sensible, and latent heats from the surface. Upon returning equatorward, that parcel will have lost moist static energy through radiative cooling to space. Hence, completing a meridional circuit (e.g., the one symbolized in Fig. 6.7) results in a net poleward transfer of moist static energy, which compensates radiative heating at low latitude and cooling at middle and high latitudes. Integrating (15.2) over the entire atmosphere and averaging over time obtains the mean change of moist static energy, which must vanish under thermal equilibrium. It follows that mean heat transfer into the atmosphere must likewise vanish

$$\overline{\dot{q}_{rad}} + \overline{\dot{q}_{sen}} + \overline{\dot{q}_{lat}} = 0, \tag{15.4}$$

where the overbar denotes global and time mean. Radiative cooling of the atmosphere (Fig. 8.24) must then be balanced by transfers of sensible and latent heat from the earth's surface (Fig. 1.27).

---

[1] This differs from the development in Sec. 10.7, which focuses on the dry air component and therefore must account for transfer of latent heat from the vapor component.

Inside the boundary layer, turbulent diffusion of heat and absorption of longwave (LW) radiation from the surface increase an air parcel's sensible heat content. Absorption of water vapor from the surface increases its latent heat content. Both increase the moist static energy of the parcel. Outside the boundary layer and convection, turbulent diffusion is small enough to be ignored. Radiative cooling then takes over as the primary diabatic process and gradually depletes the parcel's store of moist static energy, which is subsequently replenished when it returns to regions of positive heat transfer.

15.1.2   Total Potential Energy

Thermal energy is reflected in the internal energy of dry air. Heat transfer from the surface increases an atmospheric column's temperature and hence its internal energy. The hypsometric relationship (6.12) then implies vertical expansion of the column. Since it elevates the column's center of mass, a process that increases the column's internal energy must also increase its potential energy.

Consider a dry atmosphere in hydrostatic equilibrium and in which air motion occurs on a timescale short enough to be regarded adiabatic. An incremental volume of unit cross-sectional area (Fig. 15.1) has internal energy

$$d\mathscr{U} = \rho c_v T dz. \tag{15.5.1}$$

Integrating upward from the surface gives the column internal energy

$$\mathscr{U} = c_v \int_0^\infty \rho T dz. \tag{15.5.2}$$

The potential energy of the incremental volume is

$$d\mathscr{P} = \rho g z dz$$
$$= -z dp. \tag{15.6.1}$$

Then the column potential energy up to a height z is given by

$$\mathscr{P}(z) = \int_{p(z)}^{p_s} z' dp', \tag{15.6.2}$$

where $z' = z'(p')$ and the integral is evaluated in the direction of increasing pressure. Integrating by parts transforms this into

$$\mathscr{P}(z) = z'p'\big|_{p(z)}^{p_s} + \int_0^z p dz'$$
$$= -zp(z) + R\int_0^z \rho T dz',$$

which reduces to

$$\mathscr{P} = R\int_0^\infty \rho T dz \tag{15.7}$$

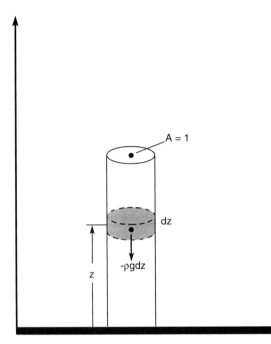

Figure 15.1   An incremental volume of unit cross-sectional area and height z inside an atmospheric column.

for $z \to \infty$. A similar expression holds for the column internal energy (15.5.2), so

$$\frac{\mathscr{P}}{\mathscr{U}} = \frac{R}{c_v}$$

$$= \gamma - 1. \tag{15.8}$$

Equation (15.8) asserts that the internal and potential energies of an atmospheric column preserve a constant ratio, which is a consequence of hydrostatic equilibrium. Energy can then be drawn from those reservoirs only in a fixed proportion. Adding the internal and potential energies of the column results in

$$\mathscr{P} + \mathscr{U} = \int_0^\infty \rho c_p T dz$$

$$= \frac{c_p}{g} \int_0^{p_s} T dp$$

$$= \mathscr{H}, \tag{15.9}$$

which is just the column enthalpy and is called the *total potential energy* (Margules, 1903).

Consider the atmosphere as a whole and, for the sake of illustration, with an absence of heat transfer at its boundaries. Because p vanishes at the top of the atmosphere and w vanishes at its bottom, no work is performed on this system as well. Then the first law for the entire atmosphere and inclusive of kinetic energy $\mathscr{K}$ asserts

$$\Delta(\mathscr{H} + \mathscr{K}) = 0, \qquad (15.10.1)$$

or

$$\Delta\mathscr{K} = -\Delta\mathscr{H}$$
$$= -\Delta(\mathscr{P} + \mathscr{U}), \qquad (15.10.2)$$

where column properties are understood to be globally integrated. Under adiabatic conditions, the atmosphere's kinetic energy is drawn from its reservoir of total potential energy.

15.1.3 Available Potential Energy

Only a small fraction of the atmosphere's total potential energy is actually available for conversion to kinetic energy. Consider an atmosphere that is initially motionless and barotropically stratified (Sec. 12.2). Isobaric surfaces then coincide with isentropic surfaces, as shown in Fig. 12.5a. Because air parcels must move along θ surfaces under adiabatic conditions, no pressure gradient exists in the directions of possible motion. Consequently, a barotropically stratified atmosphere possesses no means of generating motion internally, so none of its potential energy is available for conversion to kinetic energy.

Suppose this atmosphere is now heated at low latitude and cooled at middle and high latitudes, in such a manner that net heating integrated over the globe vanishes. By the first law, the atmosphere's total energy is unchanged. So is its total potential energy if the atmosphere remains motionless (15.10.1). But, locally, total potential energy clearly does change. Increased temperature increases total potential energy at low latitude, whereas reduced temperature diminishes total potential energy at middle and high latitudes. By the hypsometric relationship, the vertical spacing of isobaric surfaces is expanded at low latitude and compressed at middle and high latitudes, which tilts isobaric surfaces clockwise into the positions assumed in Fig. 12.5b. Just the reverse occurs for isentropic surfaces under hydrostatically stable conditions. Heating increases potential temperature at low latitude, which, since θ increases upward, is achieved by bringing down greater values from above. Conversely, cooling reduces θ at middle and high latitudes, which is achieved by bringing up smaller values from below. Isentropic surfaces are thus tilted counterclockwise (Fig. 12.5b). In this fashion, nonuniform heating drives isobaric surfaces out of coincidence with isentropic surfaces and hence drives the thermal structure into baroclinic stratification.

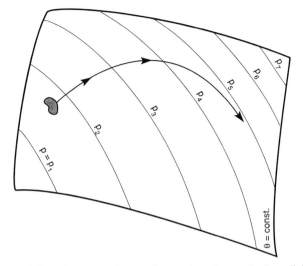

Figure 15.2 Variation of pressure along an isentropic surface under baroclinic stratification. The pressure variation vanishes under barotropic stratification.

The atmosphere now possesses a pressure gradient along isentropic surfaces (Fig. 15.2). An air parcel can then accelerate under the action of the pressure gradient force, so potential energy is available for conversion to kinetic energy. Even though no potential energy has been added to the atmosphere as a whole, that energy reservoir has been tapped by introducing a horizontal variation of buoyancy. Similar reasoning shows that uniformly heating the atmosphere, which does increase its store of total potential energy, introduces none that is available for conversion to kinetic energy.

From hydrostatic equilibrium, the variation of pressure along a θ surface reflects a nonuniform distribution of mass above that surface. The situation is analogous to two immiscible fluids of different densities that have been

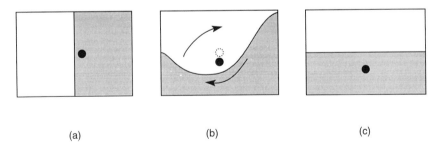

(a)   (b)   (c)

Figure 15.3 Schematic illustrating the rearrangement of mass in a system of two immiscible fluids of different densities which are initially juxtaposed horizontally. In the absence of rotation, heavier fluid (shaded) undercuts and comes to rest underneath lighter fluid, which renders the system hydrostatically stable and lowers its center of mass (dot). Rotation modifies this adjustment by deflecting motion into the page. Adapted from Wallace and Hobbs (1977).

juxtaposed horizontally (Fig. 15.3). Hydrostatic equilibrium implies a pressure gradient force directed from the heavier fluid to the lighter one. Motion will then develop internally to alleviate the mechanical imbalance. The nonrotating system in Fig. 15.3 accomplishes this by rearranging mass so that heavier fluid undercuts and eventually comes to rest underneath lighter fluid, so the system's final state is hydrostatically stable. By lowering the center of gravity of the system, this process releases potential energy, which is converted into kinetic energy and eventually dissipated by viscosity to increase the system's internal energy.

In the atmosphere, horizontal rearrangement of mass is inhibited by rotation, which deflects air motion parallel to isobars (e.g., into the page in Fig. 15.3). Nevertheless, the pressure gradient along θ surfaces enables air motion to develop and, although more complex, to neutralize the mechanical imbalance. The Coriolis force makes those motions highly rotational, which favors horizontal mixing (Fig. 12.4). This process is illustrated by interleaving swirls of tropical and polar air in the cyclone off the coast of Africa in Fig. 1.15. By mixing air horizontally, extratropical cyclones homogenize the distribution of mass along isentropic surfaces, which drives isobaric surfaces back into coincidence with isentropic surfaces and restores the thermal structure to barotropic stratification. Those motions also result in a net poleward transfer of heat and moisture because air drawn poleward from low latitudes has greater moist static energy than air drawn equatorward from high latitudes. In making the horizontal mass distribution uniform, air motions lower the overall center of gravity of the atmosphere. Potential energy is therefore converted into kinetic energy, which in turn is dissipated by viscosity, converted into internal energy, and finally rejected to space as heat.

Air motions responsible for this redistribution of mass are fueled by a conversion of potential energy to kinetic energy. They are termed *baroclinic instability* because their source of energy is directly related to the baroclinicity of the stratification. Since temperature then varies along isobaric surfaces, thermal wind balance implies that those motions are also related to vertical shear of the flow.

ADIABATIC ADJUSTMENT

The potential energy available for conversion to kinetic energy is reflected in the departure from barotropic stratification. Consider an adiabatic redistribution of mass from a given baroclinic state. Because air must move along θ surfaces, horizontal mixing will eventually render the distributions of mass and pressure uniform over those surfaces, restoring the thermal structure to barotropic stratification. In that limiting state, the atmosphere has no more potential energy available for conversion, so $\mathcal{H} = \mathcal{H}_{min}$.

The *available potential energy* $\mathcal{A}$ (Lorenz, 1955) is defined as the difference between the total potential energy for the state under consideration and the minimum that would result through an adiabatic rearrangement of mass

$$\mathscr{A} = \mathscr{H} - \mathscr{H}_{\min}. \tag{15.11}$$

Available potential energy has the following properties:

1. $\mathscr{A} + \mathscr{K}$ is preserved under adiabatic rearrangement, so $\Delta\mathscr{K} = -\Delta\mathscr{A}$.
2. $\mathscr{A}$ is positive for baroclinic stratification and zero for barotropic stratification.
3. $\mathscr{A}$ is uniquely determined by the distribution of mass.

To derive an expression for $\mathscr{A}$, consider an isentropic surface of area S. The average pressure on this surface

$$\overline{p} = \frac{1}{S} \int_S p \, dS \tag{15.12}$$

is conserved under adiabatic rearrangement (Problem 15.7). The atmosphere as a whole has total potential energy

$$\begin{aligned}
\mathscr{H} &= \int_S dS \int_0^\infty \rho c_p T \, dz \\
&= \frac{c_p}{g} \int_S dS \int_0^{p_s} T \, dp.
\end{aligned} \tag{15.13}$$

In terms of potential temperature, this becomes

$$\begin{aligned}
\mathscr{H} &= \frac{c_p}{g p_0^\kappa} \int_S dS \int_0^{p_s} \theta p^\kappa \, dp \\
&= \frac{1}{p_0^\kappa (1 + \kappa) \Gamma_d} \int_S dS \int_0^{p_s} \theta \, dp^{1+\kappa},
\end{aligned} \tag{15.14}$$

where $\Gamma_d = g/c_p$ is the dry adiabatic lapse rate. Because the atmosphere is hydrostatically stable, θ increases monotonically with height, so it may be interchanged with p for the vertical coordinate. Integrating (15.14) by parts results in

$$\mathscr{H} = \frac{S}{p_0^\kappa (1 + \kappa) \Gamma_d} \left[\overline{\theta p^{1+\kappa}} \Big|_0^{p_s} + \int_{\theta_s}^\infty \overline{p^{\kappa+1}} \, d\theta \right], \tag{15.15}$$

with θ_s denoting the potential temperature at the surface. The limit $p = 0$ in the first term in brackets vanishes (Problem 15.9). Defining $p = p_s$ for $\theta < \theta_s$ allows us to express (15.15) as

$$\mathscr{H} = \frac{S}{p_0^\kappa (1 + \kappa) \Gamma_d} \int_0^\infty \overline{p^{\kappa+1}} \, d\theta. \tag{15.16}$$

When the stratification has been driven barotropic, $\mathscr{H} = \mathscr{H}_{\min}$ and p has become uniform over isentropic surfaces. Since p then equals $\overline{p}$,

$$\mathscr{H}_{\min} = \frac{S}{p_0^\kappa (1 + \kappa) \Gamma_d} \int_0^\infty \overline{p}^{\kappa+1} \, d\theta. \tag{15.17}$$

Hence, the state under consideration has available potential energy

$$\mathscr{A} = \frac{S}{p_0^\kappa (1 + \kappa)\Gamma_d} \int_0^\infty \left[\overline{p^{\kappa+1}} - \overline{p}^{\kappa+1} \right] d\theta. \tag{15.18}$$

For small departures p' and T' from their areal averages, (15.18) is approximated by

$$\mathscr{A} = \frac{RS}{2gp_0^\kappa} \int_0^\infty \overline{p}^{1+\kappa} \left(\frac{\overline{p'^2}}{\overline{p}^2} \right) d\theta \tag{15.19.1}$$

and

$$\mathscr{A} = \frac{gS}{2} \int_0^{p_s} \frac{1}{\overline{N}^2} \left(\frac{\overline{T'^2}}{\overline{T}^2} \right) dp, \tag{15.19.2}$$

the demonstration of which is left as an exercise. Thus, the variance of pressure on an isentropic surface and the variance of temperature on an isobaric surface, which reflect the degree of baroclinicity, are direct measures of available potential energy. Calculations based on typical conditions reveal that only about 0.1% of the atmosphere's total potential energy is actually available for conversion to kinetic energy (Problem 15.12).

15.2   Heat Transfer in an Axisymmetric Circulation

Available potential energy is released by meridional air motion, which transfers heat poleward and is influenced importantly by the Coriolis force. To explore how rotation influences poleward heat transfer, we consider a simple model within the framework of the Boussinesq approximation (Sec. 12.5.3). A spherical atmosphere of finite vertical extent H is bounded below by a rigid surface and above by a free surface. The lower boundary is maintained at a zonally symmetric temperature $T_0(\phi)$, with T_0 decreasing poleward, whereas the upper boundary is maintained at $T_0(\phi) + \Delta T$, with $\Delta T = \text{const}$. If $\Delta T < 0$, this corresponds to heat being supplied at the atmosphere's lower boundary and rejected at its upper boundary. It is through imposed temperature contrast that laboratory simulations are driven thermally, which will be seen in the next section to provide a mechanical analogue of this theoretical model.

Following Charney (1973), we consider zonally symmetric motion $\overline{\mathbf{v}} = (\overline{u}, \overline{v}, \overline{w})$ in the presence of turbulent diffusion. Scaling variables as in Sec. 11.4 casts the governing equations into nondimensional form, which allows the prevailing balances to be identified in terms of the dimensionless Rossby number and *Ekman number*:

$$Ro = \frac{U}{2\Omega a} \qquad E = \frac{K}{2\Omega H^2}, \tag{15.20}$$

where $K = K_M = K_H$ is a constant eddy diffusivity (Sec. 13.1). The limit of fast rotation and small friction makes $E \leq O(Ro) \ll 1$. To streamline the development, we follow the convention adopted in Sec. 12.5. Other than Ro and E, dimensionless factors are absorbed into the dependent variables, which allows the nondimensional equations to be expressed in a form nearly identical to their dimensional counterparts. Prevailing balances can then be identified in dimensional form because, at a specified order of Ro and E, those dimensionless parameters drop out. Proceeding as we did earlier casts the governing equations in log-pressure coordinates into

$$Ro\left\{\frac{\overline{v}}{a}\frac{\partial \overline{u}}{\partial \phi}+\overline{w}\frac{\partial \overline{u}}{\partial z}\right\}-f\overline{v}=EK\frac{\partial^2 \overline{u}}{\partial z^2},\tag{15.21.1}$$

$$Ro\left\{\frac{\overline{v}}{a}\frac{\partial \overline{v}}{\partial \phi}+\overline{w}\frac{\partial \overline{v}}{\partial z}\right\}+f\overline{u}=-\frac{1}{a}\frac{\partial \overline{\Phi}}{\partial \phi}+EK\frac{\partial^2 \overline{v}}{\partial z^2},\tag{15.21.2}$$

$$\frac{\partial \overline{\Phi}}{\partial z}=\frac{RT}{H},\tag{15.21.3}$$

$$\frac{1}{a\cos\phi}\frac{\partial}{\partial \phi}(\cos\phi\overline{v})+\frac{\partial \overline{w}}{\partial z}=0,\tag{15.21.4}$$

$$Ro\left\{\frac{\overline{v}}{a}\frac{\partial \overline{T}}{\partial \phi}+\overline{w}\frac{\partial \overline{T}}{\partial z}\right\}=EK\frac{\partial^2 \overline{T}}{\partial z^2},\tag{15.21.5}$$

where factors such as a, $f=\sin\phi$, and K are understood to be dimensionless and $O(1)$. At the rigid lower boundary, no-slip and the imposed temperature require

$$\begin{aligned}\overline{u}=\overline{v}=\overline{w}=0\\ \overline{T}=T_0(\phi)\end{aligned}\qquad z=0.\tag{15.21.6}$$

At the free surface, the stresses $\tau_{xz}=K(\partial \overline{u}/\partial z)$ and $\tau_{yz}=K(\partial \overline{v}/\partial z)$ must vanish. Further, for motion that is nearly geostrophic, vertical deflections of the free surface are small enough to replace the kinematic condition (Sec. 14.1) by the requirement of no vertical motion. Then the upper boundary conditions are

$$\begin{aligned}\frac{\partial \overline{u}}{\partial z}=\frac{\partial \overline{v}}{\partial z}=0\qquad \overline{w}=0\\ \overline{T}=T_0(\phi)+\Delta T\end{aligned}\qquad z=H.\tag{15.21.7}$$

The dominant balances in (15.21) are geostrophic and hydrostatic equilibrium, with advective acceleration and friction introducing a small ageostrophic component into the momentum balance. As in Chapter 12, we consider an asymptotic series solution, one that is valid in the limit $E\to 0$. Expanding dependent variables in power series of $E^{\frac{1}{2}}$, for example,

$$\overline{v}=\overline{v}^{(0)}+E^{\frac{1}{2}}\overline{v}^{(1)}\dots,\tag{15.22}$$

allows the motion to be determined recursively to successively higher accuracy.[2]

To $O(E^0)$, the governing equations reduce to

$$f\overline{v}^{(0)} = 0, \tag{15.23.1}$$

$$f\overline{u}^{(0)} = -\frac{1}{a}\frac{\partial \overline{\Phi}^{(0)}}{\partial \phi}, \tag{15.23.2}$$

$$\frac{\partial \overline{\Phi}^{(0)}}{\partial z} = \frac{R}{H}\overline{T}^{(0)}, \tag{15.23.3}$$

$$\frac{1}{a\cos\phi}\frac{\partial}{\partial\phi}\left[\cos\phi\overline{v}^{(0)}\right] + \frac{\partial\overline{w}^{(0)}}{\partial z} = 0, \tag{15.23.4}$$

which are identical to their dimensional counterparts. Equations (15.23) imply that, to lowest order, the motion is zonal and in geostrophic balance. By cross-differentiating (15.23.2) and (15.23.3), we obtain the thermal wind relation

$$\frac{\partial \overline{u}^{(0)}}{\partial z} = -\frac{R}{aHf}\frac{\partial \overline{T}^{(0)}}{\partial\phi}. \tag{15.24}$$

With (15.23.1) and the boundary conditions, (15.23.4) gives

$$\overline{w}^{(0)} \equiv 0. \tag{15.25}$$

Then the thermodynamic equation (15.21.5) implies

$$\frac{\partial^2 \overline{T}^{(0)}}{\partial z^2} = 0. \tag{15.26.1}$$

The vertical heat flux is nondivergent, so the circulation is in diffusive equilibrium. Heat absorbed at the atmosphere's lower boundary is transferred vertically to its upper boundary, where it is rejected. The solution of (15.26.1) satisfying the boundary conditions of imposed temperature is

$$\overline{T}^{(0)}(\phi, z) = T_0(\phi) + \Delta T\left(\frac{z}{H}\right). \tag{15.26.2}$$

Then (15.24) implies that the E^0 motion increases linearly with height

$$\overline{u}^{(0)}(z) = -\frac{R}{aHf}\frac{\partial T_0}{\partial\phi}z. \tag{15.27}$$

---

[2] Powers of $E^{\frac{1}{2}}$ are the appropriate form for the expansion because K enters the equations with second-order derivatives.

The $O(E^{\frac{1}{2}})$ equations have the same form as (15.23). Proceeding along similar lines leads to

$$\overline{v}^{(1)} \equiv 0, \tag{15.28.1}$$

$$\overline{w}^{(1)} \equiv 0, \tag{15.28.2}$$

$$\overline{T}^{(1)} \equiv 0. \tag{15.28.3}$$

Thermal wind balance (15.24) then implies that the $O(E^{\frac{1}{2}})$ zonal flow is independent of height

$$\overline{u}^{(1)} \neq \overline{u}^{(1)}(z). \tag{15.29}$$

The solution accurate to $O(E^{\frac{1}{2}})$ has constant vertical shear, so it violates the upper boundary conditions of no stress on the free surface. This deficiency follows from the neglect of viscous terms, which reduces the order of the governing equations. Instead, the full solution must possess a boundary layer, in which the motion adjusts sharply to meet boundary conditions. Viscous terms are then nonnegligible in a shallow neighborhood of the free surface, where friction drives the motion out of geostrophic balance and renders the preceding equations invalid.

To obtain an $O(E^{\frac{1}{2}})$ solution that is uniformly valid, we introduce a stretching transformation that accounts for sharp changes near the upper boundary. Boundary conditions can be satisfied by augmenting the first-order geostrophic motion with ageostrophic corrections

$$\boldsymbol{v}^{(1)} = \overline{u}_g^{(1)}\boldsymbol{i} + \overline{\boldsymbol{v}}_a^{(1)}, \tag{15.30}$$

where $\overline{u}_g^{(1)}$ is the inviscid motion described by (15.29). The ageostrophic velocity $\boldsymbol{v}_a^{(1)}$ imposes a secondary circulation onto the geostrophic flow, analogous to Ekman pumping (Sec. 13.4), which transports heat poleward by driving the motion across isotherms. Introducing the stretching transformation

$$H - z = E^{\frac{1}{2}}\zeta \tag{15.31}$$

inside the upper boundary layer magnifies sharp changes to make viscous terms comparable to others in the momentum balance. Then the $O(E^{\frac{1}{2}})$ momentum equations reduce to the Ekman balance

$$-f\overline{v}_a^{(1)} = K\frac{\partial^2 \overline{u}_a^{(1)}}{\partial \zeta^2}, \tag{15.32.1}$$

$$f\overline{u}_a^{(1)} = K\frac{\partial^2 \overline{v}_a^{(1)}}{\partial \zeta^2}, \tag{15.32.2}$$

with the geostrophic contribution automatically satisfied by $\overline{u}_g^{(1)}$. Requiring the total solution accurate to $O(E^{\frac{1}{2}})$ to satisfy (15.21.7) provides boundary conditions at the free surface

$$\frac{\partial \overline{u}^{(0)}}{\partial z} - \frac{\partial \overline{u}_a^{(1)}}{\partial \zeta} = 0 \qquad \frac{\partial \overline{v}_a^{(1)}}{\partial \zeta} = 0 \qquad \zeta = 0. \tag{15.32.3}$$

For the solution to reduce to the inviscid behavior outside the boundary layer, the ageostrophic corrections must also satisfy

$$\overline{u}_a^{(1)} \sim 0 \qquad \overline{v}_a^{(1)} \sim 0 \qquad \zeta \to \infty. \tag{15.32.4}$$

The boundary value problem (15.32) has solution

$$\overline{u}_a^{(1)}(\zeta) = \frac{1}{2\gamma} \frac{\partial \overline{u}^{(0)}}{\partial z} e^{-\gamma\zeta} \left(\sin \gamma\zeta - \cos \gamma\zeta\right), \tag{15.33.1}$$

$$\overline{v}_a^{(1)}(\zeta) = \frac{1}{2\gamma} \frac{\partial \overline{u}^{(0)}}{\partial z} e^{-\gamma\zeta} \left(\sin \gamma\zeta + \cos \gamma\zeta\right) \tag{15.33.2}$$

(Figs. 15.4a and b), where

$$\gamma = \sqrt{\frac{f}{2K}} \tag{15.33.3}$$

is the same parameter defining the Ekman spiral in Sec. 13.2.1.

Because $\overline{v}$ vanishes outside the boundary layer, meridional mass transfer is proportional to

$$\int_{\substack{H-\zeta \\ \zeta \to \infty}}^{H} E^{\frac{1}{2}} \overline{v}^{(1)} dz = E \int_0^\infty \overline{v}_a^{(1)} d\zeta$$

$$= -E \frac{KR}{aHf^2} \frac{\partial T_0}{\partial \phi}. \tag{15.34}$$

Hence, meridional mass transfer along the upper boundary is $O(E)$ and poleward (Fig. 15.4c). Integrating the continuity equation (15.21.4) gives the vertical motion just beneath the boundary layer

$$\overline{w}(H) = \int_{\substack{H-\zeta \\ \zeta \to \infty}}^{H} \frac{1}{a \cos \phi} \frac{\partial}{\partial \phi} \left(\cos \phi v\right) dz$$

$$= E \int_0^\infty \frac{1}{a \cos \phi} \frac{\partial}{\partial \phi} \left(\cos \phi \overline{v}_a^{(1)}\right) d\zeta,$$

which is likewise $O(E)$. Thus, to leading order, vertical motion beneath the upper boundary layer is

$$\overline{w}^{(2)}(H) = -\frac{KR}{Ha^2 \cos \phi} \frac{\partial}{\partial \phi} \left(\frac{\cos \phi}{f^2} \frac{\partial T_0}{\partial \phi}\right). \tag{15.35}$$

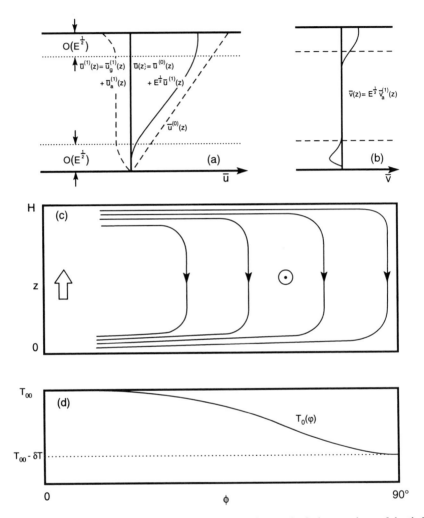

Figure 15.4 Zonally symmetric circulation in a Boussinesq spherical atmosphere of depth H that is driven thermally by imposing different temperatures along its rigid lower boundary and upper free surface. Motion is valid in the limit of fast rotation and small friction, which makes the *Ekman number E* small. The solution accurate to $O(E^{\frac{1}{2}})$: $\overline{\boldsymbol{v}} = \overline{\boldsymbol{v}}^{(0)} + E^{\frac{1}{2}}\overline{\boldsymbol{v}}^{(1)}$, is characterized by a strong zonal jet, with meridional motion confined to shallow boundary layers along the bottom and top. (a) Zonal motion. (b) Meridional motion. (c) Streamfield in a meridional cross section. (d) Meridional profile of imposed temperature. Adapted from Charney (1973) with permission of Kluwer Academic Publishers.

At $O(E)$, the zonal momentum equation reduces to

$$-f\overline{v}^{(2)} = K\frac{\partial^2 \overline{u}^{(0)}}{\partial z^2}$$
$$= 0. \tag{15.36}$$

Then continuity implies

$$\frac{\partial \overline{w}^{(2)}}{\partial z} = 0,$$

so $\overline{w}^{(2)}$ is independent of height in the interior and equal to $\overline{w}^{(2)}(H)$. Air expelled from the upper boundary layer drives a gentle $O(E)$ subsidence in the interior, which must be absorbed in another boundary layer at the bottom.

The lower boundary is a rigid surface, so motion there has the form of a standard Ekman layer. Introducing the stretching transformation

$$z = E^{\frac{1}{2}}\xi \tag{15.37}$$

inside the lower boundary layer magnifies sharp changes to make viscous terms comparable to others in the momentum balance. Then, at $O(E^{\frac{1}{2}})$, the momentum equations reduce to the Ekman balance

$$-f\overline{v}_a^{(1)} = K\frac{\partial^2 \overline{u}_a^{(1)}}{\partial \xi^2}, \tag{15.38.1}$$

$$f\overline{u}_a^{(1)} = K\frac{\partial^2 \overline{v}_a^{(1)}}{\partial \xi^2}, \tag{15.38.2}$$

with the geostrophic contribution again automatically satisfied by $\overline{u}_g^{(1)}$. Requiring the total solution accurate to $O(E^{\frac{1}{2}})$ to satisfy (15.21.6) and to reduce to the inviscid behavior outside the boundary layer yields

$$\overline{u}_g^{(1)} + \overline{u}_a^{(1)} = 0 \qquad \overline{v}_a^{(1)} = 0 \qquad \xi = 0 \tag{15.38.3}$$
$$\overline{u}_a^{(1)} \sim 0 \qquad \overline{v}_a^{(1)} \sim 0 \qquad \xi \to \infty. \tag{15.38.4}$$

The boundary value problem (15.38) has solution

$$\overline{u}_a^{(1)}(\xi) = -\overline{u}_g^{(1)}e^{-\gamma\xi}\cos\gamma\xi, \tag{15.39.1}$$

$$\overline{v}_a^{(1)}(\xi) = \overline{u}_g^{(1)}e^{-\gamma\xi}\sin\gamma\xi \tag{15.39.2}$$

(Figs. 15.4a and b), which is analogous to the Ekman structure in Fig. 13.3.

Because $\overline{v}$ vanishes in the interior, meridional mass transfer inside the lower boundary layer must be exactly compensated by meridional mass transfer inside the upper boundary layer. This requires $\bar{v}_a^{(1)}$ to satisfy the condition

$$
E \int_0^\infty \bar{v}_a^{(1)} d\xi = -E \int_0^\infty \bar{v}_a^{(1)} d\zeta
$$

$$
= E \frac{KR}{aHf^2} \frac{\partial T_0}{\partial \phi}. \tag{15.40.1}
$$

Thermal wind (15.24) then implies

$$
\bar{u}_g^{(1)} = \frac{R}{aH\gamma f} \frac{\partial T_0}{\partial \phi} \tag{15.40.2}
$$

for the $O(E^{\frac{1}{2}})$ geostrophic motion, which completes the solution to this order.

Figure 15.4 illustrates the motion for the profile $T_0(\phi) = T_{00} - \delta T \sin^4 \phi$ (Fig. 15.4d). This temperature structure may be thought to result from the distribution of radiative heating (Fig. 1.29c) and is sufficiently flat at low latitude for the solution to remain valid up to the equator. The motion characterizes a zonally symmetric Hadley cell. Air heated near the equator rises, gradually spirals poleward in a strong zonal jet along the upper surface, sinks in the extratropics, and then returns along the lower boundary. Because motion in the interior is zonal and parallel to isotherms, meridional heat flux is concentrated inside the boundary layers, which have thickness $O(E^{\frac{1}{2}})$ and $v = O(E^{\frac{1}{2}})$. As Ω increases, $E \to 0$. The zonally symmetric circulation then becomes increasingly geostrophic, with meridional motion confined to shrinking Ekman layers along the walls in which poleward heat transfer collapses like E. Hence, rotation acts to choke the poleward transfer of heat by the zonally symmetric circulation.

In the atmosphere, the zonal–mean meridional circulation (Fig. 15.5) qualitatively resembles the idealized Hadley cell—but only in the tropics. Rising motion is found near the equator in the summer hemisphere, coincident with the Inter Tropical Convergence Zone (ITCZ) and organized latent heat release (Fig. 9.38). This is compensated almost entirely by sinking motion in the winter hemisphere. Strong meridional motion appears near the tropopause and surface, but only equatorward of 30°. Like subsidence, that ageostrophic motion is found chiefly in the winter hemisphere, where the meridional temperature gradient is steepest (15.40).

The observed Hadley cell can, in fact, be understood from considerations of angular momentum conservation like those discussed at the beginning of this chapter (see, e.g., James, 1993). Air moving poleward in the upper troposphere accelerates via the Coriolis force to produce strong westerlies (see Fig. 15.8a later in this chapter). This acceleration is observed, however, only out to the poleward edge of the Hadley cell, which happens to coincide with the subtropical jet (compare Fig. 1.8). At higher latitudes, the thermally direct Hadley cell in Figure 15.5 is replaced by a thermally indirect *Ferrell cell*,

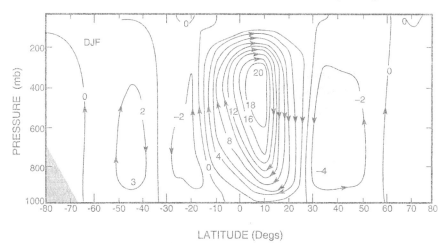

Figure 15.5 Zonal–mean meridional circulation observed during December–February. Adapted from Peixoto and Oort (1992).

in which air sinks at low latitude and rises at high latitude.[3] Contrary to the Hadley cell, which is driven thermally by heating at low latitude, the Ferrell cell is driven mechanically by zonally asymmetric eddies that develop along the poleward flank of the subtropical jet and dominate the extratropical circulation.

15.3 Heat Transfer in a Laboratory Analogue

Insight into why the observed meridional circulation differs from the zonally symmetric model is provided by laboratory simulations. A rotating cylindrical annulus (Fig. 15.6) is driven thermally by imposing high temperature along its outer surface, which represents the equator, and low temperature along its inner surface, which represents the pole. Figure 15.7 shows the motion observed in the rotating frame, as a function of its rotation rate Ω. At slow rotation, the motion is zonally symmetric and has the form of a thermally direct Hadley cell. Fluid along the surface gradually spirals inward in a zonal jet, converges, and then sinks along the interior wall. As Ω increases, fluid moving radially experiences an intensified Coriolis force that deflects the motion increasingly parallel to isotherms and spins up a strong zonally symmetric jet in the interior. Heat transfer then becomes concentrated inside the Ekman layers along the walls, diminishing like E.

---

[3] This thermally indirect motion is actually an artifact of the Eulerian mean along latitude circles, disappearing in other representations (see, e.g., Holton, 1992). For this reason, the Ferrell cell is not the principal mechanism for meridional transport in extratropical regions.

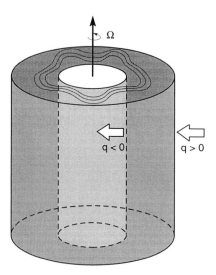

Figure 15.6 Schematic of a rotating cylindrical annulus that is driven thermally by imposing high temperature along its outer surface and low temperature along its inner surface.

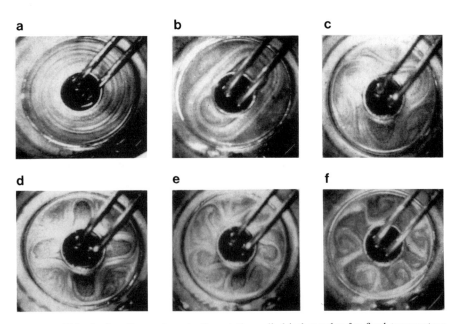

Figure 15.7 Surface flow pattern in the rotating cylindrical annulus for fixed temperature contrast and as a function of increasing rotation rate. Adapted from Hide (1966).

Beyond a critical rotation rate, zonally asymmetric waves appear. These waves develop from a steep temperature gradient that forms in the interior when radial heat transfer by the zonally symmetric circulation is choked by rotation. The zonally symmetric stratification is then strongly baroclinic, possessing available potential energy. By thermal wind balance (12.11), the radial temperature gradient is accompanied by strong vertical shear, which renders the zonally symmetric flow *baroclinically unstable* (Chapter 16). Unsteady disturbances then amplify by extracting available potential energy from the zonal–mean state and converting it to eddy kinetic energy. Baroclinic waves accomplish this by transferring heat radially in *sloping convection*. Warm inward-moving fluid overrides heavier fluid and ascends, whereas cool outward-moving fluid undercuts lighter fluid and descends (compare Fig. 15.3). The result is net heat transfer inward. This asymmetric heat transfer occupies much of the interior, so it is far more efficient than the shallow $O(E)$ heat transfer of the zonally symmetric circulation that has been confined along the walls by rotation. Although more complex, sloping convection likewise lowers the overall center of gravity to release available potential energy.

The wavelength of the dominant baroclinic disturbance decreases with increasing rotation rate (Fig. 15.7b–e). In certain ranges of Ω, the annulus circulation resembles observed flows in the troposphere. The pentagonal structure in Fig. 15.7e is similar to patterns observed in the Southern Hemisphere (Fig. 2.10). Because its storm track is almost zonally symmetric, cyclones often appear uniformly spaced about the pole. At sufficiently fast rotation, the dominant wavelength becomes small enough to make the wave itself unstable.[4] Wave motion then breaks down into isolated vortices that transfer heat efficiently by rolling up and eventually mixing fluid across the annulus. Behavior similar to that in Fig. 15.7 occurs if the imposed temperature contrast exceeds a critical value, which likewise renders the zonally symmetric circulation baroclinically unstable.

In the troposphere, nonuniform heating (Fig. 1.29c) continually makes the zonal-mean stratification baroclinic, producing available potential energy. Baroclinicity is strongest at midlatitudes, between regions of radiative heating and cooling. Unstable eddies develop on the strong vertical shear of the zonal–mean jet, where they transfer heat poleward through sloping convection. Compared to the troposphere, heating in the stratosphere is fairly uniform. Therefore, baroclinicity remains weak, as does the potential energy available to generate baroclinic eddies.

By transferring heat poleward, baroclinic motions along with the Hadley circulation make the general circulation of the troposphere behave as a heat engine (Fig. 6.7). Air is heated at high temperature while it is near the equator

---

[4] Vertical shear of the wave reinforces that of the zonal-mean flow, which makes shear strong enough locally to render the motion baroclinically unstable. A parallel exists in the Northern Hemisphere, where planetary waves reinforce zonal-mean shear in the North Pacific and North Atlantic storm tracks to provide conditions favorable for the development of baroclinic eddies.

and surface, whereas it is cooled at low temperature after being displaced upward and poleward. The second law then implies that an individual air parcel performs net work when executing a thermodynamic cycle. Kinetic energy produced in this fashion maintains the circulation against frictional dissipation. About half of the organized kinetic energy of the troposphere is dissipated inside the planetary boundary layer through turbulent mixing (Sec. 13.1.2). The remainder is dissipated in the interior through turbulence generated by dynamical instability (Sec. 14.7) and convection.

15.4 Tropical Circulations

At low latitude the Coriolis force is weak, which allows the circulation to possess a greater contribution from divergence and be thermally direct. The relatively flat temperature distribution (Fig. 1.7) implies little available potential energy to drive large-scale motion. Instead, latent heat release inside organized convection provides the primary source of energy for the tropical circulation, which in turn is derived from evaporative cooling of the oceans.

Precipitation inside the ITCZ (Fig. 9.38) exceeds evaporation directly beneath it. Thus, moisture must be imported into convection from surrounding areas. Much of this converges in the easterly equatorward *trade winds* near the surface (Fig. 15.8b). The trades comprise the lower branch of the zonal–mean Hadley cell in Fig. 15.5 and develop when equatorward-moving air is deflected westward by the Coriolis force. Air subsiding from the upper branch of the Hadley cell dries the troposphere and stabilizes it through mechanisms discussed in Sec. 7.4.4. By inhibiting convection, subsidence in the Hadley circulation maintains the deserts common at subtropical latitudes.

In addition to the zonal–mean Hadley cell, the tropics contain two other classes of circulation. These are also thermally direct, but zonally asymmetric. Monsoon circulations are established by horizontal gradients of surface temperature, which vary annually. During summer, subtropical landmasses such as India and northern Australia become warmer than neighboring maritime regions.[5] During winter, the horizontal temperature gradient reverses, with continents becoming colder than the surrounding maritime regions.

Surface heating during summer favors enhanced convection over land, which is fueled by moist inflow at low levels. Latent heat release then warms a deep layer, which leads to increased thickness between 1000 and 100 mb by the hypsometric relationship (Fig. 1.30). Isobaric surfaces that have been depressed in the lower troposphere introduce a surface low over land, which reinforces convection through frictional convergence inside the boundary layer.

---

[5] Higher temperatures over land result from the comparatively small heat capacity of soil and its immobility, wherein only diffusion is available to transfer heat away from the surface. By comparison, water has a much greater heat capacity and can efficiently transfer heat away from the surface through fluid motion. Both effects make seasonal swings in ocean temperature much smaller than corresponding swings over continents.

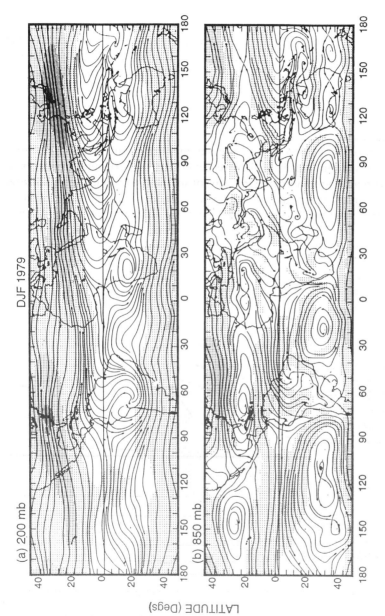

DJF 1979

(a) 200 mb

(b) 850 mb

LATITUDE (Degs)

LONGITUDE (Degs)

Figure 15.8 Streamlines and wind speed (shaded) during December–February 1979 at (a) 200 mb and (b) 850 mb. *Trade winds* are manifested at 850 mb by southwestward flow across the tropical western Pacific, Atlantic, and Indian oceans. Outflow at 200 mb over tropical Africa, South America, and Indonesia assumes anticyclonic curvature as it diverges away from those convective centers. Note the reversal of curvature to conserve potential vorticity as streamlines cross the equator. Strong jets at 200 mb east of North America and Asia mark the *north Atlantic* and *north Pacific storm tracks* (compare Fig. 9.38), where the mean temperature gradient is steep and cyclone development is favored. Stipling levels: $|v_h| = 15$ to 30 m s^{-1}, 30 to 45 m s^{-1}, > 45 m s^{-1}. Adapted from Lau (1984).

The influx of water vapor then increases θ_e to make the lower troposphere potentially unstable (Fig. 7.12). Convergence, which extends well above the boundary layer, also reduces the moist static stability through mechanisms described in Sec. 7.4.3. These changes provide conditions favorable to deep convection by allowing air to penetrate more easily to the level of free convection (LFC). Low-level air that converges into the convective region spins up cyclonically to conserve potential vorticity (e.g., over northern Australia in Fig. 15.8b). Convected to the tropopause, that air then diverges from the convective region and spins up anticyclonically (e.g., also over tropical Africa and South America in Fig. 15.8a).

The onshore circulation during the summer monsoon is analogous to a seabreeze circulation, which is likewise generated by horizontal gradients of heating. Both may be understood in terms of solenoidal production of horizontal vorticity, which occurs under baroclinic stratification (Fig. 15.9). Nonuniform heating introduces a horizontal variation of temperature that drives isochoric surfaces (α = const) out of coincidence with isobaric surfaces. A solenoidal production term proportional to $\nabla p \times \nabla \alpha$ then appears in the

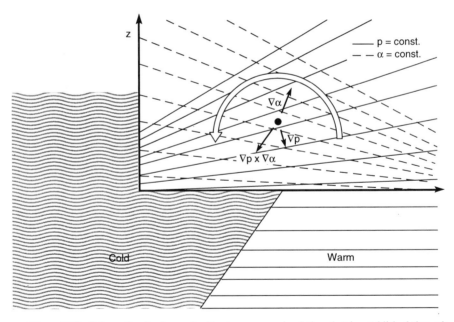

Figure 15.9 Schematic of a thermally direct monsoon circulation that is established through solenoidal production of horizontal vorticity under baroclinic stratification. Heating over land expands the vertical spacing of isobaric surfaces (solid lines), whereas it draws down isochoric surfaces (dashed lines) of larger α from aloft. The reverse occurs from cooling over ocean. The nonuniform distribution of heating drives isochoric surfaces out of coincidence with isobaric surfaces to produce baroclinic stratification. Solenoidal production of horizontal verticity $\nabla p \times \nabla \alpha$ then spins up an onshore circulation.

budget of horizontal vorticity, analogous to the one in the budget of vertical vorticity (12.36). This spins up horizontal vorticity as an onshore circulation, with rising motion over the heated region and sinking motion over the cooled region.

During winter, the monsoon circulation is reversed. Land, colder than neighboring water, then cools the overlying air to produce subsidence. This process inhibits convection by drying the environment with air from aloft, which in turn reduces θ_e and stabilizes the stratification. The compressed thickness between 1000 and 100 mb forms a surface high (Fig. 1.30). Frictional divergence at low levels then drives anticyclonic motion (e.g., as appears over the south Indian ocean in Fig. 15.8b) and subsidence, which strengthens the moist static stability of the lower troposphere (Sec. 7.4.3).

The other class of circulation important in the tropics is the Walker cell, which is depicted in Fig. 1.30. A zonally asymmetric overturning in the equatorial plane, the Walker circulation is characterized by rising motion over regions of heating and sinking motion to the east and west. The Pacific Walker circulation is forced by latent heating in the western Pacific. Low surface pressure in the western Pacific supports deep convection, in the same way it does monsoons. Strong surface easterlies along the equator fuel convection with moisture that has been absorbed by air during its traversal of the western Pacific, where sea surface temperature (SST) approaching 30°C provides an abundant source of water vapor (Fig. 5.1).

Latent heating that drives zonal overturning in the Walker cell also drives meridional overturning, which is likewise zonally asymmetric. Figure 15.10a shows surface motion forced by a steady heating anomaly (shaded area) over the equator. Surface easterlies along the equator that converge into the heating resemble the Pacific Walker cell and are mirrored in the upper troposphere by westerlies (Fig. 15.10b). Flanking ascending motion inside the heating are two cyclonic gyres in the subtropics, which reinforce inflow along the equator to the west. Those gyres represent a Rossby wave mode that is trapped about the equator (see, e.g., Gill, 1982). They are evident in the 200-mb streamfield (Fig. 15.8b), along with zonal flow to the east that comprises the Walker cell.

The Pacific Walker circulation changes markedly on interseasonal timescales. *El Niño*, which occurs with a frequency of 3 to 5 years, is marked by warming of SST in the eastern and central Pacific (Fig. 15.11).[6] The equatorial Pacific is especially sensitive to temperature perturbations because SST exceeds 25°C there. The exponential dependence in the Clausius–Clapeyron equation (4.39) then implies that small variations of SST produce large changes in evaporation and latent heat transfer. When the temperature anomaly in Fig. 15.11 is introduced, the warmest SST and convection shift eastward from their normal position (Fig. 9.38) to the central Pacific.

---

[6]The name refers to "the child" because warm waters off the coast of Peru, which signal its onset, appear around Christmas.

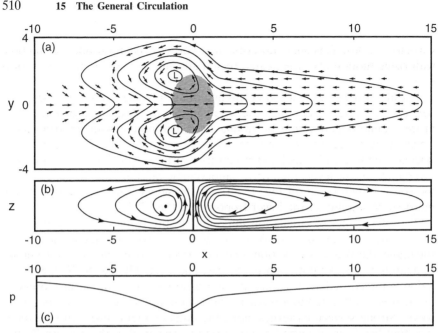

Figure 15.10 Zonally asymmetric circulation produced by a deep heating anomaly over the equator (stippled). (a) Planform view of surface motion. (b) Streamfield in a vertical cross section over the equator. (c) Surface pressure distribution. Adapted from Gill (1980).

El Niño is accompanied by a seesaw in surface pressure between the western and eastern Pacific. Known as the *Southern Oscillation*, this seesaw shifts low pressure from the western Pacific and Indian ocean to the eastern Pacific, as is evidenced by the anticorrelation of changes in those regions (Fig. 15.12). These phenomena are referred to jointly as El Niño Southern Oscillation or *ENSO*. Because it reconfigures the convective pattern, ENSO introduces a major perturbation to the tropical circulation.

ENSO is also felt at extratropical latitudes. By upsetting the distribution of heating in the equatorial troposphere, ENSO alters planetary waves excited thermally at low latitude that radiate poleward. Figure 14.20 shows the anomalous upper-tropospheric height for winters during El Niño, when convective heating is positioned in the equatorial central Pacific. A planetary wavetrain radiates poleward from anomalous convection, being refracted equatorward along a great circle route. Known as the *Pacific North American (PNA) pattern*, this perturbation to the normal height field alters the extratropical circulation, in particular, by introducing a prominent ridge over western Canada and trough to its east. The jet stream (wavy trajectory in Fig. 14.20a) is then deflected northward over the western United States and Canada and southward over the eastern United States. Since the jet stream marks the boundary between tropical and polar air (Sec. 12.2.2), these changes produce abnor-

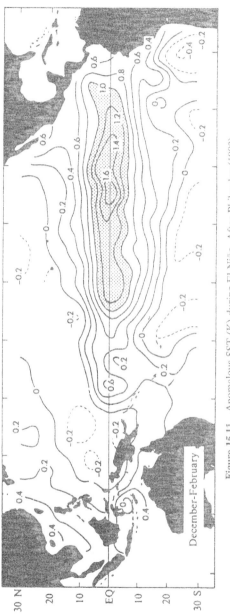

Figure 15.11 Anomalous SST (K) during El Niño. After Philander (1983).

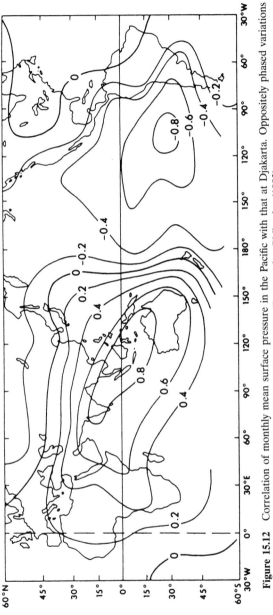

Figure 15.12 Correlation of monthly mean surface pressure in the Pacific with that at Djakarta. Oppositely phased variations in the eastern and western Pacific characterize the *Southern Oscillation*. After Philander (1983).

mally warm temperatures over the western United States and Canada and abnormally cold temperatures over the eastern United States. By altering the locations of the storm tracks, they also modify the development and movement of baroclinic systems.

Suggested Reading

Atmospheric Science: An Introductory Survey (1977) by Wallace and Hobbs contains a nice introduction to moist static energy and other concepts figuring in the global energy budget.

An historical overview of available potential energy is given in *Atmosphere-Ocean Dynamics* (1982) by Gill, along with applications to wave motions. It also includes a comprehensive discussion of tropical circulations and their interaction with the oceans.

An Introduction to Dynamic Meteorology (1992) by Holton contains a formal treatment of the zonal–mean circulation and its relationship to eddies.

Introduction to Circulating Atmospheres (1993) by James develops the observed Hadley circulation from first principles and in light of the three-dimensional structure of the general circulation.

Global Physical Climatology (1994) by Hartmann includes a thorough description of the general circulation and its relationship to eddy transports of heat and momentum and to regional climates.

Physics of Climate (1992) by Peixoto and Oort is a comprehensive treatment of the general circulation that surveys a wide range of observations.

Problems

15.1. See Problem 1.18.

15.2. A national research announcement for a precipitation monitoring satellite asserts that latent heating inside convection is the primary energy source driving the atmospheric circulation. Discuss the accuracy of this assertion (a) for the globe as a whole, in light of Problem 15.1, and (b) inside regions of organized deep convection, in light of the discussion in Sec. 9.5.1.

15.3. Derive expression (15.2.2) for conservation of moist static energy.

15.4. Relate the change of moist static energy $\epsilon = c_p T + l_v r + \Phi$ for an air parcel to its change of equivalent potential temperature θ_e.

15.5. Consider an air parcel that moves equatorward at sea level, ascends adiabatically in the tropics, returns poleward at a constant altitude of 15 km, and then descends in the extratropics—as symbolized by the idealized thermodynamic circuit in Fig. 6.7. If heat transfer is restricted to upper and lower legs of the circuit and if, along the lower leg, the parcel's initial temperature and mixing ratio in the extratropics are 250 K and 1 g kg^{-1}, respectively, whereas it assumes a temperature and mixing ratio in the tropics of 300 K and 20 g kg^{-1}, calculate the radiative heat transfer that occurs along the upper leg of the circuit.

15.6. Demonstrate that no work is performed on the atmosphere as a whole.

15.7. Show that the pressure averaged over an isentropic surface $\overline{p}$ is preserved under an adiabatic rearrangement of mass.

15.8. Discuss why uniformly heating an atmosphere increases its store of total potential energy, but makes none available for conversion to kinetic energy.

15.9. Show that the integration limit $p = 0$ in (15.15) vanishes.

15.10. Derive expressions (15.19) for available potential energy.

15.11. Consider the following thermal structure, representative of zonal–mean stratification in the troposphere:

$$T(\phi, p) = \overline{T}(p) - T'(\phi, p),$$

where

$$\overline{T}(p) = \overline{T}_0 - \Gamma H \ln\left(\frac{p_0}{p}\right) + \frac{\Delta T}{4}\left(\frac{p}{p_0}\right),$$

$$T'(\phi, p) = \Delta T |\sin \phi|^3 \left(\frac{p}{p_0}\right) - \frac{\Delta T}{4}\left(\frac{p}{p_0}\right),$$

and an overbar denotes global average. For $\overline{T}_0 = 268$ K, $\Gamma = 4$ K km^{-1}, $H = 8$ km, $\Delta T = 70$ K, $\phi_0 = 45°$, and $p_0 = 1000$ mb, (a) plot the thermal structure and zonal wind as functions of latitude and $\xi = \ln(p_0/p)$, (b) calculate the available potential energy $\mathscr{A}$ of the atmosphere, and (c) plot the vertical concentration of available potential energy $\partial \mathscr{A}/\partial \xi$ as a function of ξ.

15.12. For the atmosphere in Problem 15.11, determine the fraction of total potential energy that is available for conversion to kinetic energy.

15.13. For the atmosphere in Problem 15.11, determine a characteristic eddy velocity if thermal structure is driven adiabatically into barotropic stratification and if the kinetic energy produced is concentrated between $\pm 30°$ and $\pm 60°$ latitude. What conclusion can you draw?

15.14. Show that $E^{\frac{1}{2}}$ is the appropriate power of Ekman number in which equations (15.21) should be expanded.

15.15. Verify that equations (15.21) at $O(E^{\frac{1}{2}})$ have the same form as (15.23).

15.16. For the spherical atmosphere treated in Sec. 15.2, (a) verify that (15.33) describes the $O(E^{\frac{1}{2}})$ behavior inside the lower Ekman layer and (b) verify that (15.39) describes the $O(E^{\frac{1}{2}})$ behavior inside the upper Ekman layer.

15.17. Show that the temperature profile

$$T_0(\phi) = T_{00} - \delta T \sin^4 \phi$$

makes the solution in Sec. 15.2 valid up to the equator.

15.18. Observed zonal–mean flow in the troposphere is easterly at low latitudes and westerly at middle and high latitudes (Fig. 1.8). Surface drag must then represent a sink of easterly momentum at low latitudes, which is equivalent to a source of westerly momentum there, and a sink of westerly momentum at middle and high latitudes. On average, the atmosphere's angular momentum remains constant, so the preceding source and sink of momentum must be compensated by a poleward transfer of westerly momentum. (a) Demonstrate that transport of angular momentum by the zonal–mean Hadley circulation produces a momentum flux $\overline{uv}$ of the correct sense to accomplish this transfer of westerly momentum. (b) Not all of the momentum transfer required to maintain equilibrium is accomplished by the zonal–mean Hadley circulation. The remainder occurs through large-scale eddy transport. In terms of horizontal phase structure and group velocity, describe the meridional propagation of Rossby waves needed to accomplish the remaining momentum transfer between tropical and extratropical regions.

15.19. The length of day can vary through exchanges of momentum between the atmosphere and solid earth. (a) Estimate the velocity fluctuation over the equator corresponding to observed fluctuations in the length of day of order 10^{-3} s, if the earth has a mean density of order 5.0×10^3 kg m^{-3} and if the atmosphere responds through uniform changes of angular velocity. (b) More generally, where would velocity fluctuations most effectively introduce changes in the length of day?

15.20. Describe how sloping convection lowers the center of gravity of air to release available potential energy.

15.21. Discuss how horizontal rearrangement of air by baroclinic eddies drives thermal structure toward barotropic stratification.

15.22. Relate heat transfer inside convective and cloud-free regions to the distribution of surface pressure in the tropics.

15.23. (a) Use observed distributions of precipitation rate (Fig. 9.38) and total precipitable water vapor (Fig. 1.16) to calculate a characteristic timescale for the column abundance of water vapor in the tropics. (b) Discuss this timescale in relation to the efficiency of dehydration inside individual convective cells, their fractional coverage, and the efficiency with which water vapor is produced at the earth's surface.

15.24. Describe the disturbed Walker circulation and trade winds in the Pacific anticipated during El Niño, when convection is found near the dateline.

15.25. The variation with latitude of planetary vorticity (i.e., β) is difficult to simulate directly in the laboratory. Use the vorticity budget for an incompressible fluid to construct an analogue of β in terms of the variation of fluid depth inside a rotating cylindrical annulus.

15.26. The height of convective towers in the tropics is controlled by the convective available potential energy (CAPE) of surface air (Sec. 7.4.1), which in turn is controlled by fluxes of sensible and latent heat from the surface. A surge of extratropical air into the tropics produces the temperature profile

$$T(z) = \begin{cases} T_0 - \Gamma z & z < z_T \\ \\ T_T & z \geq z_T, \end{cases}$$

where $T_0 = 290$ K, $\Gamma = 7.5$ K km^{-1}, $z_T = 12$ km, and $T_T = 200$ K. (a) Calculate the CAPE of surface air if, through contact with an isolated landmass, it attains a temperature of 300 K and a mixing ratio of 25 g kg^{-1}. (b) As in part (a) but if the air is dry. (c) Estimate the height to which convection would develop under the conditions of parts (a) and (b).

15.27. In terms of the first law, hydrostatic equilibrium, and the budget of horizontal vorticity, (a) describe the development of stratification over a continental and neighboring maritime region in the tropics under monsoonal conditions and (b) describe how the accompanying thermally direct circulation spins up.

15.28. Many components of the global energy budget, like convection, cloud albedo, and fluxes of sensible and latent heat, vary diurnally. Discuss how systematic diurnal variations of such properties contribute to the time–mean energetics illustrated in Fig. 1.27.

Chapter 16 | Hydrodynamic Instability

Wave propagation is supported by a positive restoring force, which opposes air displacements by driving parcels back toward their undisturbed positions (Chapter 14). Under certain conditions, the sense of the restoring force is reversed. Air displacements are then reinforced by a negative restoring force, one which accelerates parcels away from their undisturbed positions. Instability was encountered earlier in connection with hydrostatic stratification (Chapter 7). If distributions of temperature and moisture violate the conditions for hydrostatic stability, small vertical displacements produce buoyancy forces that accelerate parcels away from their undisturbed positions. Unlike the response under stable stratification, this reaction leads to finite displacements of air. Fully developed convection then drives the stratification toward neutral stability by rearranging air.

Two classes of instability are possible. *Parcel instability* follows from reinforcement of air displacements by a negative restoring force, such as that occurring in the development of convection. *Wave instability* occurs in the presence of a positive restoring force, but one that amplifies parcel oscillations inside wave motions. Unstable waves amplify by extracting energy from the mean circulation (e.g., from available potential energy associated with baroclinic stratification and vertical shear; Chapter 15). Strong zonal motion that results from the nonuniform distribution of heating and geostrophic equilibrium makes this class of instability the one most relevant to the large-scale circulation. Like parcel instability, it develops to neutralize instability in the mean flow, which it achieves by rearranging air.

16.1 Inertial Instability

The simplest form of large-scale instability relates to the inertial oscillations described in Sec. 12.4. Consider disturbances to a geostrophically balanced zonal flow $\bar{u}$ on an f-plane. If the disturbances introduce no pressure perturbation, the total motion is governed by the horizontal momentum balance

$$\frac{du}{dt} - fv = 0, \tag{16.1.1}$$

$$\frac{dv}{dt} + f(u - \bar{u}) = 0, \tag{16.1.2}$$

where geostrophic equilibrium has been used to eliminate pressure in favor of $\bar{u}$.

Since a parcel's motion satisfies

$$v = \frac{dy}{dt},\tag{16.2}$$

(16.1.1) implies

$$\frac{du}{dt} = f\frac{dy}{dt}.\tag{16.3}$$

Integrating from the parcel's initial position y_0 to its displaced position $y_0 + y'$ gives

$$u(y_0 + y') - \bar{u}(y_0) = fy'.$$

To first order in parcel displacement, this can be expressed

$$u(y_0) + \frac{\partial \bar{u}}{\partial y}y' - \bar{u}(y_0) = fy'$$

or

$$(u - \bar{u})\big|_{y_0} - \left(f - \frac{\partial \bar{u}}{\partial y}\right)y' = 0.$$

Then incorporating (16.1.2) yields

$$\frac{d^2y'}{dt^2} + f\left(f - \frac{\partial \bar{u}}{\partial y}\right)y' = 0.\tag{16.4}$$

If the mean flow is without shear, (16.4) reduces to a description of the inertial oscillations considered previously. In the presence of shear, displacements either oscillate, decay exponentially, or grow without bound. The system possesses unstable solutions if the absolute vorticity of the mean flow

$$f + \bar{\zeta} = f - \frac{\partial \bar{u}}{\partial y}\tag{16.5}$$

has sign opposite to the planetary vorticity f. Displacements then amplify exponentially and the zonal flow is *inertially unstable*. These circumstances make the specific restoring force $f(f + \bar{\zeta})y'$ negative, so the ensuing instability is of the parcel type. Because $f + \bar{\zeta}$ usually has the same sign as f, the criterion for inertial instability is tantamount to the absolute vorticity reversing sign somewhere.

Inertial instability does not play a major role in the atmosphere. Extratropical motions tend to remain inertially stable, even locally in the presence of synoptic and planetary wave disturbances. However, the criterion for inertial instability is violated more easily near the equator, where f is small. Evidence of inertial instability exists in the tropical stratosphere, where horizontal shear flanking the strong zonal jets (Fig. 1.8) can violate the criterion for inertial stability.

16.2   Shear Instability

More relevant to the large-scale circulation is instability associated directly with shear. Shear instability is of the wave type, so it requires a more involved analysis than that applying to an individual parcel. Like the treatment of wave motions, describing shear instability requires the solution of partial differential equations that govern perturbation properties. Closed-form solutions can be found only for very idealized profiles of zonal–mean flow $\bar{u}(y, z)$. However, an illuminating criterion for instability, due originally to Rayleigh, can be developed under fairly general circumstances.

16.2.1   Necessary Conditions for Instability

Consider quasi-geostrophic motion on a beta plane in an atmosphere that extends upward indefinitely and is bounded below and laterally at $y = \pm L$ by rigid walls. Disturbances to the zonal–mean flow $\bar{u}(y, z)$ are governed by first-order conservation of potential vorticity (14.75)

$$\frac{DQ'}{Dt} + v'\beta_e = 0 \tag{16.6.1}$$

where $D/Dt = \partial/\partial t + \bar{u}\partial/\partial x$,

$$Q' = \nabla^2 \psi' + \frac{1}{\rho_0}\frac{\partial}{\partial z}\left(\frac{f_0^2}{N^2}\rho_0\frac{\partial \psi'}{\partial z}\right), \tag{16.6.2}$$

$$\beta_e = \beta - \frac{\partial^2 \bar{u}}{\partial y^2} - \frac{1}{\rho_0}\frac{\partial}{\partial z}\left(\frac{f_0^2}{N^2}\rho_0\frac{\partial \bar{u}}{\partial z}\right)$$

$$= \frac{\partial \bar{Q}}{\partial y}, \tag{16.6.3}$$

and z denotes log–pressure height. Requiring vertical motion to vanish at the ground (which is treated as an isobaric surface) gives, via the thermodynamic equation and thermal wind balance, the lower boundary condition

$$\frac{D}{Dt}\left(\frac{\partial \psi'}{\partial z}\right) - \frac{\partial \bar{u}}{\partial z}\frac{\partial \psi'}{\partial x} = 0 \qquad z = 0. \tag{16.6.4}$$

Physically meaningful solutions must also have bounded column energy, which provides the upper boundary condition

$$\text{finite energy condition} \qquad z \to \infty. \tag{16.6.5}$$

At the lateral walls, v' must vanish, so ψ'=const. It suffices to prescribe

$$\psi' = 0 \qquad y = \pm L. \tag{16.6.6}$$

Equations (16.6) define a second-order boundary value problem for the disturbance streamfunction $\psi'(x, t)$—one that is homogeneous. Containing no imposed forcing, (16.6) describes a system that is self-governing or autonomous. Nontrivial solutions (i.e., other than $\psi' \equiv 0$) exist only for certain *eigenfrequencies* that enable boundary conditions to be satisfied. Determined by solving the homogeneous boundary value problem for a given zonal flow $\bar{u}(y, z)$, those eigenfrequencies are, in general, complex.

Consider solutions of the form

$$\psi' = \Psi(y, z)e^{ik(x-ct)}, \tag{16.7}$$

where Ψ and $c = c_r + ic_i$ can assume complex values. Substituting (16.7) transforms (16.6) into

$$(\bar{u} - c)\left[\frac{\partial^2 \Psi}{\partial y^2} + \frac{1}{\rho_0}\frac{\partial}{\partial z}\left(\frac{f_0^2}{N^2}\rho_0\frac{\partial \Psi}{\partial z}\right) - k^2\Psi\right] + \beta_e\Psi = 0. \tag{16.8.1}$$

The lower boundary condition (16.6.4) becomes

$$(\bar{u} - c)\frac{\partial \Psi}{\partial z} - \frac{\partial \bar{u}}{\partial z}\Psi = 0 \qquad z = 0. \tag{16.8.2}$$

For c real, (16.8) is singular at a critical line where $\bar{u} = c$. The singularity disappears if $c_i \neq 0$, in which case (16.7) contains an exponential modulation in time. If the flow is stable, wave activity incident on the critical line is absorbed when dissipation is included (Sec. 14.3). The wave field then decays exponentially. If the flow is unstable, wave activity can be produced at the critical line. The wave field can then amplify exponentially.

Multiplying the conjugate of (16.8) by Ψ and (16.8) by the conjugate of Ψ and then subtracting yields

$$\left[\Psi\frac{\partial^2 \Psi^*}{\partial y^2} - \Psi^*\frac{\partial^2 \Psi}{\partial y^2}\right] + \left[\Psi\frac{1}{\rho_0}\frac{\partial}{\partial z}\left(\frac{f_0^2}{N^2}\rho_0\frac{\partial \Psi^*}{\partial z}\right)\right.$$
$$\left. - \Psi^*\frac{1}{\rho_0}\frac{\partial}{\partial z}\left(\frac{f_0^2}{N^2}\rho_0\frac{\partial \Psi}{\partial z}\right)\right] - 2ic_i\frac{|\Psi|^2}{|\bar{u} - c|^2}\beta_e = 0 \tag{16.9.1}$$

and

$$\Psi\frac{\partial \Psi^*}{\partial z} - \Psi^*\frac{\partial \Psi}{\partial z} + 2ic_i\frac{|\Psi|^2}{|\bar{u} - c|^2}\frac{\partial \bar{u}}{\partial z} = 0 \qquad z = 0. \tag{16.9.2}$$

The chain rule allows terms in the first set of square brackets in (16.9.1) to be expressed

$$\Psi\frac{\partial^2 \Psi^*}{\partial y^2} - \Psi^*\frac{\partial^2 \Psi}{\partial y^2} = \frac{\partial}{\partial y}\left[\Psi\frac{\partial \Psi^*}{\partial y} - \Psi^*\frac{\partial \Psi}{\partial y}\right].$$

Terms in the second set of square brackets can be expressed in a similar manner. Then (16.9.1) can be written

$$\frac{\partial}{\partial y}\left[\Psi\frac{\partial\Psi^*}{\partial y}-\Psi^*\frac{\partial\Psi}{\partial y}\right]+\frac{1}{\rho_0}\frac{\partial}{\partial z}\left[\frac{f_0^2}{N^2}\rho_0\left(\Psi\frac{\partial\Psi^*}{\partial z}-\Psi^*\frac{\partial\Psi}{\partial z}\right)\right]$$

$$-2ic_i\frac{|\Psi|^2}{|\bar{u}-c|^2}\beta_e=0. \qquad (16.10)$$

Integrating over the domain unravels the exterior differentials in (16.10) to give

$$\int_0^\infty\left[\Psi\frac{\partial\Psi^*}{\partial y}-\Psi^*\frac{\partial\Psi}{\partial y}\right]_{y=-L}^{y=L}\rho_0 dz$$

$$+\int_{-L}^L\left[\frac{f_0^2}{N^2}\rho_0\left(\Psi\frac{\partial\Psi^*}{\partial z}-\Psi^*\frac{\partial\Psi}{\partial z}\right)\right]_{z=0}^{z=\infty}dy \qquad (16.11)$$

$$-2ic_i\int_{-L}^L\int_0^\infty\frac{|\Psi|^2}{|\bar{u}-c|^2}\beta_e dy\rho_0 dz=0.$$

By (16.6.6), the first integral vanishes. The finite energy condition makes the upper limit inside the second integral also vanish. Then incorporating (16.9.2) for the lower limit yields the identity

$$c_i\left\{\int_0^\infty\int_{-L}^L\beta_e\frac{\rho_0|\Psi|^2}{|\bar{u}-c|^2}dydz-\int_{-L}^L\left[\frac{f_0^2}{N^2}\frac{\rho_0|\Psi|^2}{|\bar{u}-c|^2}\frac{\partial\bar{u}}{\partial z}\right]_{z=0}dy\right\}=0, \qquad (16.12)$$

which must be satisfied for $\Psi(y,z)$ to be a solution of (16.6).

Advanced by Charney and Stern (1962), the preceding identity provides "necessary conditions" for instability of the zonal–mean flow $\bar{u}(y,z)$. If c is complex, (16.7) describes a disturbance whose amplitude varies in time exponentially

$$e^{ik(x-ct)}=e^{kc_it}\cdot e^{ik(x-c_rt)},$$

with the growth/decay rate kc_i. Without loss of generality, k can be considered positive, so the existence of unstable solutions requires $c_i>0$. Unstable solutions are then possible only if the quantity inside braces in (16.12) vanishes. If it does not, (16.12) implies $c_i=0$ and solutions to (16.6) are stable.

16.2.2   Barotropic and Baroclinic Instability

Requiring (16.12) to be satisfied with $c_i>0$ provides two alternative criteria for instability:

1. If $\partial \bar{u}/\partial z$ vanishes at the lower boundary, so does the temperature gradient by thermal wind balance. Then $\beta_e = \partial \overline{Q}/\partial y$ must reverse sign somewhere in the interior. Since β_e is normally positive, a region of negative potential vorticity gradient, $\partial \overline{Q}/\partial y < 0$, is identified as an unstable region of the mean flow.

2. If $\beta_e > 0$ throughout the interior, $\partial \bar{u}/\partial z$ must be positive somewhere on the lower boundary. By thermal wind balance, this implies the existence of an equatorward temperature gradient at the surface.

Other combinations are also possible, but these are the ones most relevant to the atmosphere. Neither represents a "sufficient condition" for instability. Satisfying criterion (1) or (2) does not ensure the existence of unstable solutions.

Criterion (1) defines a necessary condition for free-field instability (e.g., instability for which boundaries do not play an essential role). From (16.6.3), the mean gradient of potential vorticity can reverse sign through strong horizontal curvature of the mean flow or through strong (density-weighted) vertical curvature of the mean flow. It is customary to distinguish these contributions to $\partial \overline{Q}/\partial y$. If the necessary condition for instability is met through horizontal shear, amplifying disturbances are referred to as *barotropic instability*. If it is met through vertical shear (which is proportional to the horizontal temperature gradient and the departure from barotropic stratification), amplifying disturbances are referred to as *baroclinic instability*. Realistic conditions often lead to criterion (1) being satisfied by both contributions, in which case amplifying disturbances are combined barotropic–baroclinic instability.

In the absence of rotation, criterion (1) reduces to Rayleigh's (1880) necessary condition for instability of one-dimensional shear flow (Problem 16.12). Criterion (1) is then equivalent to requiring the mean flow profile to possess an inflection point. Since β is everywhere positive, rotation is stabilizing. It provides a positive restoring force that inhibits instability and supports stable wave propagation.

Recall that (16.8) is singular at a critical line $\bar{u} = c_r$ if $c_i = 0$. Exponential amplification removes the singularity by making $c_i > 0$. When boundaries do not play an essential role, amplifying solutions usually possess a critical line inside the unstable region where $\partial \overline{Q}/\partial y < 0$ (see, e.g., Dickinson, 1973). Rather than serving as a localized sink of wave activity, as its does under conditions of stable wave propagation ($\partial \overline{Q}/\partial y > 0$), the critical line then functions as a localized source of wave activity. Wave activity flux then diverges out of the critical line, where it is produced by a conversion from the mean flow. Alternatively, incident wave activity that encounters the critical line is "overreflected": More radiates away than is incident on the unstable region.

Criterion (2) describes instability that is produced through the direct involvement of the lower boundary. This criterion applies to baroclinic instability because it requires a temperature gradient at the surface and hence baroclinic stratification. Since air must move parallel to it, the boundary can then drive motion across mean isotherms, which transfers heat meridionally (e.g., in sloping convection). By weakening the temperature gradient, eddy heat transfer drives the mean thermal structure toward barotropic stratification and releases available potential energy (Sec. 15.1), which in turn is converted to eddy kinetic energy. This situation underlies the development of extratropical cyclones. Temperature gradients introduced by the nonuniform distribution of heating make the stratification baroclinic and produce available potential energy, on which baroclinic instability feeds.

16.3 The Eady Problem

The simplest model of baroclinic instability is that of Eady (1949). Consider disturbances to a mean flow that is invariant in y, bounded above and below by rigid walls at $z = 0, H$ on an f plane, and within the Boussinesq approximation (Sec. 12.5). A uniform meridional temperature gradient is imposed, which, by thermal wind balance, corresponds to constant vertical shear (Fig. 16.1)

$$\bar{u} = \Lambda z \qquad \Lambda = \text{const.} \tag{16.13}$$

Under these circumstances, $\partial \bar{Q}/\partial y$ vanishes identically in the interior, so instability can follow solely from the temperature gradient along the boundaries.

Disturbances to this system are governed by the perturbation potential vorticity equation in log–pressure coordinates

$$\frac{D}{Dt}\left(\nabla^2 \psi' + \frac{f_0^2}{N^2}\frac{\partial^2 \bar{u}}{\partial z^2}\right) = 0, \tag{16.14.1}$$

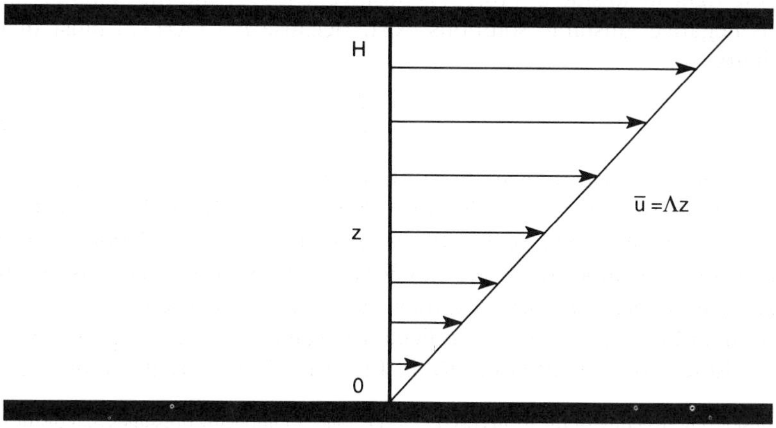

Figure 16.1 Geometry and mean zonal flow in the Eady problem of baroclinic instability.

with the boundary conditions

$$\frac{D}{Dt}\left(\frac{\partial \psi'}{\partial z}\right) - \frac{\partial \overline{u}}{\partial z}\frac{\partial \psi'}{\partial x} = 0 \qquad z = 0, H. \tag{16.14.2}$$

Considering solutions of the form

$$\psi' = \Psi(z)\cos(ly)e^{ik(x-ct)} \tag{16.15}$$

reduces (16.14) to the one-dimensional boundary value problem

$$\frac{d^2\Psi}{dz^2} - \alpha^2\Psi = 0, \tag{16.16.1}$$

$$(\overline{u} - c)\frac{d\Psi}{dz} - \Lambda\Psi = 0 \qquad z = 0, H, \tag{16.16.2}$$

where

$$\alpha = \frac{N}{f_0}|k_h| \tag{16.16.3}$$

with $|k_h|^2 = k^2 + l^2$, is a weighted horizontal wavenumber.[1] Solutions of (16.16.1) are of the form

$$\Psi = A\cosh(\alpha z) + B\sinh(\alpha z). \tag{16.17}$$

Substituting (16.17) into the boundary conditions (16.16.2) leads to a homogeneous system of two equations for the coefficients A and B. Nontrivial solutions exist only if the determinant of that system vanishes, which yields the dispersion relation for Eady modes (Problem 16.14)

$$\left(c - \frac{\Lambda H}{2}\right)^2 = \Lambda^2 H^2\left[\frac{1}{4} - \frac{\coth(\alpha H)}{\alpha H} + \frac{1}{(\alpha H)^2}\right]. \tag{16.18}$$

If the right-hand side of (16.18) is positive, c is real and the system is stable. If it is negative, unstable solutions exist. Because $c - (\Lambda H/2)$ must then be imaginary,

$$c_r = \frac{\Lambda H}{2}. \tag{16.19}$$

Thus, amplifying disturbances have phase speeds equal to the mean flow at the middle of the layer. Unstable disturbances are advected eastward by the mean flow with its speed at the the *steering level*: $z = H/2$. While influenced by rotation, baroclinic waves are not Rossby waves in a strict sense because, as the Eady model demonstrates, they can exist in the absence of β.

The quantity $k^2(c - \Lambda H/2)^2$ reflects the square of the complex frequency and is plotted as a function of αH in Fig. 16.2. For α greater than a critical

---

[1] Considering structure of the form $\cos(ly) = \frac{1}{2}\left(e^{ily}+e^{-ily}\right)$ implicitly presumes that disturbances are trapped meridionally (e.g., by rigid walls at $y = \pm\pi/2l$).

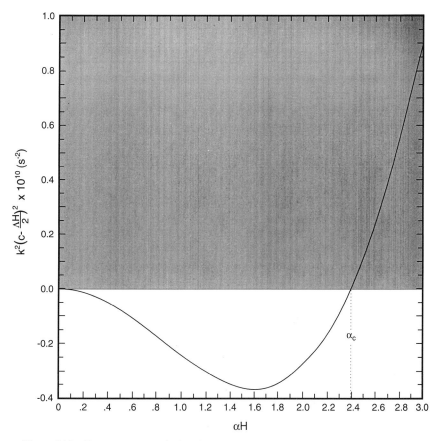

Figure 16.2 Frequency squared of Eady modes as a function of scaled horizontal wavenumber. Instability occurs for wavenumbers smaller than the cutoff α_c.

value $\alpha_c \cong 2.4$, $(c - \Lambda H/2)^2 > 0$. Thus, the system possesses a "shortwave cutoff" for instability, amplifying solutions existing only for smaller α (larger scales). Wavenumbers $\alpha < \alpha_c$ are all unstable: $(c - \Lambda H/2)^2 < 0$. A maximum growth rate kc_i is achieved at $\alpha H \cong 1.6$ for $l = 0$. This value of αH also maximizes the growth rate for waves of fixed aspect ratio l/k, which have smaller kc_i. For square waves $(k = l)$ and values representative of the troposphere, $\alpha H \cong 1.6$ predicts a wavenumber of order 5, which is typical of extratropical cyclones. The maximum growth rate is proportional to Λ and therefore to the meridional temperature gradient. Representative values give an e-folding time for amplification of a couple of days, consistent with the observed development of extratropical cyclones.

Insight into how instability is achieved follows from the structure of the most unstable disturbance. The lower boundary condition (16.16.2) implies the following relationship between the coefficients in (16.17):

$$\frac{B}{A} = -\frac{\Lambda}{\alpha c},\qquad(16.20)$$

which yields the structure shown in Fig. 16.3 for $k = l$. The fastest growing Eady mode is characterized by a westward tilt with height that places geopotential anomalies about $90°$ out of phase between the lower and upper boundaries. This can be inferred from the eddy meridional velocity $v' = ik\psi'$, which is contoured (solid/dashed lines) in Fig. 16.3a on a vertical section passing zonally through the center of the disturbance ($y = 0$).

Eddy geopotential and motion maximize at $z = 0$ and H in the form of two edge waves that are sandwiched between the upper and lower boundaries. In addition to driving air across mean isotherms, those boundaries trap instability, which allows the disturbance to amplify. Each edge wave decays vertically with the Rossby height scale $H_R = \alpha^{-1}$. In the limit $|k_h| \to \infty$, H_R is short enough for the two edge waves to be isolated from one another. Because neither has vertical phase tilt, hydrostatic balance implies a temperature perturbation that is in phase with the geopotential perturbation (Problem 16.15). The perturbation velocity v' is then everywhere in quadrature with θ', so the eddy heat flux averaged over a wavelength $\overline{v'\theta'}$ vanishes. Consequently, eddy motion does not alter the baroclinic stratification of the mean state and no available potential energy is released. It is for this reason that short wavelengths in the Eady problem are neutral.

For $\alpha < \alpha_c$, H_R is long enough for the two edge waves to influence one another. Phase-shifted, they produce the westward tilt apparent in Fig. 16.3a. The temperature perturbation is then no longer in phase with the geopotential perturbation. Isentropes (dotted lines in Fig. 16.3a) actually tilt slightly eastward with height to make θ' positively correlated with v': Cold air moves equatorward (out of the page) and warm air moves poleward (into the page).[2] Both contribute positively to $\overline{v'\theta'}$ to produce a poleward eddy heat flux, which releases available potential energy. The mode's amplification is thus directly related to its westward tilt with height.

The streamfunction in a horizontal section at $z = H/2$ (Fig. 16.3b) and in a meridional section at $kx = 3\pi/2$ (Fig. 16.4) illustrates that poleward-moving air ascends, whereas, half a wavelength away, equatorward-moving air descends. Characteristic of sloping convection, this is the same sense meridional motion would have were it to move along zonal-mean isentropic surfaces in Fig. 16.4 (dotted lines), which slope upward toward the pole. However, meridional displacements of the fastest growing Eady mode have a slope at midlevel which is only about half that of mean isentropic surfaces (Problem 16.18). Eddy motion directed across those surfaces advects high θ poleward and low θ equatorward in adjacent branches of a cell, as is evident from isentropes at midlevel (Fig. 16.3b). This asymmetric motion produces a net heat

---

[2] In fact, v' and θ' are perfectly in phase at the steering level: $z = H/2$.

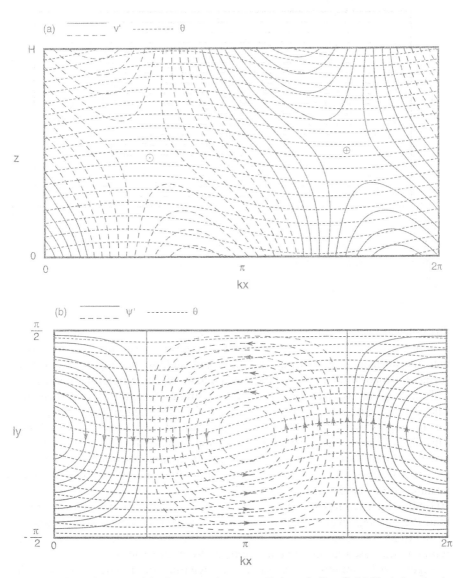

Figure 16.3 Structure of the fastest growing square Eady mode ($k = l$). (a) Vertical section in the zonal plane at $y = 0$ of eddy meridional velocity v' (solid/dashed lines) and isentropic surfaces $\theta = $ const (dotted lines). Potential temperature increases upward. ⊙ marks equatorward motion ($v' < 0$) and ⊕ marks poleward motion ($v' > 0$). (b) Horizontal section at $z = H/2$ of eddy streamlines (solid/dashed lines) and isentropes (dotted lines). Potential temperature increases equatorward.

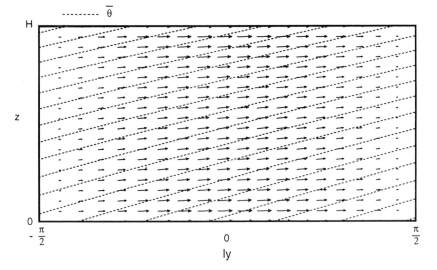

Figure 16.4   Vertical section in the meridional plane at $kx = 3\pi/2$ of mean isentropic surfaces (dotted lines) and motion for the fastest growing square Eady mode. Potential temperature increases upward and equatorward (compare Fig. 12.5).

flux poleward, which releases available potential energy by driving the thermal structure toward barotropic stratification. Poleward heat flux acts to eliminate the horizontal temperature gradient and shallow the slope of isentropic surfaces, which in turn reduces the zonal–mean available potential energy.

Extratropical cyclones have qualitatively similar structure during their development. Figure 16.5 shows distributions of 700-mb height and temperature for an amplifying cyclone situated off the coast of Africa on March 2, 1984. This disturbance is the precursor to the cyclone apparent in Figs. 1.15 and 1.24 two days later. During amplification, the system tilts westward, which transfers heat poleward and releases available potential energy—analogous to an unstable Eady mode with $\alpha < \alpha_c$. Eddy heat flux tends to maximize near 700 mb, which typifies the steering level of observed cyclones. This is lower than the steering level predicted by the Eady model. However, an unbounded model treated by Charney (1947) reproduces the observed steering level, while retaining the essential ingredients captured by Eady's solution.

Figure 16.5 contains the characteristic signature of sloping convection: A tongue of warm air is drawn poleward ahead of the closed low, while cold air is advected equatorward behind it. Those bodies of air have disparate histories, which are reflected in contemporaneous infrared (IR) and water vapor imagery (Figs. 16.6a and b). A tongue of high cloud cover and moisture that extends northwestward from the African coast defines the *warm sector* ahead of the cyclone. A complementary tongue of cloud-free conditions and low moisture is being drawn equatorward behind it. Sharp gradients separating those bodies

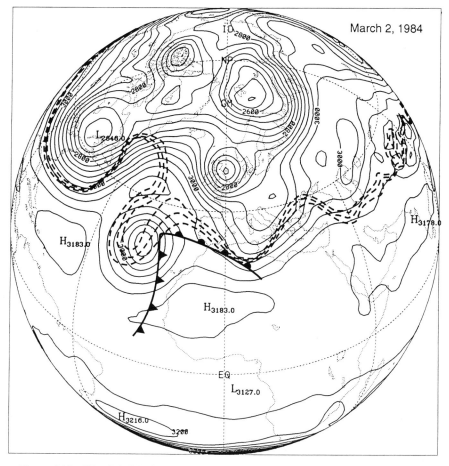

Figure 16.5 700-mb height (solid lines) and selected isotherms (dashed lines) on March 2, 1984. A surface frontal analysis is superposed.

of air mark warm and cold fronts at the surface, which are superposed in Fig. 16.5 and delineate the warm sector. Moisture inside the warm sector can be traced back in water vapor imagery (e.g., Fig. 16.6d) to its source: tropical convection over the Amazon basin (compare Fig. 1.24). Air advancing behind the cold front undercuts the warm sector in sloping convection, lifting moist air to produce the extensive cloud shield in Fig. 16.6a.

16.4 Nonlinear Considerations

While capturing the development of extratropical cyclones, Eady's solution provides only a hint of their behavior at maturity. Conversion of available

Infrared   Water Vapor

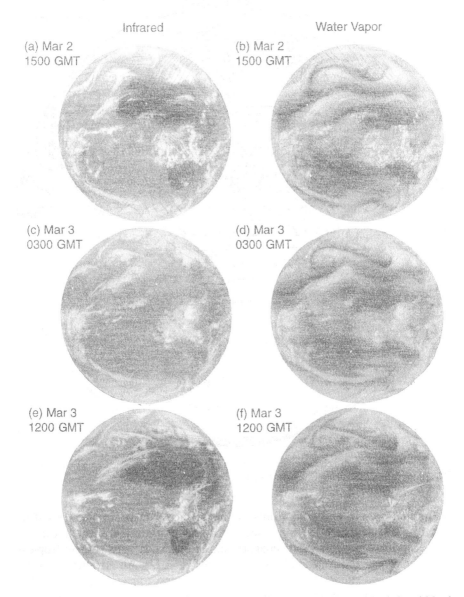

(a) Mar 2
1500 GMT

(b) Mar 2
1500 GMT

(c) Mar 3
0300 GMT

(d) Mar 3
0300 GMT

(e) Mar 3
1200 GMT

(f) Mar 3
1200 GMT

Figure 16.6 Infrared and water vapor imagery from Meteosat between March 2 and March 4, 1984, that reveal the evolution of an extratropical cyclone off the northwest coast of Africa. The warm sector ahead of the cyclone is marked by a tongue of high cloud cover and moisture, which slopes northwestward on March 2. That air mass subsequently overturns near the juncture of cold and warm fronts delineating it to form an occlusion, in which cold and warm air are entrained and mixed horizontally (compare Fig. 9.21).

potential energy into eddy kinetic energy enables a baroclinic system to intensify and eventually attain finite amplitude. Finite horizontal displacements then invalidate the linear description on which Eady's model is based and which predicts exponential amplification to continue indefinitely—analogous to vertical displacements under hydrostatically unstable conditions (Sec. 7.3). Second-order effects then modify the zonal–mean state, which in turn limits subsequent amplification of the baroclinic system.

As a baroclinic disturbance amplifies, horizontal displacements (e.g., Fig. 16.3b) become increasingly exaggerated. Warm tropical air is eventually folded north of cold polar air, as is revealed by the distribution of potential vorticity on March 2, 1984 (Fig. 12.10). Deformations experienced by those air masses steepen potential temperature gradients separating warm and cold air (e.g., Fig. 12.4), which intensifies the accompanying fronts. Continued advection leads to warm and cold air eventually encircling the low, with the warm sector overturning at the junction of the warm and cold fronts. (Cloud cover in Fig. 9.21 provides a textbook example.) The cyclone in Fig. 16.5 then *occludes*, with warm and cold fronts overlapping in the center of the system. The occlusion actually develops when the surface trough (not shown) separates from the junction of warm and cold fronts and deepens farther back into the cold air mass (see, e.g., Wallace and Hobbs, 1977). This structure marks the mature stage of the cyclone's life cycle because the surface trough is then positioned beneath the upper-level trough, which eliminates the system's westward tilt and hence its release of available potential energy—now analogous to a neutral Eady mode with $\alpha > \alpha_c$. Cold and warm air drawn into the occlusion are then wound together and mixed horizontally.

Figure 16.6 shows sequences of IR and water vapor imagery while the disturbance in Fig. 16.5 matures. At 1500 GMT on March 2 (Figs. 16.6a and b), the cold front is approaching the warm front near their junction, with the warm sector clearly defined. Twelve hours later, the 700-mb trough has deepened (not shown). The warm sector in IR and water vapor imagery (Figs. 16.6c and d) has then been sheared to the northwest, where cold dry air is being entrained with warm moist air in the occlusion that has formed. This process culminates in cold and warm air at the occlusion winding up into a spiral—similar to behavior in a cylindrical annulus at high rotation (compare Fig. 15.7f). By 1200 GMT on March 3 (Figs. 16.6e and f), air inside the occlusion has wound up into a vortex, which is seen to separate from the remaining warm sector to its south and east. Interleaving bands of cold and warm air that are apparent in both cloud cover and moisture symbolize efficient horizontal mixing. One day later (Figs. 1.15 and 1.23), that mass of air has become nearly homogeneous. The 700-mb trough has then weakened and high cloud cover that developed earlier through sloping convection is dissipating. Only a broad spiral of equatorward-moving air remains, drawn cyclonically around the now-diffuse anomaly of potential vorticity, in which cold and warm air have been mixed.

Since θ (more generally, θ_e) is conserved for individual air parcels, horizontal mixing makes the potential temperature uniform and restores the thermal structure to barotropic stratification. Baroclinic instability that was present initially has then been neutralized and no more potential energy is available for conversion to eddy kinetic energy. This process is a direct counterpart of vertical mixing by convection, which neutralizes hydrostatic instability (Sec. 7.3). Strong modification of the mean state by unstable eddies contrasts sharply with wave propagation under stable conditions. Parcel displacements are then bounded, so the mean state remains largely unaffected outside regions of dissipation, where wave activity is absorbed (Chapter 14).

The preceding treatment applies formally to a zonally symmetric mean state. In practice, the results also lend insight into isolated regions of instability. Amplified planetary waves in the Northern Hemisphere reinforce zonal–mean vertical shear to produce a broken storm track that is marked by localized jets east of Asia and North America (refer to Figs. 15.8a and 9.38). Strongly baroclinic, the north Atlantic and north Pacific storm tracks are preferred sites for cyclone development. Weaker planetary waves in the Southern Hemisphere leave stratification there nearly uniform in longitude to produce a continuous storm track. This enables cyclones there to assume a more regular distribution than is observed in the Northern Hemisphere, one that occasionally resembles baroclinic modes in a rotating annulus (see Figs. 2.10 and 15.7e).

The criteria for instability also have implications important to planetary waves that have attained large amplitude. The situation is analogous to the breaking of gravity waves when isentropic surfaces are overturned (Fig. 14.25). High θ is then folded underneath low θ to make $\partial\theta/\partial z < 0$ and render the region hydrostatically unstable. Convective mixing that ensues absorbs organized wave motion and neutralizes the instability by driving $\partial\theta/\partial z$ to zero. Planetary waves of sufficient amplitude overturn the distribution of the conserved property Q (Fig. 14.26). High potential vorticity folded poleward of low potential vorticity then reverses $\partial Q/\partial y$ locally to render the motion dynamically unstable. At that point, the planetary wave field breaks. Small-scale eddies that develop in the region of instability mix Q horizontally, which absorbs organized wave motion and neutralizes the instability by driving $\partial Q/\partial y$ to zero.

Suggested Reading

Atmospheric Science: An Introductory Survey (1977) by Wallace and Hobbs contains a synoptic analysis of the life cycle of extratropical cyclones.

Atmosphere-Ocean Dynamics (1982) by Gill includes a nice comparison between Charney's (1947) model of baroclinic instability and the Eady model.

An Introduction to Dynamic Meteorology (1992) by Holton provides a complete description of baroclinic extratropical disturbances, their energetics, and aspects surrounding their prediction. It also discusses the process of frontal formation.

Middle Atmosphere Dynamics (1987) by Andrews *et al.* discusses inertial instability in the tropical stratosphere and includes a description of planetary wave breaking.

Problems

16.1. Derive equations (16.1).

16.2. Carry out the Lagrangian integration of (16.3) to relate an air parcel's meridional displacement to its change of zonal velocity.

16.3. Provide an expression for the specific restoring force inside an inertially unstable layer.

16.4. (a) At what latitudes is inertial instability favored? (b) Identify those regions of the zonal–mean flow in Fig. 1.8 that would be most susceptible to inertial instability.

16.5. Consider westerly flow that corresponds to an angular velocity $A(z) = \bar{u}/(a \cos \phi)$, which varies with height but not with latitude. (a) Within the framework of quasi-geostrophic motion and the Boussinesq approximation, what sign of curvature must the velocity profile $A(z)$ have for shear instability to develop? (b) At what latitudes is shear instability favored most if $A(z) = \epsilon(z) \cdot 2\Omega$, with $\epsilon << 1$?

16.6. Obtain the lower boundary condition (16.6.4).

16.7. A wave packet approaches its critical line with phase speed c_r equal to the real part of c in (16.8.1). Describe the wave packet's evolution in the presence of weak dissipation if, under inviscid adiabatic conditions, (a) c in (16.8.1) is real and (b) c in (16.8.1) is complex.

16.8. Obtain equations (16.9) and (16.10).

16.9. Recover the necessary condition (16.12).

16.10. Derive a necessary condition for instability of a quasi-geostrophic zonal flow that is unbounded above and bounded below by a rigid surface of constant height.

16.11. Derive a necessary condition for instability of a quasi-geostrophic zonal flow that is bounded vertically at $z = 0$ and H by rigid walls.

16.12. (a) Show that, in the absence of rotation, criterion (1) in Sec. 16.2.2 recovers Rayleigh's condition for instability: a barotropic flow must possess an inflection point in its interior. (b) Discuss the influence rotation has on shear instability.

16.13. Why do midlatitude cyclones intensify during winter?

16.14. Obtain the dispersion relation (16.18) for Eady modes.

16.15. Demonstrate that the limiting structure of Eady modes for $H \to \infty$ has v' in quadrature with θ'.

16.16. Derive an expression for the shortwave cutoff α_c for instability in the Eady problem.

16.17. (a) Show that the maximum growth rate of Eady modes is achieved for $l = 0$. (b) Express the growth rate of the fastest growing square Eady mode ($k = l$) in terms of that of the fastest growing Eady mode ($l = 0$).

16.18. Show that, for the fastest growing Eady mode, the maximum slope of motion in the meridional plane is only about half the slope of mean isentropic surfaces.

16.19. Calculate the e-folding times of square Eady modes ($k = l$) for an f plane at $45°$, $N^2 = 10^{-4}$ s^{-2}, vertical shear of 3 m s^{-1} km^{-1}, a rigid lid at 10 km, and for zonal wavenumbers 1–8.

16.20. Within the framework of an initial value problem, describe the structure and evolution predicted by the Eady model if the flow in Problem 15.19 is initialized with random structure having a broad wavenumber spectrum (a) under inviscid adiabatic conditions, (b) in the presence of linear dissipation[3] with a timescale equal to the shortest e-folding time of zonal wavenumber 2 under inviscid adiabatic conditions and (c) in the presence of linear dissipation with a timescale equal to the shortest e-folding time of zonal wavenumber 4 under inviscid adiabatic conditions.

16.21. Meridional components of the eddy momentum flux $\overline{u'v'}$ and group velocity c_{gy} vanish for Eady modes. Why?

16.22. Show that the eddy heat flux $\overline{v'\theta'}$ is positive and independent of height for Eady modes.

16.23. Use the Eady problem to explain (a) why midlatitude cyclones develop and track along the jet stream, (b) how baroclinic instability would be altered if the jet were displaced equatorward and (c) where the storm tracks should be positioned in relation to the time–mean motion in Fig. 1.9.

16.24. Precipitation inside cyclones often assumes a banded structure. Discuss this feature in relation to the evolution in Fig. 16.6 and the corresponding distribution of potential vorticity Q_g.

16.25. Consider an idealized atmosphere in which the circulation is described by the barotropic zonal flow

$$\overline{u}(y) = -U \cos(2\phi),$$

---

[3] Rayleigh friction and Newtonian cooling of the same timescale.

where $U = 0.55\Omega a$. (a) Characterize the inertial stability of this flow as a function of latitude. (b) Describe the zonal–mean circulation toward which inertial instability will drive the flow.

16.26. Discuss the relationship of barotropic and baroclinic instability to stratification (Sec. 12.2.1).

16.27. Unlike extratropical cyclones, tropical cyclones are driven by latent heat release. Suppose a typhoon drifts poleward from the ITCZ. (a) Describe its evolution as it migrates over colder sea surface temperatures. (b) What structure of midlatitude westerlies will allow it to sustain itself?

Chapter 17 | The Middle Atmosphere

Above the tropopause, the horizontal gradient of heating is generally weak enough to leave the thermal structure close to barotropic stratification. Therefore, baroclinic instability generated through local conversion of available potential energy does not play an essential role in the stratosphere. Further, strong westerlies in the winter stratosphere and easterlies in the summer stratosphere trap baroclinic disturbances generated in the troposphere, limiting their presence in the stratosphere to a neighborhood of the tropopause. These features leave the circulation in the stratosphere much smoother than in the troposphere (Fig. 1.10). However, planetary waves propagate through strong westerlies of the polar-night jet (Sec. 14.5.2). Vertical growth enables planetary waves to disturb the polar vortex and, during strong amplifications, to disrupt the circumpolar flow completely. Air then moves freely between tropical and polar regions, where it experiences sharply different radiative environments. Global-scale disturbances like the one in Fig. 1.10a play a key role in establishing mean distributions of motion and chemical constituents.

With convection inhibited by strong static stability, the energy budget of the stratosphere is dominated by radiative transfer, so it generally remains closer to radiative equilibrium than the troposphere (Fig. 8.24). The radiative energy budget is controlled by shortwave (SW) heating due to ozone absorption in the Hartley and Huggins bands (Sec. 8.3.1) and longwave (LW) cooling due to CO_2 emission to space at 15 μm. Ozone thus determines the thermal structure of the middle atmosphere. Since this controls the circulation through hydrostatic and geostrophic equilibrium, ozone figures indirectly in much of the observed behavior of the middle atmosphere.

17.1 Ozone Photochemistry

The simplest treatment of ozone photochemistry is attributed to Chapman (1930). Consider a pure oxygen atmosphere, the composition of which is governed by the following reactions

$$O_2 + h\nu \rightarrow O + O \qquad (J_2), \qquad (17.1.1)$$

$$O + O_2 + M \rightarrow O_3 + M \qquad (k_2), \qquad (17.1.2)$$

$$O + O_3 \rightarrow 2O_2 \qquad (k_3), \qquad (17.1.3)$$

$$O_3 + h\nu \rightarrow O_2 + O \qquad (J_3), \qquad (17.1.4)$$

where the rate coefficients indicated parenthetically to the right character-ize the speeds of individual reactions.[1] Reactions (17.1.1) and (17.1.4) de-scribe photodissociation or *photolysis* of O_2 by ultraviolet (UV) radiation in the Herzberg continuum near 242 nm and of O_3 in the Hartley and Huggins bands near 310 nm, (17.1.4), operating at all wavelengths shorter than 1 μm. Reactions (17.1.2) and (17.1.3) describe recombination of O_2 and O_3 with O. The molecule M represents a third body needed to conserve momentum and energy in the recombination of O and O_2. Atomic oxygen produced by pho-tolysis of ozone in (17.1.4) recombines immediately with molecular oxygen in (17.1.2) to reform ozone. Hence, those reactions constitute a closed cycle in O and O_3 that absorbs UV radiation efficiently.

17.1.1 The Chemical Family

The rate coefficients determine the photochemical lifetimes of these species (Problem 17.1), which are shown in Fig. 17.1 as functions of altitude. Lifetimes of the odd oxygen species O and O_3 are short by comparison with that of O_2, which is present in fixed proportion. Moreover, the lifetimes of O and O_3 differ by several orders of magnitude, which complicates their treatment in numerical calculations. For this reason, it is convenient to introduce the

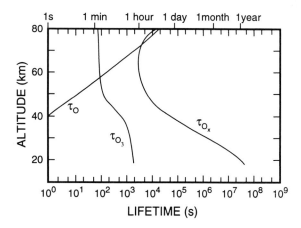

Figure 17.1 Photochemical lifetimes of oxygen species. Source: Brasseur and Solomon (1986).

[1] Dimensions of $\left(\frac{\text{molecules}}{\text{volume}}\right)^{-2}$ time^{-1} for k_2, $\left(\frac{\text{molecules}}{\text{volume}}\right)^{-1}$ time^{-1} for k_3, and time^{-1} for J_2 and J_3.

odd oxygen family:

$$O_x = O + O_3. \tag{17.2}$$

Individual members of the family have lifetimes much shorter than that of O_x, so they can be thought of as remaining in photochemical equilibrium with one another. Reactions (17.1) then describe comparatively slow changes in O_x that are attended by immediate adjustments of O and O_3 within the family to maintain photochemical equilibrium. Such changes are introduced whenever an air parcel is displaced from one radiative environment to another.

The relative abundance or *partitioning* of family members is determined by reactions (17.1.2) and (17.1.4), which operate much faster than the others. Those reactions preserve odd oxygen, so they represent a simple redistribution between O and O_3. The rate of destruction of O_3 in (17.1.4) is expressed by

$$\left.\frac{d[O_3]}{dt}\right|_{\text{destruction}} = -J_3[O_3], \tag{17.3.1}$$

where [ ] denotes the number density (molecules/volume) and d/dt reflects the Lagrangian derivative. [The chemical budget should formally be expressed in terms of mixing ratio to account for expansion and compression of an air parcel (Sec. 1.2), but, for the moment, we ignore motion.] The rate of production of O_3 in (17.1.2) is given by

$$\left.\frac{d[O_3]}{dt}\right|_{\text{production}} = k_2[O_2][O][M]. \tag{17.3.2}$$

Adding yields the net rate of production of O_3 in (17.1.2) and (17.1.4):

$$\left.\frac{d[O_3]}{dt}\right|_{\text{net}} = k_2[O_2][O][M] - J_3[O_3]. \tag{17.3.3}$$

Photochemical equilibrium within the family requires the left-hand side to vanish, so

$$J_3[O_3] = k_2[O_2][O][M], \tag{17.4}$$

which describes the partioning of its member species. Below 60 km, O_3 is the dominant member of O_x. Consequently, ozone tracks the behavior of odd oxygen. The lifetime of O_x (Fig. 17.1) is several weeks in the lower stratosphere, where it is an order of magnitude longer than the timescale of advection (~1 day). But it decreases rapidly with height, falling to under 1 day above about 30 km.

17.1.2 Photochemical Equilibrium

The remaining reactions (17.1.1) and (17.1.3) describe the production and destruction of O_x, processes which operate on timescales much longer than reactions for the individual member species. For photochemical equilibrium of the family as well, the net rate of production of $[O_x]$ must also vanish. The rate at which O_x molecules are produced in (17.1.1) is expressed by

$$\left.\frac{d[O_x]}{dt}\right|_{\text{production}} = 2J_2[O_2], \tag{17.5.1}$$

whereas the rate they are destroyed in (17.1.3) is given by

$$\left.\frac{d[O_x]}{dt}\right|_{\text{destruction}} = -2k_3[O][O_3]. \tag{17.5.2}$$

Adding yields the net rate of production of $[O_x]$

$$\left.\frac{d[O_x]}{dt}\right|_{\text{net}} = 2J_2[O_2] - 2k_3[O][O_3]. \tag{17.5.3}$$

Photochemical equilibrium of O_x then requires

$$J_2[O_2] = k_3[O][O_3]. \tag{17.6}$$

Eliminating $[O]$ from (17.4) and (17.6) results in the equilibrium ozone number density

$$[O_3] = [O_2]\left(\frac{k_2 J_2}{k_3 J_3}[M]\right)^{\frac{1}{2}} \tag{17.7}$$

in terms of the number densities of molecular oxygen and air.

The profile of $[O_3]$ predicted by (17.7) is plotted in Fig. 17.2. Chapman chemistry (solid line) reproduces the essential structure of the observed vertical ozone distribution, which is superposed for tropical and extratropical regions. Ozone number density decreases downward due to increasing $[O_2]$ and photolysis in (17.1.1). A maximum in $[O_3]$ is achieved near 30 km, below which UV flux and J_2 fall off sharply due to extinction by photolysis.

Despite this general agreement, the behavior predicted by (17.7) deviates importantly from the observed distribution of ozone. Ozone number density is generally underpredicted in the lower stratosphere and overpredicted in the upper stratosphere. In extratropical regions, the observed profile of $[O_3]$ is displaced downward from that in the tropics, with the maximum concentration lying below 20 km (refer to Fig. 1.17). When the distribution of ozone column abundance Σ_{O_3} is considered (Figs. 1.18 and 1.19), the discrepancy with observed behavior is even more serious. Observed total ozone at $60°$ is almost double that in the tropics. By contrast, (17.7) predicts total ozone to minimize at high latitude because UV flux and hence J_2 are small there. Both vanish in the polar night, where some of the largest values of Σ_{O_3} are, in fact, found.

These discrepancies are attributable to two key ingredients that are not accounted for by (17.7). First, ozone chemistry involves species other than oxygen. Catalytic cycles involving free radicals of hydrogen, nitrogen, and chlorine deplete odd oxygen and make ozone dependent upon a wide array of chemical species. The second factor not accounted for is motion. Below 30 km, where

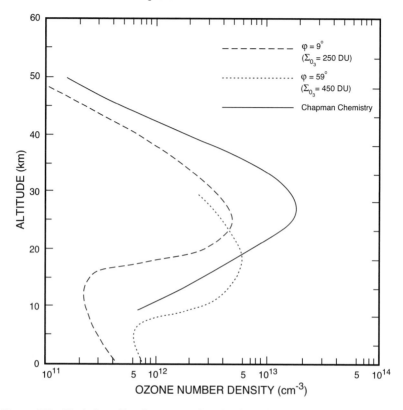

Figure 17.2 Vertical profile of ozone number density under photochemical equilibrium, as calculated from Chapman chemistry (solid line), and observed in tropical (dashed line) and extratropical (dotted line) regions. Chapman chemistry was calculated with rate coefficients from Brasseur and Solomon (1986) and Nicolet (1980). It yields realistic vertical structure, but a column abundance of $\Sigma_{O_3} \cong 1000$ DU. Observed data are from Hering and Borden (1965) and Krueger (1973).

the ozone column is concentrated, the photochemical lifetime is long enough for O_3 to be transported by the circulation and therefore to pass between widely differing photochemical environments. Even at higher altitudes, other species participating in the complex photochemistry of ozone are influenced importantly by transport.

17.2   Involvement of Other Species

Photodissociation by UV radiation produces a number of free radicals from less reactive reservoir species of tropospheric origin. Photochemically active, these free radicals can then go on to destroy ozone in catalytic cycles that leave the free radical unchanged.

17.2.1 Nitrous Oxide

The free radical nitric oxide, NO, represents an important link to human activities. Nitric oxide is a by-product of inefficient combustion (e.g., in aircraft exhaust). In the stratosphere, the principal source of NO is dissociation of nitrous oxide N_2O through reaction with atomic oxygen:

$$N_2O + O \rightarrow 2NO. \tag{17.8}$$

Nitrous oxide is produced in the troposphere by natural as well as anthropogenic means (Chapter 1). Away from the surface, N_2O is long-lived, having a photochemical lifetime of order 100 years (Fig. 17.3). Further, nitrous oxide is not water soluble, so it is immune to normal scavenging mechanisms associated with precipitation (Sec. 9.5.2). These properties allow N_2O to become well-mixed in the troposphere and make it useful as a tracer of air motion in the stratosphere.

After entering the stratosphere, N_2O is photodissociated according to the reaction

$$N_2O + h\nu \rightarrow N_2 + O, \tag{17.9}$$

which constitutes the primary destruction mechanism for nitrous oxide and is responsible for the decrease of its mixing ratio r_{N_2O} above the tropopause (Fig. 1.20). The mixing ratio of N_2O is largest above the tropical tropopause (Fig. 17.4), which, in view of its long lifetime, illustrates how air enters the stratosphere. A dome of high values reflects upwelling motion in the tropical stratosphere that carries nitrous oxide across the tropical tropopause from its reservoir below.

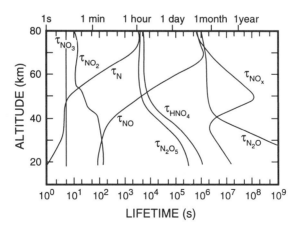

Figure 17.3 Photochemical lifetimes of nitrogen species. Source: Brasseur and Solomon (1986).

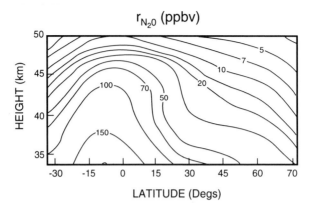

Figure 17.4 Zonal–mean mixing ratio of nitrous oxide N_2O on January 9, 1992, as observed by the ISAMS instrument on board the Upper Atmosphere Research Satellite (UARS). Adapted from Ruth *et al.* (1994).

The free radical NO, once produced via (17.8), can efficiently destroy ozone through the catalytic cycle

$$NO + O_3 \rightarrow NO_2 + O_2, \tag{17.10.1}$$

$$NO_2 + O \rightarrow NO + O_2, \tag{17.10.2}$$

which has the net effect

$$O_3 + O \rightarrow 2O_2 \quad \text{(net)}. \tag{17.10.3}$$

This closed cycle leaves $NO + NO_2$ unchanged, so one molecule of nitric oxide can destroy many molecules of ozone.

Reactions with other nitrogen compounds make NO dependent upon a number of species. Because reactions among those species are fast, it is convenient to introduce the *odd nitrogen family*:

$$NO_x = N + NO + NO_2 + NO_3 + 2N_2O_5 + HNO_4, \tag{17.11}$$

the members of which may be thought to remain in photochemical equilibrium with one another. The lifetime of NO_x is much longer than the timescale for advection (Fig. 17.3), despite the much shorter lifetimes of some of its members. The abundance of NO then follows from $[NO_x]$ and from partitioning within the family, which reacts on a slower timescale with other families such as O_x.

17.2.2 Chlorofluorocarbons

Free radicals of chlorine represent another important link to human activities. Atomic chlorine is produced naturally in the stratosphere through photodissociation of methyl chloride CH_3Cl, which is a product of ocean processes, and through its oxidation with reactive species. Atomic chlorine is also produced by photodissociation of chlorofluorocarbons (CFCs), the sole source of which is industrial. Like N_2O, CFC-11 ($CFCl_3$) and CFC-12 (CF_2Cl_2) are long-lived (Fig. 17.5). Photochemical lifetimes of several years along with their water insolubility, which circumvents normal scavenging mechanisms, allow CFCs to become homogenized in the troposphere (Fig. 1.20).

In the stratosphere, CFC-11 and CFC-12 decrease upward due to photodissociation, which releases free chlorine in the reactions

$$CFCl_3 + h\nu \rightarrow CFCl_2 + Cl, \tag{17.12.1}$$

$$CF_2Cl_2 + h\nu \rightarrow CF_2Cl + Cl, \tag{17.12.2}$$

which operate at wavelengths shorter than 225 nm. Once produced, free Cl reacts with ozone in the catalytic cycle involving chlorine monoxide ClO

$$Cl + O_3 \rightarrow ClO + O_2, \tag{17.13.1}$$

$$ClO + O \rightarrow Cl + O_2, \tag{17.13.2}$$

which has the net effect

$$O_3 + O \rightarrow 2O_2 \quad \text{(net)}. \tag{17.13.3}$$

This closed cycle leaves Cl + ClO unchanged, so one atom of chlorine can destroy many atoms of ozone.

Like free radicals of nitrogen, atomic chlorine can react with other species on widely varying timescales. It is therefore convenient to introduce the *odd chlorine family*:

$$Cl_x = Cl + ClO + HOCl, \tag{17.14}$$

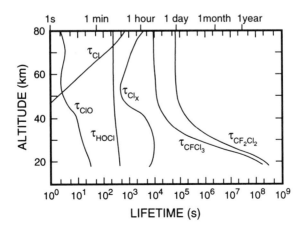

Figure 17.5 Photochemical lifetimes of chlorine species, including CFC-11 ($CFCl_3$) and CFC-12 (CF_2Cl_2). Source: Brasseur and Solomon (1986).

from which the abundance of Cl follows with the partitioning of member species. Relative to air motion, Cl_x is short-lived (Fig. 17.5), so it is not passively advected. Nevertheless, Cl_x is much longer-lived than its members, which facilitates the numerical treatment of chlorine species. Odd chlorine can interact with odd nitrogen in the reaction

$$ClO + NO \rightarrow Cl + NO_2, \tag{17.15}$$

which regenerates free Cl and couples the families NO_x and Cl_x. Other interactions among O_x, NO_x, and Cl_x bear importantly on ozone depletion in the polar stratosphere (Sec. 17.7).

17.2.3 Methane

Methane enters the equation for ozone because it interacts with NO_x and O_x. In addition to being radiatively active, CH_4 represents an important link between chemical constituents in the stratosphere and water vapor, both of which have the troposphere as a reservoir. Like N_2O, methane has a photochemical lifetime of several years below 30 km (Fig. 17.6) and is water insoluble. These properties allow CH_4 to become well mixed in the troposphere (Fig. 1.20) and make it useful as a tracer in the stratosphere. Figure 17.7 shows the zonal–mean distribution of r_{CH_4} in the middle atmosphere. Mirroring behavior in the N_2O distribution, a dome of high values above the tropical tropopause reflects upwelling motion in the equatorial stratosphere.

Decreasing mixing ratio in the stratosphere follows from chemical destruction of CH_4. Methane is oxidized by the hydroxyl radical in the reaction

$$CH_4 + OH \rightarrow CH_3 + H_2O. \tag{17.16.1}$$

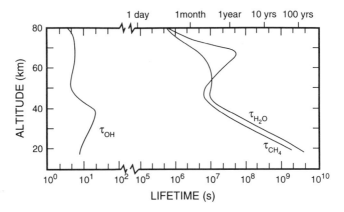

Figure 17.6 Photochemical lifetimes of hydrogen species. Source: Brasseur and Solomon (1986).

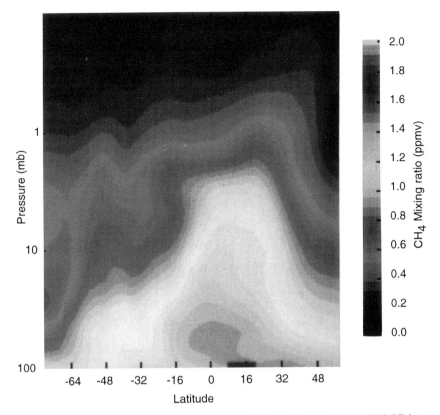

Figure 17.7 Zonal–mean mixing ratio of methane CH_4, as observed by the HALOE instrument on board UARS during September 21–October 15, 1992. Adapted from Bithell *et al.* (1994).

It is also destroyed by atomic oxygen and chlorine in the reactions

$$CH_4 + O \rightarrow CH_3 + OH, \qquad (17.16.2)$$

$$CH_4 + Cl \rightarrow CH_3 + HCl. \qquad (17.16.3)$$

The latter represents an important sink of reactive Cl_x. The free radical CH_3 produced in reactions (17.16) immediately combines with molecular oxygen to form CH_3O_2. Two reactions involving NO_x,

$$CH_3O_2 + NO \rightarrow CH_3O + NO_2 \qquad (17.17.1)$$

$$NO_2 + h\nu \rightarrow NO + O, \qquad (17.17.2)$$

then constitute a closed cycle that has the net effect

$$CH_3O_2 + h\nu \rightarrow CH_3O + O \qquad \text{(net)}. \qquad (17.17.3)$$

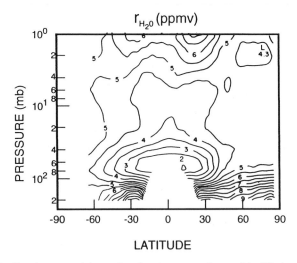

Figure 17.8   Zonal–mean mixing ratio of water vapor observed by Nimbus-7 LIMS during December, 1978. Adapted from Remsberg *et al.* (1984).

Methane oxidation thus leads to production of O_x. Since the partitioning of O_x fixes the abundance of O_3, reactions (17.17) provide a potentially important source of ozone in the lower stratosphere.

The free radical OH that initiates the foregoing process is produced by dissociation of water vapor

$$H_2O + O \rightarrow 2OH. \tag{17.18}$$

Above the tropopause, the lifetime of H_2O is several years (Fig. 17.6), so it too behaves as a tracer at these heights. Water vapor mixing ratio decreases sharply in the troposphere to a minimum of just a few ppmv at the *hygropause* (Fig. 17.8). The steep decrease with height in the troposphere reflects precipitation processes that dehydrate air reaching the tropopause by limiting r_{H_2O} to saturation values. Above the hygropause, r_{H_2O} increases due to oxidation of methane, as r_{CH_4} decreases. How much of the observed increase is actually explained by reactions (17.16) remains unclear.

17.3   Air Motion

Photochemical considerations alone predict maximum column abundances of ozone at low latitude. This shortcoming is unchanged by the addition of other chemical species because UV radiation and hence photolysis maximize in the tropics. Consequently, the maximum of observed Σ_{O_3} at middle and high latitudes can be explained only by incorporating dynamical considerations.

17.3.1 The Brewer–Dobson Circulation

The latitudinal gradient of heating in combination with geostrophic equilibrium favors motion that is chiefly zonal, with only a small meridional component to transfer heat and chemical species between the equator and poles (Chapter 15). In the winter hemisphere, ozone heating establishes an equatorward temperature gradient over a deep layer (Fig. 1.7), which, by thermal wind balance, produces strong westerly flow. Stratospheric westerlies intensify upward along the *polar-night terminator*, where SW absorption vanishes, to produce the polar-night jet (Fig. 1.8) and the circumpolar vortex (Fig. 1.10b). In the summer hemisphere, a poleward gradient of heating, which follows from the distribution of daily insolation (Fig. 1.28), produces a deep temperature gradient of the opposite sense and strong easterly circumpolar flow.

Under radiative equilibrium, the circulation is zonally symmetric and experiences no net heating. Then, by the first law, individual air parcels do not cross isentropic surfaces. This implies no net vertical motion and, by continuity, no net meridional motion—analogous to the zonally symmetric circulation in Sec. 15.2. Therefore, mechanical disturbances that drive the circulation out of radiative equilibrium play a key role in producing the gradual meridional overturning that accompanies strong zonal motion in the middle atmosphere.

The behavior of long-lived tracers (Figs. 17.4 and 17.7) implies upwelling in the tropics, where tropospheric air enters the stratosphere. Downwelling at middle and high latitudes, which is required by continuity, can then explain the large abundances of ozone observed in extratropical regions. Air drawn poleward and downward from the chemical source region of ozone in the tropical stratosphere (Fig. 17.9) would then undergo compression to yield the greatest absolute concentrations ρ_{O_3} (Fig. 1.17) and hence the greatest column abun-

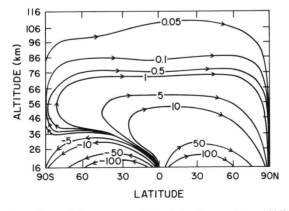

Figure 17.9 Streamlines of the mean meridional circulation of the middle atmosphere, in a quasi-Lagrangian representation. Adapted from Garcia and Solomon (1983), copyright by the American Geophysical Union.

dances at middle and high latitudes. Suggested by Dobson (1930) and Brewer (1949) to explain tracer observations, this gradual meridional overturning is responsible for observed distributions of ozone and other chemical species in the middle atmosphere. The *Brewer–Dobson circulation* also underlies chemical production because, by displacing air out of photochemical equilibrium, it enables chemical reactions to occur.

17.3.2   Wave Driving of the Mean Meridional Circulation

Although it may be envisioned as a zonally symmetric overturning in the meridional plane, the Brewer–Dobson circulation is in reality accomplished by zonally asymmetric processes. A clue to its origin lies in the radiative-equilibrium state of the middle atmosphere (Fig. 17.10). During solstice, the sharp gradient of heating across the polar-night terminator produces radiative-equilibrium temperatures T_{RE} colder than 150 K (Fig. 17.10a). These are much colder than observed zonal–mean temperatures (Fig. 1.7), which are more uniform latitudinally and remain above 200 K in the Northern Hemisphere. Radiative-equilibrium winds implied by the deep layer of sharp temperature gradient (Fig. 17.10b) intensify to more than 300 m s^{-1}—much stronger than observed winds (Fig. 1.8).

The observed polar-night vortex is warmer than radiative equilibrium, so it must experience net radiative cooling (Sec. 8.5). Observed temperatures and ozone imply cooling rates exceeding 8 K day^{-1} in the polar night (Fig. 8.27). By the first law, that cooling must reduce the potential temperature of air parcels inside the vortex. Positive stability ($\partial\theta/\partial z > 0$) then implies downwelling across isentropic surfaces (e.g., parcels sink to lower θ). Calculations in which this diabatic cooling is prescribed (Murgatroyd and Singleton, 1961) qualitatively reproduce the meridional circulation inferred from tracer behavior by Brewer and Dobson.

In the Southern Hemisphere, polar temperatures as cold as 180 K are observed. Zonal–mean winds are then commensurately faster. Colder and stronger than its counterpart over the Arctic, the Antarctic polar-night vortex remains closer to radiative-equilibrium behavior. According to the preceding arguments, radiative cooling and downwelling in extratropical regions should then be weaker. Similar considerations apply to the summertime circulation (compare Fig. 8.27), in which strong easterlies block planetary wave propagation from below (Sec. 14.5).[2]

The key to understanding interhemispheric differences of the circumpolar vortex lies in understanding how the circulation is maintained out of radiative equilibrium. Observed motion during winter is rarely in a quiescent state of zonal symmetry, especially in the Northern Hemisphere, where the circumpolar

[2]The absence of eddy motion enables the cloud of volcanic debris in Fig. 9.6 to remain confined meridionally through the summer season. Wher the zonal flow reverses near equinox and planetary waves emerge from below, the volcanic cloud is quickly dispersed over the hemisphere.

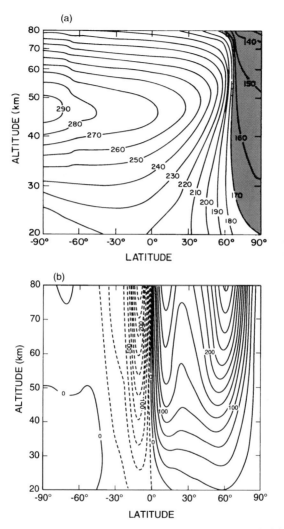

Figure 17.10 Radiative-equilibrium (a) temperature and (b) zonal wind in the middle atmosphere during solstice, as calculated in a radiative–convective–photochemical model. Thermal structure adapted from Fels (1985). Zonal wind calculated from thermal wind balance and from climatological motion at 20 km in Fig. 1.8.

flow is continually disturbed by planetary waves that propagate upward from the troposphere. Planetary waves in the Northern Hemisphere are stronger than those in the Southern Hemisphere because of larger orographic forcing at the earth's surface. Able to penetrate into the strong westerlies of the winter stratosphere, planetary waves drive the circulation out of zonal symmetry and away from radiative equilibrium. As illustrated by Fig. 1.10a, air then flows meridionally from one radiative environment to another (e.g., from sunlit regions, where it experiences SW heating, into polar darkness, where LW cooling prevails). Irreversible heat transfer can then lead to net cooling of individual parcels (Sec. 3.6), which is manifested in a drift of air to lower θ.

Figure 17.11a shows the horizontal trajectory of an air parcel inside the polar-night vortex, superposed on the instantaneous distributions of motion and potential vorticity Q (Sec. 12.5) in a numerical integration. During its evolution, the vortex is sporadically displaced out of zonal symmetry and distorted by planetary waves. The air parcel shown then passes between different latitudes and therefore different radiative-equilibrium temperatures as it orbits about the displaced vortex. In its thermodynamic state space (Fig. 17.11b), the parcel's temperature oscillates between excursions of T_{RE}. This prevents the parcel from adjusting to radiative equilibrium, instead trailing behind the lo-

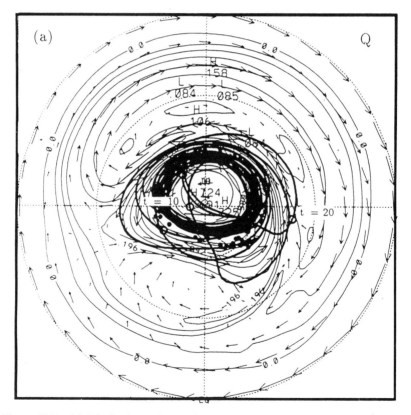

Figure 17.11   (a) Distributions of potential vorticity Q and horizontal motion in a two-dimensional calculation representative of the polar-night vortex (compare Fig. 14.26). The bold line shows the trajectory of one of an ensemble of air parcels (dots) initialized inside the vortex. The folding of contours along the edge of the vortex and regions where the Q distribution has been homogenized by secondary eddies symbolizes horizontal mixing. (*continues*)

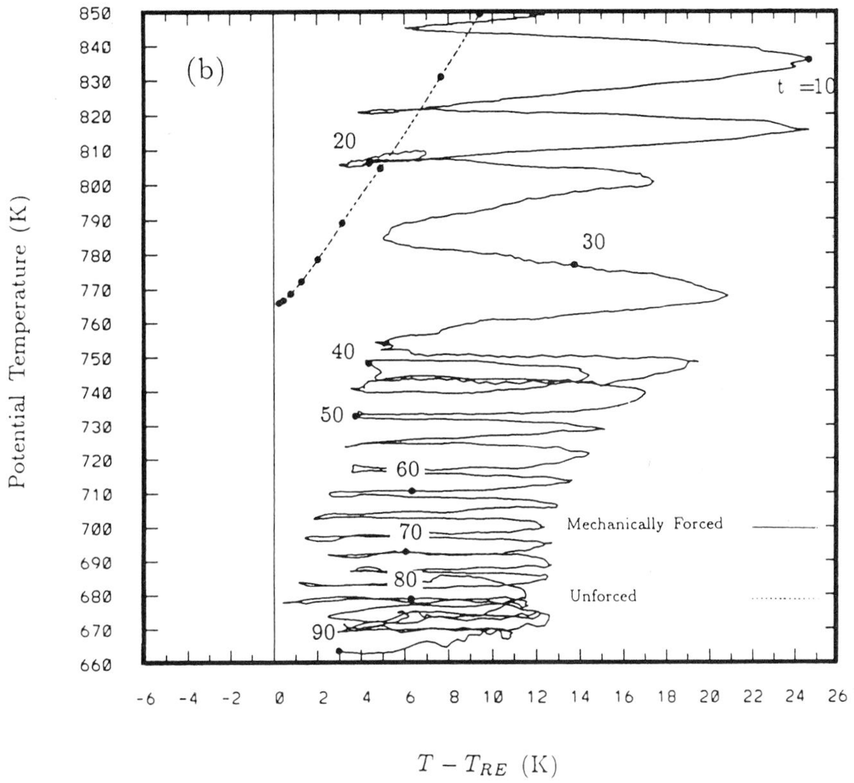

Figure 17.11 (*Continued*) (b) Evolution of the parcel whose trajectory is shown in part (a) in its thermodynamic state space (solid line), represented in terms of the parcel's departure from local radiative equilibrium. Irreversible heat transfer introduces a hysteresis into the parcel's state with each circuit about the displaced vortex, which produces a steady drift of air to lower θ. In the same calculation, but without planetary waves to maintain the circulation out of radiative equilibrium (dotted line), the parcel lapses into radiative equilibrium, which brings to a halt subsequent cooling and descent inside the vortex. After Fusco and Salby (1994), copyright by the American Geophysical Union.

cal value of T_{RE} by a finite temperature difference. The air parcel is therefore out of thermodynamic equilibrium with its surroundings (Chapter 3). According to the second law, the parcel then experiences irreversible heat transfer, which introduces a hysteresis into its thermodynamic state with each circuit about the vortex. Net cooling during successive circuits causes the parcel to drift across isentropic surfaces toward lower θ (refer to Fig. 3.6). Following ultimately from work performed by planetary waves, net rejection of heat makes the parcel and the circulation as a whole behave as a radiative refrigerator.

Irreversible heat transfer leads to thermal dissipation of planetary waves because it destroys large-scale temperature anomalies. Diabatic behavior introduced when the vortex is driven out of zonal symmetry corresponds to Newtonian cooling (Sec. 8.6), which acts to restore the circulation to radiative equilibrium. Thermal relaxation operates on a timescale of order a week (Fig. 8.29)—still much longer than the characteristic timescale of advection. As air relaxes toward radiative equilibrium, temperature anomalies associated with the wave field collapse, and with them so does eddy motion. Momentum carried by planetary waves is then transferred to the zonal–mean flow.

Planetary waves also experience mechanical dissipation when the polar-night vortex is subjected to quasi-horizontal mixing, which destroys large-scale motion anomalies. Such behavior is evident in Fig. 17.11a from the folding of tracer contours along the edge of the vortex and in regions where the distribution of Q has been homogenized. Those features neighbor the critical line of stationary planetary waves (see Fig. 14.23), where large eddy displacements overturn the distribution of Q and render the local motion dynamically unstable (Sec. 16.4). Air is then wound up in secondary eddies that cascade large-scale tracer structure to small dimensions, where it is eventually absorbed by diffusion. Secondary eddies generated through instability produce an irreversible dispersion of air along isentropic surfaces, because advection operates much faster than diabatic processes. Under amplified conditions, dispersive eddy motions can entrain and mix low-latitude air well into the body of the vortex. Like thermal dissipation, mechanical dissipation of planetary waves acts to maintain the Arctic polar-night vortex warmer and weaker than the radiative-equilibrium circulation.

Weaker planetary waves in the Southern Hemisphere introduce smaller meridional displacements and hence smaller departures of individual parcels from local radiative equilibrium. The Antarctic polar-night vortex also experiences anomalous radiative cooling at high latitudes through exchange with the underlying surface. The Antarctic plateau, which is as high as 700 mb, is much colder than the surrounding maritime region. Radiative exchange between the cold surface and the lower stratosphere in the 9.6-μm band of ozone represents anomalous cooling not experienced by the Arctic vortex. Together, these factors leave the Antarctic polar-night vortex colder and closer to radiative equilibrium than its counterpart over the Arctic. The foregoing considerations then imply that the Antarctic vortex should possess a weaker meridional circulation and reduced transfer of ozone out of its photochemical source region.

Even though the actual circulation is more complex, it is convenient to consider motion two-dimensionally in terms of the zonal–mean circulation $\overline{\mathbf{v}} = (\overline{u}, \overline{v}, \overline{w})$. This representation is especially useful in light of the numerous chemical species that must be accounted for in stratospheric photochemistry. Within this framework, zonal–mean distributions of motion and chemical species interact with the wave field. Wave fluxes of momentum, heat, and chemical mixing ratio enter the equations governing zonal-mean properties (see, e.g., Andrews *et al.*, 1987), analogous to the convergence of eddy momentum flux appearing in the zonal–mean budget (14.5). Transfers of momentum accompanying dissipation of planetary waves then force the zonal–mean flow by exerting an eddy drag. Deceleration of the zonal–mean flow must be attended by a change in zonal–mean thermal structure to maintain thermal wind balance. Wave dissipation thus drives the circulation out of radiative equilibrium, which in turn forces a mean meridional circulation like the one inferred earlier from Lagrangian considerations.

The zonal–mean circulation in Fig. 17.9 was calculated in this manner. In the lower stratosphere, it consists of two equator-to-pole Hadley cells that transport ozone-rich air from the photochemical source region in the tropics to extratropical regions. Meridional transport and subsequent downwelling explain the large column abundances observed at high latitudes (Fig. 1.19). Most active when planetary waves are amplified, that transport also explains the strong seasonality of total ozone (Fig. 1.18). Maximum values of Σ_{O_3} appear during late winter and spring, just following the period of strongest planetary wave activity.

In the mesosphere, the meridional circulation is transformed into a pole-to-pole cell, with air passing from the radiatively heated summer hemisphere to the radiatively cooled winter hemisphere. Despite the distribution of net heating (Fig. 8.27), the summer mesosphere is actually colder than the winter hemisphere (Fig. 1.7), which is only weakly illuminated. Like the winter stratosphere, the mesosphere is driven out of radiative equilibrium by mechanical disturbances that propagate up from below. Planetary waves play a role at these altitudes, but gravity waves are thought to be the primary agent responsible for driving the mesosphere out of radiative equilibrium. This conclusion is supported by the fact that, unlike the stratosphere, both hemispheres of the mesosphere are far from radiative equilibrium. Westward-propagating gravity waves can propagate through westerlies of the winter stratosphere without encountering a critical level (Sec. 14.3.4), whereas eastward-propagating gravity waves can do the same in the summer stratosphere. Upon reaching mesospheric heights, gravity waves excited in the troposphere have amplified sufficiently to break (Fig. 14.25). Momentum carried by the waves is then transferred to the zonal–mean flow, which, on being decelerated, alters the thermal structure to maintain thermal wind balance.

17.4 Sudden Stratospheric Warmings

Occasionally during northern winter, the circulation becomes highly disturbed. Accompanied by a marked amplification of planetary waves, the disturbed motion is characterized by a marked deceleration of zonal–mean westerlies or even a reversal into zonal–mean easterlies. At the same time, temperatures over the polar cap increase sharply—by as much as 50 K, so the dark winter pole actually becomes warmer than the illuminated tropics. This dramatic sequence of events takes place in just a few days and is known as a *sudden stratospheric warming*.

A stratospheric warming is under way on the day shown in Fig. 1.10a. Zonal–mean wind on the same day (Fig. 17.12b) has reversed across much of the winter hemisphere at 30 mb and above. Zonal–mean temperature (Fig. 17.12a) is anomalously warm from 70 mb upward. Polar temperatures there are some 40 K warmer than climatological values (Fig. 1.7) and exceed those over the equator, so the meridional temperature gradient is reversed. These dramatic changes in the zonal–mean circulation result from the absorption of planetary waves, which have amplified and transmit upward anomalously strong momentum flux.

Even though it is formally defined in terms of zonal–mean properties, the stratospheric warming is actually a phenomenon that is inherently zonally asymmetric. Figure 1.10a reveals that zonal–mean deceleration on this day really follows from a marked displacement of the cyclonic low out of zonal symmetry and its replacement over the polar cap by an anticyclonic high that

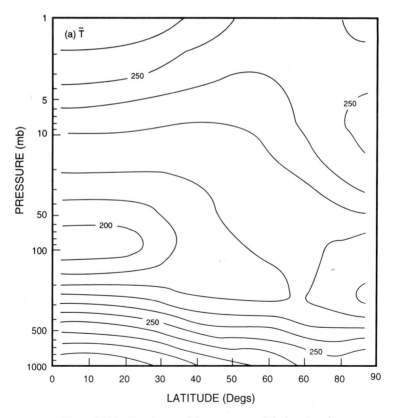

Figure 17.12 Zonal–mean (a) temperature (K). (*continues*)

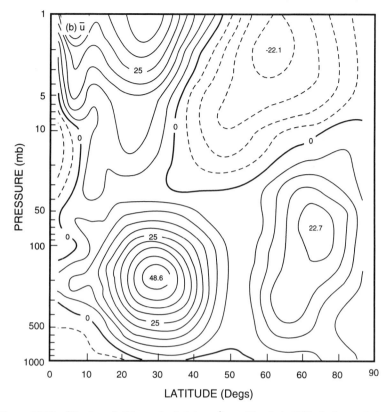

Figure 17.12 (*Continued*) (b) zonal wind (m s^{-1}) on March 4, 1984, during a stratospheric sudden warming (compare Fig. 1.10a).

has built in from midlatitudes. This asymmetric flow pattern corresponds to an amplification of planetary wavenumber 1. It permits air to flow freely between tropical and extratropical regions—on the timescale of a day, rather than on the much longer timescale of the zonal–mean meridional circulation. Figure 17.13 shows the distributions of potential vorticity and motion near 10 mb on the same day. The cyclonic vortex has been displaced well off the pole and has undergone a complex distortion, with high-latitude air (blue) being drawn equatorward around the anticyclone that has invaded the polar cap. Large eddy displacements have overturned the potential vorticity distribution [e.g., near (135°W,45°N)] to make $\partial Q/\partial y < 0$. That motion is then dynamically unstable (Sec. 16.2), so the planetary wave field breaks (McIntyre and Palmer, 1983).

Secondary eddies that amplify in the region of instability bring about an irreversible rearrangement of air. Midlatitude air (red) drawn poleward spins up anticyclonically to conserve potential vorticity. It then forms easterly flow over the polar cap that prevails in the zonal mean (Fig. 17.12). At the same time, polar air drawn equatorward degenerates into small anomalies of cyclonic vorticity that are only suggested in the analyzed distribution of Q.[3] The foregoing behavior resembles an expanded critical layer that engulfs much of the hemisphere when the wave field amplifies and overturns the Q distribution (compare Fig. 14.26).

Motions during a stratospheric warming bring about a major rearrangement of air and chemical species. Ozone then flows freely from its site of chemical production at low latitude to middle and high latitudes. Air advected poleward moves approximately along isentropic surfaces. In the lower stratosphere, where the ozone column is concentrated, isentropic surfaces slope downward toward the pole, so air descends as it streams poleward. On a longer timescale, poleward-moving air also experiences radiative cooling, which introduces a net downward displacement across isentropic surfaces. These are the same mechanisms responsible for generating the Brewer–Dobson circulation, which is temporarily accelerated due to the exaggerated meridional displacements accompanying a stratospheric warming.

Ozone-rich air advected poleward descends and undergoes compression, which increases the absolute concentration ρ_{O_3} and magnifies the column abundance of ozone. Stratospheric warmings are often accompanied by a marked increase of total ozone at high latitudes. Figure 17.14 shows the distributions of Σ_{O_3} and 400 K isentropic motion on February 25, 1984, when the zonal–mean flow was reversed down to 70 mb. A tongue of enhanced total ozone ($\Sigma_{O_3} > 500$ DU) coincides with the cross-polar flow, where air from lower latitudes has descended and experienced compression. In the Eastern Hemisphere, air that has been displaced equatorward leads to anomalously low Σ_{O_3}.

Stratospheric warmings occur primarily in the Northern Hemisphere, where planetary waves are strong. However, behavior like that described here invariably closes the winter season of both hemispheres. Spring witnesses a weakening of the equatorward temperature gradient that supports strong zonal flow. Circumpolar westerlies then collapse in the *final warming* and are eventually replaced during summer by circumpolar easterlies.

---

[3] The Q distribution must be derived from meteorological analyses, which have limited ability to resolve such features because they are based largely on temperature observations.

Figure 17.13 Distributions of potential vorticity Q and horizontal motion on the 850 K isentropic surface (near 10 mb) for March 4, 1984. High-Q polar air (blue), which marks the polar-night vortex, has been displaced off the pole by a large-amplitude planetary wave and has undergone pronounced distortion, with secondary eddies appearing along its tail. Low-Q air from equatorward (red), which is advected into the polar cap, spins up anticyclonically to form a reversed circulation, with easterly circumpolar flow at high latitudes (compare Fig. 14.26).

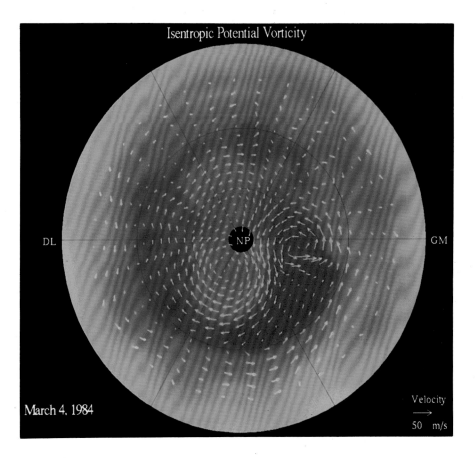

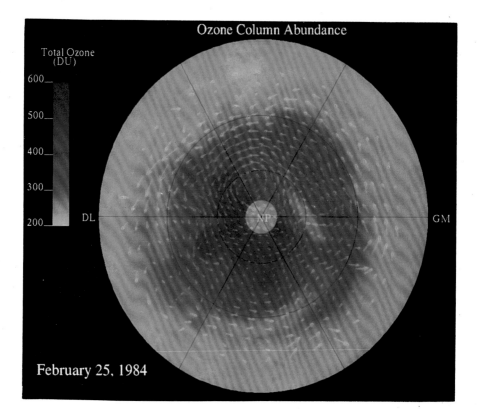

17.5   The Quasi-Biennial Oscillation

The circulation in the tropics is also chiefly zonal. However, it is modified importantly by interannual variations. Figure 17.15 shows zonal wind averaged over equatorial stations as a function of time and height. An oscillation of 26 to 28 months prevails over seasonal variations at heights of 15 to 30 km. Known as the *quasi-biennial oscillation* (QBO), this variation is seen to descend with time in an alternating series of easterlies and westerlies that attain speeds of 20 to 30 m s^{-1}. At higher levels, the QBO gives way to the semi-annual oscillation, which is a harmonic of the seasonal cycle. The QBO is symmetric about the equator and is confined to latitudes of less than about 15°.

Even though it is nearly periodic, the QBO is not a harmonic of the seasonal cycle. Therefore, it cannot be explained simply in terms of seasonality that is imparted to the lower stratosphere from other regions. Latent heating in the troposphere contains no preferred period on timescales of the QBO. Likewise, diabatic heating in the equatorial stratosphere is unable to explain the oscillation (Wallace, 1967). This leaves momentum transfer. During the westerly phase of the QBO, air over the equator moves faster than the earth, so the specific angular momentum exceeds that found at other latitudes. This rules out horizontal advection of zonal–mean momentum as an explanation.

The QBO is thought to be driven mechanically by vertically propagating waves that transfer momentum upward from the troposphere. The particular waves involved remain a matter of some discussion, but there is no debate over their origin. Unsteady latent heating inside tropical convection excites a spectrum of wave activity, much of which propagates vertically into the middle atmosphere. Upon being absorbed, that wave activity transfers momentum to the zonal–mean flow, which is thereby accelerated toward the phase speeds of individual waves.

As illustrated in Fig. 17.16, eastward- and westward-propagating waves have transmission and absorption characteristics that differ between easterly and westerly layers of the QBO (e.g., as controlled by their intrinsic frequencies; Sec. 14.6). These differences allow wave absorption to establish a westerly critical line, where the zonal flow matches the phase speed and therefore absorbs westerly waves, flanked aloft by an easterly critical line, where the zonal flow matches the phase speed and therefore absorbs easterly waves. Transfers of westerly and easterly momentum at those sites of wave absorption

---

Figure 17.14   Total ozone Σ_{O_3} and horizontal motion on the 400 K isentropic surface in the Northern Hemisphere on February 25, 1984, during a stratospheric warming when zonal flow was disturbed down to 70 mb. A tongue of enhanced Σ_{O_3} coincides with cross-polar flow between longitudes of 150°E and 30°W, where air descends along isentropic surfaces and undergoes compression. Air displaced equatorward in the Eastern Hemisphere ascends isentropically and undergoes expansion to introduce anomalously low Σ_{O_3}.

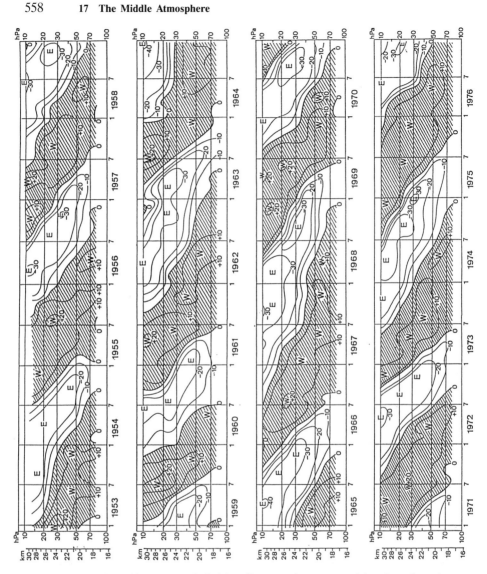

Figure 17.15 Monthly mean zonal wind (m s^{-1}) averaged over equatorial rawinsonde stations. Updated from Naujokat (1986); courtesy of B. Naujokat (Free U. Berlin). (*continues*)

then cause both critical lines to descend (Lindzen and Holton, 1968; Holton and Lindzen, 1972), as is observed in Fig. 17.15.

The tropical QBO has an extratropical counterpart. During years when equatorial winds are easterly, the polar-night vortex is warmer and more disturbed than during years when equatorial winds are westerly (Labitzke, 1982; Holton and Tan, 1982). Owing to the close relationship between ozone and

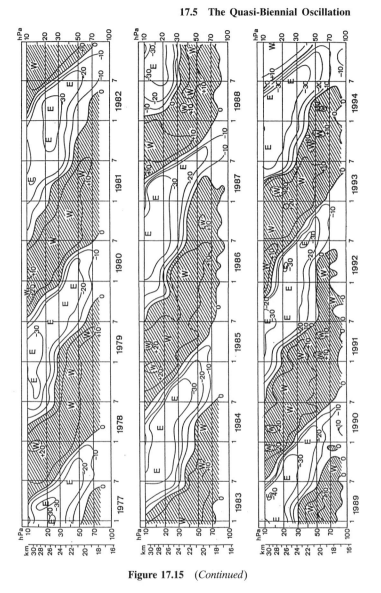

Figure 17.15   (*Continued*)

disturbances to the circumpolar vortex, similar interannual variability is manifested by total ozone in extratropical regions. For instance, interannual variations of Antarctic Σ_{O_3} (see Fig. 17.23 later in this chapter) are strongly correlated with equatorial winds (Garcia and Solomon, 1987). The extratropical
QBO appears to result from a repositioning of the critical line for planetary
waves in relation to the polar-night vortex. During the easterly phase of the
QBO, the critical line advances into the winter hemisphere, where it introduces
large eddy displacements of the vortex (Fig. 17.17a). Conversely, the critical

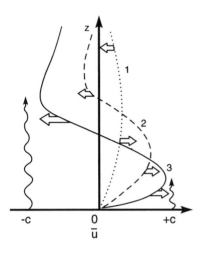

Figure 17.16 Schematic illustrating the differential transmission and absorption characteristics of vertically propagating eastward- and westward-traveling waves with phase speed c. Absorption of momentum from the eastward-traveling wave accelerates the zonal–mean flow toward the phase speed of that wave. Westerlies then form a critical level, where eastward-traveling wave activity is fully absorbed. The westward-traveling wave propagates through those westerlies without encountering a critical level, but it experiences absorption at higher levels in easterly shear. Absorption of its momentum accelerates the zonal–mean flow toward the phase speed of the westward-traveling wave to produce a critical level where that wave is fully absorbed. Continued absorption of eastward and westward momentum at the critical levels of the two waves then causes the layers of easterlies and westerlies to descend. Adapted from Plumb (1982).

line is removed into the summer hemisphere during the westerly phase of the QBO, leaving the polar-night vortex comparatively isolated (Fig. 17.17b). The vortex is therefore more disturbed during the easterly phase, which implies intensified diabatic motion, than during the westerly phase, when it remains closer to zonal symmetry and radiative equilibrium.

17.6 Direct Interactions with the Troposphere

For the most part, the middle atmosphere is driven indirectly through vertically propagating waves that carry momentum upward from the troposphere. However, mean vertical motion implied by the Brewer–Dobson circulation allows stratospheric air to interact directly with tropospheric air. Chemical tracers such as N_2O and CH_4 imply that air enters the stratosphere across the tropical tropopause. By continuity, that air must eventually return to the troposphere at middle and high latitudes. Estimates of mass transfer suggest a residence time in the middle atmosphere of 2.5 years (Holton, 1990). Exchanges of air between the troposphere and stratosphere are complex and not well understood. However, tracer observations like those in Figs. 17.4 and 17.7 point to the involvement of tropical convection and synoptic disturbances at midlatitudes.

Water vapor mixing ratio reaches a minimum at the hygropause due to moisture in tropospheric air being limited to saturation values. In the tropics, the hygropause is found as high as 18 km—several kilometers above the mean elevation of the tropopause. Figure 17.18 shows the profile of water vapor mixing ratio over Darwin, Australia, along with the mixing ratio of ozone

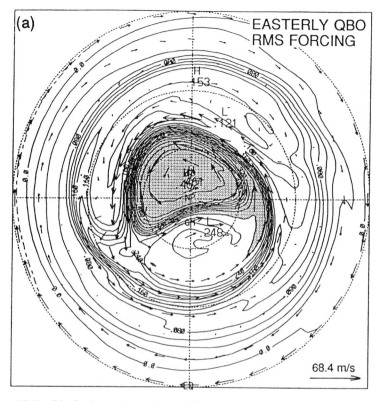

Figure 17.17 Distributions of potential vorticity and motion in a two-dimensional calculation representative of the polar-night vortex during (a) the easterly phase of the QBO, when the critical line for stationary planetary waves is situated in the winter subtropics, and (*continues*)

and potential temperature. Well-mixed air below 17 km is marked by nearly uniform r_{O_3} and slowly increasing θ that reflects neutral moist static stability. Having been processed by deep convection, that air is characterized by r_{H_2O} which decreases sharply to a broad minimum between 16 and 18 km. The minimum in water vapor mixing ratio extends about a kilometer above the local tropopause. This feature is suggestive of dehydrated tropospheric air that has entered the stratosphere in overshooting convection. The observed minimum of $r_{H_2O} < 2$ ppmv is significantly drier than saturation values corresponding to the mean tropical tropopause. These departures from mean conditions indicate that air entering the stratosphere at low latitude is confined to deep convective towers that have temperatures of 190 K and colder. Penetrative convection culminates in mixing of tropospheric and stratospheric air, which maintains the tropopause against radiative drive and temporarily elevates its position (Fig. 7.8).

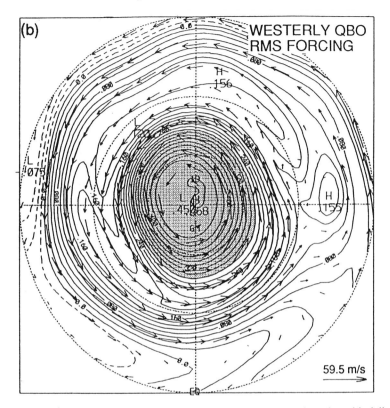

Figure 17.17 (*Continued*) (b) the westerly phase of the QBO, when the critical line for stationary planetary waves is removed into the summer subtropics. Adapted from O'Sullivan and Salby (1990).

Vertical mixing that occurs when cumulonimbus towers encounter stable air overhead and subsequent heat transfer can lead to upwelling across the mean elevation of the tropical tropopause. Convective turrets that overshoot their equilibrium level (Fig. 7.7) eventually collapse into extensive anvils, in which tropospheric and stratospheric air are mixed.[4] Mixing results in irreversible heat transfer between air parcels, which reduces the potential temperature of the stable layer overhead. Radiative heating, which acts to restore the layer to radiative equilibrium, then increases θ, which in turn produces upwelling across isentropic surfaces in the tropical lower stratosphere. How much this air is taken up by large-scale upwelling in the middle atmosphere must, to some degree, be controlled by large-scale downwelling at middle and high latitudes (Figs. 17.9 and 17.11). Since downwelling motion carries stratospheric

---

[4] Vertical mixing is an inevitable consequence of cumulus convection because a buoyant thermal comprised of tropospheric air advances only by entraining environmental air (Fig. 9.16), which, for an overshooting turret, is stratospheric in origin.

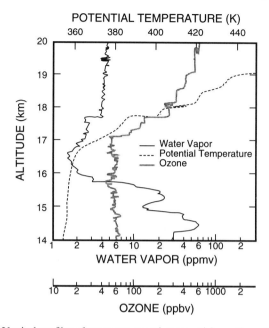

Figure 17.18   Vertical profiles of water vapor and ozone mixing ratio and potential temperature over Darwin, Australia, on January 13, 1987. Source: Kelly *et al.* (1993).

air across the extratropical tropopause, by continuity, it must act to pump the troposphere and, therefore, promote large-scale ascent in the tropics (see Holton *et al.*, 1995, for an overview of the possibilities).

For convective transport to explain the observed distribution of water vapor, air introduced into the tropical stratosphere must be relatively free of water in the condensed phase, which would subsequently evaporate and increase r_{H_2O} over the very lean values observed (see, e.g., Holton, 1984). Cumulonimbus anvils, in which tropospheric and stratospheric air have been mixed, are destabilized by LW heating below and cooling aloft (Sec. 9.4). Vertical overturning then allows condensate to precipitate out (Danielsen, 1982). Airborne observations point to a sink of total water in the neighborhood of convective anvils, where measured ice particles are large enough to undergo sedimentation (Kelly *et al.*, 1993). This process dehydrates tropospheric air that has been mixed with stratospheric air.

Air that enters the stratosphere in the tropics must eventually return to the troposphere in extratropical regions. Vertical mixing is believed to be involved in this transfer of air as well, but on the larger scale of synoptic disturbances. Although confined to a neighborhood of the tropopause, baroclinic systems disturb the stratospheric circulation (see Fig. 2.10). Their importance is underscored by the fact that the ozone column is concentrated between 10 and

20 km (Fig. 1.17). It is also suggested by daily variations of Σ_{O_3}, which are as large as seasonal variations.

The daily distribution of total ozone (Fig. 1.19a) is punctuated by anomalies of high Σ_{O_3} that circumscribe the poles. Comparison with the 500-mb circulation in Fig. 1.9a reveals that those anomalies are positively correlated with midlatitude cyclones. According to the pressure on the 375 K isentropic surface, which is contoured in Fig. 1.19, midlatitude cyclones coincide with depressions of θ surfaces in the lower stratosphere. Isentropic surfaces are deflected downward to greater pressure over mature baroclinic systems, in which static stability is reduced. Ozone-rich air descending along those surfaces undergoes compression, which magnifies Σ_{O_3} locally—by as much as 100%. Analogous behavior occurs on planetary dimensions during a stratospheric warming, when air is advected meridionally along isentropic surfaces (refer to Fig. 17.14). Conversely, expansion and adiabatic cooling when air ascends isentropically over an anticyclone have been found to support the formation of extensive polar stratospheric clouds (PSCs) (Jones *et al.*, 1990; Pitts *et al.*, 1990).

Meteorological analyses suggest that isentropic surfaces undergo vertical deflections of a couple of kilometers. However, airborne measurements record anomalous concentrations of stratospheric species as low as 700 mb. Stratospheric air enters the troposphere inside *tropopause folds*, in which isopleths of stratospheric tracers are overturned and drawn downward into the troposphere along frontal zones that intensify during the development of baroclinic systems. Their impact is manifested in the zonal–mean ozone mixing ratio (Fig. 1.17), which dips downward and enhances ozone number density at the poleward edge of the subtropical jet—precisely where baroclinic systems develop (Sec. 15.2).

Figure 17.19 shows a cross section of a tropopause fold. Isentropic surfaces (Fig. 17.19a), which reflect stratospheric air in strong vertical stability (e.g., large $\partial\theta/\partial z$), have been drawn downward and concentrated into a tilting frontal zone that separates cold and warm air. Strongly baroclinic, that region develops from deformations accompanying the amplification of a baroclinic disturbance (Hoskins, 1982). According to the distribution of ozone mixing ratio (Fig. 17.19b), stratospheric air drawn into the frontal zone is diluted by small-scale mixing that diminishes r_{O_3} along the intrusion. Once it has entered the troposphere, the water-soluble species O_3 quickly encounters convective systems, in which it is absorbed, precipitated to the surface, and destroyed through oxidation processes.

17.7   Heterogeneous Chemical Reactions

The chemical reactions considered in Secs. 17.1 and 17.2 involve species only in gas phase. During the 1970s and 1980s, these and other elements of gas-phase

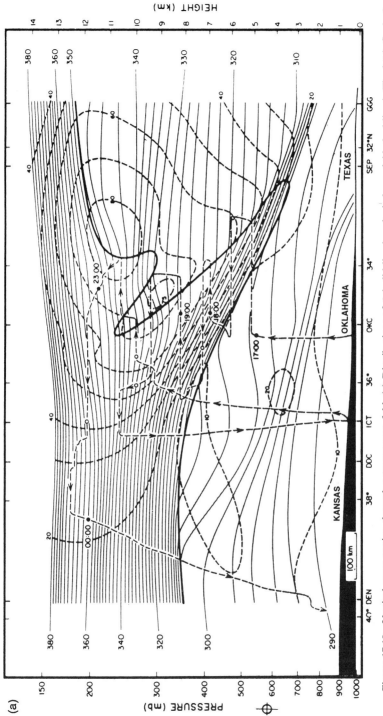

Figure 17.19 Vertical cross section through a tropopause fold. (a) Distributions of θ (solid lines) and wind speed (dashed lines). The jet is flanked by a baroclinic frontal zone in which isentropic surfaces descend into the troposphere from stratospheric levels. (*continues*)

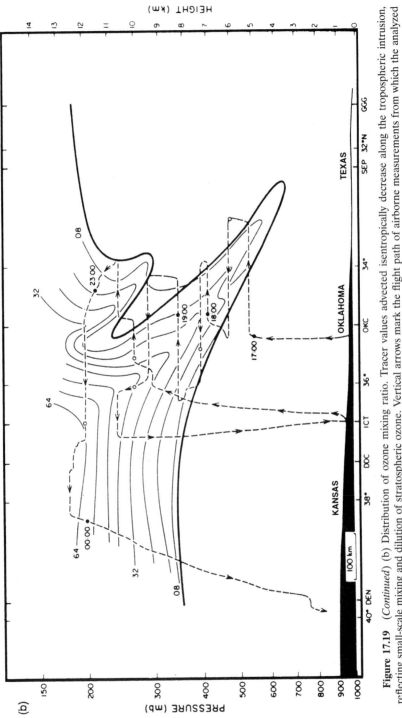

Figure 17.19 (*Continued*) (b) Distribution of ozone mixing ratio. Tracer values advected isentropically decrease along the tropospheric intrusion, reflecting small-scale mixing and dilution of stratospheric ozone. Vertical arrows mark the flight path of airborne measurements from which the analyzed fields were inferred. After Shapiro (1982).

photochemistry were the focus of intensive investigation. Although an impact from anthropogenic species was suggested, it was not expected to emerge clearly before the twenty-first century and even then it was expected to reduce the column abundance of ozone by only about 5%. The timing and magnitude of anticipated changes were set by two considerations: (1) Reactive species had to increase to relatively large concentrations before catalytic destruction of ozone via (17.10) and (17.13) became sufficiently fast. (2) Those reactions were favored at higher altitudes, where only a small fraction of the ozone column is affected.

For these reasons, the discovery of the Antarctic ozone hole in 1985 caught much of the scientific community by surprise. Moreover, the order-of-magnitude greater depletions observed (e.g., Fig. 1.19b) were found at latitudes where ozone was thought to be photochemically inert—because UV fluxes there are small. The key ingredient not considered in earlier investigations was the presence of solid phase in the stratosphere, which is normally excluded by very low mixing ratios of water vapor.

PSCs (Sec. 9.3) were already recognized to form over the Antarctic due to its very cold temperatures, but they were regarded largely as a curiosity. PSCs appear far more frequently in the Antarctic stratosphere than in the warmer Arctic stratosphere (Fig. 17.20). During late Austral winter, when temperatures are coldest, fractional coverage over the Antarctic exceeds 50%, chiefly by very tenuous clouds in the PSC I category. It is now widely accepted that PSCs provide the surfaces on which certain reactions proceed much faster than they can in gas phase alone. Moreover, the presence of PSCs shifts catalytic destruction of ozone from the upper stratosphere, where only a small fraction of the ozone column resides, to the lower stratosphere, where Σ_{O_3} is concentrated.

Figure 17.21 shows profiles of ozone concentration over Antarctica during Austral winter (solid line) and shortly after equinox (dashed line), when the sun rises above the horizon. A marked reduction of ozone has occurred between 10 and 20 km, where most of the ozone column normally resides. Decreases in Σ_{O_3} of 50% are observed at this time year, with column abundances as low as 100 DU having been recorded. Superposed in Fig. 17.21 is the profile of ClO (shaded line), which is produced by destruction of ozone (17.13). Consistent with the observed ozone depletion, r_{ClO} maximizes between 10 and 25 km—precisely where PSCs are sighted (Fig. 17.20). A correspondence between reduced O_3 and increased ClO is also apparent across the edge of the polar-night vortex (Fig. 17.22). Ozone decreases sharply and chlorine monoxide increases sharply where temperature becomes colder than 196 K, which is close to the threshold temperature for the formation of type I PSCs.

The reactions now recognized to be primarily responsible for the ozone losses in Figs. 1.19b, 17.21, and 17.22 involve two stages: First, inactive chlorine species such as HCl and $ClONO_2$ are converted to reactive forms of Cl_x

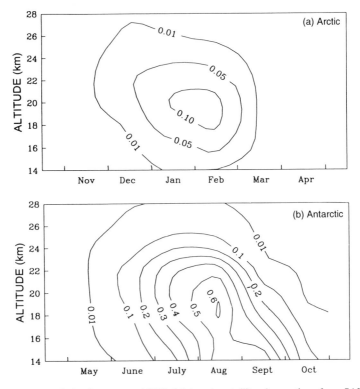

Figure 17.20   Relative frequency of PSC sightings in satellite observations from SAM II, as a function of height and month, over (a) the Arctic and (b) the Antarctic. After WMO (1991).

through heterogeneous reactions like

$$HCl(s) + ClONO_2(g) \rightarrow Cl_2(g) + HNO_3(s) \tag{17.19.1}$$

and

$$H_2O(s) + ClONO_2(g) \rightarrow HNO_3(s) + HOCl(g), \tag{17.19.2}$$

which involve solid (s) as well as gas (g) phases. These reactions proceed rapidly on ice, but slowly in the gas phase alone. Once produced, Cl_2 and HOCl are readily photolyzed by sunlight to release free chlorine

$$Cl_2 + h\nu \rightarrow 2Cl, \tag{17.19.3}$$

$$HOCl + h\nu \rightarrow OH + Cl, \tag{17.19.4}$$

which is a reactive form of Cl_x. Then the sequence of reactions

$$Cl + O_3 \rightarrow ClO + O_2, \tag{17.20.1}$$

$$2ClO + M \rightarrow Cl_2O_2 + M, \tag{17.20.2}$$

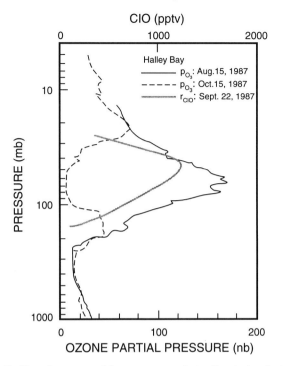

Figure 17.21   Profiles of ozone partial pressure over Antarctica during Austral winter (solid line) and shortly after spring equinox (dashed line). A profile of the chlorine monoxide mixing ratio (shaded line) after equinox is superposed. Sources: WMO (1988), Solomon (1990).

$$Cl_2O_2 + h\nu \rightarrow Cl + ClO_2, \qquad (17.20.3)$$

$$ClO_2 + M \rightarrow Cl + O_2 + M \qquad (17.20.4)$$

destroys ozone catalytically with the net effect of

$$2O_3 + h\nu \rightarrow 3O_2 \qquad (net). \qquad (17.20.5)$$

With established reaction rates, observed mixing ratios of $r_{ClO} \cong 1$ pptv are adequate to explain the observed ozone depletion rate of about 2% per day. This concentration of ClO is two orders of magnitude greater than that predicted by gas-phase photochemistry alone. However, it is consistent with calculations that include heterogeneous reactions like (17.19).

The catalytic sequence (17.20) is initiated by free chlorine, which originates largely from photolysis of CFCs (17.12). Produced exclusively by industry, CFCs have led to steadily increasing levels of atmospheric chlorine (Fig. 17.23). Above a natural background level of about 0.6 ppbv associated with ocean processes (Sec. 17.2.2), atmospheric chlorine has increased fivefold since the 1950s—shortly after the introduction of CFCs in industrial applications. Tending in the opposite sense, the signature of ozone depletion over

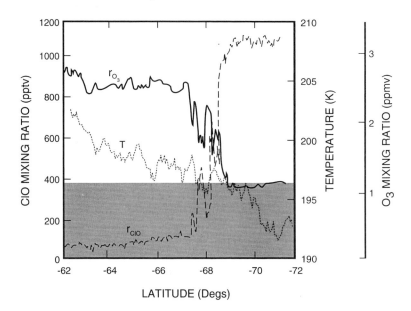

Figure 17.22 Mixing ratios of ozone (solid line) and chlorine monoxide (dashed line) and temperature (dotted line) along a flight path into the Antarctic polar-night vortex. Temperatures colder than about 196 K (shaded) coincide with the formation of type I PSCs. Source of O_3 and ClO profiles: Anderson *et al.* (1989).

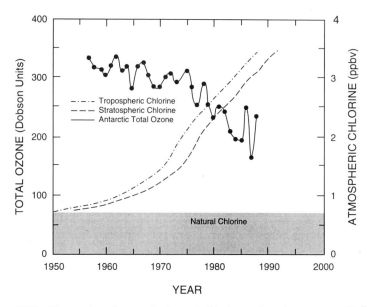

Figure 17.23 Time series of atmospheric total chlorine and total ozone over Halley Bay, Antarctica. Adapted from Solomon (1990).

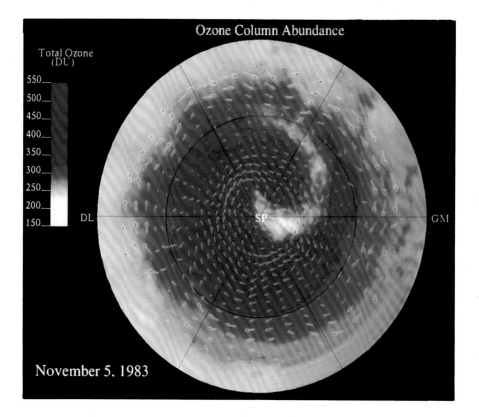

Antarctica emerges clearly after 1980, presumably when chlorine levels exceeded a threshold for reactions (17.20) to become an important sink of O_3 (Solomon, 1990).

While converting inert forms of chlorine into reactive Cl_x, heterogeneous reactions (17.19) have the opposite effect on reactive nitrogen. They convert NO_x into relatively inactive nitric acid. This bears importantly on ozone depletion because NO_x regulates the abundance of reactive chlorine. The principal means by which Cl_x is converted back to inactive forms is via reaction with nitrogen dioxide

$$ClO + NO_2 + M \rightarrow ClONO_2 + M. \qquad (17.21)$$

The abundance of NO_2 thus controls the duration over which reactive chlorine is available to destroy ozone in (17.20). Since PSCs are composed of hydrated forms of nitric acid (Sec. 9.3), their formation removes NO_x from the gas phase. Should PSC particles become large enough to undergo sedimentation, NO_x is removed entirely. The stratosphere is then denitrified, leaving reactive chlorine available much longer to destroy ozone.

Figure 17.24 compares the seasonal cycle of Σ_{O_3} over Antarctica based on the historical record against that based on years since the appearance of the ozone hole. The two evolutions diverge near Austral spring, when solar radiation triggers reactions (17.19.3) and (17.19.4), which release reactive chlorine and set the stage for catalytic destruction of ozone in (17.20). Minimum column abundances are observed in mid-October. By November, increasing values restore Σ_{O_3} toward historical levels, when ozone-rich air is imported from low latitudes during the final warming. But even then, values remain below historical levels due to the dilution of subpolar air with ozone-depleted air from inside the polar-night vortex. As illustrated by Fig. 17.25, the breakdown of the vortex during Austral spring involves a complex rearrangement of air. Meridional displacements when the circumpolar flow weakens allow ozone-depleted air that had been confined inside the vortex to escape from the polar cap. Anomalies of mixing ratio created in this fashion can survive for several weeks before they are eventually destroyed by deformation and small-scale diffusion (see Hess and Holton, 1985).

Airborne observations have established that the same reactions operating over the Antarctic occur in the Arctic stratosphere. But, owing to the warmer temperature of the Arctic polar-night vortex, PSCs are a relatively infrequent phenomenon. Moreover, free chlorine that is produced in isolated PSCs via (17.19) is quickly acted on by dynamical effects that, by elevating temperature, reverse the process through other chemical reactions (see, e.g., Garcia, 1994). This limits the amount of reactive chlorine available when the sun rises over the Arctic, which in turn limits catalytic destruction of ozone via (17.20). Similar considerations apply to midlatitudes, where ozone depletions of 5 to 10% have been documented (WMO, 1991). Occurring at temperatures too warm to support cloud formation, those depletions may also follow from

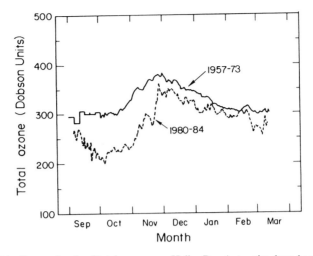

Figure 17.24   Seasonal cycle of total ozone over Halley Bay, Antarctica, based on the historical record since 1957 and on years since the appearance of the Antarctic ozone hole. After Solomon (1990).

heterogeneous reactions, but involving sulfuric acid aerosols of volcanic origin (Hoffman and Solomon, 1989).

The limiting factor in ozone depletions appears to be temperature, which controls the formation of stratospheric cloud (Sec. 9.3). Indeed, the deepest reductions of Antarctic Σ_{O_3} are observed during the coldest winters, in agreement with the sharp correspondence between temperature and perturbed photochemistry (Fig. 17.22). Consequently, dynamical disturbances that control temperature inside the vortex through diabatic effects (Sec. 17.3) are a key ingredient that regulates ozone depletion at high latitudes.

Intriguingly, recent temperatures over Antarctica during October are cooler than those in the historical record. Evidence suggests that those depressed temperatures follow from diminished ozone heating associated with anomalously low ozone concentrations (Poole *et al.*, 1989). By steepening the meridional temperature gradient, this response reinforces the vortex, which in turn acts to postpone the final warming that eventually restores the circulation toward normal conditions. It therefore represents a positive feedback on ozone depletion, one that involves interactions between chemistry, radiation, and dynamics. An improved understanding of such interactions will be important to achieving

Figure 17.25   Distributions of total ozone and horizontal motion on the 325 K isentropic surface over the Southern Hemisphere on November 5, 1983, during the final warming. Ozone-depleted air, which had previously been confined inside the polar-night vortex, escapes in a tongue of anomalously lean values (white) that spirals anticyclonically into midlatitudes to conserve potential vorticity.

further progress on the ozone hole as well as on other complex problems facing atmospheric science.

Suggested Reading

An Introduction to Dynamic Meteorology (1992) by Holton provides a nice introductory development of stratospheric waves and a quantitative description of how they interact with the mean meridional circulation.

An advanced treatment of these topics is presented in *Middle Atmosphere Dynamics* (1987) by Andrews *et al.*

Holton (1984) discusses the historical view of the Brewer–Dobson circulation in relation to transport of water vapor across the tropical tropopause. WMO (1987) gives a more general development of stratosphere–troposphere exchange in light of aircraft and satellite observations. Recent observations of stratosphere–troposphere exchange and their interpretation are developed in Holton *et al.* (1995).

Aeronomy of the Middle Atmosphere (1986) by Brasseur and Solomon is a comprehensive treatment of gas phase photochemistry. Heterogeneous chemical processes are reviewed in WMO (1991).

Solomon (1990) provides an excellent overview of the Antarctic ozone hole and its relationship to polar stratospheric clouds.

Problems

17.1. Within the framework of Chapman chemistry, derive expressions for the photochemical lifetimes τ_{O_3} and τ_{O_x} for O_3 and O_x, respectively, where $\tau_X^{-1} = -(1/X)dX/dt$ is defined from reactions that destroy species X. Given the rate coefficients in cgs units: $k_2 = 6.0 \times 10^{-34}(300/T)[M]$, $k_3 = 8.0 \times 10^{-12}e^{-2060/T}$, where T is in Kelvin, and $[M]$ is the number density of air, $J_2 = 4.35 \times 10^{-15}, 7.44 \times 10^{-11}, 1.01 \times 10^{-9}$, and $J_3 = 4.47 \times 10^{-4}, 7.80 \times 10^{-4}$, and 6.96×10^{-3}, at 100, 10, and 1 mb, respectively, evaluate τ_{O_3} and τ_{O_x} at (a) 100 mb, (b) 10 mb, and (c) 1 mb.

17.2. Ozone is useful as a tracer in the lower stratosphere. Yet, O_3 has a photochemical lifetime there of only hours (see Fig. 17.1). Resolve this discrepancy.

17.3. (a) Construct an analytical approximation to the radiative-equilibrium temperature in Fig. 17.10. (b) Construct an analytical approximation to the zonal wind at 100 mb in Fig. 1.8. (c) Use the results of parts (a) and (b) to plot the radiative-equilibrium wind in the middle atmosphere.

17.4. Use observed temperature (Fig. 1.7), radiative-equilibrium temperature (Fig. 17.10), and the coefficient of Newtonian cooling (Fig. 8.29) to estimate the radiative cooling rate $(K\ day^{-1})$ inside the polar-night vortex at (a) 100 mb, (b) 10 mb, and (c) 1 mb.

17.5. Describe how the polar-night vortex is maintained warmer than radiative equilibrium by (a) thermal dissipation of planetary waves and (b) mechanical dissipation of planetary waves.

17.6. Use the fact that upward momentum flux carried by planetary waves is absorbed to describe how vertically propagating wave activity makes the circulation of the middle atmosphere behave as a refrigerator.

17.7. If the middle atmosphere remains in thermodynamic equilibrium, net heat transfer vanishes for individual air parcels, even though they may temporarily sustain heat transfer during their motion about the pole (see, e.g., Fig. 3.6). Likewise, if the middle atmosphere remains in photochemical equilibrium, no ozone is produced. (a) Discuss net production of ozone in relation to air being driven out of thermodynamic equilibrium and photochemical equilibrium and relate those concepts to the Brewer–Dobson circulation. (b) Extend the foregoing ideas to explain the seasonal cycle of total ozone in Fig. 1.18.

17.8. Discuss how changes of radiative transfer introduced by depleted ozone at high latitudes can reinforce the polar-night vortex to represent a positive feedback.

17.9. According to the Brewer–Dobson circulation (Fig. 17.9), air enters the stratosphere from the tropical troposphere. Estimate the water vapor mixing ratio in ppmv (Problem 1.2) for the tropical lower stratosphere if tropospheric air entering the stratosphere (a) evolves through a thermodynamic state corresponding to the mean temperature and height of the tropical tropopause (Fig. 1.7), (b) were actually introduced through isolated convective overshoots that have a mean height and temperature of 16.5 km and 185 K (see Fig. 1.25a), and (c) as in part (b), but if, after entering the stratosphere, tropospheric air inside convective overshoots were to mix with subtropical stratospheric air that has a mixing ratio of 5 ppmv.

17.10. The Brewer–Dobson circulation (Fig. 17.9) gives no indication of equatorward transport in the lower stratosphere. (a) How could subtropical air in Problem 17.9, part (c), be supplied to the equator? (b) Why does such transport not appear in the Brewer–Dobson circulation?

17.11. Methane oxidation is thought to be responsible for about 25% of stratospheric water vapor. Given the reactions

$$CH_4 + OH \rightarrow H_2O + CH_3 \qquad (k_a)$$

$$H_2O + O \rightarrow 2OH \qquad\qquad (k_b),$$

with the rate coefficients in cgs units: $k_a = 2.4 \times 10^{-12} e^{-1710/T}$ and $k_b = 2.2 \times 10^{-10}$, calculate the photochemical-equilibrium mixing ratio of water vapor produced by methane oxidation at 20 km. Compare this value with observed mixing ratios at this altitude.

17.12. See Problem 4.8.

17.13. Consider Chapman chemistry, but augmented by the catalytic cycle (17.13) involving Cl_x. Determine the photochemical lifetime of O_x at 100 mb for (a) rate coefficients for (17.13.1) and (17.13.2) in cgs units of $k_a = 8.5 \times 10^{-12}$ and $k_b = 3.7 \times 10^{-11}$, respectively, and a ClO mixing ratio of 100 pptv, which are representative of gas-phase chemistry under unperturbed conditions, and (b) rate coefficients for (17.13.1) and (17.13.2) in cgs units of $k_a = 8.5 \times 10^{-10}$ and $k_b = 3.7 \times 10^{-9}$, respectively, and a ClO mixing ratio of 1000 pptv, which are representative of heterogeneous chemistry involving (17.20) under chemically perturbed conditions. (c) Compare the photochemical lifetimes of O_x in parts (a) and (b) with that under pure oxygen chemistry calculated in Problem 17.1.

Appendix A
Conversion to SI Units

Physical quantity	Unit	SI (MKS) equivalent
Length	ft	0.305 m
	μm	10^{-6} m
	nm	10^{-9} m
Time	day	8.64×10^4 s
Mass	lb	0.454 kg
Temperature	°F	$273 + (°F - 32)/1.8$ K
Volume	liter	10^{-3} m^3
Velocity	mph	0.447 m s^{-1}
	knots	0.515 m s^{-1}
	km hr^{-1}	0.278 m s^{-1}
	fps	0.305 m s^{-1}
Force	kg m s^{-2}	1 N
	lb	0.138 N
	dyne	10^{-5} N
Pressure	N m^{-2}	1 Pa
	bar	10^5 Pa
	mb	10^2 Pa = 1 hPa
Energy	kg m^2 s^{-2}	1 J
	Nm	1 J
	erg	10^{-7} J
	cal	4.187 J
Power	kg m^{-2} s^{-3}	1 W
	J s^{-1}	1 W
	Langley day^{-1}	4.84×10^{-1} W m^{-2}
Specific heat	cal gm^{-1}	4.184×10^3 J kg^{-1}
Energy flux	cal cm^{-2} min^{-1}	6.97×10^2 W m^{-2}

Appendix B

Thermodynamic Properties of Air and Water

Dry Air

Mean molecular weight	$M_d = 28.96 \text{ g mol}^{-1}$
Specific gas constant	$R = 287.05 \text{ J kg}^{-1} \text{K}^{-1}$
Density	$\rho = 1.293 \text{ kg m}^{-3}$ (at STP*)
Number density (Loschmidt number)	$n = 2.687 \times 10^{25} \text{ m}^{-3}$ (at STP)
Isobaric specific heat capacity	$c_p = 1.005 \times 10^3 \text{ J kg}^{-1} \text{K}^{-1}$ (at 273 K)
Isochoric specific heat capacity	$c_v = 7.19 \times 10^2 \text{ J kg}^{-1} \text{K}^{-1}$ (at 273 K)
Ratio of specific heats	$\gamma = c_p/c_v = 1.4$
	$\kappa = (\gamma - 1)/\gamma = R/c_p = 0.286$
Coefficient of viscosity	$\mu = 1.73 \times 10^{-5} \text{ kg m}^{-1} \text{s}^{-1}$ (at STP)
Kinematic viscosity	$\nu = 1.34 \times 10^{-5} \text{ m}^2 \text{s}^{-1}$ (at STP)
Coefficient of thermal conductivity	$k = 2.40 \times 10^{-2} \text{ W m}^{-1} \text{K}^{-1}$ (at STP)
Sound speed	$c_s = 331 \text{ m s}^{-1}$ (at 273 K)

Water

Mean molecular weight	$M_v = 18.015 \text{ g mol}^{-1}$
	$\epsilon = M_v/M_d = 0.622$
Specific gas constant	$R = 461.51 \text{ J kg}^{-1} \text{K}^{-1}$
Density (liquid water)	$\rho = 10^3 \text{ kg m}^{-3}$ (at STP)
Density (ice)	$\rho = 9.17 \times 10^2 \text{ kg m}^{-3}$ (at STP)
Isobaric specific heat capacity (vapor)	$c_p = 1.85 \times 10^3 \text{ J kg}^{-1} \text{K}^{-1}$ (at 273 K)
Isochoric specific heat capacity (vapor)	$c_v = 1.39 \times 10^3 \text{ J kg}^{-1} \text{K}^{-1}$ (at 273 K)
Ratio of specific heats (vapor)	$\gamma = c_p/c_v = 1.33$
Specific heat capacity (liquid water)	$c = 4.218 \times 10^3 \text{ J kg}^{-1} \text{K}^{-1}$ (at 273 K)
Specific heat capacity (ice)	$c = 2.106 \times 10^3 \text{ J kg}^{-1} \text{K}^{-1}$ (at 273 K)
Specific latent heat of fusion	$l_f = 3.34 \times 10^5 \text{ J kg}^{-1}$
Specific latent heat of vaporization	$l_v = 2.50 \times 10^6 \text{ J kg}^{-1}$
Specific latent heat of sublimation	$l_s = l_f + l_v$

---

*Standard temperature and pressure (STP) = 1013 mb and 273 K.

Appendix C

Physical Constants

Avogadro's number	$N_A = 6.022 \times 10^{23}\ \mathrm{mol}^{-1}$
Universal gas constant	$R^* = 8.314\ \mathrm{J\,mol}^{-1}\,\mathrm{K}^{-1}$
Boltzmann constant	$k = 1.381 \times 10^{-23}\ \mathrm{J\,K}^{-1}$
Planck constant	$h = 6.6261 \times 10^{-34}\ \mathrm{J\,s}^{-1}$
Stefan-Boltzmann constant	$\sigma = 5.67 \times 10^{-8}\ \mathrm{W\,m}^{-2}\,\mathrm{K}^{-4}$
Speed of light	$c = 2.998 \times 10^{8}\ \mathrm{m\,s}^{-1}$
Solar constant	$F_s = 1.372 \times 10^{3}\ \mathrm{W\,m}^{-2}$
Radius of the earth	$a = 6.371 \times 10^{3}\ \mathrm{km}$
Standard gravity	$g_0 = 9.806\ \mathrm{m\,s}^{-2}$
Earth's angular velocity	$\Omega = 7.292 \times 10^{-5}\ \mathrm{s}^{-1}$

Appendix D
Vector Identities

$$A \times (B \times C) = (A \cdot C)B - (A \cdot B)C \tag{D.1}$$

$$A \cdot (B \times C) = (A \times B) \cdot C = B \cdot (C \times A) \tag{D.2}$$

$$(A \times B) \cdot (C \times D) = (A \cdot C)(B \cdot D) - (A \cdot D)(B \cdot C) \tag{D.3}$$

$$\nabla(fg) = f\nabla g + g\nabla f \tag{D.4}$$

$$\nabla \cdot (fA) = \nabla f \cdot A + f\nabla \cdot A \tag{D.5}$$

$$\nabla \times (fA) = \nabla f \times A + f\nabla \times A \tag{D.6}$$

$$\nabla \cdot (A \times B) = B \cdot (\nabla \times A) - A \cdot (\nabla \times B) \tag{D.7}$$

$$\nabla \cdot \nabla \times A = 0 \tag{D.8}$$

$$\nabla \times \nabla f = 0 \tag{D.9}$$

$$\nabla \cdot \nabla f = \nabla^2 f \tag{D.10}$$

$$\nabla \times \nabla \times f = \nabla(\nabla \cdot f) - \nabla^2 f \tag{D.11}$$

$$\nabla(A \cdot B) = A \cdot \nabla B + B \cdot \nabla A + A \times \nabla B + B \times \nabla \times A \tag{D.12}$$

$$\nabla(A \times B) = B \cdot \nabla A - A \cdot \nabla B + A(\nabla \cdot B) - B(\nabla \cdot A) \tag{D.13}$$

$$A \cdot \nabla A = \frac{1}{2}\nabla(A \cdot A) - A \times (\nabla \times A) \tag{D.14}$$

Appendix E
Curvilinear Coordinates

Spherical Coordinates (λ, ϕ, r)

$$\nabla \psi = \frac{1}{r \cos \phi} \frac{\partial \psi}{\partial \lambda} e_\lambda + \frac{1}{r} \frac{\partial \psi}{\partial \phi} e_\phi + \frac{\partial \psi}{\partial r} e_r \tag{E.1}$$

$$\nabla \cdot A = \frac{1}{r \cos \phi} \frac{\partial A_\lambda}{\partial \lambda} + \frac{1}{r \cos \phi} \frac{\partial}{\partial \phi} (\cos \phi A_\phi) + \frac{1}{r^2} \frac{\partial}{\partial r} (r^2 A_r) \tag{E.2}$$

$$\nabla \times A = \frac{1}{(r^2 \cos \phi)} \left\{ r \cos \phi \left[\frac{\partial A_r}{\partial \phi} - \frac{\partial (r A_\phi)}{\partial r} \right] e_\lambda \right.$$

$$+ r \left[\frac{\partial}{\partial r} (r \cos \phi A_\lambda) - \frac{\partial A_r}{\partial \lambda} \right] e_\phi$$

$$\left. + \left[\frac{\partial (r A_\phi)}{\partial \lambda} - \frac{\partial}{\partial \phi} (r \cos \phi A_\lambda) \right] e_r \right\} \tag{E.3}$$

$$\nabla^2 \psi = \frac{1}{r^2 \cos^2 \phi} \frac{\partial^2 \psi}{\partial \lambda^2} + \frac{1}{r^2 \cos \phi} \frac{\partial}{\partial \phi} \left(\cos \phi \frac{\partial \psi}{\partial \phi} \right) + \frac{1}{r^2} \frac{\partial}{\partial r} \left(r^2 \frac{\partial \psi}{\partial r} \right) \tag{E.4}$$

$$\nabla^2 A = \left[\nabla^2 A_\lambda - \frac{A_\lambda}{r^2 \cos^2 \phi} + \frac{2}{r^2 \cos \phi} \frac{\partial A_r}{\partial \lambda} + \frac{2}{r^2} \frac{\sin \phi}{\cos^2 \phi} \frac{\partial A_\phi}{\partial \lambda} \right] e_\lambda$$

$$+ \left[\nabla^2 A_\phi - \frac{A_\phi}{r^2 \cos^2 \phi} + \frac{2}{r^2} \frac{\partial A_r}{\partial \phi} - \frac{2}{r^2} \frac{\sin \phi}{\cos^2 \phi} \frac{\partial A_\lambda}{\partial \lambda} \right] e_\phi$$

$$+ \left[\nabla^2 A_r - \frac{2}{r^2} A_r - \frac{2}{r^2 \cos \phi} \frac{\partial}{\partial \phi} (\sin \phi A_\phi) - \frac{2}{r^2 \cos \phi} \frac{\partial A_\lambda}{\partial \lambda} \right] e_r \tag{E.5}$$

Cylindrical Coordinates (r, ϕ, z)

$$\nabla \psi = \frac{\partial \psi}{\partial r} e_r + \frac{1}{r} \frac{\partial \psi}{\partial \phi} e_\phi + \frac{\partial \psi}{\partial z} e_z \tag{E.6}$$

$$\nabla \cdot A = \frac{1}{r} \frac{\partial}{\partial r} (r A_r) + \frac{1}{r} \frac{\partial A_\phi}{\partial \phi} + \frac{\partial A_z}{\partial z} \tag{E.7}$$

$$\nabla \times A = \left[\frac{1}{r} \frac{\partial A_z}{\partial \phi} - \frac{\partial A_\phi}{\partial z} \right] e_r + \left[\frac{\partial A_r}{\partial z} - \frac{\partial A_z}{\partial r} \right] e_\phi$$

$$+ \left[\frac{1}{r} \frac{\partial (r A_\phi)}{\partial r} - \frac{1}{r} \frac{\partial A_r}{\partial \phi} \right] e_z \tag{E.8}$$

$$\nabla^2 \psi = \frac{1}{r} \frac{\partial}{\partial r} \left(r \frac{\partial \psi}{\partial r} \right) + \frac{1}{r^2} \frac{\partial^2 \psi}{\partial \phi^2} + \frac{\partial^2 \psi}{\partial z^2} \tag{E.9}$$

$$\nabla^2 A = \left[\nabla^2 A_r - \frac{A_r}{r^2} - \frac{2}{r^2} \frac{\partial A_\phi}{\partial \phi} \right] e_r + \left[\nabla^2 A_\phi - \frac{A_\phi}{r^2} + \frac{2}{r^2} \frac{\partial A_r}{\partial \phi} \right] e_\phi$$

$$+ \nabla^2 A_z e_z \tag{E.10}$$

Appendix F
Pseudo-Adiabatic Chart

Figures on the following pages show adiabats (solid lines), which are characterized by constant values of potential temperature θ, pseudo-adiabats (dashed lines), which are characterized by constant values of equivalent potential temperature θ_e, and isopleths of saturation mixing ratio with respect to water r_w (thin solid lines), all as functions of temperature and pressure with the ordinate proportional to $(p_0/p)^\kappa$.

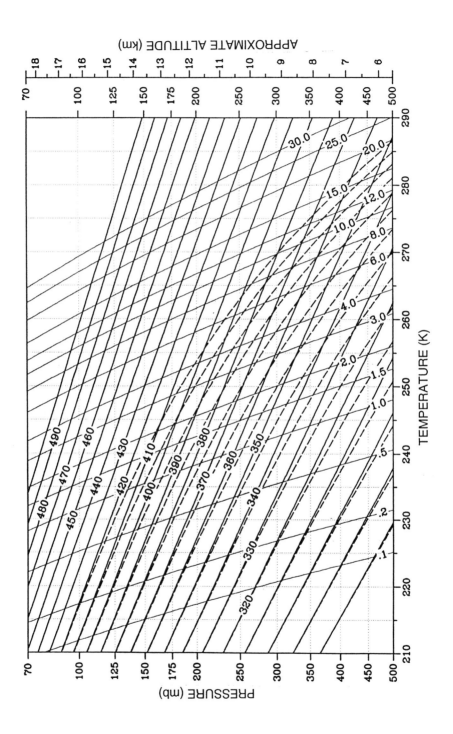

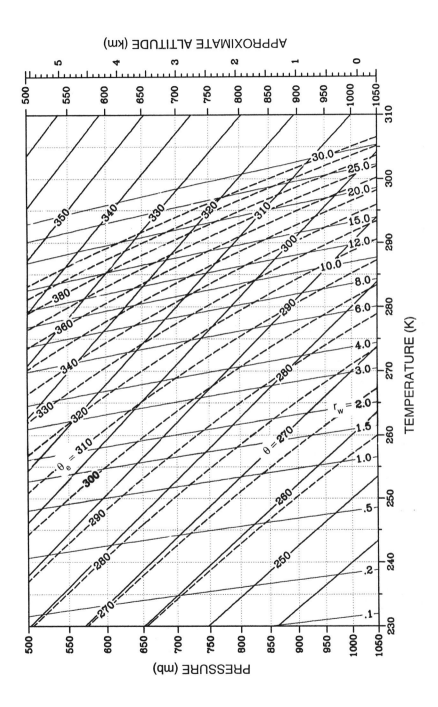

APPROXIMATE ALTITUDE (km)

TEMPERATURE (K)

PRESSURE (mb)

References

Abramowitz, M. and I. Stegun, 1972: *Handbook of Mathematical Functions*. Dover, New York, 1046 pp.

Ackerman, T., 1988: Aerosols in climate modeling. *Aerosol and Climate*, P. Hobbs and P. McCormick, Eds., A. Deepak Publishing, Hampton, 335–348.

Anderson, J., W. H. Bruce, and M. H. Proffitt, 1989: Ozone destruction by chlorine radicals within the Antarctic vortex: The spatial and temporal evolution of ClO—O_3 anticorrelation based on *in situ* ER-2 data. *J. Geophys. Res.*, **94**, 11,465–11,479.

Andrews, D., J. Holton, and C. Leovy, 1987: *Middle Atmosphere Dynamics*. Academic Press, San Diego, 489 pp.

Aris, R., 1962: *Vectors, Tensors, and the Basic Equations of Fluid Mechanics*. Prentice Hall, Englewood Cliffs, NJ, 286 pp.

Banks, P., and G. Kocharts, 1973: *Aeronomy*. Academic Press, New York, 430 pp.

Barnola J., D. Raynaud, Y. Korotkevitch, and C. Lorius, 1987: Vostok ice core: A 160,000 year record of atmospheric CO_2. *Nature*, **329**, 408–414.

Batchelor, G., 1977: *An Introduction to Fluid Dynamics*. Cambridge University Press, Cambridge, 615 pp.

Bithell, M., L. Gray, J. Harries, J. Russell, and A. Tuck, 1994: Synoptic interpretation of measurements from HALOE. *J. Atmos. Sci.*, **51**, 2942–2956.

Bleck, R., 1978a: Finite difference equations in generalized vertical coordinates. Part I: Total energy conservation. *Beiträge zur Physik deer Atmosphäre*, **51**, 360–372.

Bleck, R., 1978b: Finite difference equations in generalized vertical coordinates. Part II: Potential vorticity conservation. *Beiträge zur Physik deer Atmosphäre*, **52**, 95–105.

Brasseur, G., and S. Solomon, 1986: *Aeronomy of the Middle Atmosphere*. Reidel, Dordrecht, 2nd ed., 452 pp.

Brewer, A., 1949: Evidence for a world circulation provided by the measurements of helium and water vapor distributions in the stratosphere. *Quart. J. Roy. Met. Soc.*, **75**, 351.

Brillouin, L., 1926: Remarques sur la mecanique ondulatoire. *J. Phys. Radium*, **7**, 353–368.

Carlson, T., and S. Benjamin, 1982: Radiative heating rates for a desert aerosol (Saharan dust). In *Atmospheric Aerosols: Their Formation, Optical Properties and Effects*, A. Deepak, Ed., Spectrum Press, Hampton, 435–457.

Carrier, L. Cato, G., and K. von Essen, 1967: The backscattered and extinction of visible and infrared radiation by selected major cloud models. *Appl. Optics*, **6**, 1209–1216.

Chapman, S., 1930: On ozone and atomic oxygen in the upper atmosphere. *Phil. Mag.*, **10**, 369–383.

Charney, J., 1947: The dynamics of long waves in a baroclinic westerly current. *J. Meteorol.*, **4**, 135–163.

Charney, J., 1973: Planetary fluid mechanics. In *Dynamical Meteorology*, P. Morel, ed., Reidel, Dordrecht, 97–351.

Charney, J., and P. Drazin, 1961: Propagation of planetary scale disturbances from the lower into the upper atmosphere. *J. Geophys. Res.*, **66**, 83–109.

Charney, J., and M. Stern, 1962: On the stability of internal baroclinic jets in a rotating atmosphere. *J. Atmos. Sci.*, **19**, 159–172.

Coulson, K., 1975: *Solar and Terrestrial Radiation*, Academic Press, New York, 322 pp.

Cox, S., 1981: Radiation characteristics of clouds in the solar spectrum. In *Clouds: Their Formation, Optical Properties, and Effects*, P. Hobbs and A. Deepak, Eds., Academic Press, New York, 241–280.

Danielsen, E., 1982: A dehydration mechanism for the stratosphere. *Geophys. Res. Lett.*, **9**, 605–608.

A. Deepak, Ed., 1982: *Atmospheric Aerosols: Their Formation, Optical Properties, and Effects*. Spectrum Press, Hampton, VA 480 pp.

Denbigh, K., 1971: *The Principles of Chemical Equilibrium*. Cambridge University Press, London, 494 pp.

Dickinson, R., 1968: Planetary Rossby waves propagating through weak westerly wind wave guides. *J. Atmos. Sci.*, **25**, 984–1002.

Dickinson, R., 1970: Development of a Rossby wave critical level. *J. Atmos. Sci.*, **27**, 627–633.

Dickinson, R. E., 1973: Baroclinic instability of an unbounded zonal shear flow in a compressible atmosphere. *J. Atmos. Sci.*, **30**, 1520–1527.

Dikii, L., 1968: The terrestrial atmosphere as an oscillating system. *Izv. Acad. Sci. USSR Atmos. Oceanic Phys.*, Engl. Transl., **1**, 469–489.

Dobson, G., 1930: Observations of the amount of ozone in the Earth's atmosphere and its relation to other geophysical conditions. *Proc. Roy. Soc. London*, Sec. A, **129**, 411–433.

Dowling, D., R. Radke, and F. Lawrence, 1990: A summary of the physical properties of cirrus clouds. *J. Applied Meteorology*, **29**, 970–978.

Drazin, P., and W. Reid, 1985: *Hydrodynamic Instability*. Cambridge University Press, Cambridge, 527 pp.

Dutton, J., 1986: *The Ceaseless Wind*. Dover, New York, 617 pp.

Eady, E., 1949: Long waves and cyclone waves. *Tellus*, **1**, 33–52.

Eckart, C., 1960: *Hydrodynamics of Oceans and Atmospheres*. Pergamon, New York, 290 pp.

Eisberg, R., 1967: *Fundamentals of Modern Physics*. John Wiley, New York, 729 pp.

Elsasser, W., 1938: Mean absorption and equivalent absorption coefficient of a band spectrum. *Phys. Rev.*, **54**, 126–129.

Ertel, H. 1942: Ein neuer hydrodynamischer Wirbelsatz. *Meteorol. Z.*, **59**, 271–281.

Fels, S., 1982: A parameterization of scale-dependent radiative damping rates in the middle atmosphere. *J. Atmos. Sci.*, **39**, 1141–1152.

Fels, S., 1985: Radiative-dynamical interactions in the middle atmosphere. *Adv. Geophys.*, **28A**, 277–300.

Fleming, E., S. Chandra, M. Schoeberl, and J. Barnett, 1988: Monthly-mean climatology of temperature, wind, geopotential height, and pressure for 0–120 km. NASA TM-100697. Available from NASA Goddard Space Flight Center, Greenbelt, MD.

Fujita, T., 1992: *The Mystery of Severe Storms*. WRL Research Paper No. 239, NTIS No. PB 92-182021, 298 pp.

Fusco, A., and M. Salby, 1994: Relationship between horizontal eddy motions and mean meridional motions in the stratosphere. *J. Geophys. Res.*, **99**, 20,633–20,695.

Garcia, R., 1994: Causes of ozone depletion. *Physics World*, **7**, 49–55.

Garcia, R., and M. Salby, 1987: Transient response to localized episodic heating in the tropics. Part II: Far-field behavior. *J. Atmos. Sci.*, **44**, 499–530.

Garcia, R., and S. Solomon, 1983: A numerical model of zonally-averaged dynamical and chemical structure of the middle atmosphere. *J. Geophys. Res.*, **88**, 1379–1400.

Garcia, R., and S. Solomon, 1987: A possible interaction between interannual variability in Antarctic ozone and the quasi-biennial oscillation. *Geophys. Res. Lett.*, **14**, 848–851.

Gill, A., 1980: Some simple solutions for heat-induced tropical circulation. *Quart. J. Roy. Met. Soc.*, **106**, 447–462.

Gill, A., 1982: *Atmosphere–Ocean Dynamics*. Academic Press, San Diego, 662 pp.

Goody, R., 1952: A statistical model for water vapor absorption. *Quart. Roy. Meteorol. Soc.*, **78**, 165–169.

Goody, R., and Y. Yung, 1989: *Atmospheric Radiation*, Oxford University Press, New York, 2nd ed., 528 pp.

Gossard, E., and W. Hooke, 1975: *Waves in the Atmosphere*. Elsevier, Amsterdam, 456 pp.

Green, G., 1838: On the motion of waves in variable canal of small depth and width. *Trans. Cambridge Philos. Soc.* In *Mathematical Papers (1871)*, Macmillan, London, 223–230.

Greenspan, H., 1968: *Theory of Rotating Fluids*, Cambridge University Press, London, 327 pp.

Haltiner, G., and R. Williams, 1980: *Numerical Prediction and Dynamic Meteorology*. Wiley, New York, 477 pp.

Handbook of Geophysics and Space Environment 1965: S. Valley, Ed., Air Force Cambridge Research Laboratories, Bedford, MA.

Hansen, J., and L. Travis, 1974: Light scattering in planetary atmospheres. *Space Sci. Rev.*, **16**, 527–610.

Hartmann, D., 1993: The radiative effect of clouds on climate. In *Aerosol-Cloud-Climate Interactions*, P. Hobbs, Ed., Academic Press, San Diego, 151–170.

Hartmann, D., 1994: *Global Physical Climatology*. Academic Press, San Diego, 408 pp.

Hartmann, D., and D. Doelling, 1991: On the net radiative effectiveness of clouds. *J. Geophys. Res.*, **96**, 869–891.

Hartmann, D., V. Ramanathan, A. Berroir, and G. Hunt, 1986: Earth radiation budget data and climate research. *Rev. Geophys.*, **24**, 439–468.

Hendon, H., and K. Woodberry, 1993: The diurnal cycle of tropical convection. *J. Geophys. Res.*, **98**, 16,623–16,637.

Hering, W., and T. Borden, 1965: Ozone sonde observations over North America, Vol. 3, AFCRL Report AFCRL-64-30. Air Force Cambridge Research Laboratories, Bedford MA.

Herzberg, G., 1945: *Molecular Spectra and Molecular Structure. II: Infrared and Raman Spectra of Polyatomic Molecules*. Van Nostrand, New York, 616 pp.

Herzberg, L., 1965: In *Physics of the Earth's Upper Atmosphere*, C. Hines, I. Paghis, T. Hartz, and J. Fejer, Eds., Prentice-Hall, Englewood Cliffs, NJ, 31–45.

Hess, P., and J. Holton, 1985: The origin of temporal variance in long-lived trace constituents in the summer stratosphere. *J. Atmos. Sci.*, **42**, 1455–1463.

Hide, R., 1966: On the dynamics of rotating fluids and related topics in geophysical fluid dynamics. *Bull. Amer. Meteorol. Soc.*, **47**, 873–885.

Hobbs, P., and P. McCormick, Eds., 1988: *Aerosol and Climate*. Deepak Pub., Hampton VA, 486 pp.

Hoffman, D., 1988: Aerosols from past and present volcanic emissions. In *Aerosol and Climate*, P. Hobbs and P. McCormick, Eds., A. Deepak Publishing, Hampton, VA, 195–214.

Hoffman, D., and S. Solomon, 1989: Ozone destruction through heterogeneous chemistry following the eruption of El Chichon. *J. Geophys. Res.*, **94**, 5029–5041.

Holton, 1984: Troposphere-stratosphere exchange of trace constituents: the water vapor puzzle. In *Dynamics of the Middle Atmosphere*, J. Holton and T. Matsuno, Eds., Terrapub, Tokyo, 369–385.

Holton, J., 1990: On the global exchange of mass between the stratosphere and troposphere. *J. Atmos. Sci.*, **47**, 392–395.

Holton, J., 1992: *An Introduction to Dynamic Meteorology*. Academic Press, San Diego, 511 pp.

Holton, J., and R. Lindzen, 1972: An updated theory for the quasi-biennial cycle of the tropical stratosphere. *J. Atmos. Sci.*, **29**, 1076–1080.

Holton, J., and H. Tan, 1982: The quasi-biennial oscillation in the Northern Hemisphere lower stratosphere. *J. Meteorol. Soc. Jpn.*, **60**, 140–148.

Holton, J., Haynes, P., McIntyre, M., Douglass, A., Rood, R., and L. Pfister, 1995: Stratosphere–Troposphere Exchange. *Revs. Geophys.* (in press).

Horel, J., and J. Wallace, 1981: Planetary scale atmospheric phenomena associated with the interannual variability of sea-surface temperature in the equatorial Pacific. *Mon. Wea. Rev.*, **109**, 813–829.

Hoskins, B., 1982: A mathematical theory of frontogenesis. *Annu. Rev. Fluid Mech.*, **14**, 131–151.

Hoskins, B., and D. Karoly, 1981: The steady linear response of a spherical atmosphere to thermal and orographic forcing. *J. Atmos. Sci.*, **38**, 1179–1196.

Hoskins, B., M. McIntyre, and A. Robertson, 1985: On the use and significance of isentropic potential vorticity maps. *Quart. J. Roy. Meteorol. Soc.*, **111**, 877–946.

Houze, R. A., 1982: Cloud clusters and large-scale vertical motions in the tropics. *J. Meteor. Soc. Jpn.*, **60**, 396–410.

Houze, R., 1993: *Cloud Dynamics*. Academic Press, San Diego, 573 pp.

Humphreys, W., 1964: *Physics of the Air*. Dover, New York, 661 pp.

IPCC, 1990: *Climate Change*. Intergovernmental Panel on Climate Change, J. Houghton, G. Jenkins, and J. Ephrams, Eds., Cambridge University Press, 365 pages.

Iribarne, J., and W. Godson, 1981: *Atmospheric Thermodynamics*. Reidel, Dordrecht, 259 pp.

Jackson, J., 1975: *Classical Electrodynamics*. Wiley, New York, 848 pp.

James, I., 1993: *Introduction to Circulating Atmospheres*. Cambridge University Press, New York, 422 pp.

Jin, F., and B. Hoskins, 1995: The direct response to tropical heating in a baroclinic atmosphere. *J. Atmos. Sci.*, **52**, 307–319.

Johnson, R., and D. Kriete, 1982: Thermodynamic circulation characteristics of winter monsoon tropical mesoscale convection. *Mon. Wea. Rev.*, **110**, 1898–1911.

Jones, D., S. McKenna, L. R. Poole, and S. Solomon, 1990: On the influence of PSC formation on chemical composition during the 1988/89 Arctic winter. *Geophys. Res. Lett.*, **17**, 545–548.

Keeling, C., R. Bacastow, A. Carter, S. Piper, T. Whorf, M. Heimann, W. Mook, and H. Rheloffzen, 1989: A three dimensional model of atmospheric CO_2 based on observed winds: 1. Analysis of observational data. *Aspects of Climate Variability in the Western Pacific and Western Americas*, D. Peterson, Ed., *Geophysical Monographs*, **55**, American Geophysical Union, Washington DC, 165–236.

Keenan, J., 1970: *Thermodynamics*. The MIT Press, Cambridge, 507 pp.

Kelly, K., M. Proffitt, K. Chan, M. Lowenstein, J. Podolske, S. Strahan, J. Wilson, and D. Kley, 1993: Water vapor and cloud water measurements over Darwin during the STEP 1987 tropical mission. *J. Geophys. Res.*, **98**, 8713–8723.

Kelvin, Lord, 1880: On a disturbing infinity in Lord Rayleigh's solution for waves in a plane vortex stratum. *Nature*, **23**, 45–46.

Kent, G., and P. McCormick, 1984: SAGE and SAM II measurements of global stratospheric aerosol optical depth and mass loading. *J. Geophys. Res.*, **89**, 5303–5314.

Kiehl, J., and S. Solomon, 1986: On the radiative balance of the stratosphere, *J. Atmos. Sci.*, **43**, 1525–1534.

Knollenberg, R. G., K. Kelly, and J. C. Wilson, 1993: Measurements of number densities of ice crystals in the tops of tropical cumulonimbus. *J. Geophys. Res.*, **98**, 8639–8664.

Kocharts, G., 1971: Penetration of solar radiation in the Schumann-Runge bands of molecular oxygen. In *Mesospheric Models and Related Experiments*, G. Fiocco, Ed., Reidel, Dordrecht, 160–176.

Koschmieder, E., and S. Pallas, 1974: Heat transfer through a shallow horizontal convecting fluid layer. *Int. J. Heat Mass Transfer*, **17**, 991–1002.

Kramer, H., 1926: Wellenmechanik und halbzhalige Quantisierung. *Z. Phys.*, **39**, 828–840.

Krueger, A., 1973: The mean ozone distributions from several series of rocket soundings to 52 km at latitudes from 58 S to 64 N. *Pure Appl. Geophys.*, **106–108**, 1272–1280.

Labitzke, K., 1982: On interannual variability of the middle stratosphere during northern winters. *J. Meteorol. Soc. Jpn.*, **60**, 124–139.

Landau, X., Y. Lifshitz, and Z. Pitaevskii, 1980: *Statistical Physics*, Pergamon Press, New York, 3rd ed., Part 1, 544 pp.

Lamb, H., 1910: On atmospheric oscillations. *Proc. Roy. Soc. London*, **84**, 551–572.

Lau, N., 1984: *Circulation Statistics Based on FGGE Level III-B Analyses*. NOAA Data Report ERL GFDL-5. NOAA Environmental Research Laboratories, Boulder, CO.

Lebedev, N., 1972: *Special Functions and Their Applications*, R. Silverman, Ed. Dover, New York, 308 pp.

Lee, J., F. Sears, and D. Turcotte, 1973: *Statistical Thermodynamics*. Addison-Wesley, Reading, MA, 371 pp.

Lighthill, J., 1978: *Waves in Fluids*. Cambridge University Press, Cambridge, 504 pp.

Lilly, D., 1978: A severe downslope windstorm and aircraft turbulence event induced by a mountain wave. *J. Atm. Sci.*, **35**, 59–77.

Lincoln, C., 1972: *On Quiet Wings: A Soaring Anthology*. Northland Press, Flagstaff, AZ, 397 pp.

Lindzen, R., and J. Holton, 1968: A theory of the quasi-biennial oscillation. *J. Atmos. Sci.*, **25**, 1095–1107.

Liou, K., 1980: *An Introduction to Atmospheric Radiation*. Academic Press, San Diego, 392 pp.

Liou, K., 1990: *Radiation and Cloud Processes in the Atmosphere*. Oxford University Press, New York, 487 pp.

Liouville, J., 1837: Sur le développment des fonctions ou parties de fonctions en séries. *J. Math. Pure Appl.*, **2**, 16–35.

List, R., 1958: *Smithsonian Meteorological Tables*. Smithsonian Institute Press, Random House, New York, 6th ed.

London, J., 1980: Radiative energy sources and sinks in the stratosphere and mesosphere. *Proceedings of the NATO Advanced Study Institute on Atmospheric Ozone: Its Variation and Human Influences*. Report FAA-EE-80-20, A. Aiken, Ed., 703–721.

Lorenz, E., 1955: Available potential energy and the maintenance of the general circulation. *Tellus*, **7**, 157–167.

Manabe, S., and R. Strickler, 1964: Thermal equilibrium of the atmosphere with convective adjustment. *J. Atmos. Sci.*, **21**, 361–385.

Manabe, S., and R. Wetherald, 1967: Thermal equilibrium of the atmosphere with a given distribution of relative humidity. *J. Atmos. Sci.*, **24**, 241.

Margules, M., 1903: Über die energie der sturme. *Jahrb. Zentralanst. Meteorol. Wien*, **40**, 1–26.

Matveev, L., 1967: *Physics of the Atmosphere*. Israel Program for Scientific Translations, Jerusalem, 699 pp.

McClatchey, R., and J. Selby, 1972: Atmospheric transmittance, 7–30 μm: Attenuation of CO_2 laser radiation. *Environmental Research Paper* No. 419, AFCRL-72-0611.

McIntyre, M., and T. Palmer, 1983: Breaking planetary waves in the stratosphere. *Nature*, **305**, 593–600.

Mie, G., 1908: Beigrade zur optik trüber medienspeziell kolloidaler metallösungen. *Ann. Physik*, **25**, 377–445.

Miles, T., and W. Grose, 1986: Transient medium-scale wave activity in the summer stratosphere. *Bull. Amer. Meteor. Soc.*, **67**, 674–686.

Minnis, P., and E. Harrison, 1984: Diurnal variability of regional cloud and clear-sky radiative parameters derived from GOES data. III. November 1978 radiative parameters. *J. Climate Appl. Meteor.*, **23**, 1032–1051.

Monin, A., and A. Yaglom, 1973: *Statistical Fluid Mechanics*. The MIT Press, Cambridge, 769 pp.

Morse, P., and H. Feschbach, 1953: *Methods of Theoretical Physics*, Vols. I and II. McGraw–Hill, New York, 1978 pp.

Murgatroyd, R., and F. Singleton, 1961: Possible meridional circulations in the stratosphere and mesosphere. *Q. J. Roy. Meteorol. Soc.*, **87**, 125–135.

Nicolet, M., 1980: The chemical equations of stratospheric and mesospheric ozone. *Proceedings of the NATO Advanced Study Institute on Atmospheric Ozone (Portugal)*. U.S. Dept. of Transportation, FAA, Washington, D.C., Report FAA-EE-80-20.

Naujokat, B., 1986: An update of the observed quasi-biennial oscillation of he stratospheric winds over the tropics. *J. Atmos. Sci.*, **43**, 1873–1877.

Oort, A., and J. Peixoto, 1983: Global angular momentum and energy balance requirements from observations. *Adv. Geophys.*, **25**, 355–490.

O'Sullivan, D., and M. Salby, 1990: Coupling of the quasi-biennial oscillation and the extratropical circulation in the stratosphere through planetary wave transport. *J. Atmos. Sci.*, **47**, 650–673.

Paeschke, W., 1937: Experimentelle Untersuchungen zum Rauhigkeits-und Stabilitaetsproblem in der freien Atmosphaere. *Beitr. Phys. Atmos.*, **24**, 163–189.

Paltridge, G., and C. Platt, 1981: Aircraft measurements of solar and infrared radiation and the microphysics of cirrus clouds. *Quart. J. Roy. Meteorol. Soc.*, **107**, 367–380.

Patterson, E., 1982: Size distributions, concentrations, and composition of continental and marine aerosols. In *Atmospheric Aerosols: Their Formation, Optical Properties, and Effects*, A. Deepak, Ed., Spectrum Press, Hampton, 1–23.

Pedlosky, J., 1979: *Geophysical Fluid Dynamics*. Springer-Verlag, New York, 624 pp.

Peixoto, J., and A. Oort, 1992: *Physics of Climate*. American Institute of Physics, New York, 520 pp.

Philander, S., 1983: El Niño Southern Oscillation phenomena. *Nature*, **302**, 295–301.

Phillips, N., 1966: The equations of motion for a shallow rotating atmosphere and the "Traditional Approximation". *J. Atmos. Sci.*, **23**, 626–628.

Pitts, M. C., L. R. Poole, and M. P. McCormick, 1990: SAGE observations of PSCs near 50 degrees north January 31–February 2, 1989. *Geophys. Res. Lett.*, **17**, 405–408.

Platt, C., A. Dilley, J. Scott, I. Barton, and G. L. Stephens, 1984a: Remote sounding of high clouds. V: Infrared properties and structures of tropical thunderstorm anvils. *J. Climate Appl. Meteor.*, **23**, 1296–1308.

Platt, C., J. Scott, and A. Dilley, 1984b: Remote sounding of high clouds. IV: Optical properties of midlatitude and tropical cirrus. *J. Atmos. Sci.*, **44**, 729–747.

Plumb, R., 1982; The circulation of the middle atmosphere. *Aust. Meteorol. Mag.*, **30**, 107–121.

Poole, L., S. Solomon, M. McCormick, and M. Pitts, 1989: *Geophys. Res. Lett.*, **16**, 1157–1160.

Pruppacher, H., 1981: The microstructure of atmospheric clouds and precipitation. In *Clouds: Their Formation, Optical Properties, and Effects*. P. Hobbs and A. Deepak, Eds., Academic Press, San Diego, 93–186.

Pruppacher, H., and J. Klett, 1978: *Microphysics of Clouds and Precipitation*. Reidel, Dordrecht, 714 pp.

Queney, P., 1948: The problem of air flow over mountains: A summary of theoretical studies. *Bull. Amer. Meteorol. Soc.*, **29**, 16–26.

Ramanathan, V., 1987: Atmospheric general circulation and its low frequency variance: radiative influences. *J. Meteor. Soc. Jpn.*, **65**, 151–175.

Ramanathan, V., B. Barkstrom, and E. Harrison 1989: Climate and the Earth's Radiation Budget. *Physics Today*, **42**, 22–32.

Rayleigh, Lord, 1871: On the light from the sky, its polarization and colour. *Phil. Mag.*, **41**, 107–120.

Rayleigh, Lord, 1880: On the stability or instability of certain fluid motions. *Proc. London Math. Soc.*, **9**, 57–70.

Rektorys, K., 1969: *Survey of Applicable Mathematics*. The MIT Press, Cambridge, 1369 pp.

Remsberg, E., J. Russell, L. Gordley, J. Gille, and P. Bailey, 1984: Implications of the stratospheric water vapor distribution as determined from the Nimbus-7 LIMS experiment. *J. Atmos. Sci.*, **41**, 2934–2945.

Reynolds, D., T. Vonder Haar, and S. Cox, 1975: The effect of solar radiation absorption in the tropical troposphere. *J. Appl. Meteor.*, **14**, 433–443.

Rodgers, C., and C. Walshaw, 1966: The computation of infrared cooling rate in planetary atmospheres. *Quart. J. Roy. Meteor. Soc.*, **92**, 67–92.

Roper, R., 1977: Turbulence in the lower thermosphere, Chap. 7 in *The Upper Atmosphere and Magnetosphere*, F. Johnson, Ed., U.S. National Academy of Science, 117–129.

Rossby, C., *et al.* 1939: Relations between variations in the intensity of the zonal circulation of the atmosphere and the displacements of the semipermanent centers of action. *J. Mar. Res.*, **2**, 38–55.

Ruth, S., J. Remedios, B. Lawrence, and F. Taylor, 1994: Measurements of N_2O by the Improved Stratospheric and Mesospheric Sounder during early northern winter 1991/92. *J. Atmos. Sci.*, **51**, 2818–2833.

Salby, M., D. O'Sullivan, R. Garcia, and P. Callaghan, 1990: Air motions accompanying the development of a planetary wave critical layer. *J. Atmos. Sci.*, **47**, 1179–1204.

Schlicting, H, 1968: *Boundary Layer Theory*. McGraw–Hill, New York, 744 pp.

Scorer, R., 1978: *Environmental Aerodynamics*. Ellis Horwood Ltd., Chichester, 488 pp.

Scorer, R., 1986: *Cloud Investigation by Satellite*. Ellis Horwood Ltd., Chichester, 23 chaps.

Shapiro, M., 1982: Nowcasting the position and intensity of jet streams using a satellite-borne total ozone mapping spectrometer. In *Nowcasting*, K. A. Browning, Ed., Academic Press, San Diego, 256 pp.

Shapiro, M., and J. Hastings, 1973: Objective cross-section analysis by Hermite polynomial interpolation on isentropic surfaces. *J. Appl. Meteorol.*, **12**, 753–762.

Shea, D., K. Trenberth, and R. Reynolds, 1990: *A Global Monthly Sea Surface Temperature Climatology*. Technical Note TN 345+STR, National Center for Atmospheric Research, Boulder, CO, 167 pp.

Slingo, A., 1989: A GCM parameterization for the SW radiative properties of water clouds. *J. Atmos. Sci.*, **46**, 1419–1427.

Slinn, W., 1975: Atmospheric aerosol particles in surface-level air. *Atmos. Env.*, **9**, 763–764.

Smith, W., and D. Gottlieb, 1974: Solar flux and its variations. *Space Sci. Rev.*, **16**, 771–802.

Solomon, S., 1990: Progress towards a quantitative understanding of Antarctic ozone depletion. *Nature*, **347**, 347–354.

Sorbjan, Z., 1989: *The Structure of the Atmospheric Boundary Layer*. Prentice Hall, Englewood Cliffs, NJ, 317 pp.

Stephens, G., 1978a: Radiation profiles in extended water clouds I: Theory. *J. Atmos. Sci.*, **35**, 2111–2122.

Stephens, G., 1978b: Radiation profiles in extended water clouds II: Parameterization schemes. *J. Atmos. Sci.*, **35**, 2123–2132.

Stephens, G., 1984: The parameterization of radiation for numerical weather prediction and climate models. *Mon. Wea. Rev.*, **112**, 826–867.

Stephens, G., G. Paltridge, and C. Platt, 1978: Radiation profiles in extended water clouds III: Observations. *J. Atmos. Sci.*, **35**, 2133–2141.

Stull, R., 1988: *An Introduction to Boundary Layer Meteorology*. Kluwer, Boston, 666 pp.

Sun, W., and C. Chang, 1986: Diffusion model for a convective layer. *J. Climate Appl. Meteorol.*, **25**, 1445–1453.

Taylor, G., 1929: Waves and tides in the atmosphere. *Proc. Roy. Soc. London Ser. A*, **126**, 169–183.

Tennekes, H., and J. Lumley, 1972: *A First Course in Turbulence*. The MIT Press, Cambridge, 300 pp.

Thorpe, S., 1969: Experiments on the instability of stratified flows: Immiscible fluids. *J. Fluid Mech.*, **39**, 25–48.

Thorpe, S., 1971: Experiments on the instability of stratified flows: Miscible fluids. *J. Fluid Mech.*, **46**, 299–319.

Turco, R., K. Drdla, A. Tabazadeh, and P. Hamill, 1993: Heterogeneous chemistry of polar stratospheric clouds and volcanic aerosols. *The Role of the Stratosphere in Global Change*, NATO ASI Series I, **8**, M. Chanin, Ed., Springer-Verlag, Heidelberg.

Twomey, S., 1977: *Atmospheric Aerosols*. Elsevier, Amsterdam, 302 pp.

Understanding Climate Change, 1975: U.S. National Academy of Sciences, Washington, DC, 239 pp.

U.S. Standard Atmosphere, 1976: National Oceanic and Atmospheric Administration, National Aeronautics and Space Administration, U.S. Air Force, U.S. Government Printing Office, NOAA-S/T 76-1562, Washington, DC, 228 pp.

Van Dyke, M., 1982: *An Album of Fluid Motion*. Parabolic Press, Stanford, CA, 174 pp.

Vinnichenko, N., 1970: The kinetic energy spectrum in the free atmosphere—1 second to 5 years. *Tellus*, **22**, 158–166.

Wallace, J., 1967: A note on the role of radiation in the biennial oscillation. *J. Atmos. Sci.*, **24**, 598–599.

Wallace, J., and D. Gutzler, 1981: Teleconnections in the geopotential height field during the Northern Hemisphere winter. *Mon. Wea. Rev.*, **109**, 784–812.

Wallace, J., and P. Hobbs, 1977: *Atmospheric Science: An Introductory Survey*. Academic Press, San Diego, 467 pp.

Warner, J., 1969: The microstructure of cumulus cloud. Part I. General features of the droplet spectrum. *J. Atmos. Sci.*, **26**, 1049–1059.

Webster. P., 1983: The large-scale structure of the tropical atmosphere. In *Large Scale Dynamical Processes in the Atmosphere*, B. Hoskins and R. Pierce, Eds., Academic Press, San Diego, 235–276.

Webster, P., and R. Lukas, 1992: TOGA-COARE: The coupled ocean-atmosphere response experiment. *Bull. Amer. Meteor. Soc.*, **73**, 1377–1416.

Webster, P., and G. Stephens, 1980: Tropical upper-tropospheric extended clouds: Inferences from winter MONEX. *J. Atmos. Sci.*, **37**, 1521–1541.

Welander, P., 1955: Studies of the general development of motion in a two-dimensional ideal fluid. *Tellus*, **7**, 141–156.

Wentzel, G., 1926: Eine Verallgemeinerung der Quantenbedingung fur die Zwecke der Wellen-mechanik. *Z. Phys.*, **38**, 518–529.

Whipple, 1930: The great Siberian meteor, and the waves, seismic and aerial, which it produced. *Quart. J. Roy. Meteorol. Soc.*, **56**, 287–298.

Willson, R. C., H. S. Hudson, C. Frohlich, and R. W. Brusa, 1986: Long-term downward trend in total solar irradiance. *Science*, **234**, 1114–1117.

WMO, 1969: *International Cloud Atlas*. World Meteorological Organization, Geneva, 62 pp., 72 plates.

WMO, 1986: *Atmospheric Ozone: Assessment of Our Understanding of the Processes Controlling Its Present Distribution and Change*. Report 16, World Meteorological Organization, Global Ozone Research and Monitoring Project, NASA, Washington, DC.

WMO, 1988: *Report of the International Ozone Trends Panel 1988*. Report 18, World Meteorological Organization, Global Ozone Research and Monitoring Project, NASA, Washington, DC.

WMO, 1991: *Scientific Assessment of Ozone Depletion: 1991*. Report 25, World Meteorological Organization, Global Ozone Research and Monitoring Project, NASA, Washington, DC.

Yanai, M., S. Esbensen, and J. Chu, 1973: Determination of the bulk properties of tropical cloud clusters from large-scale heat and moisture budgets. *J. Atmos. Sci.*, **30**, 11–27.

Young, K., 1993: *Microphysical Processes in Clouds*. Oxford University Press, New York, 427 pp.

Answers to Selected Problems

Chapter 1

2. $\frac{V_i}{V_d} = \frac{r_i}{\epsilon_i}$.

3. $\frac{V_i}{V_d} = \frac{n_i}{n_d}$.

6. (a) $N_{H_2O} = 0.0231$, (b) $\overline{M} = 28.71$ g mol^{-1}, (c) $\overline{R} = 289.58$ J kg^{-1} K^{-1},
   (d) $\rho_{H_2O} = 0.0173$ kg m^{-3}, (e) $r_{H_2O} = 0.0147$, (f) $\frac{V_i}{V_d} = 0.0236$.

8. $\Sigma_{H_2O} = \frac{1000}{\rho_w} \int_0^\infty r_{H_2O}\rho_d dz$ (mm).

9. (a) 47 mm, (b) 9.7 mm.

12. $\Delta T_e = 4.4$ K.

13. (a) 0.

16. $T(z) = \frac{gh^2}{2Rz}\left[1 + \left(\frac{z}{h}\right)^2\right]$.

17. 236 K.

Chapter 2

1. (a) 5 km, (b) 0.9 km.

2. (a) $\theta_{950} = 30°$ C.

3. (a) 3.06×10^3 J, (b) 3.10×10^3 J, (c) 4.06×10^4 J.

6. 15.85.

Chapter 3

1. (a) 298 K, (b) -6.3 J K^{-1}, (c) 7.05 J K^{-1}, (d) 0.75 J K^{-1}.

2. (a) -20.7 J K^{-1}, (b) 21.5 J K^{-1}.

3. 0.90.

4. (a) 0.967.

5. (a) 277 K, (b) 285 K.

7. 199 J K^{-1} kg^{-1}.

9. (a) 1.3×10^2 J K^{-1} mol^{-1}, (b) 5.2×10^4 J mol^{-1}.

Chapter 4

6. (a) 95° C, (b) 86° C, (c) 71° C.
7. (a) 599 K.
8. (a) 188 K, (b) in the wintertime Antarctic stratosphere.

Chapter 5

1. (a) 33° C, (b) 10.2 g kg^{-1}, (c) 26%.
3. (a) $T(z, t; x) = T_\infty[z - h(x - ct)] - \Gamma_d h(x - ct)$, (b) $r(z, t; x) = r_\infty[z - h(x - ct)]$.
7. (a) 75%, (b) 25° C, (c) 325 K, (d) 770 mb, (e) 400 K, (f) 22.5 g kg^{-1}, (g) 450 mb.
8. 53%.
9. (a) 53%, (b) 100%.
11. (a) 20.1° C, (b) 32.6° C.
15. (e) 328 K.
16. 9° C.
18. (a) 920 mb, (b) 610 mb.
19. 9%.
20. (a) 5.4%, (b) 0.51 kg, (c) 7.22 × 10^6 J.
22. (a) 13.6 km.
23. 3.4 km.
25. (d) 1.3 km.

Chapter 6

3. $w = -R \oint T d \ln p = -RA.$
4. $w = c_p(T_2 - T_1)\left[1 - \left(\frac{p_{34}}{p_{12}}\right)\right] > 0.$
7. (a) $z = \frac{R}{g}\left\{(\overline{\overline{T}} + \overline{T}\cos\phi)\xi - T'\left(\frac{\xi_T}{\pi}\right)\exp\left[-\frac{\lambda^2 + (\phi - \phi_0)^2}{L^2}\right]\sin\left(\frac{\pi\xi}{\xi_T}\right)\right\}$ $(\xi < \xi_T);$
   $= \frac{R}{g}\left[(\overline{\overline{T}} + \overline{T}\cos\phi)\xi\right]$ $(\xi \geq \xi_T).$
8. 23 K.
9. (a) $z = \frac{R\overline{T}}{g}(\xi - \xi_s)\left\{1 - \frac{\xi_s[1 - e^{\alpha(\xi_s - \xi)}]}{(\xi_s - \xi)}\right\}.$
12. (a) 158 m.

Chapter 7

2. (a) unstable: $z < z_c$, stable: $z > z_c$; $z_c = h_1 \ln \left(\frac{h_1}{h_2} \right)$, (b) $N^2 = g \frac{\frac{1}{h_1} e^{z/h_1} - \frac{1}{h_2}}{e^{z/h_1} - \frac{z}{h_2}}$.

4. $1.24 \ 10^{-4}$ kg s^{-2}, (b) $5.41 \ 10^{-4}$ kg s^{-2}.

5. (a) $\frac{w_0}{N}$.

6. (a) $\theta = T_0 \frac{a}{a+z} \exp\left[\frac{\Gamma_d}{aT_0} \left(az + \frac{z^2}{2} \right) \right]$, (b) unstable: $z < z_c$, stable: $z > z_c$;

$z_c = \sqrt{\frac{aT_0}{\Gamma_d}} - a$.

8. (a) $T_{\text{trigger}} = T_0 + \frac{3\Gamma_d}{2}$, (b) $\Delta t \cong 5$ hr.

10. (a) stable: $z < a$, unstable: $z > a$; (b) $\Delta z = 2a$ (exclusive of instability).

13. 27 km.

14. 15 km.

15. $\frac{d\theta_e}{dz} > 0$ (stable), $\frac{d\theta_e}{dz} = 0$ (neutral), $\frac{d\theta_e}{dz} < 0$ (unstable).

18. 57 m s^{-1}.

19. (a) $TI(z) = 4z - 20$, (b) 5 km, (c) $w_{\text{max}} = 58$ m s^{-1} at 5 km, $z_{\text{max}} = 9.5$ km.

20. (a) $K' > 0$ for $z < 9.8$ km, (b) $K' > 0$ for $z < 5.7$ km, (c) $K' > 0$ for $z < 5.3$ km.

24. (a) 244 K at 13 km.

25. (a) potentially unstable (stable) for $z < 1.6$ km ($z > 1.6$ km), (b) potentially unstable (stable) for $z < 4.8$ km ($z > 4.8$ km).

Chapter 8

1. (a) 254 K, (b) 290 K, (c) 270 K, (d) 240 K.

3. -32.4 K.

4. (a) 4.6 K.

5. (a) $u(z) = \rho_0 rH \exp\left[-\left(\frac{\lambda - \lambda_0}{\alpha} \right)^2 - \frac{z}{H} \right]$.

9. (a) 227 K.

10. (a) 95%, (b) 3.4.

11. (a) 393 K, (b) 393 K.

12. (a) $T_{\text{RE}} = \left(\frac{a_{\text{SW}}}{a_{\text{LW}}} \frac{F_s}{\sigma} \right)^{\frac{1}{4}}$, (b) 278 K.

13. $T(t) = \left(0.12\sigma t + \frac{1}{T_0^3} \right)^{-\frac{1}{3}}$.

14. $\frac{2 - a_{\text{LW}}^{\text{atm}} + a_{\text{LW}}^{\text{atm}} a_{\text{LW}}^{\text{hood}}}{2 - a_{\text{LW}}^{\text{atm}}} a_{\text{SW}}^{\text{hood}} F_s \cos \theta_s = a_{\text{LW}}^{\text{hood}} \sigma T_{\text{hood}}^4$; (a) 298 K, (b) 392 K.

15. (a) 324 K, (b) 212 K.

16. (a) $[a_{\text{SW}} + (1 - a_{\text{SW}})a_{\text{LW}}] \sum_{n=0}^{\infty} \left(\frac{a_{\text{LW}}}{2} \right)^n$, (b) 252 K.

17. (a) 291 K, (b) 252 K, (c) 294 K.

18. (a) 5.0 mb, (b) 6.4 mb.

19. (a) $F_s = F_0 \frac{2}{2-a}$ $(N = 1)$, $F_0 \frac{2+a}{2-a}$ $(N = 2)$, ... $F_0 \frac{2+(N-1)a}{2-a}$
    $= F_0 \left[1 + N \frac{a}{2-a} \right] (N)$.

22. -97 K.

24. (a) $\sigma T_{RE}^4(z) = \frac{F_0}{2} \left\{ \frac{3k}{2g} p_0 \exp \left[-\int_0^z \frac{gdz'}{RT_{RE}(z')} \right] + 1 \right\}$, (b) $T_{RE}(z) \cong$
    $\left[\frac{3kF_0}{4\sigma g} p_0 \exp \left(-\frac{z}{H_s} \right) \right]^{\frac{1}{4}}$, (c) $\left[\frac{3kF_0}{4\sigma g} p_0 \exp \left(-\frac{z_T}{H_s} \right) \right]^{\frac{1}{4}} = T_s - \Gamma_s(0) z_T$.

32. (a) 16 days, (b) 0.

33. (a) 4.6 K, (b) 17.6 K.

Chapter 9

3. (a) $\frac{dS}{d(\log a)} = 4\pi a^2 \frac{dn}{d(\log a)}$, (b) $\frac{dM}{d(\log a)} = \frac{4}{3}\pi a^3 \rho \frac{dn}{d(\log a)}$.

8. $\Delta G = \dfrac{16\pi\sigma^3}{3\left[n_w KT \ln\left(\frac{e}{e_w} \right) \right]^2}$

9. (a) 0.60 μm, (b) 0.19 μm.

10. 1.6 g m^{-3}.

11. (a) ~ 100 m.

15. (a) $w_l = \frac{2}{9} \frac{\rho_w g a^2}{\mu}$, (b) 87 μm, (c) 88 μm.

17. (a) 0.76 years, (b) 76 years.

19. (a) 150 m, (b) 1500 m.

26. $\cos \Theta = \cos \theta \cos \theta' + \sin \theta \sin \theta' \cos(\phi - \phi')$.

27. (a) 0.013, (b) 0.051, (c) 0.301.

28. $\beta_s(\pi) = \frac{64\pi^5}{\lambda^4} \frac{n}{\eta^2} a^6 \left(\frac{m^2-1}{m^2+2} \right)^2$.

35. (a) $a_e = 5.7$ μm, (b) $\Sigma_l = 785$ g m^{-2} (c) $\tau_c = 208$.

36. $\tau_c = 410$.

38. (a) $\Sigma_l = 240$ g m^{-2}, (b) $\rho_l = 0.24$ g m^{-3}.

39. (9.40), but with $C = \dfrac{E_-\alpha_- e^{-\gamma \tau_c^*} - E_+\alpha_+ e^{-\frac{\bar{\mu}}{\mu_s} \tau_c^*} - (1-a)\alpha_+ E_- \left(e^{-\gamma \tau_c^*} - e^{-\frac{\bar{\mu}}{\mu_s} \tau_c^*} \right)}{\alpha_+^2 e^{\gamma \tau_c^*} - \alpha_-^2 e^{-\gamma \tau_c^*} + (1-a)\alpha_+\alpha_- \left(e^{-\gamma \tau_c^*} - e^{\gamma \tau_c^*} \right)}$,

$D = \dfrac{E_+\alpha_- e^{-\frac{\bar{\mu}}{\mu_s} \tau_c^*} - E_-\alpha_+ e^{\gamma \tau_c^*} + (1-a)\alpha_+ E_- \left(e^{\gamma \tau_c^*} - e^{-\frac{\bar{\mu}}{\mu_s} \tau_c^*} \right)}{\alpha_+^2 e^{\gamma \tau_c^*} - \alpha_-^2 e^{-\gamma \tau_c^*} + (1-a)\alpha_+\alpha_- \left(e^{-\gamma \tau_c^*} - e^{\gamma \tau_c^*} \right)}$.

41. $C_{SW} = -131$ W m^{-2}, $C_{LW} = 125$ W m^{-2}, $C = -6$ W m^{-2}.

42. 3.1 mm day^{-1}.

47. Strong forward scattering implies that the reduction of direct transmission is offset by an enhancement of diffuse transmission.

Chapter 10

6. (a) two thermodynamic degrees of freedom, (b) two independent families of surfaces, (c) the families coincide.

10. (c) For a given time interval, the deformation experienced by a material volume approaches zero with its dimension.

14. (a) $\frac{d\rho}{dt} = \nabla \cdot \mathbf{v} = 0$, (b) $\frac{d\theta}{dt} = 0$.

15. $\mathbf{v}$ orthogonal to $\nabla\psi$.

18. (a) $3.5°$, (b) $34.8°$, (c) $348°$, (d) $0.019, 0.19, 1.9$.

Chapter 11

6. $\frac{d}{dt}[(u + \Omega r \cos\phi)r\cos\phi] = -r\cos\phi\left[\frac{1}{\rho r\cos\phi}\frac{\partial p}{\partial\lambda} + D_\lambda\right]$.

11. $\frac{d\mathbf{v}_h}{dt} + f\mathbf{k}\times\mathbf{v}_h = -\nabla_\pi\Phi - \mathbf{D}_h$; $\frac{\partial\Phi}{\partial\pi} = -\theta$; $\frac{\partial}{\partial\pi}\left(\pi^{\frac{1}{\kappa}-1}\dot\pi\right) + \pi^{\frac{1}{\kappa}-1}\nabla_\pi\cdot\mathbf{v}_h = 0$; $\frac{d\theta}{dt} = \frac{\dot q}{\pi}$; $\frac{d}{dt} = \frac{\partial}{\partial t} + \mathbf{v}_h\cdot\nabla_\pi + \dot\pi\frac{\partial}{\partial\pi}$.

19. (a) $u_g = -\frac{1}{fa}\frac{\partial\Phi}{\partial\phi}$; $v_g = \frac{1}{fa\cos\phi}\frac{\partial\Phi}{\partial\lambda}$, (b) $u_g = -\frac{1}{fa}\frac{\partial\Psi}{\partial\phi}$; $v_g = \frac{1}{fa\cos\phi}\frac{\partial\Psi}{\partial\lambda}$.

20. $-\frac{\partial\mathbf{v}_g}{\partial\ln p} = \frac{R}{f}\mathbf{k}\times\nabla_p T$.

21. (a) $w = \zeta h\left\{1 - e^{-z^*/h}\right\}$, (b) convergence inside the boundary layer must be compensated by divergence aloft.

Chapter 12

3. (a) $\mathbf{v}_g = \frac{1}{\rho f}\mathbf{k}\times\nabla_z p$, (b) $\mathbf{v}_g = \frac{1}{f}\mathbf{k}\times\nabla_\theta\Psi$.

4. (a) $u_g = -\frac{1}{f}\frac{\partial\Phi}{\partial y}$; $v_g = \frac{1}{f}\frac{\partial\Phi}{\partial x}$, (b) $u_g = -\frac{1}{fa}\frac{\partial\Phi}{\partial\phi}$; $v_g = \frac{1}{fa\cos\phi}\frac{\partial\Phi}{\partial\lambda}$,
   (c) $u_{g\phi} = \frac{\partial\Phi}{\partial r}$; $u_{gr} = -\frac{1}{r}\frac{\partial\Phi}{\partial\phi}$.

6. (a) $\mathbf{k}\cdot(\nabla\times\mathbf{v}_g) = \frac{1}{f}\nabla^2\Phi + \frac{\cot\phi}{a}u_g$, (b) $\nabla\cdot\mathbf{v}_g = -\frac{\cot\phi}{a}v_g$.

7. 105 m s^{-1}.

10. (a) $v(r) = \frac{\text{const}}{r}$, (b) $\Gamma(r) = 2\pi\cdot\text{const} \neq f(r)$, (c) vorticity is concentrated at $r = 0$ in a point vortex.

11. (b) Diffusion.

13. (a) $v_\parallel = \frac{f^2 v_g}{K^2+f^2}$, $v_\perp = \frac{Kf v_g}{K^2+f^2}$, (b) $\delta = \tan^{-1}\left(\frac{K}{f}\right)$, (c) $6.4°$, (d) $17.1°$.

14. $\nabla\cdot\mathbf{v}_h = -\frac{K}{K^2+f^2}\nabla^2\Phi$.

15. $4.6°$.

16. (a) $\mathbf{v} = \frac{2}{L^2}\exp\left(-\frac{x^2+y^2}{L^2}\right)[-(\Psi y + Xx)\mathbf{i} + (\Psi x - Xy)\mathbf{j}]$,
   (b) $\nabla\cdot\mathbf{v} = \frac{4}{L^2}X\exp\left(-\frac{x^2+y^2}{L^2}\right)\left[\frac{x^2+y^2}{L^2} - 1\right]$,
   (c) $\mathbf{k}\cdot(\nabla\times\mathbf{v}) = -\frac{4}{L^2}\Psi\exp\left(-\frac{x^2+y^2}{L^2}\right)\left[\frac{x^2+y^2}{L^2} - 1\right]$.

17. (a) $\omega = -\rho g w$, (b) $w = -18$ mm s^{-1}.

18. (a) $v_{500} = \frac{g}{f}\left\{\left[\frac{\overline{z}_{500}}{a} - \frac{2\pi y}{L^2} z'_{500} \sin\left(\pi \frac{x^2+y^2}{L^2}\right)\right]i + \left[\frac{2\pi x}{L^2} z'_{500} \sin\left(\pi \frac{x^2+y^2}{L^2}\right)\right]j\right\}$,

    (c) $(\zeta + f) = \frac{g}{f}\frac{4\pi}{L^2} z'_{500}\left[\sin\left(\pi \frac{x^2+y^2}{L^2}\right) + \pi \frac{(x^2+y^2)}{L^2} \cos\left(\pi \frac{x^2+y^2}{L^2}\right)\right] + f_0$.

24. (a) 16 m s^{-1}, (b) 14.3 m s^{-1}.

25. (a) 768 m s^{-1}, (b) 45 m s^{-1}.

27. (b) $\zeta = \frac{2y}{Y^2} \exp\left[-\left(\frac{y}{Y}\right)^2\right] \exp[i(kx - \sigma t)]$;

    $\nabla \cdot v_h = ik \exp\left[-\left(\frac{y}{Y}\right)^2\right] \exp[i(kx - \sigma t)]$;

    (c) $\nabla^2 \chi = ik \exp\left[-\left(\frac{y}{Y}\right)^2\right] \exp[i(kx - \sigma t)]$; $\psi = \frac{i}{k}\frac{\partial \chi}{\partial y}$.

30. (a) 950–850: $\frac{\partial \overline{T}}{\partial x} = 0$, $\frac{\partial \overline{T}}{\partial y} = -0.032$ K km^{-1}; 850–750: $\frac{\partial \overline{T}}{\partial x} = 0$,

    $\frac{\partial \overline{T}}{\partial y} = +.028$ K km^{-1}, (b) 950–850: 32 K day^{-1}; 850–750:

    -28 K day^{-1},

    (c) 63 K km^{-1}.

32. Positive (negative) feedback between changes of $-\frac{\partial \Phi}{\partial n}$ and $\frac{v^2}{R}$.

35. $p = p_0 \exp\left(\frac{\Omega^2 r^2}{2RT}\right)$.

36. (a) $v = \pm \left(\frac{r}{L}\right)\sqrt{2\epsilon \overline{\Phi}} e_\phi$.

37. (a) $z_{500} = \frac{R}{g} \ln(2)\left\{\left[\overline{T} - T\left(\frac{y}{a}\right)\right] - \exp\left(-\frac{x^2+y^2}{L^2}\right)\left[A + \frac{B}{2} \ln(2)\right]\right\}$,

    (b) $u_{500} = -\frac{R}{f} \ln(2)\left\{-\frac{\overline{T}}{a} + \frac{2y}{L^2} \exp\left(-\frac{x^2+y^2}{L^2}\right)\left[A + \frac{B}{2} \ln(2)\right]\right\}$,

    $v_{500} = \frac{R}{f} \ln(2)\left\{\frac{2x}{L^2} \exp\left(-\frac{x^2+y^2}{L^2}\right)\left[A + \frac{B}{2} \ln(2)\right]\right\}$, (c) $p \cong 430$ mb.

38. 33°.

39. (a) $\frac{\partial^2 y}{\partial x^2} + \frac{\beta}{u} y = 0$, (b) $y(x) = Ae^{i\sqrt{\frac{\beta}{u}}x} + Be^{-i\sqrt{\frac{\beta}{u}}x}$.

45. $\frac{d\zeta}{dt} \cong -f\nabla \cdot v_h$.

47. (a) $\zeta = -7.5 \times 10^{-5}$ s^{-1}, (b) $\zeta = -8.3 \times 10^{-5}$ s^{-1}.

48. (b) Solenoidal production is absent in isobaric coordinates as well, but
tilting remains.

50. (a) $dM = -\frac{d\theta}{g}\left(\frac{\partial p}{\partial \theta}\right) dA = \text{const} \cdot \left(\frac{\partial p}{\partial \theta}\right) dA$, (b) $\frac{d\Gamma}{dt} = -\frac{d(fdA)}{dt}$.

54. (c) Diffusion.

Chapter 13

1. (a) 0.003, (b) 10^8, (c) 5.0×10^{11}.

3. 87 km for $H = 7.3$ km.

5. 10^5 MW (!).

15. 8.9 mm s^{-1}

16. 0.08 N m^{-2}.

17. (a) $K\bar{u} - f\bar{v} = 0$; $K\bar{v} + f(\bar{u} - u_g) = 0$; $K = \frac{c_d|\bar{v}|}{h}$, (b) $\bar{v} = \frac{u_g}{1+\frac{K^2}{f^2}}(i + \frac{K}{f}j)$,

   (c) 0.46 days.

18. (a) 20 m s^{-1}km^{-1}, (b) 50 m s^{-1} km^{-1}.

19. 4.5 days.

20. (a) $\frac{dr_t}{dx} + \frac{1}{L}r_t = \frac{1}{L}r_a$, (b) $r_t(x) = r_0 e^{-x/L} + r_a$; $T(x) = T_0 e^{-x/L} + T_a$,

   (c) $r_w(x) = \frac{0.622}{200} 10^{9.4 - \frac{2354}{T_0 \exp\left(-\frac{x}{L} + T_a\right)}}$.

Chapter 14

9. (a) $k = \left[k_0^2, \frac{\sigma^2}{gH(y)} - k_0^2\right]$.

19. (a) k and c_g are both horizontal, (b) k points downward and c_g is horizontal.

24. (a) vertical propagation: $z < z_t$, external: $z \geq z_t$; $z_t = h\sqrt{1 - \frac{k^2\bar{u}^2}{N_0^2}}$,

   (c) $\bar{u} > \frac{N_0}{k}$.

28. (a) $z_t = 18$ km, (c) wave activity disperses for $\Sigma \to \infty$.

30. (a) 0.0004 mb, (b) 5.1 m s^{-1}.

32. $|P_k^\sigma|^2 = \frac{\kappa^2}{H^2}\frac{|W_k^\sigma|^2 e^{z/H}}{\sigma^2\kappa(\kappa-1)+N^2k^2H^2} + \frac{\kappa^2}{\sigma^2 H^2}\frac{|W_k^\sigma|^2 e^{\frac{\dot{z}}{H}-2\hat{m}z}}{[\frac{1}{2}-\kappa-\hat{m}H]^2}$; $-\hat{m}^2 + \frac{1}{4H^2} = \frac{N^2}{\sigma^2}k^2$.

41. (a) Through divergence, the mountain acts as a vorticity source,

   (b) $\frac{D}{Dt}\nabla^2\psi' + \beta\frac{\partial\psi'}{\partial x} = \frac{Q_0}{2\pi L^2}\exp\left(-\frac{x^2+y^2}{2L^2}\right)$, (c) $\Psi_{sl} = \frac{i}{k}Q_0\frac{\exp\left[-\frac{(k_s^2+l^2)L^2}{2}\right]}{\bar{u}(k_s^2+l^2)-\beta}$;

   $k_s l > 0, y > 0$; $k_s l < 0, y < 0$.

46. (a) $\frac{1}{a^2\cos\phi}\frac{\partial}{\partial\phi}\left(\cos\phi\frac{\partial\psi'}{\partial\phi}\right) + \frac{1}{\bar{\rho}}\frac{\partial}{\partial z}\left(\frac{f^2}{N^2}\bar{\rho}\frac{\partial\psi'}{\partial z}\right) + \nu^2\psi' = 0$;

   $\nu^2 = \frac{\beta_e}{\bar{u}-c} - k^2$; $\beta_e = \beta - \frac{1}{a^2\cos\phi}\frac{\partial}{\partial\phi}\left(\cos\phi\frac{\partial\bar{u}}{\partial\phi}\right) - \frac{1}{\bar{\rho}}\frac{\partial}{\partial z}\left(\frac{f^2}{N^2}\bar{\rho}\frac{\partial\bar{u}}{\partial z}\right)$, (b) 74°,

   (c) 45°.

47. With $\beta_e = \frac{\partial\bar{Q}}{\partial y}$ driven to zero, reflection can occur at both high and low latitudes, supporting the existence of horizontal modes. Since propagation is also trapped vertically, normal modes can exist as well.

51. (a) $m_b = \frac{\bar{\theta}}{\Theta}\frac{N^2}{g}$, (b) $\frac{\bar{\theta}}{g\Theta} + \frac{k^2}{N^2} - \frac{1}{(c-\bar{u})^2} = 0$.

Chapter 15

4. $d\epsilon = c_p T d\ln\theta_e$.

5. -9.8×10^4 J kg^{-1}.

11. (b) 2.6×10^{21} J.

12. 0.0013.

13. 81 m s^{-1}.

18. (b) Phase lines tilt SW–NE in the Northern Hemisphere, so that wave activity propagates equatorward with $c_{gy} < 0$.

19. (a) 2.9 m s^{-1}.

23. (a) 6 days.

26. (a) 9.2×10^3 J kg^{-1}, (b) 7.8×10^2 J kg^{-1}, (c) 13.3 km; 4.4 km.

Chapter 16

3. $-f\left(f - \frac{\partial \bar{u}}{\partial y}\right) y'$.

4. (a) low latitudes.

5. (a) positive, (b) $\phi = \cos^{-1}\left(\sqrt{\dfrac{\epsilon(z)}{1 - \frac{f_0^2}{N^2} a^2 \frac{\partial^2 \epsilon}{\partial z^2}}}\right)$.

16. $\frac{\alpha_c H}{2} = \coth\left(\frac{\alpha_c H}{2}\right)$.

17. (b) $(kc_i)_{k=l} = \frac{1}{\sqrt{2}} (kc_i)_{l=0}$.

19. 6.1 days, 3.2 days, 2.25 days, 1.86 days, 1.71 days, 1.77 days, 2.22 days, ∞.

25. (a) unstable (stable) equatorward (poleward) of $25°$.

Chapter 17

1. $\tau_{O_3} = \frac{1}{J_3 + k_3[O]}$; $\tau_{O_x} = \frac{1}{2k_3[O]}$, (a) $\tau_{O_3} = 37$ min; $\tau_{O_x} = 32$ years, (b) $\tau_{O_3} = 21$ min; $\tau_{O_x} = 15$ days, (c) $\tau_{O_3} = 2.4$ min; $\tau_{O_x} = 3.8$ hr.

9. (a) 16.4 ppmv, (b) 1.6 ppmv, (c) 3.3 ppmv.

11. 1.26 ppmv.

13. (a) $\tau_{O_x} = \frac{1}{2k_3[O] + 2k_a[Cl]}$; $[O] = \frac{J_2[O_2]}{k_3[O_3] + k_b[ClO]}$, (a) 8 years, (b) 3.1 days.

Index

Absolute concentration, 6–7
Absolute humidity, 26, 118
Absolute vorticity, 388, 389–392
Absorption
  atmospheric waves, 468–472
  by gases, characteristics, 216–224; *see also*
    specific gases
  graybody, 211–212
  heat, 65–66, 81–87
  plane parallel atmosphere, 226, 228–230,
    232
  radiation, LW, 46, 200, 469, 472
  radiation, SW, 48–49, 200, 243, 244, 263,
    307
  radiative, 199–200, 207–208, 243–244, 246
  strong, 230, 232
Absorption coefficient, 207
Absorption cross section, 207
Absorptivity, 208, 211, 216
  cloudy atmosphere, 298, 300
  plane parallel atmosphere, 229
Acceleration, 335–336
  centrifugal, 143, 144–145, 336
  centripetal, 384–385
  Coriolis, 351, 355, 371, 385
  material, 347–348, 355, 382–384, 409
Accumulation particle, 261
Acid rain, 41
Acoustic cutoff frequency, 441
Acoustic-gravity wave, 441, 444
Acoustic wave, 439–446
Activated droplet, 268
Adiabat, 69, 71, 128
  pseudo-adiabatic process, 132, 134, 135
Adiabatic cooling, 283
Adiabatic layer, 155
Adiabatic process, 58, 63, 68–75, 128–129
  Carnot cycle, 82–84
  control surface, 58

entropy, 87, 93
equation of motion, 352
equilibrium conditions, 91
isentropic coordinates, 364, 367
lapse rate, 171
  dry, 74, 135, 151
  saturated, 132–133, 151
path independence, 63
potential energy, 493–495
potential temperature, 71–73, 74, 93, 352
pseudo-adiabatic, 130–132, 134–148
stratification, 130–131, 156–160
Advection, 27, 552
  material derivative, 328, 338
Aerosol, 33, 35, 46; *see also* Clouds, droplet
    growth
  chemical processes, role, 314–315
  cloudy atmosphere, 295–305
  global energy budget, role, 312–314
  ice nuclei, 275–276
  LW and SW radiation, 244, 263, 312–313
  Mie scattering, 291–295
  morphology, 258–264, 270–271
  thermal structure, 275–305
  as trace constituent, 35
Ageostrophic motion, 394, 422, 502
Air, *see also* Dry air; Moist air
  compressibility, 2, 6–7, 152–153
  constituents, 3–4, 12, 14; *see also specific*
    constituents
  internal energy, 44
Air parcel, 1, 32
Aitken nuclei, 261
Albedo, 305
  aerosol, 312
  cloud, 298–299, 300, 307
  global energy budget, 48–49, 307
  Mie scattering, 293
  radiative equilibrium, 41, 43

Albedo (*continued*)
  Rayleigh scattering, 290
  single scattering, 214
Aleutian high, 374
Altitude, *see also* Geopotential height;
    Height
  coordinate systems, 146
  pseudo-adiabatic chart, 134
Amazon basin, global energy budget, 307
Angle
  scattering, 288–289
  zenith, *see* Zenith angle
Angular momentum
  absolute vorticity, 391
  coordinate systems, 350, 352
  general circulation, 486, 502
  quasi-biennial oscillation, 557
Angular velocity
  earth, 579
  reference frame, 334
Answers, to selected problems, 593–600
Antarctic plateau, 552
Anthropogenic impact
  aerosol, 35, 258, 262, 271
  albedo, 312
  carbon dioxide, 23–24
  chlorofluorocarbons, 33, 542
  greenhouse effect, 252
  heterogeneous chemistry, 567, 569
  methane, 32
  nitric oxide, 35, 541
  nitrous oxide, 34–35
Anticyclone, 187, 192; *see also* Cyclone
Anticyclonic motion, 374; *see also* Cyclonic
    motion
  geostrophic equilibrium, 380, 386–387, 388
  stratospheric warming, 554–556
  tropical circulation, 508
Anvil cirrus cloud
  global energy budget, 307
  radiative cooling, 304
Archimedes' principle, 167
Arctic haze, 312
Asymmetry factor, scattering, 297
Asynoptic satellite observation, 404
Atacama desert, LW radiation, 211
Atmospheric behavior, *see also* Atmospheric
    motion
  influencing mechanisms, 2–3
  perspectives, 1–2, 321–323
Atmospheric energy, *see* General circulation,
    atmospheric energy

Atmospheric motion, *see also* Equations of
    motion; Horizontal motion; Large-scale
    motion; Vertical motion
  Cauchy's equations, 331–334
  conservation of mass, 330–331
  descriptions, 321–323
  first law of thermodynamics, 336–338
  fluid motion, kinematics, 323–328
  material derivative, 328
  momentum budget, 331–336
  Reynolds' transport theorem, 329–330
  rotating reference frame, 334–336
Atmospheric window, 200, 203
  absorption characteristics, 216
Auxiliary function, 88
Avogadro's number, 579
Axisymmetric circulation, 495–503

Band, spectrum, 218–220
Band absorptivity, 229
Band model, 217, 229–230
Band transmissivity, 228
Baroclinic instability, 493, 505, 521–523; *see
    also* Eady problem
Baroclinic stratification, 378–379, 389–393
  potential energy, 491, 494
Barotropic instability, 521–523
Barotropic nondivergent motion, 387–388
Barotropic nondivergent vorticity equation,
    388
Barotropic nondivergent wave, 459–461
Barotropic stratification, 378
Beta effectiveness, 466
Beta-plane approximation, 394
Blackbody radiation theory, 208–212
Blackbody spectrum, 209
Blackbody temperature, 43, 44, 208–212
Boltzmann constant, 579
Boundary layer, *see* Planetary boundary
    layer
Boussinesq approximation, 394–395
Break temperature, 193
Brewer–Dobson circulation, 547–548
Brightness temperature, 209
Brunt-Väisäillä frequency, 175
Bulk viscosity, 340
Buoyancy
  cloud dissipation, 286–287
  cloud formation, 277–281
  conditional instability, 178–179

Buoyancy (*continued*)
 entrainment, 181, 184
 vertical stability, 166–170, 174–176, 189
Buoyancy wave, 440–456

CAPE, *see* Convective available potential
 energy
Carbonaceous aerosol, 258, 260, 262
Carbon dioxide (CO_2)
 absorption by, 200, 201, 203, 216–217, 232,
  244
 greenhouse effect, 249–250, 252
 line broadening, 224
 LW radiation, 46, 242
 as trace constituent, 22–25
Carbon monoxide (CO), absorption by, 216
Carnot cycle, 82–84, 159
Carnot's theorem, 84–85
Cartesian coordinates
 rectangular, 341
 transformations, 341–342, 344–346
Cauchy's equations of motion, 331–334
CCN, *see* Cloud condensation nuclei
Centrifugal acceleration, 143, 144–145, 336
Centripetal acceleration, 384–385
CFCs, *see* Chlorofluorocarbons
Chapman layer, 243, 246, 304
Chappuis band, 219, 220
Chemical equilibrium, 118, 120; *see also*
 Photochemical equilibrium
 Gibbs function, 102–103, 105–106
 heterogeneous system, 99, 102–104
Chemical potential, 102–104, 105, 109
Chinook wind, 138
Chlorine chemistry, 35, 41; *see also*
 Chlorofluorocarbons; Heterogeneous
 chemistry
 odd chlorine family, 543–544
Chlorofluorocarbons (CFCs)
 absorption by, 216, 220
 greenhouse effect, 252
 heterogeneous chemistry, 569
 increased levels, 33–34
 middle atmosphere, 542–544
 mixing ratio, 33
 ozone destruction, 34
 radiative equilibrium, 43
 as trace constituents, 33–34
Chromosphere, 201
Circulation, 2, 41; *see also* Atmospheric
 motion; General circulation;

Large-scale motion; Meridional
 circulation; Middle atmosphere
Brewer–Dobson, 547–548
Hadley, 39, 51, 505
secondary, 395, 420
structure, 17–22
thermally direct, 51
time-mean, 19–21, 51
Walker, 51
zonal-mean, 17–18
Circumpolar flow, 18–19, 21, 51, 150
Cirriform cloud, 277, 307
Cirrus cloud, 300, 304, 305
 dissipation, 287
 global energy budget, 307
 microphysical properties, 285–286
 Mie scattering, 295
Cirrus uncinus cloud, 286
Clausius–Clapeyron equation, 112–114
 distribution of water vapor, 122
 greenhouse effect, 250
 saturation properties, 120
Clausius inequality, 85–86
Closed system, 55
Cloud condensation nuclei (CCN), 268,
 270–271, 275
 chemical species, 314
Cloud cover, 250, 306–312
Cloud radiative forcing, 308–310
Clouds, 39, 41, 259; *see also specific cloud
 types*
 climate, role in
  chemical processes, 314–315
  global energy budget, 306–312
 droplet growth
  collision, 273–275
  condensation, 264–273, 274
 ice particle growth, 275–276
 IR radiation, 36–38, 43
 LW radiation, 46, 280, 281, 287, 300–302,
  304–305
 macroscopic characteristics
  dissipation, 286–287
  formation and classification, 277–285
  microphysical properties, 285–286
 microphysics, 264–276
 optical properties, 299–305
 radiative transfer
  cloudy atmosphere, 295–305
  Mie scattering, 291–295
 structure, 36–41
Coagulation, ice particle growth, 276

Coefficient
  absorption, 207
  drag, 424
  extinction, 214, 293
  isobaric expansion, 125
  scattering, 293
  viscosity, 333
Cold cloud, 275
Collection efficiency, 273
Collector drop, 273–274
Collision
  droplet growth by, 273–275
  ice particle growth by, 276
  molecule, critical level, 13
Collisional broadening, 223, 224
Combustion aerosol, 258, 260, 261
Compressibility, air, 2, 6–7, 152–153
Concentration
  absolute, 6–7
  relative, 7
Condensation, 126–130; *see also* Lifting
    condensation level; Moist air; Water
    vapor
  aerosol formation, 261
  droplet growth, 264–273, 274
  ice particle growth, 276
  pseudo-adiabatic chart, 137–138
  pseudo-adiabatic process, 130–132
Conditional instability, 173, 176–181, 189
Conservation of mass, 330–331
Conserved property, 7–8
Constants, physical, 579
Continental aerosol, 258, 260–263, 270–271
Continental cloud
  droplet growth, 271
  glaciation, 275
  reflective properties, 300
Continuity equation, 331, 338
  incompressible motion, 352
  isentropic coordinates, 366–367
  isobaric coordinates, 360, 361
  log-pressure coordinates, 363
  quasi-geostrophic motion, 396
  turbulence, 409
Control surface, 55, 58
Convection, 27
  boundary layer, 419
  cumulus, 39, 190–191
  entrainment, 181, 183–184
  free, 178, 179, 189, 415, 508
  Rayleigh–Bénard, 282
  sloping, 39, 505

stratification, 159
  vertical stability, 176, 189
Convective available potential energy
    (CAPE), 179, 189
Convective cloud, 38, 127, 136
Convective mixing, 189–191, 192–193
Convective overshoot, 180–181
Convective overturning, 236–240
Conversion, to SI units, 577
Cooling, *see also* Longwave radiation;
    Shortwave radiation
  adiabatic, 283
  cirrus, 304
  evaporative, 45
  Newtonian, 76, 248, 469, 472, 552
Cooling rate, 160–161
Cooling to space, 241, 246–248
Cool-to-space approximation, 246
Coordinates
  Cartesian, 341–342, 344–346
  curvilinear, 341–344, 581–582
  cylindrical, 581–582
  fluid motion, kinematics, 323–326
  geopotential, 145–147, 336
  isentropic, 364–368, 372
  isobaric, 151, 357–362
  log-pressure, 362–364
  orthogonal, 343, 344, 346
  spherical, 146, 344–352, 581
  thermodynamic, 357–368
  trajectory, 383
Coriolis acceleration, 335–336, 351, 355, 371
  cyclostrophic motion, 385
  inertial motion, 385
Coriolis force, 336
  general circulation, 502
  geostrophic equilibrium, 372–374
  Rossby waves, 459
Coriolis parameter, 351
Critical level, 13–15
  gravity waves, 473
  Rossby waves, 465
  wave activity, 450, 454–456
Critical line
  quasi-biennial oscillation, 557–560
  shear instability, 522
Critical point, water, 108
Critical radius, droplet formation, 268
Critical supersaturation, 268
Crustal aerosol, 258, 260, 261, 262, 275
Cumuliform cloud, 277; *see also* Cumulus
    cloud

Cumulonimbus cloud, 180–181, 277,
    280–281
  global energy budget, 307
  microphysical properties, 285–286
  troposphere-stratosphere interaction, 562,
    563
Cumulus cloud, 39, 128; *see also*
    Stratocumulus cloud
  albedo, 299
  dissipation, 286, 287
  droplet growth, 271, 274
  formation, 277–281
  global energy budget, 307, 310, 311
  microphysical properties, 285–286
  Mie scattering, 293–295
  transmissivity, 299
Cumulus convection, 39, 190–191
Cumulus tower, 180–181
Curl, motion field, 326
Curvature effect, droplet growth, 268
Curvilinear coordinates, 341–344, 581–582
Curvilinear motion, 382–387
Cyclic process, 59, 62; *see also* Circuit, heat
    transfer
  first law, 64–65
  second law, 80–82, 85
Cyclic work, 57
Cyclone, 20, 72; *see also* Anticyclone
  Eady model, 525
  potential energy, 493
  vertical instability, 188
Cyclonic motion, *see also* Anticyclonic
    motion
  geostrophic equilibrium, 374, 380,
    386–387, 388
  Rossby waves, 460
Cyclostrophic equilibrium, 385–386
Cylindrical coordinates, 581–582

Dalton's law, 5
Deforestation, 191
Deformation tensor, 326
Degrees of freedom, 59
  barotropic stratification, 378
  heterogeneous system, 99, 106–107
  homogeneous system, 99
  moist air, 123
  water, 108–109
Density, number, *see* Number density
Deposition, 276
Desert dust, 312–314
Dew point, 121, 135–136

Diabatic process, 75–76, 156
  changes of state, 63–65
  entrainment, 181
  momentum budget, 332
  stratification, 160–162
Diagnostic equation, 387
Diffuse flux transmission function, 229
Diffuse radiation, 205, 295–296; *see also*
    Scattering
Diffusive transport, 11–12
Diffusivity, 232, 333, 411, 412
Dilatation, motion field, 326
Dipole moment, scatterer, 287
Direction, *see also* Reversibility
  thermodynamic process, 85, 87, 90
Dispersion
  turbulent, 174, 189–193
  wave, 434–438
Dissipation, 392
  cloud, 286–287
  frictional, 50, 52
Dobson units, ozone abundance, 31
Doppler broadening, 224
Doppler shift, 432, 447, 472
Drag
  frictional, 381
  Stokes, 317
  turbulent, 420
Drag coefficient, 424
Drag force, 334
Droplet growth, 264–275
Dry adiabatic lapse rate, 74, 135, 151
Dry air, 4, 7
  equation of state, 118
  thermodynamic properties, 578
Dynamical structure, 16–22
Dynamic condition, free surface, 428–429

Eady problem, 523–529, 531
Earth
  angular velocity, 579
  radius, 579
  rotation, *see* Rotation, earth
Eddy diffusivity, boundary layer, 411, 412
Eddy momentum, 409–411, 428
Edge wave, 434, 442, 526
Effective gravity, 143–145, 336; *see also*
    Height
Eigenfrequency, shear instability, 520
Einstein notation, 325
Ekman balance, 498, 501
Ekman layer, 412–413, 501, 502, 503

Ekman number, 495–496
Ekman pumping, 420–423
Ekman spiral, 413, 499
El Chichon volcano, 263, 315
Elevation, *see* Altitude; Geopotential height;
    Height
El Niño, 467–468, 509–510
El Niño Southern Oscillation (ENSO), 510
Emission
    Lyman-α, 201
    radiative, 208–212
        plane parallel atmosphere, 226, 228, 232
Emissivity, 211
Energy transfer, *see* General circulation;
    Global energy budget; Heat transfer;
    Radiation; Radiative transfer
ENSO, *see* El Niño Southern Oscillation
Enthalpy, 65, 110–111
    ideal gas law, 67
    specific heat capacity, 66
Entrainment, 181, 183–184
    cloud dissipation, 286–287
    cloud formation, 280
Entropy
    changes, 87–88
    potential temperature, 91–96, 337
    saturated behavior, 126
    second law, 84–87, 91–96
Environmental lapse rate, 151–153, 168
Environmental temperature, 169–170
Equations of motion, 331–334, 338; *see also*
    Perturbation equation
    adiabatic motion, 352
    balances
        scale analysis, 354–357
        stratification, 353–354
    Cauchy's, 331–334
    coordinates
        curvilinear, 341–344
        spherical, 344–352
        thermodynamic
            isentropic, 364–368
            isobaric, 357–362
            log-pressure, 362–364
    incompressible motion, 352
    traditional approximation, 350–352
    turbulence, 407–408, 409
Equations of radiative transfer, 215, 233, 295
Equations of state, 5–7, 59
    dry air, 118
    heterogeneous system, 106, 112–113
    moist air, 119
    water, 107–108, 112, 113

Equilibrium
    stable, 90
    true, 90, 91
Equilibrium phase transformation, 110–114
Equilibrium supersaturation, 268, 269
Equilibrium temperature, 236
Equilibrium vapor pressure, 109, 112–114
Equivalent potential temperature, 131–132,
    136
Ertel potential vorticity, 392
Escape velocity, 13–15
Eulerian derivative, 328
Eulerian description
    versus Lagrangian, 1–2, 322–323, 324
        material derivative, 328, 329–330
Evanescent wave, 441–442
Evaporation
    cloud dissipation, 286–287
    droplet growth, 265–268
Evaporative cooling, 45
Exact differential theorem, 61
Exchange integral, 241
Exner function, 369
Exosphere, 13–15
Expansion work, 56–58
    adiabatic conditions, 63, 73
    diabatic conditions, 75
Exponential kernel approximation, 232
Extensive property, 55–56, 99–101
External wave, 441–442
Extinction, 206
Extinction coefficient, 214, 293
Extinction cross section, 213–214
Extinction efficiency, 292

Fair weather cumulus cloud, 281
Feedback, greenhouse effect, 250–252
Ferrell cell, 502–503
Field variable, 322, 331, 338; *see also*
    Atmospheric motion
Final warming, 557
Finite-energy condition, 457
First law of thermodynamics, 63–65, 85
    atmospheric motion, 336–338
    forms, 66
        adiabatic process, 68–69, 132
        diabatic process, 75, 76
        isochoric process, 87
Fluid motion, 1–2, 321–328; *see also*
    Atmospheric motion
Flux Richardson number, 415
f-plane motion, 374–375

Forcing
  cloud radiative, 308–310, 311
Fourier's law of heat conduction, 337
Fourier synthesis, 429–432
Fourier transform, 431
Fractus cloud, 282
Free atmosphere, 405
Free convection, 178, 179, 189, 415, 508
Free energy, 89, 265–267
Free surface, 426–429
Freezing level, 136–137
Freezing nucleus, 275
Frictional convergence, 420, 421
Frictional dissipation, 50, 52
Frictional geostrophic equilibrium, 381
Friction velocity, 413
Frost point, 121
Fundamental relations, 88–90
  heterogeneous system, 104–106

Gases, thermodynamics, *see*
    Thermodynamics of gases
Gas phase properties, 117–120
Gas-to-particle conversion, 35
General circulation, 50–52, 486–487; *see also*
    Circulation
  atmospheric energy
    moist static, 487–489
    potential, available, 491–495
    potential, total, 489–491
  heat transfer
    axisymmetric circulation, 495–503
    laboratory simulations, 503–506
  radiative transfer, 198
  tropical, 506–513
General circulation model (GCM), 24, 362
Geological evidence
  carbon dioxide, 24–25
  ozone, 29–30
Geometric altitude, *see* Altitude
Geometric optics method, 451–456
Geopotential coordinates, 145–147, 336
Geopotential height, 146–147; *see also*
    Gravity; Potential energy
Geostrophic balance, 377
Geostrophic equilibrium
  curvilinear motion, 382–384, 386–387
  frictional, 381
  large-scale motion, 371–377
  thermal wind balance, 380
  weakly divergent motion, 388, 394–395

Geostrophic motion, 372, 374, 377–381; *see
    also* Quasi-geostrophic motion
Geostrophic paradox, 372, 486
Geostrophic streamfunction, 376, 461
Geostrophic velocity, 371–373, 374, 378–379
  quasi-geostrophic motion, 394
Geostrophic wind, 370, 377–381
Gibbs–Dalton law, 117, 120
Gibbs free energy, 265–267
Gibbs function, 88–89
  chemical equilibrium, 102–103, 105–106
  equilibrium conditions, 91
Gibbs' phase rule, 107, 109, 123
Glaciated cloud, 275
Global energy budget, *see also* Momentum
    budget; Vorticity budget
  aerosol, role, 312–314, 315
  clouds, role, 306–312
  global-mean energy balance, 44–46
  radiative transfer, 198
Global-mean energy balance, 44–46
Global warming, 23–24
Governing equations, *see* Continuity
    equation; Equations of motion;
    Equations of state; Perturbation
    equation
Gradient equilibrium, 386
Gradient motion, 386–387
Gradient Richardson number, 417
Gradient wind speed, 386
Grauple, 276
Gravity, 2, 8, 44; *see also* Height; Hydrostatic
    equilibrium; Potential energy
  centrifugal acceleration, 336
  effective, 143–145, 336
  escape velocity, 13–15
  standard, 579
Gravity wave, 440–450
  critical level, 473
  dynamical structure, 22
  nonlinear conditions, 477
Gray atmosphere, *see also* Plane parallel
    atmosphere
  greenhouse effect, 248–249
  radiative equilibrium, 233–236
Graybody absorption, 211–212
Greenhouse effect, 43, 203, 248–252
  global energy budget, 45, 306
  runaway, 250
Group speed, 435
Group velocity, 436
  acoustic waves, 439–440
  gravity waves, 441

Group velocity (*continued*)
Lamb waves, 458
Rossby waves, 462

Hadley cell
general circulation, 486–487, 501, 502
tropical circulation, 506
Hadley circulation, 39, 51, 505
Hartley band, 219, 220, 243, 537
Haze, 270, 312
Heat engine, 64, 82, 83
Heat transfer, 3, 58; *see also* Circulation;
General circulation; Global energy
budget; Radiation; Radiative transfer;
Thermal equilibrium
axisymmetric circulation, 495–503
diabatic stratification, 160–161
meridional, 486
noncompensated, 89–90
polytropic processes, 156
simulations, 503–506
vertical stability, 188–189
Height, 150–155, 161
entrainment, 183, 184
geopotential, 146–147
scale, 9, 526
as vertical coordinate, 349, 350, 362
Helmholtz function, 88–89, 91
Helmholtz theorem, 375–377
Hermitian property, 431, 435
Herzberg continuum, 218
Heterogeneous chemistry
aerosol, 35, 314–315
clouds, 41
middle atmosphere, 564, 567–573
Heterogeneous nucleation, 268, 275
Heterogeneous state, 110
Heterogeneous system, 56, 99–102
chemical equilibrium, 99, 102–104
degrees of freedom, 99, 106–107
equilibrium phase transformations,
110–114
fundamental relations, 104–106
thermodynamic characteristics, 107–109
Heterosphere, 12–13
Historical record
carbon dioxide, 24–25
ozone, 29–30, 571, 572
Homogeneous nucleation, 264–268, 269, 275
Homogeneous state, 66
Homogeneous system, 56, 99
characteristics, 108–109
heterogeneous subsystem, 106

Homopause, 11, 12
Homosphere, 11, 13, 147
Horizontal distribution, radiative transfer,
46–50
Horizontal motion, *see also* Atmospheric
motion; Large-scale motion; Vertical
motion
frictional drag, 381
geostrophic balance, 377
isobaric coordinates, 151
quasi-, 351, 352, 389
scale analysis, 354–355
traditional approximation, 350–351
Horizontal shear, 376, 391
Horizontal velocity, 371
Huggins band, 219, 220, 537
Human activity, *see* Anthropogenic impact
Humidity
absolute, 26, 118
relative, 121, 135, 262
specific, 118, 120
Hydrocarbons, aerosol, 258, 262
Hydrodynamic instability, *see* Instability
Hydrogen
depletion, 15
escape velocity, 15
Lyman-α emission, 201
Hydrostatic balance, 9, 147–151
Hydrostatic equation, 360, 362, 366
Hydrostatic equilibrium
balance, 147–151
geopotential coordinates, 145–147
gravity, effective, 143–145
gravity waves, 440–441
primitive equations, 357
stratification, 151–155
Lagrangian interpretation, 155–162
wave propagation, 427–428, 434
Hydrostatic stability, 166–172, 174, 189; *see
also* Vertical stability
Hygropause, 546, 560
Hygroscopic particle, 268–269
Hypsometric equation
hydrostatic balance, 149, 150–151
potential energy, 489, 491
thermal wind balance, 378, 380
tropical circulation, 506
Hysteresis, 62, 96

Ice particle growth, 275–276, 305
shapes, 276, 295
Ideal gas law, 5, 67

Incompressibility, Boussinesq approximation, 406–407
Incompressible fluid, 426–427
Incompressible motion, 352
Index
  refractive, 289–290, 291, 449–450
  thermal, 196
Inertial motion, 385
Inertial oscillation, 385, 517–518
Inertial reference frame, 334–336, 351
Inertio-gravity wave, 446
Infrared (IR) radiation
  aerosol, 35
  global energy budget, 44
  greenhouse effect, 43
  Newtonian cooling, 76
Infrared (IR) wavelength
  absorption at, 216, 217, 244
  clouds, 36–38
  line broadening, 224
  spectra, 199–200, 201, 216–217
Insolation, 46–47, 200
Instability, *see also* Stability; Vertical stability
  baroclinic, 493, 505, 521–523
  barotropic, 521–523
  conditional, 173, 176–181, 189
  criteria, 521–522
  Eady problem, 523–529
  inertial, 517–519
  nonlinear considerations, 529, 531–532
  parcel, 517
  potential, 184–186, 189
  shear, 519–523
  wave, 517
Instantaneous circulation, 50–51, 149
Intensity, 205–206
Intensive property, 55–56, 69
  heterogeneous system, 99, 106
  homogeneous system, 106
Internal energy, 44, 63, 67, 91; *see also* Temperature
  general circulation, 489–490
  material element, 336–337
  saturated behavior, 126
  specific heat capacity, 66
Internal wave, 441, 444
International Geophysical Year, 464–465
Inter-Tropical Convergence Zone (ITCZ), 39
  axisymmetric circulation, 502
  cloud radiative forcing, 309
  general circulation, 51
  precipitation, 114, 311, 506
Intrinsic frequency, 432

Inversion, 191–193
  nocturnal, 192–193, 410, 419
  subsidence, 187, 192
Invertibility principle, 392
Ionization, 16, 200, 201, 217
Irradiance, 206
Irreversibility, *see* Reversibility
Irrotational component, 375
Irrotational vector field, 62
Isentropic coordinates, 364–368, 372
Isentropic potential vorticity, 392
Isentropic process, 87, 93–96, 131
Isentropic surface, 129, 352, 378
  potential energy, 491–492, 495
Isobaric coordinates, 151, 357–362
Isobaric expansion coefficient, 125
Isobaric process, 68
  heterogeneous states, 110
  heterogeneous system, 101
  homogeneous states, 66
Isobaric surface
  potential energy, 491, 495
  stratification, 378
  temperature gradient, 147–149
Isochoric process, 66, 68, 87
Isochoric surface, 352
Isopleth, 129, 132, 134
Isothermal layer, 155
Isothermal process, 80–84, 89
Isotropic radiation, 206, 208, 212
Isotropy, hemispheric, 233
ITCZ, *see* Inter Tropical Convergence Zone

Jacobian
  curvilinear coordinates, 341–343
  isobaric coordinates, 359
  spherical coordinates, 346
Jet stream, 17–18, 21
Joule's experiment, 67

Kelvin's cat's-eye pattern, 473
Kelvin's formula, 268
Kelvin wave, 400
Kinematic condition, free surface, 429
Kinematics, fluid motion, 323–328
Kinematic viscosity, 333
Kinetic energy
  general circulation, 487, 488, 491–493
  simulations, 505–506
Kirchhoff's equation, 111
Kirchhoff's law, 211–212

Köhler curve, 269–270, 272
Krakatoa volcano, 191, 263, 458–459

Lagrangian derivative, 328, 338
    isentropic coordinates, 364–365
    isobaric coordinates, 358–359
    log-pressure coordinates, 363
    spherical coordinates, 347–348
Lagrangian description, 155–162, 321–323
    versus Eulerian, 1–2, 322–323, 324
        material derivative, 328, 329–330
Lambert's law, 207–208
Lamb mode, 458
Lamb wave, 444, 456–459
Laminar air flow, 282
Lapse rate, 16
    constant, 153–155, 156
    dry adiabatic, 74, 135, 151
    environmental, 151–153, 168
    saturated adiabatic, 132–133, 151
    superadiabatic, 171
Large-scale motion, *see also* Atmospheric
        motion; Equations of motion
    curvilinear, 382–385
        cyclostrophic, 385–386
        gradient, 386–387
        inertial, 385
    geostrophic equilibrium, 371–374
        f plane, 374–375
        frictional, 381
        Helmholtz theorem, 375–377
        thermal wind balance, 378–381
        vertical shear, 377–378
    weakly divergent
        barotropic nondivergent, 387–388
        quasi-geostrophic, 394–397
        vorticity budget, 389–393
Latent heat, 44, 45–46
    condensation, 126–130
    entrainment, 184
    general circulation, 51–52
    global energy budget, 310–311
    heterogeneous system, 110–112
    thermal energy, 487–488
    vertical motion, 130
LCL, *see* Lifting condensation level
Lenticular cloud, 281–282, 448
Level of free convection (LFC), 178, 179,
        189
    tropical circulation, 508
LFC, *see* Level of free convection
Lifting condensation level (LCL), 127–130

cloud formation, 277, 280, 281
    conditional instability, 178
    potential instability, 186
    pseudo-adiabatic chart, 136
    vertical stability, 189
Light speed, 579
Linearized equation, 428
Line broadening, 220–224, 229
Line-by-line calculation, 217, 229
Line strength, 221, 229
Liquid water content, cloud, 271–272, 285,
        300
Liquid water path, 300
Local derivative, 328
Local thermodynamic equilibrium (LTE),
        212
Log-pressure coordinates, 362–364
Longwave (LW) radiation, *see also* Outgoing
        longwave radiation
    aerosol, 244, 312–313
    boundary layer, 489
    cloudy atmosphere, 300–302, 304–305
    cooling, 241–242, 244, 302, 304, 550
        dissipation, 287
        wave absorption, 469, 472
    cumulonimbus cloud, 280, 281
    emission, 209, 210–211, 212
    global energy budget, 44–46, 49–50,
        306–309
    gray atmosphere, 233–235
    heating, 241–242, 244, 258, 302, 304,
        306–307
    Mie scattering, 295
    plane parallel atmosphere, 227, 228, 230,
        232, 233
    radiative equilibrium, 41–43
    spectra, 198–203
    vertical stability, 188
Lorentz line, 223, 224, 229
LTE, *see* Local thermodynamic equilibrium
LW radiation, *see* Longwave radiation
Lyman-α emission, 201

Mammatus cloud, 280–281
Maritime aerosol, 263, 271
Maritime cloud
    albedo, 307
    droplet growth, 271–272, 274–275
    glaciation, 275
    global energy budget, 309
    reflective properties, 300
Maritime continent, 189, 307

Mass
    atmospheric column, 50
    conservation, 330–331
    diffusion, 104, 109
    stratification, 2, 8–15
Mass absorption coefficient, 207
Mass mixing ratio, *see* Mixing ratio
Material acceleration, 335–336, 355
    geostrophic equilibrium, 382–384
    spherical coordinates, 347–348
    turbulence, 409
Material coordinate, 323
Material derivative, 328, 329–330, 338
Material element, 323–328, 336–337
Material line, 128–129
Material surface, 352–353
Material tracer, 8
Material velocity, 324–328, 335–336
Material volume, 329–330
    conservation of mass, 330–331
    first law of thermodynamics, 336–337
    momentum budget, 331–333
Maxwell relations, 89
Maxwell's equations, Mie theory, 291
Mean free path, 9–11
Mechanical equilibrium, 57, 59, 90; *see also*
        Hydrostatic equilibrium
    heterogeneous system, 102, 106
    homogeneous system, 99
Meridional circulation, 3, 50–51, 150, 486
    axisymmetric circulation, 495, 499–502
    laboratory simulation, 503
    tropical circulation, 509
    wave driving, 548–553
Meridional overturning, 548
Mesosphere, 16, 17
    collisional broadening, 224
    Lyman-α emission, 201
    meridional circulation, 553
    Newtonian cooling, 248
    ozone formation, 220
    radiative heating, 243–244
Metastable equilibrium, 90, 91
Methane (CH_4)
    absorption by, 200, 216, 220
    greenhouse effect, 252
    middle atmosphere, 544–546
    mixing ratio, 33, 544
    as trace constituent, 32–33
Metric scale factor, 342
Metric terms, 348
Microphysical properties, cloud, 285–286
Microphysics, cloud, 264–276

Microwave radiation, 199
Middle atmosphere, *see also* Stratosphere
    air motion, 546
        Brewer–Dobson circulation, 547–548
        meridional circulation, 548–553
    chlorofluorocarbons, 542–544
    heterogeneous chemistry, 564, 567–573
    methane, 544–546
    nitrous oxide, 541–542
    ozone photochemistry, 536–540
    quasi-biennial oscillation, 557–560
    structure, 16
    sudden stratospheric warming, 553–557
    troposphere interactions, 560–564
Mie scattering, 291–295, 297
Mixing ratio, 7, 352–353
    chlorofluorocarbons, 33
    cloud formation, 283
    methane, 33, 544
    moist air, 118
    nitric acid, 72
    nitrogen, 10
    nitrous oxide, 35, 541
    oxygen, 10
    ozone, 7, 30, 31, 564
    potential instability, 185, 186
    pseudo-adiabatic chart, 137–138
    saturation, 120, 126–130, 134
    vertical stability, 189
    water vapor, 26, 27, 29, 546, 560–561
Moist air
    adiabatic lapse rate, 132–133
    adiabatic process, 75
    condensation, 126–130
    equation of state, 119
    latent heat, 126–130
    properties
        gas phase, 117–120
        saturation, 120–121, 124–126
    pseudo-adiabatic chart, 134–138
    pseudo-adiabatic process, 130–132
    state variables, 120, 123–126, 132
    unsaturated behavior, 123–124
    vertical motion, 126–133, 134–138
Moist static energy, 487–489
Moisture dependence, 173–174
Molar abundance, 5–6, 7
Molar fraction, 7
Molar property, 101–102, 105
Molar weight, 5–6
Molecule–radiation interaction, 216–220
Momentum balance, 167, 168
Momentum budget, 331–336, 407

Momentum equation, 334–336, 351; *see also* Horizontal momentum equation; Vertical momentum equation
Monsoon, 39, 191, 506, 508–509
Montgomery streamfunction, 365–366, 372
Motion, *see* Atmospheric motion; Equations of motion; Horizontal motion; Large-scale motion; Vertical motion; Wave
Motion field, 322, 323–328
Mountain wave, 447
Mountain wave cloud, 282, 283
Mount St. Helens volcano, 264

Nacreous cloud, 282, 283
Natural broadening, 223
Natural process, 79
Navier Stokes equations, 334
Net radiation, 49–50
Net work, 65
Neutral stability, 415
Newtonian cooling, 76, 248, 469, 472, 552
Newtonian fluid, 333
Newton's second law, 143, 331–332, 333; *see also* Hydrostatic equilibrium
Nimbostratus cloud, 285–286
Nitrate aerosol, 261–262
Nitric acid, 72, 571
Nitric oxide (NO), 34, 35, 541–542
Nitrogen, 3, 10
   odd nitrogen family, 542
Nitrogen compounds, as trace constituents, 34–35
Nitrous oxide (N_2O), 34–35
   absorption by, 200, 216, 220
   greenhouse effect, 252
   middle atmosphere, 541–542
   mixing ratio, 35, 541
Nocturnal inversion, 192–193, 410, 419
Noncompensated heat transfer, 89–90
Nondispersive wave, 436
Nondivergent motion, 387–388
Nondivergent Rossby wave, 459–461
Nonlinearity
   instability, 529, 531–532
   waves, 472–479
Nonuniform heating, 47–48, 51–52, 505
Normal stress, 332
No-slip condition, 356
Nuclear detonation, 191
Nucleation, 261
   heterogeneous, 268, 275
   homogeneous, 264–268, 269, 275

Number density
   aerosol, 260, 263
   clouds, 285–286
   droplet, 271
   equations of state, 6
   ozone, 539
   scatterers, 290

Oceans
   evaporative cooling, 45
   heat exchange, 39
   SW absorption, 48–49
   water vapor distribution, 122, 130
Odd chlorine family, 543–544
Odd nitrogen family, 542
Odd oxygen family, 538–540
OLR, *see* Outgoing longwave radiation
Omega block, 400
Open system, 55
Optical path length, 208
Optical properties, *see also* Scattering
   cloudy atmosphere, 299–305
   Mie scattering, 291–295
   radiative transfer, 46–50
   Rayleigh scattering, 288–290
Optical thickness, 215, 231
Orographic cloud, 281–283
Orthogonal coordinates, 343, 344, 346
Oscillation
   El Niño Southern, 510
   inertial, 385, 517–518
   quasi-biennial, 557–560
Outgoing longwave radiation (OLR), 43, 305, 307; *see also* Longwave radiation
   global energy budget, 48–50
Overturning, 51
   boundary layer, 419
   convective, 236–240
   meridional, 548
   potential temperature, 73
Oxygen, 3, 10, 15
   absorption, 201, 220
   lifetime, 537
   odd oxygen family, 538–540
   photodissociation, 12, 218, 220
Ozone, 3–4, 7, 29–32
   absorption
      characteristics, 216, 219, 220
      radiative heating, 243, 244, 246
      spectra, 200, 201, 203
   chemical processes, 314–315, 567–573
   chlorofluorocarbons, 34

Ozone (*continued*)
  circulation, 31–32
  general circulation, 51
  greenhouse effect, 249–250
  LW heating, 242
  mixing ratio, 30, 31, 564
  nitrogen compounds, 34–35
  number density, 539
  photochemistry, 352–353, 536–546, 567
  photodissociation, 219–220
  quasi-biennial oscillation, 558–559
  radiative-convective equilibrium, 240
  radiative equilibrium, 43
  stratospheric warming, 556
  troposphere–stratosphere interaction,
      563–564
  variable distribution, 25
Ozone heating, 16
  geostrophic equilibrium, 374
  vertical stability, 189, 191
Ozone hole, 32, 34, 35, 41
  heterogeneous chemistry, 567, 569, 571
  PSCs, 314–315

Paddle work, 58, 65
Parallel-beam radiation, 205
Parcel, 1
  instability, 517
  trajectory, 323–324
Parseval's theorem, 431
Partial molar property, 101–102, 105
Partial pressure, 5
Partial specific property, 101–102
Partial volume, 5
Partitioning, odd oxygen family, 538
Path, mean free, 9–11
Path dependence, 64
Path function, 62
Path independence, 59, 62, 63
Pencil of radiation, *see* Absorption;
      Emission; Radiative transfer; Scattering
Perpetual motion machine, 85
Perturbation equation, *see also* Equations of
      motion.
  acoustic waves, 439
  gravity waves, 440
  Lamb waves, 456–457
  Rossby waves, 459
Phase function, 214, 296–297
Phase rule, Gibbs', 107, 109, 123
Phases, single versus multiple, *see*
      Heterogeneous system; Homogeneous
      system

Phase speed, 430–431, 432–433
Phase transformation, 79, 110–114
Photochemical equilibrium, *see also* Ozone;
      Photodissociation
  Brewer–Dobson circulation, 548
  odd oxygen family, 538–540
Photodissociation, 12, 29–30, 34, 200
  absorption, 217–218, 219–220
  Lyman-α emission, 201
  middle atmosphere, 537, 540, 541, 543,
      546
  plane parallel atmosphere, 232
  vertical motion, 130
Photoionization, 200, 201, 217
Photolysis, 537; *see also* Photochemical
      equilibrium; Photodissociation
Photons, 209, 214
Photosphere, 200
Photosynthesis, 29–30
Physical constants, 579
Planck's constant, 209, 579
Planck's law, 209
Planck spectrum, 209
Plane parallel atmosphere, 224–228; *see also*
      Gray atmosphere
  transmission function, 228–230
  two-stream approximation, 230–233
Planetary boundary layer, 52
  Ekman pumping, 420–423
  stratification, 415, 417–419
  structure
      Ekman layer, 412–413
      surface layer, 413–415
  turbulence, 405–408, 420
      diffusion, 410–411
      Reynolds decomposition, 408–410
  well-mixed, 424
Planetary vorticity, 336, 355
  geostrophic motion, 374
  Rossby waves, 460
  traditional approximation, 350–351
Planetary wave, 465–468; *see also* Rossby
      wave
  dynamical structure, 19–21
  El Niño Southern Oscillation, 510
  general circulation, 51–52
  instability, 532
  meridional circulation, 549–553
  nonlinear considerations, 478
  thermodynamic cycle, 65
Plant life, 29–30
Point function, 61–62
Poisson's equations, 69, 71

Polar front, 149, 151
Polarization, 287–290, 290
Polar-night jet, 17, 21, 151
Polar-night terminator, 47, 547
Polar-night vortex, 374, 478, 479, 548–552
    heterogeneous chemistry, 571
    quasi-biennial oscillation, 558–560
Polar stratospheric cloud (PSC), 41, 283–284
    global energy budget, 314–315
    heterogeneous chemistry, 567, 571
    microphysical properties, 285–286
    troposphere-stratosphere interaction, 564
Pollution, *see also* Aerosol; Anthropogenic
        impact; Volcano
    inversion, 187, 192, 193
    precipitation, 314
    turbulence, 410
Polytropic process, 75–76, 156
Polytropic specific heat capacity, 76
Potential energy, *see also* Gravity
    available, 491–495
    baroclinic system, 491, 494, 531
    conditional instability, 178–179
    general circulation, 487, 489–495
    gravitational field, 44
    tropical circulation, 506
    vertical stability, 189
Potential function, 62, 375
Potential instability, 184–186, 189
Potential temperature
    adiabatic process, 71–73, 74, 93, 352
    diabatic process, 75, 76
    entropy, 91–96, 337
    equivalent, 131–132, 136
    inversions, 191, 192
    isobaric coordinates, 361
    potential energy, 494
    potential instability, 185
    pseudo-adiabatic process, 131–132
    stratification, 153–155
    surface, 135, 136
    thermodynamic equation, 337
    vertical coordinate, 364
    vertical displacement, 186, 187–188
    vertical stability, 171–173
    virtual, 124
Potential vorticity, 392, 396
Potential well, 179
Power spectrum, 431, 437, 438
Precipitable water vapor, 28
Precipitation, 27, 41; *see also* Condensation;
        Monsoon
    cloud absorptivity, 300

cloud dissipation, 287
droplet growth, 272–273, 275
global energy budget, 310, 311
Inter Tropical Convergence Zone, 114,
        311, 506
pollution, 314
pseudo-adiabatic chart, 137–138
vertical motion, 129
Pressure, 5, 9; *see also* Log-pressure
        coordinates
    hydrostatic balance, 147–151
    potential temperature, 71–72
    pseudo-adiabatic chart, 134–137
    vapor, 120, 121–122, 250
        equilibrium, 109, 112–114
    as vertical coordinate, 357
Pressure broadening, 223
Pressure gradient force, 371
Primitive equations
    isobaric coordinates, 357
    scale analysis, 356
Prognostic equation, 387
Prognostic system, 393, 394
PSC, *see* Polar stratospheric cloud
Pseudo-adiabatic chart, 97, 134–138, 583–584
Pseudo-adiabatic process, 130–132, 134–138
    adiabats, 132, 134, 135
    stratification, 158
Pure substance, 58–59, 69

QBO, *see* Quasi-biennial oscillation
Quasi-biennial oscillation (QBO), 557–560
Quasi-geostrophic motion, 394–397
    boundary layer, 420, 422
    shear instability, 519
Quasi-geostrophic potential vorticity, 396
Quasi-geostrophic Rossby wave, 461
Quasi-geostrophic vorticity, 461
Quasi-horizontal motion, 351, 352, 389

Radiance, 205
Radiation, 198–200; *see also* Infrared
        radiation; Longwave radiation;
        Shortwave radiation; Ultraviolet
        radiation
    absorption characteristics, 216–224
    blackbody, 208–212
    diffuse, 205, 295–296
    greenhouse effect, 248–252
    isotropic, 206, 208, 212
    line broadening, 220–224

Radiation (*continued*)
  microwave, 199
  net, 49–50
  outgoing longwave, 43, 48–50, 305, 307
  parallel-beam, 205
  plane parallel atmosphere, 224–233
  terrestrial, 41–43
  thermal relaxation, 245–248, 552
  top of atmosphere, 307, 311
  transfer
    absorption, 207–208
    emission, 208–212
    equation, 215
    radiometric quantities, 203, 205–207
    scattering, 213–214
  X-ray, 216
Radiation–molecule interaction, 216–220
Radiative–convective equilibrium, 236–240
  cloudy atmosphere, 302–304
  greenhouse effect, 249, 250–252
  thermal structure, 321
Radiative equilibrium, 41–43; *see also* Global
    energy budget
  gray atmosphere, 233–236
  temperature, 211–212
  thermal structure, 226, 321
Radiative heating, 241–245, 246
  cloud cover, 307–309
  cloud dissipation, 287
  thermal equilibrium, 486
Radiative transfer, 44–46, 58, 75
  absorption, 207–208
  cloudy atmosphere, 295–305
  directionality, 205–206, 214
  emission, 208–211
  gray atmosphere, 233–236
  plane parallel atmosphere, 224–233
  radiometric quantities, 203, 205–207
  scattering, 212–215, 287–295
  thermal structure, 198, 248
Radiative transfer equation, 215, 233, 295
Radioactive debris, 191
Radiometric quantities, 203, 205–207
Rain out, 41
Rate of strain, 326
Rawinsonde measurement, versus satellites,
    149
Rayleigh–Bénard convection, 282
Rayleigh friction, 381
Rayleigh scattering, 287–290, 292, 293, 297
Rectangular coordinates, 341; *see also*
    Cartesian coordinates

Reflection
  cloud properties, 300
  wave activity, 450, 454
Refractive index, 289–290, 291, 449–450
Refrigerator, 65, 84, 85
Relative concentration, 7
Relative humidity, 121, 135, 262
Relative vorticity, 387–388, 459
Restoring force, 426, 517; *see also* Instability
  negative, 170
  positive, 169
Reversibility, 79–84
  entropy, 84–87
  equilibrium conditions, 90–91
  fundamental relations, 88–90
  potential temperature, 91–93
  saturated adiabatic process, 130–131
Reynolds decomposition, 408–410
Reynolds number, 317, 405
Reynolds stress, 410
Reynolds' transport theorem, 329–330, 331,
    333, 337
Richardson number, 415, 417
Riming, 276
Rossby height scale, 526
Rossby number, 355, 371, 427
  axisymmetric circulation, 495–496
  curvilinear motion, 382, 384–385, 386
  geostrophic equilibrium, 394–395
Rossby wave, *see also* Planetary wave
  barotropic nodivergent, 459–461
  nonlinear considerations, 473
  propagation
    planetary waves, 465–468
    three-dimensional, 461–465
  tropical circulation, 509
Rotating reference frame, 334–336
Rotation
  earth, 3
    axisymmetric circulation, 495
    cyclostrophic motion, 385–386
    general circulation, 51
    geostrophic paradox, 372
    gravitational components, 143, 144–145
    meridional heat transfer, 486
    potential energy, 493
    reference frame, 334, 336
  laboratory simulation, 503–505
  sun, 203
  Venus, 386
Rotational energy, 213, 216
Rotational wave, 459; *see also* Planetary
    wave; Rossby wave

Rotor cloud, 282
Roughness length, 414

Sahara desert
    crustal aerosol, 258
    dust, 312–314
Satellite measurement, versus rawinsondes,
    149
Saturated conditions
    adiabatic lapse rate, 132–133, 151
    adiabatic stratification, 130–131, 159
    mixing ratio, 120, 126–130, 134
    moisture dependence, 173–174
    phases, 109
    plane parallel atmosphere, 230, 231
    potential temperature, 173
    properties, 120–121
    specific humidity, 120
    supersaturation, 268–270, 277
    two-component system, 124–126
    vapor pressure, 120, 121–122, 250
Saturation values, 120
Scalar potential, 375
Scale analysis, 354–357, 371
Scale height, 9, 526
Scattering, 213–215
    aerosol, 312–313
    Mie, 291–295, 297
    plane parallel atmosphere, 226
    Raleigh, 287–290, 292, 293, 297
    thermal structure, 295–305
Scattering angle, 288–289
Scattering coefficient, 293
Scattering cross section, 213
Scattering efficiency, 292–293
Schumann–Runge band, 218–219, 224, 243
Schwartzchild's equation, 215
Sea salt aerosol, 263, 271
Sea surface temperature (SST), 122
    stratification, 417, 419
    tropical circulation, 509
    vertical stability, 190
Secondary aerosol, 258, 260, 261
Secondary circulation, 395, 420
Second law of thermodynamics
    entropy, 84–87, 91–96
    forms
        Clausius inequality, 85
        differential, 86
        latent heat, 112
        restricted, 87–88
    fundamental relations, 88–90, 104–106

    natural and reversible processes, 79–84
    potential temperature, 91–96
    thermodynamic equilibrium, 90–91
Sedimentation, 262, 264, 287
Sensible heat, 44, 45–46
    entrainment, 184
    thermal energy, 487–489
Shallow atmosphere approximation, 348–349
Shear
    horizontal, 376, 391
    motion field, 326
    vertical, 377–381, 415, 493
Sheared mean flow, 465–468
Shear instability, 519–523
Shear stress, 332, 333
Shortwave cutoff, 525
Shortwave limit, 434, 447–448
Shortwave (SW) radiation, 198–200
    aerosol, 191, 244, 263, 312–313
    cloudy atmosphere, 295–296, 304, 305
    cooling, 258, 307
    emission, 209, 212
    global energy budget, 44–50, 307–309
    gray atmosphere, 233–234
    heating, 304, 307, 550
    nocturnal inversion, 192
    plane parallel atmosphere, 233
    radiative–convective equilibrium, 240
    radiative equilibrium, 41–43
    radiative heating, 242, 243, 244
    scattering, 290, 294
    solar constant, 200
    spectra, 198–203, 219
Sidereal day, 385
Sierra Wave Project, 477
Simulation, heat transfer, 503–506
Single scattering, 214
SI units, 577
Skin temperature, 236
Sloping convection, 39, 505
Slowly varying approximation, 450
Sodium chloride (NaCl), droplet growth,
    268–269
Solar constant, 200, 579
Solar inclination, 200
Solar-max, 203
Solar-min, 203
Solar spectrum, 200–201
Solar structure
    chromosphere, 201
    photosphere, 200
    rotation, 203
    sunspots, 203

Solar variability, 203
Solenoid, 391
Solenoidal component, 375
Solenoidal production, 391
Sound wave, *see* Acoustic wave
Source function, radiative emission, 212
Southern Oscillation, 510
Specific gas constant, 5–6
Specific heat, 110, 119–120
Specific heat capacity, 65–66, 67–68, 76
Specific humidity, 118, 120
Specific property, 56, 101–102
Specific volume, 5
Spectrum
    bands, 218–220
    blackbody, 209
    clouds, 286, 293–294
    IR, 201, 216–217
    line broadening, 220–224
    LW, 198–203
    Planck, 209
    power, 431, 437, 438
    radiation–molecule interactions, 216
    solar, 200–201
    SW, 198–203, 219
    UV, 200, 201, 203
Speed of light, 579
Spherical coordinates, 146, 344–352, 581
Sphericity, 144, 145, 224
SST, *see* Sea surface temperature
Stability, *see also* Instability; Vertical stability
    absolute, 173, 186
    hydrostatic, 166–172, 174, 189
    neutral, 415
    static, 449
Stability criteria, 169, 172–173, 176, 188
Stable equilibrium, 90
Standard gravity, 579
State
    equations of, *see* Equations of state
    thermodynamic, 55–56, 58–63, 99, 106
State change, 63–65
    reversible, 80–84, 86–87
State space, 59
State variable, 58–63, 65, 66
    moist air, 120, 123–126, 132
Static stability, 449
Steering level, 524
Stefan–Boltzmann constant, 210, 579
Stefan–Boltzmann law, 43, 210–211
Stokes drag, 317
Stokesian fluid, 340
Storm track, 20, 41, 51

Rossby waves, 465
    laboratory simulation, 505
Stratification, 2, 8–15; *see also* Plane parallel
    atmosphere; Vertical stability
    adiabatic, 130–131, 156–160
    baroclinic, 378–379, 389–393, 491, 494, 508
    barotropic, 378, 422, 493–494, 532
    boundary layer, 415, 417–419
    components, 353
    diabatic, 160–162
    equations of motion, 353–354
    hydrostatic equilibrium, 151–162
    Lagrangian interpretation, 155–162
    moisture, 184–186
    subadiabatic, 161
    superadiabatic, 162
    unstable, 236–238
Stratiform cloud, 38, 277, 280–281, 287
Stratocumulus cloud, 281; *see also* Cumulus
    cloud
    albedo, 307
    global energy budget, 309
    microphysical properties, 285–286
Stratopause, 16, 242, 244
Stratosphere, 16, 17, 20–21; *see also* Middle
    atmosphere
    aerosol, 260, 262, 263–264
    global energy budget, 307
    gravity waves, 448
    hydrostatic equilibrium, 160, 161
    inertial instability, 518
    LW radiation, 65, 242
    nacreous clouds, 282, 283
    ozone, 219, 220, 353
    ozone heating, 374
    ozone photochemistry, 536–546
    planetary waves, 478–479
    radiation absorption, 200
    radiative-convective equilibrium, 237, 238
    radiative emission, 209
    radiative heating, 243–244
    Rossby waves, 465
    sudden warming, 478–479, 553–557
    thermal heating, 246
    vertical stability, 166, 189, 191
    wave activity, 451, 454, 472
Stratus cloud, 285–286
Streakline, 324
Streamfunction
    geostrophic, 376, 388, 461
    Montgomery, 365–366, 372
Streamline, 324
Stress tensor, 332, 333

Stretching
   motion field, 326
   vertical, 390–391
Stretching transformation, 498, 501
Strong absorption, 230, 232
Structure
   boundary layer, 412–415
   circulation, 17–22
   cloud, 36–41
   dynamical, 16–22
   general circulation, 50–52
   global energy budget, 44–50
   radiative equilibrium, 41–43
   stratification of mass, 2, 8–15
   thermal, *see* Thermal structure
Subadiabatic process, 161, 171
Subsidence inversion, 187, 192
Sudden stratospheric warming, 478–479,
   553–557
Sulfate aerosol, 261–262, 271
Sulfur dioxide (SO_2), 35
Sun, structure, *see* Solar structure
Sunspot activity, 203
Superadiabatic layer, 419
Supercooled water, 275
Supersaturation
   cloud formation, 277
   droplet growth, 268–270
Surface layer, 413–415
Surface water wave, 426–429, 442
SW radiation, *see* Shortwave radiation
Synoptic weather system, 18, 149
   general circulation, 51
   vorticity budget, 392
System, thermodynamic, 55–63; *see*
   Thermodynamics of gases

Taylor–Goldstein equation, 449
Taylor–Proudman column, 377–378
Taylor–Proudman flow, 378
Taylor–Proudman theorem, 377–378
Temperature, 5, 14, 16; *see also* Greenhouse
   effect; Potential temperature; Sea
   surface temperature; Thermal
   equilibrium; Thermal structure
   adiabatic process, 74, 132, 133
   blackbody, 43, 44, 208–212
   break, 193
   brightness, 209
   clouds
      cold, 275
      warm, 273

   critical point, 108
   dew point, 121
   environmental, 169–170
   equilibrium vapor pressure, 112–114
   frost point, 121
   global-mean, 16, 45
   global warming, 23–24
   hydrostatic balance, 147–150
   isentropic behavior, 93–96
   ozone depletion, 572
   plane parallel atmosphere, 226
   polar stratospheric clouds, 285
   potential instability, 185
   pseudo-adiabatic chart, 134–137
   pseudo-adiabatic process, 131
   radiative-convective equilibrium, 238
   radiative equilibrium, 41, 43, 211–212, 236
   saturation mixing ratio, 126
   saturation vapor pressure, 122
   sudden stratospheric warming, 478–479,
      553–557
   trigger, 193
   vertical stability, 169–170
   virtual, 119, 120
   zonal-mean, 16–17
Temperature inversion, 152, 191–193
Tensor analysis, 324–326, 332, 333
Terminator, polar-night, 47, 547
Terrestrial radiation, 41–43; *see also* Infrared
   radiation; Longwave radiation
Thermal, 277–280, 286
Thermal damping, 472
Thermal energy, 44, 487–489
Thermal equilibrium, 41, 43, 59; *see also*
   Global energy budget
   absorption-emission, 208
   general circulation, 486
   gray atmosphere, 233–236
   heterogeneous system, 102, 106
   homogeneous system, 99
   inversions, 192
   isentropic behavior, 94–95
   radiative-convective, 236–240
   radiative emission, 211
   radiative heating, 241–245
   radiative transfer, 198
Thermal index, 196
Thermally direct circulation, 51
Thermal relaxation, 245–248, 552
Thermal structure, 16–22, 51
   barotropic stratification, 532
   cloud and aerosol influences, 295–305
   gray atmosphere, 233

Thermal structure (*continued*)
  ozone, 536
  plane parallel atmosphere, 226
  radiative-convective equilibrium, 238, 321
  radiative transfer, 198, 248
  stratification, 156–162
  vertical stability, 188–189
Thermal wind balance, 378–381
  general circulation, 486
  potential energy, 493
  shear instability, 522
Thermodynamic coordinates, 357–368
Thermodynamic cycle, 65, 157, 159, 161
Thermodynamic equation, 337–338
  adiabatic motion, 352
  isentropic coordinates, 367
  isobaric coordinates, 361
  log-pressure coordinates, 363
  quasi-geostrophic motion, 395–396
  turbulence, 407
Thermodynamic equilibrium, 59
  conditions, 90–91
  heterogeneous system, 99, 103–104, 106
  homogeneous system, 99
  irreversibility, 79
  local, 212
Thermodynamic potential, 97; *see also* Free
  energy
Thermodynamic process, 59, 61
  direction, 85, 87, 90
Thermodynamic properties, 55–56, 578
Thermodynamics of gases, *see also* First law
  of thermodynamics; Second law of
  thermodynamics
  adiabatic process, 63, 68–74, 132
  diabatic process, 75–76
    changes of state, 63–65
  entropy, 84–87, 91–96
  fundamental relations, 88–90, 104–106
  heat capacity, 65–68
  internal energy, 63
  natural and reversible processes, 79–84
  potential temperature, 71–76, 91–96
  system concepts, 55–56
    expansion work, 56–58
    heat transfer, 58
    state variables, 58–62
  thermodynamic equilibrium, 90–91
  vertical motion, 73–74, 126–133
Thermodynamic state, 55–56, 58–63
  heterogeneous system, 99, 106
  homogeneous system, 99
Thermodynamic system, 55–63

Thermosphere, 16, 201, 219
Time-mean circulation, 19–21, 51
Time-mean cloud field, 39, 41
Time-mean cloudiness, 39, 41
Time-mean thermal structure, 51
Timescale
  advection, 552
  diabatic process, 75, 156, 160
  dissipative process, 392
  Ekman pumping, 423
  heat transfer, 71, 128
  horizontal motion, 71
  radiative transfer, 75, 248
  thermal damping, 472
  thermal relaxation, 552
  turbulent exchange, 405, 408
  vertical mixing, 190
  vertical motion, 71, 128, 156
Top of atmosphere (TOA) radiation, 307,
  311
Trace constituent, 22–36; *see also specific
  constituents*
Tracer, 8, 71
Trace speed, 431
Trade wind, 506
Traditional approximation, 350–352
Trajectory, parcel, 323–324
Trajectory coordinates, 383
Transmission function, 228–230
Transmissivity, 208, 228, 298–299
Transport theorem, 329–330, 331, 333, 337
Trigger temperature, 193
Triple point, 106
Tropical circulation, 506–513; *see also* Inter
  Tropical Convergence Zone
Tropical cirrus cloud, 285–286
Tropopause, 16, 17
  cloud formation, 283
  conditional instability, 181
  cumulus formation, 280
  nitrous oxide, 541
  potential temperature, 191
  radiation spectra, 200
  radiative-convective equilibrium, 237–238,
    240
  tropical circulation, 508
  water vapor, 546
  wave activity, 451, 454
Tropopause fold, 564
Troposphere, 16, 17, 18
  absorption, 216
  aerosol, 258, 260–263, 314
  air motion timescale, 248

Troposphere (*continued*)
  cloud formation, 285
  conditional instability, 181
  cyclones, 72
  Ekman pumping, 423
  El Niño Southern Oscillation, 510
  hydrostatic equilibrium, 157, 159–161
  kinetic energy, 506
  LW radiation, 65, 232, 241, 244
  middle atmosphere interactions, 560–564
  nitrous oxide, 541
  nonuniform heating, 505
  ozone, 219
  planetary waves, 468
  radiation absorption, 200
  radiative-convective equilibrium, 236–237,
    240
  radiative emission, 209
  radiative forcing, 311
  Rossby waves, 465
  thermal heating, 246
  thermal wind, 380
  tropical circulation, 508
  vertical stability, 166, 178, 188, 189,
    190–192
  water vapor mixing ratio, 546
  wave activity, 451, 454
Turbopause, 11
Turbulence, boundary layer, 405–411, 420
Turbulent dispersion, 174, 189–193
Turbulent transport, 10–12, 52
Turning level, wave activity, 450, 451–455
Two-stream approximation, 230–233

Ultraviolet (UV) radiation
  absorption of, 219
  ozone, 29–30
  photodissociation, 12, 537, 540, 546
Ultraviolet (UV) wavelength
  absorption at, 216, 217, 219, 220
  line broadening, 224
  spectra, 200, 201, 203
Universal gas constant, 579
Unsaturated conditions
  moist air, 123–124, 132
  moisture dependence, 173–174
  potential temperature, 171–173
  stratification, 151
Unstable equilibrium, 90
Unstable stratification, 236–238

Vapor pressure, 109, 120, 121–122, 250
  equilibrium, 109, 112–114

Vector field, irrotational, 62
Vector identities, 580
Vector operations
  curvilinear coordinates, 344
  spherical coordinates, 346, 348–349
Vector potential, 375
Velocity, *see also* Geostrophic velocity;
    Group velocity
  angular, 334, 579
  friction, 413
  horizontal, 371
  material, 324–328, 335–336
  spherical coordinates, 347
Velocity potential, 376
Venus
  cyclostrophic motion, 386
  greenhouse effect, 250, 251–252
Vertical coordinate
  height, 349, 350
  log-pressure height, 362
  potential temperature, 364
  pressure, 357
Vertical momentum budget, 407
Vertical momentum equation, 354–355, 429,
    434
Vertical motion, *see also* Atmospheric
    motion; Gravity; Horizontal motion;
    Large-scale motion; Vertical stability
  adiabatic process, stratification, 156–160
  boundary layer, 420
  buoyancy, 128–130, 157, 161
  cloud formation, 277
  condensation, 126–130
  hydrostatic balance, 9
  isentropic behavior, 93–96
  pseudo-adiabatic chart, 134–138
  scale analysis, 354–355
  stability implications, 174–176
  thermodynamic behavior, 73–74, 126–133
  traditional approximation, 350–351
Vertical shear
  boundary layer, 415
  geostrophic wind, 377–381
  potential energy, 493
Vertical stability
  buoyancy, 166–170, 174–176, 189
  categories, 168–174
  finite displacements
    conditional instability, 176–181
    entrainment, 181, 183–184
    potential instability, 184–186
    unsaturated conditions, 186–188

Vertical stability (*continued*)
  hydrostatic stability, 169–172
  implications for vertical motion, 174–176
  influences, 188–189
  moisture dependence, 173–174
  potential stability, 186
  potential temperature, 171–173
  turbulent dispersion, 174
    convective mixing, 189–191
    inversions, 191–192
    nocturnal inversion, 192–193
Vertical stretching, 390–391
Vertical vorticity budget, 509
Vibrational energy, 213, 216
Virtual potential temperature, 124
Virtual process, 90, 91, 103–104
Virtual temperature, 119, 120
Viscosity
  bulk, 340
  coefficient, 333
Viscous sublayer, 414
Visible wavelength
  absorption at, 216, 217, 244
  line broadening, 224
Voigt line shape, 224
Volcano
  aerosol, 263–264
  El Chichon, 263, 315
  global energy budget, 314
  Krakatoa, 191, 263, 458–459
  Mount St. Helens, 264
Volume
  material, 329–333, 336–337
  specific heat capacity, 65–66, 67–68
Volume heating rate, 243
Volume mixing ratio, 7
von Karman constant, 414
Vorticity, 336
  absolute, 388, 389–392
  Ekman pumping, 420
  motion field, 326
  planetary, *see* Planetary vorticity
  potential, 392, 396
  quasi-geostrophic, 396, 461
  relative, 387–388, 459
  Rossby waves, 459–460, 461, 463
  wave absorption, 470
Vorticity budget, 389–393, 421, 509
Vorticity equation, 388, 396, 422

Walker cell, 509
Walker circulation, 51

Washout, 264, 314
Water
  Clausius–Clapeyron equation, 112–113
  pseudo-adiabatic chart, 137
  pure substance, 107–108, 112
  saturation vapor pressure, 120
  supercooled, 275
  thermodynamic characteristics, 107–109,
    112, 113, 578
Water vapor, 3–4; *see also* Clouds;
    Condensation; Moist air
  absorption by, 200, 201, 203, 216–217, 220,
    232, 244
  chemical equilibrium, 118
  circulation, 27, 114
  Clausius–Clapeyron equation, 113–114
  distribution, 121–122, 130
  greenhouse effect, 250–252
  heterogeneous system, 99–100
  IR absorption, 43
  line broadening, 224
  liquid, 271–272, 285, 300
  LW radiation, 46, 241
  mixing ratio, 26, 27, 29, 546, 560–561
  photodissociation, 12
  precipitable, 28
  radiative-convective equilibrium, 240
  as trace constituent, 25–29
Water wave
  deep, 434, 437–438
  shallow, 433, 436–437
  surface, 426–429, 442
Wave
  absorption, 468–472
  acoustic, 439–446
  acoustic-gravity, 441, 444
  buoyancy, 440–446
    geometric optics, 451–456
    gravity waves, 440–450
    shortwave limit, 447–448
    WKB approximation, 450–451
  dispersion, 434–438
  edge, 434, 442, 526
  evanescent, 441–442
  external, 441–442
  gravity, *see* Gravity wave
  internal, 441, 444
  Lamb, 444, 456–459
  mountain, 447
  nondispersive, 436
  nonlinear considerations, 472–479
  planetary, *see* Planetary wave

Wave (*continued*)
  propagation
    dispersion, 434–438
    Fourier synthesis, 429–432
    limiting behavior, 432–434
    surface water waves, 426–429
  refraction, 450
  Rossby, *see* Rossby wave
  water, *see* Water wave
Wave instability, 517
Wave interference, 434
Wavelength, 199–203, 208–212
Wave power spectrum, 431, 437, 438
Weakly divergent motion, 387–397
Wentzel–Kramer–Brillouin approximation,
    see WKB approximation
Wettable particle, 268
Wien's displacement law, 209
WKB (Wentzel–Kramer–Brillouin)
    approximation, 450–451
  planetary waves, 466
  wave absorption, 470
Work
  compression, 71
  cyclic, 57

  expansion, 56–58, 63, 73, 75
  net, 65
  paddle, 58, 65
  reversible processes, 79–87

X-ray radiation, 216

Zenith angle, 46–48, 200, 205
  global energy budget, 307
  plane parallel atmosphere, 224–227, 232
  radiative heating, 242
Zonal jet, 519
Zonal-mean circulation, 17–18
Zonal-mean mixing ratio
  ozone, 30
  water vapor, 26
Zonal-mean temperature, 16–17
Zonal motion
  general circulation, 486–487, 495–503
  laboratory simulation, 503, 505
  tropical circulation, 509
Zonal overturning, 51
Zonal symmetry, 19, 21, 50

International Geophysics Series

EDITED BY

RENATA DMOWSKA
Division of Applied Science
Harvard University
Cambridge, Massachusetts

JAMES R. HOLTON
Department of Atmospheric Sciences
University of Washington
Seattle, Washington

Volume 1   BEN GUTENBERG. Physics of the Earth's Interior. 1959*

Volume 2   JOSEPH W. CHAMBERLAIN. Physics of the Aurora and Airglow. 1961*

Volume 3   S. K. RUNCORN (ed.) Continental Drift. 1962*

Volume 4   C. E. JUNGE. Air Chemistry and Radioactivity. 1963*

Volume 5   ROBERT G. FLEAGLE AND JOOST A. BUSINGER. An Introduction to Atmospheric Physics. 1963*

Volume 6   L. DUFOUR AND R. DEFAY. Thermodynamics of Clouds. 1963*

Volume 7   H. U. ROLL. Physics of the Marine Atmosphere. 1965*

Volume 8   RICHARD A. CRAIG. The Upper Atmosphere: Meteorology and Physics. 1965*

Volume 9   WILLIS L. WEBB. Structure of the Stratosphere and Mesosphere. 1966*

Volume 10   MICHELE CAPUTO. The Gravity Field of the Earth from Classical and Modern Methods. 1967*

Volume 11   S. MATSUSHITA AND WALLACE H. CAMPBELL (eds.) Physics of Geomagnetic Phenomena. (In two volumes.) 1967*

Volume 12   K. YA KONDRATYEV. Radiation in the Atmosphere. 1969*

*Out of Print

Volume 13   E. PALMÉN AND C. W. NEWTON. Atmospheric Circulation Systems: Their Structure and Physical Interpretation. 1969*

Volume 14   HENRY RISHBETH AND OWEN K. GARRIOTT. Introduction to Ionospheric Physics. 1969*

Volume 15   C. S. RAMAGE. Monsoon Meteorology. 1971*

Volume 16   JAMES R. HOLTON. An Introduction to Dynamic Meteorology. 1972*

Volume 17   K. C. YEH AND C. H. LIU. Theory of Ionospheric Waves. 1972*

Volume 18   M. I. BUDYKO. Climate and Life. 1974*

Volume 19   MELVIN E. STERN. Ocean Circulation Physics. 1975

Volume 20   J. A. JACOBS. The Earth's Core. 1975*

Volume 21   DAVID H. MILLER. Water at the Surface of the Earth: An Introduction to Ecosystem Hydrodynamics. 1977

Volume 22   JOSEPH W. CHAMBERLAIN. Theory of Planetary Atmospheres: An Introduction to Their Physics and Chemistry. 1978*

Volume 23   JAMES R. HOLTON. An Introduction to Dynamic Meteorology, Second Edition. 1979*

Volume 24   ARNETT S. DENNIS. Weather Modification by Cloud Seeding. 1980

Volume 25   ROBERT G. FLEAGLE AND JOOST A. BUSINGER. An Introduction to Atmospheric Physics, Second Edition. 1980

Volume 26   KUG-NAN LIOU. An Introduction to Atmospheric Radiation. 1980

Volume 27   DAVID H. MILLER. Energy at the Surface of the Earth: An Introduction to the Energetics of Ecosystems. 1981

Volume 28   HELMUT G. LANDSBERG. The Urban Climate. 1981

Volume 29   M. I. BUDYKO. The Earth's Climate: Past and Future. 1982*

Volume 30   ADRIAN E. GILL. Atmosphere-Ocean Dynamics. 1982

Volume 31   PAOLO LANZANO. Deformations of an Elastic Earth. 1982*

Volume 32   RONALD T. MERRILL AND MICHAEL W. MCELHINNY. The Earth's Magnetic Field. Its History, Origin, and Planetary Perspective. 1983

Volume 33   JOHN S. LEWIS AND RONALD G. PRINN. Planets and Their Atmospheres: Origin and Evolution. 1983

Volume 34   ROLF MEISSNER. The Continental Crust: A Geophysical Approach. 1986

Volume 35   M. U. SAGITOV, B. BODKI, V. S. NAZARENKO, AND KH. G. TADZHIDINOV. Lunar Gravimetry. 1986

Volume 36   JOSEPH W. CHAMBERLAIN AND DONALD M. HUNTEN. Theory of Planetary Atmospheres: An Introduction to Their Physics and Chemistry, Second Edition. 1987

Volume 37   J. A. JACOBS. The Earth's Core, Second Edition. 1987

Volume 38   J. R. APEL. Principles of Ocean Physics. 1987

Volume 39   MARTIN A. UMAN. The Lightning Discharge. 1987

Volume 40   DAVID G. ANDREWS, JAMES R. HOLTON, AND CONWAY B. LEOVY. Middle Atmosphere Dynamics. 1987

Volume 41   PETER WARNECK. Chemistry of the Natural Atmosphere. 1988

Volume 42   S. PAL ARYA. Introduction to Micrometeorology. 1988

Volume 43   MICHAEL C. KELLEY. The Earth's Ionosphere. 1989

Volume 44   WILLIAM R. COTTON AND RICHARD A. ANTHES. Storm and Cloud Dynamics. 1989

Volume 45   WILLIAM MENKE. Geophysical Data Analysis: Discrete Inverse Theory, Revised Edition. 1989

Volume 46   S. GEORGE PHILANDER. El Niño, La Niña, and the Southern Oscillation. 1990

Volume 47   ROBERT A. BROWN. Fluid Mechanics of the Atmosphere. 1991

Volume 48   JAMES R. HOLTON. An Introduction to Dynamic Meteorology, Third Edition. 1992

Volume 49   ALEXANDER A. KAUFMAN. Geophysical Field Theory and Method.
Part A: Gravitational, Electric, and Magnetic Fields, 1992
Part B: Electromagnetic Fields I. 1994
Part C: Electromagnetic Fields II. 1994

Volume 50   SAMUEL S. BUTCHER, GORDON H. ORIANS, ROBERT J. CHARLSON, AND GORDON V. WOLFE. Global Biogeochemical Cycles. 1992

Volume 51   BRIAN EVANS AND TENG-FONG WONG. Fault Mechanics and Transport Properties in Rock. 1992

Volume 52   ROBERT E. HUFFMAN. Atmospheric Ultraviolet Remote Sensing. 1992

Volume 53   ROBERT A. HOUZE, JR. Cloud Dynamics. 1993

Volume 54   PETER V. HOBBS. Aerosol-Cloud-Climate Interactions. 1993

Volume 55   S. J. GIBOWICZ AND A. KIJKO. An Introduction to Mining Seismology. 1993

Volume 56   DENNIS L. HARTMANN. Global Physical Climatology. 1994

Volume 57   MICHAEL P. RYAN. Magnetic Systems. 1994

Volume 58   THORNE LAY AND TERRY C. WALLACE. Modern Global Seismology. 1995

Volume 59   DANIEL S. WILKS. Statistical Methods in the Atmospheric Sciences. 1995

Volume 60   FREDERIK NEBEKER. Calculating the Weather. 1995

Volume 61   MURRY L. SALBY. Fundamentals of Atmospheric Physics. 1996.